UNCONVENTIONAL GAS RESERVOIRS

UNCONVENTIONAL GAS RESERVOIRS

Evaluation, Appraisal, and Development

M. RAFIQUL ISLAM

Trans Canada Training
London, Ontario, Canada

ELSEVIER

AMSTERDAM • BOSTON • HEIDELBERG • LONDON
NEW YORK • OXFORD • PARIS • SAN DIEGO
SAN FRANCISCO • SINGAPORE • SYDNEY • TOKYO

Gulf Professional Publishing is an imprint of Elsevier

G P P

Gulf Professional Publishing is an imprint of Elsevier
225 Wyman Street, Waltham, MA 02451, USA
The Boulevard, Langford Lane, Kidlington, Oxford, OX5 1GB, UK

Notices

Knowledge and best practice in this field are constantly changing. As new research and experience broaden our understanding, changes in research methods, professional practices, or medical treatment may become necessary.

Practitioners and researchers must always rely on their own experience and knowledge in evaluating and using any information, methods, compounds, or experiments described herein. In using such information or methods they should be mindful of their own safety and the safety of others, including parties for whom they have a professional responsibility.

To the fullest extent of the law, neither the Publisher nor the authors, contributors, or editors, assume any liability for any injury and/or damage to persons or property as a matter of products liability, negligence or otherwise, or from any use or operation of any methods, products, instructions, or ideas contained in the material herein.

ISBN: 978-0-12-800390-9

Library of Congress Cataloging-in-Publication Data

A Catalog record for this book is available from the Library of Congress

British Library Cataloguing in Publication Data

A catalogue record for this book is available from the British Library

CONTENTS

AUTHOR BIOGRAPHY

Professor M.R. Islam is a worldwide consultant on environment and energy-related issues. He is currently director of Oil and Gas at Trans Canada Training. Dr Islam is known as one of the most published engineers in the world. He is credited to have coined terms, such as "green petroleum" and "sustainable petroleum development" at a time when "sustainable petroleum" was considered to be an oxymoron. His work has created a paradigm shift in a wide range of applications, spanning through various disciplines.

Dr Islam's most notable contribution is in the areas of sustainability, environmental integrity, and knowledge modeling, on which topic he has written dozens of books and over 700 research papers. His latest work is captured in his book: Unconventional Gas Reservoirs published by Elsevier, 2014. Dr Islam's current projects include: "greening of chemical industries," a "virtual research university," and "knowledge-based modeling." Spanning a professional career for more than 30 years, Professor Islam has held faculty positions with seven different universities and supervised over 150 research students and postdoctoral fellows. He received 30 million US dollars in research funding from various government and industrial entities. His research spun off a number of technology transfer projects that led to the development of three companies. One of them now has become one of the most innovative oil spill remediation technology development companies.

During 2000–2005, Dr Islam held the Killam Research Chair Professor in Oil and Gas, which was the only Petroleum Engineering Research Chair in Canada. Prior to this, Dr Islam was the first NRCan (Natural Resources Canada) Professor of Engineering at the University of Regina, where he was the Program Leader of Petroleum Engineering. He was instrumental in establishing the Petroleum Technology Research Center (PTRC) and the Greenhouse Gas Technology Center (GTC) in Regina. Dr Islam has written over 700 research articles on various modeling-, management-, energy-, environmental-, and communication-related topics. During the same period, he received more than 20 million US dollars in institutional and over 10 million US dollars in individual research grants, and employed over 150 postdoctoral fellows, graduate and undergraduate research students. In recognition of his research and teaching excellence, he has received a number of international awards, including the Einstein Gold Medal of Honor and the Crown and Eagle Gold Medal for both of which Dr Islam is the first recipient in Canada. He taught 30 different academic courses and several industry professional development workshops around the world. He gives numerous invited and guest speeches in multiple countries.

CHAPTER 1

Introduction

1.1 SUMMARY

Only recently it has become evident that natural gas provides one with great value with minimum impact on the environment. With the ongoing "gas crisis" in Europe and steep energy needs of China and India, the importance of natural gas is increasing tremendously in the global market. Unconventional gas reservoirs are responsible for about one-third of annual US natural gas production. In the global scale, unconventional gas reserves can add 60–250% to the proven gas reserves. This book explores the science behind unconventional gas and provides one with a basis that the realistic estimate of total available gas could easily exceed all estimates of the past. With newfound gas and technologies that harness it with environmental integrity, the world could easily see another 100 years of crisis-free energy outlook. This book paints a rosy picture that is supported by scientific basis and guides the readership to be able to discern between hype and hysteria involving petroleum resource management.

The potential of unconventional gas reserves is far greater than any enhanced oil recovery scheme designed to date. However, when the two are combined and added to that is scientific basis for revised assessment of world gas reserve, the energy outlook becomes very bright. The purpose of this book is to provide one with practical insights and tools that can be used to efficiently identify, appraise, and develop these types of reservoirs. Initially, the readership is be familiarized with both engineering and marketing aspects of natural gas. The mechanisms of gas production and reservoir engineering are discussed. Presented are the origin and mechanisms of gas production, along with the various techniques used to analyze reservoir parameters and performance, and understand how these have been applied to numerous projects. Specific features of unconventional gas reserves and the means of capitalizing on these features in order to maximize long-term benefit are discussed. Scientific characterization of gas reserves included discussion of the history of gas as well as the rock. This "bottom-up" analysis helped develop truly scientific basis for unconventional gas analysis. Thorough geological characterization was presented for US

Unconventional Gas Reservoirs
ISBN 978-0-12-800390-9
http://dx.doi.org/10.1016/B978-0-12-800390-9.00001-3

reservoirs in order to connect with potential sources of unconventional gas. This serves as a template for a new analysis that starts off with conventional reservoir characterization. It is shown that the true reserve potential of gas is much higher than the one estimated using conventional approach. Latest information on emerging technologies in typical enhanced oil recovery are presented with lucid discussion of key features of reservoir characterization and development in order to help readership identify techniques that can be used in "enhanced gas recovery" of unconventional gas. It is shown that contrary to popular belief, such techniques are economically attractive and environmentally sustainable. This unlocking of emerging technologies, coupled with a novel technique for characterization of unconventional gas reserves for which conventional techniques fail to yield sound results, creates a true paradigm shift in gas reservoir analysis.

The science of both fluid and rock is discussed using delinearized history analysis. The discussion of coal bed methane and gas hydrate is carried out with the focus on improving production rate under various production strategies. Criteria for selecting candidates for hydraulic fracturing as well as horizontal wells are discussed. For every case encountered specific suggestion is made that is custom designed for the particular application. A guideline is provided for optimization of recovery schemes under various production and enhanced gas recovery strategies.

Finally, numerical modeling approaches are discussed and challenges in modeling unconventional gas reservoirs are presented. A number of case studies involving technological application as well as numerical modeling are presented. Overall, new opportunities are highlighted and previously held beliefs and myths deconstructed.

1.2 UNCONVENTIONAL TO MAINSTREAM: A NECESSARY PARADIGM SHIFT

Newton did not know what the term "Newtonian" meant in any discipline. Scientists that followed Newton's "laws" and theories with dogmatic fervor did not know what the term non-Newtonian stood for. At present, scholarly works on "Non-Newtonian" outnumber "Newtonian" by a big margin. Over 50 years ago, US government started to invest in research in coal bed methane, shale gas, shale oil, along with other areas of petroleum development. Decades later, the term "unconventional" was coined. Even some more decades later came the success story of unconventional gas. This success would come only after Europe had been under the grip of a "gas

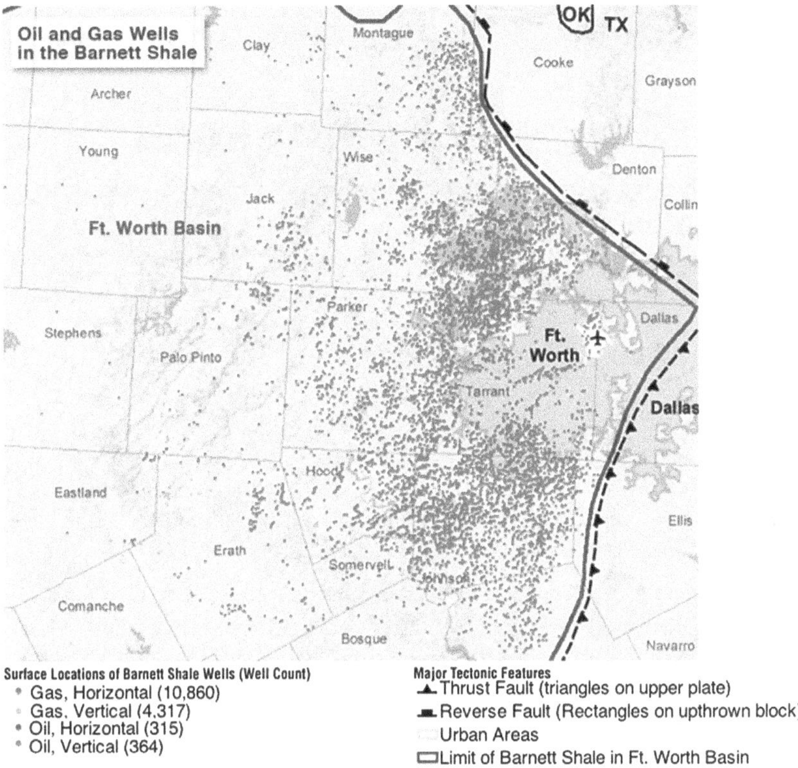

Figure 1.1 Location map for Barnett shale. *From EIA, 2011.*

crisis." Within years, "unconventional" became the biggest story in U.S. in terms of energy independence. A new type of revolution began. Figure 1.1 shows graphical representation of wells in the shale gas plays of just one basin in Texas. If 10,860 horizontal gas wells are called "unconventional" one must wonder what would the term "conventional" represent.

Chapter 2 discusses the role of unconventional gas in the big picture of petroleum. It deconstructs many myths that have perplexed modern society that has grown accustomed to vilifying carbon—the essence of life. It unravels the mysteries of "resource triangle" that is based on the false premise that natural is more expensive than artificial (Figure 1.2). It places unconventional gas in its correct position within proven gas and oil reserves and shows that this "unconventional" is anything but unconventional in conventional sense. It is the mainstream of the future.

The chapter also presents a delinearized history of energy developments in relation to petroleum production, particularly as it relates to natural gas.

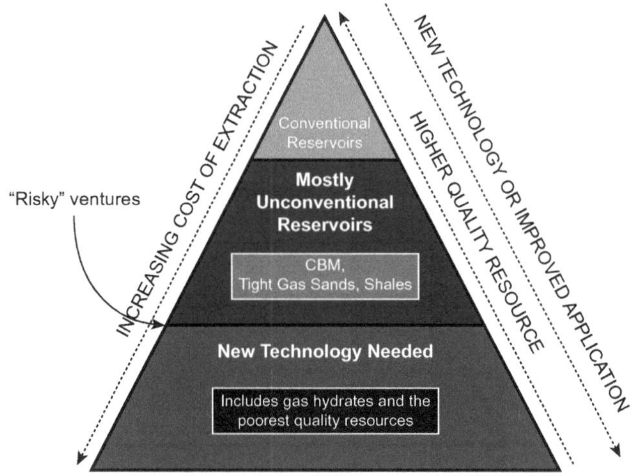

Figure 1.2 Conventional perception of resource triangle.

It shows that energy needs of the future can only be met if we excel in accessing unconventional gas. It creates the need for a paradigm shift in nomenclature as well as scientific description of unconventional reservoirs. In simple terms, it delivers the message that the science of unconventional gas cannot be understood with the conventional routines.

1.3 THE PARADOX OF UNCONVENTIONAL GAS

We live in an interesting époque. Everything surrounding us seems to point to contradictions. We made unprecedented progress in technology development, only to hear from Nobel Laureate Chemists (e.g., Robert Curl) that ours is a "technological disaster." We claim to have progressed from dark ages of savagery and lack of empathy to modern enlightenment only to hear from some of the most ardent supporters of modern European mindset of Capitalism (e.g., Pope Francis) that unfair economic structures that creates huge inequalities are actually an act of terrorism. As civilization evolved, we claim to have become wiser and more efficient, only to discover our per capita energy need has skyrocketed in the modern era. We are so sure about our superiority that we call humans as the top of the food chain and boast about our entitlement over other species, only to discover that the greatest crisis is the population growth of humans—the "best creation" of nature. We have identified "greening" as the key to resolving the environmental disaster that has befallen on us only to target carbon as the worst enemy to environmental integrity. This is the same carbon that is the essence of "greening." We praise nature for its unique ability to turn bad into good, for its tenacity and sustainability, only to target petroleum as the number one enemy of the environment. Which petroleum? The one that was perfected by nature—the very nature we love to be in tune with.

Chapter 2 started the paradigm shift and Chapter 3 continues it in the realm of unconventional technology development, its geological nature, and the economics that can support its development and growth. It shows conventional engineering analysis is not the only one that does not apply to unconventional reservoirs; conventional accounting theories as well as energy pricing models do not apply also. It shows with details why essential features of unconventional gas are such that none of the conventional techniques and principles should apply. This ranges from geology to chemistry to engineering for coal bed methane, shale gas, tight gas as well as methane gas hydrate.

1.4 GREENING THE FUTURE OF UNCONVENTIONAL GAS

Humans seek status quo. Somehow, we have great difficulty in changing our style of analysis. It starts with our inability to change the first premise. Today's first premise in energy sectors seems to be based on the misconception that global warming is because of petroleum production. This may be illogical coming from fallacious arguments, but the moment it becomes a political issue, humanity loses its ability to think from outside the box.

Even at the pinnacle of unconventional gas recovery, we have managed to use only hydraulic fracturing and horizontal wells as the only technological tool to produce gas. Talks are all abound how to increase the resource base, where to drill new wells, how to fracture more wells, how to penetrate more difficult-to-access formations, but few talk about using numerous other technologies that are readily available today. Chapter 4 introduces a section "Lessons learned from Enhanced Oil Recovery" (EOR) and summarizes success stories of last 50 years of EOR. Why learn from EOR? It is because they are much easier to apply in gas reservoirs, particularly the ones that contain huge amount of gas in shallower formations.

The chapter demystifies the global warming debate and shows that natural gas is the most environmentally appealing energy source known today. Then, it shows how the unconventional resource base can be expanded multifold by adopting EOR technologies that have been perfected by the same petroleum sector that produces unconventional gas.

These do not need any research investment for future development. They only need careful custom designing for a particular application.

Added to the equation is the tremendous resource locked in methane gas hydrates. Techniques capable of unlocking these formations are discussed.

The concept of using novel techniques for enhanced gas recovery is introduced. This would add to the current estimates of unconventional reserves. Discussion is carried out on reservoirs that are not commonly characterized as "unconventional," such as volcanic reservoirs.

1.5 PARADIGM SHIFT IN RESERVOIR CHARACTERIZATION

Recent years have brought a rapid expansion of activities around unconventional gas. The successes in unconventional gas exploitation have been mainly attributed to advances in engineering. In particular, hydraulic fracturing and horizontal wells are considered to be the game changer. To date, geophysics and geology have yet to assume a major role in the development of this important resource.

Islam et al. (2012) recently argued any "game changer" must accompany real change, which means change from the first premise to the process itself. If previous chapters established that unconventional gas reservoirs are indeed uniquely different from conventional ones, why should we assume the reservoir characterization tools and procedures of conventional reservoirs would apply to unconventional ones?

Chapter 5 introduces a new reservoir characterization technique that uses paradigm shift in its real sense. The outcome is a refreshing new approach that is practical and prepares one for in-depth understanding of the mechanisms involved in unconventional gas reservoirs. The technique relies heavily on whatever is available during the course of the reservoir development and minimizes spending on expensive tools that have little validity in unconventional reservoirs. At the end the chapter introduces a screening tool to identify different types of formations so that development strategies can be chalked out.

1.6 THE SCIENCE OF UNCONVENTIONAL GAS

Albert Einstein famously said, "Any intelligent fool can make things bigger, more complex, and more violent. It takes a touch of genius—and a lot of courage—to move in the opposite direction." Sadly, this touch of genius is missing from the New Science. The contradictions pointed out in earlier

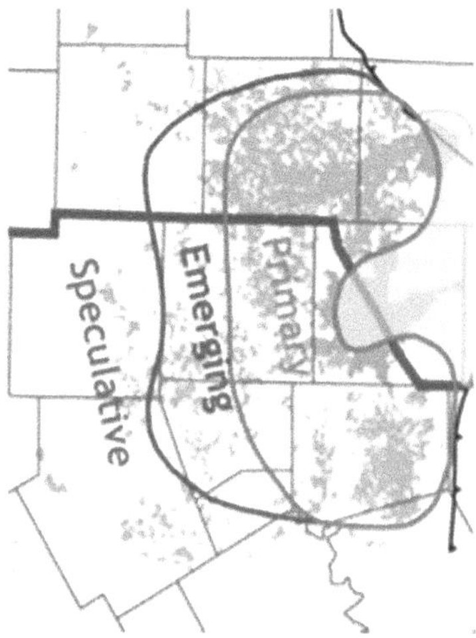

Figure 1.3 Conventional approach involves delineating unconventional proved reserve first, following by emerging and speculative zones.

sections would not occur if we did not invoke assumptions that are so complex that no one dares question their absurdity. Chapter 6 takes a dogma-free approach and eliminates all spurious assumptions in mass and energy balance equations. The results are refreshing. A new theory emerges that can explain all the events, ranging from global warming to light pollution, from digenesis to formation of gas traps. With this description a fundamental change in reservoir analysis is invoked.

Previous techniques included reservoir characterization based on production or data of the present. Even in exploration, unconventional gas reservoirs are first identified and "proved," followed by the search of additional resources that are marked as "emerging" and "speculative" (Figure 1.3). This defies the logic that all unconventional gas emerges from conventional sources. The new technique reversed that and introduced reservoir characterization based on geology first. Based on this technique, all US oil and gas fields are characterized based on origin, then migration of the fluid. This results in a truly scientific characterization of unconventional resources. Consider the significance when applied on worldwide reserves (Figure 1.4).

1.7 DEPICTING THE FUTURE

Even though not commonly perceived, modeling comes before execution of any engineering project. The petroleum engineering industry is a champion of modeling. Today, it has the ability to model a reservoir with

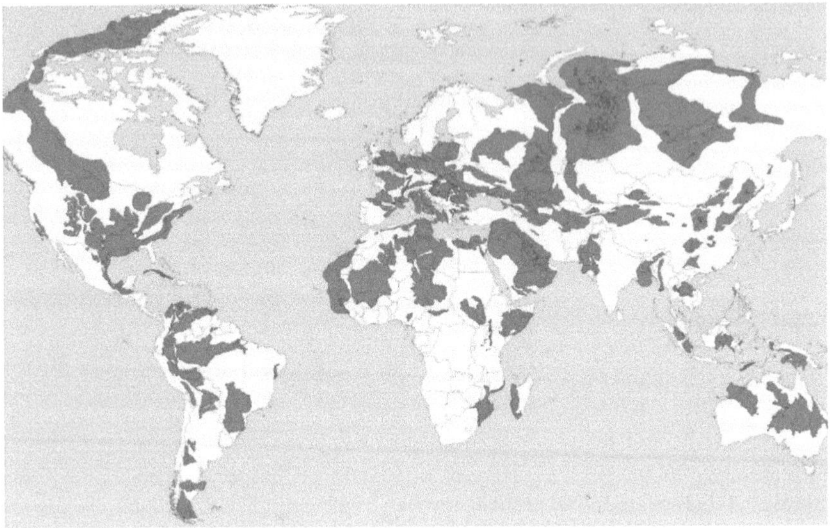

Figure 1.4 The locations of current proven reserves.

no less than trillion cells and solve the governing equation within minutes. Only a decade ago, such performance was unthinkable even from the super computers. The problem is not with the speed or the ability to model billions of grid blocks. The problem is not in collecting data and processing them. In fact, Saudi Arabia has set up massive 4D seismic in reservoirs that do not even employ waterflooding. The problem lies within the fact that none of the unconventional reservoirs follow Darcy's law—the only equation petroleum engineers have used in last 155 years. Chapter 7 shows how people tried to cope with this crisis and presents a way out, with practical advice on how to model unconventional reservoirs.

1.8 THE END GAME

Chapter 8 presents the economics of unconventional gas reservoirs. By knowing that economics is the driver of any engineering project, the chapter introduces fundamental changes in economic analysis and uses that to consolidate the conclusions that emerge from previous chapters.

Bibliography lists some 60 pages of references and bibliography that serve as a documentary evidence of the usefulness of and interest in the topic covered in this book.

CHAPTER 2

World Gas Reserve and the Role of Unconventional Gas

2.1 SUMMARY

The role of petroleum products in shaping human energy needs is undeniable. Even though the word "natural gas" is synonymous with petroleum gas, few understand the natural traits of "natural gas" and even fewer appreciate the role of "natural gas" in a natural setting. While there had been no ambiguity in terms what constitutes natural in both material and spiritual senses before modern age, modern age all the way to information is rife with confusion regarding sustainability of energy as well as mere existence of the human race. This chapter presents a delinearized history of energy developments in relation to petroleum production, particularly as it relates to natural gas. The discussion then revolves around energy needs of the future and how unconventional natural gas will play the most significant role in meeting the energy needs of the future generation. It is shown that unconventional natural gas is the most environment friendly and the cleanest of all natural resources.

2.2 PETROLEUM IN THE BIG PICTURE

Captain Drake is usually credited to have drilled the first-ever oil well in Titusville, Pennsylvania in 1859. Even if one discards the notion that petroleum was in use for thousands of years, there is credible evidence that the first well in modern age was drilled in Canada. Canadian, Charles Nelson Tripp was the first in North America to have recovered commercial petroleum products. The drilling was completed in 1851 at Enniskillen Township, near Sarnia, in present-day Ontario, which was known as Canada West at that time. Soon after the "mysterious gum" bed was discovered, first oil company was incorporated in Canada through a parliamentary charter. Tripp became the president of this company on December 18, 1854. The charter empowered the company to explore for asphalt beds and oil and salt springs, and to manufacture oils, naphtha paints,

Unconventional Gas Reservoirs
ISBN 978-0-12-800390-9
http://dx.doi.org/10.1016/B978-0-12-800390-9.00002-5

burning fluids. Even though this company (International Mining and Manufacturing) was not a financial success, the petroleum products received an honorable mention for excellence at the Paris Universal Exhibition in 1855. Failure of the company can be attributed to several factors contributed to the downfall of the operation. Lack of roads in the area made the movement of machinery and equipment to the site extremely difficult. And after every heavy rain the area turned into a swamp and the gum beds made drainage extremely slow. This added to the difficulty of distributing finished products.

In subsequent years, James Miller Williams became interested and visited the site in 1856; Tripp unloaded his hopes, his dreams, and the properties of his company, saving for himself a spot on the payroll as landman. The former carriage builder formed J.M. Williams & Company in 1857 to develop the Tripp properties. Besides asphalt, he began producing kerosene. This "refined" product, kerosene is a combustible hydrocarbon liquid. The name is derived from Greek: κ μ ρ ό ς (keros) meaning wax. The word "kerosene" was registered as a trademark by Abraham Gesner in 1854, and for several years, only the North American Gas Light Company and the Downer Company (to which Gesner had granted the right) were allowed to call their lamp oil "Kerosene" in the United States. Gesner was a medical doctor by training (from London) but took special interest in geology. Gesner's research in minerals resulted in his 1846 development of a process to refine a liquid fuel from coal, bitumen, and oil shale. His new discovery, which he named kerosene, burned more cleanly and was less expensive than competing products, such as whale oil. In 1850, Gesner created the Kerosene Gas Light Company and began installing lighting in the streets in Halifax and other cities. By 1854, he had expanded to the United States where he created the North American Kerosene Gas Light Company at Long Island, New York. Demand grew to where his company's capacity to produce became a problem, but the discovery of petroleum, from which kerosene could be more easily produced, solved the supply problem. This was the first time in recorded history artificial processing technique was introduced in refining petroleum products. Gesner did not use the term "refined" but made fortune out of the sale of this artificial processing. In 1861, he published a book titled: *A Practical Treatise on Coal, Petroleum and Other Distilled Oils*, which became a standard reference in the field. As Gesner's company was absorbed into the petroleum monopoly, Standard Oil, he returned to Halifax, where he was appointed a professor of natural history at Dalhousie University. It is this university that was founded on pirated money while other pirates continued to be hanged by the Royal Navy at Point Pleasant Park's Black Rock Beach as late as 1844.[1]

In the mean time, a parallel event took place elsewhere in Canada. In 1858, Williams dug a well in search of cleaner drinking water and came across oil at a depth of 15.5 m. It became the first commercial oil well in North America, remembered as the Williams No. 1 well at Oil Springs, Ontario. The Sarnia Observer and Lambton Advertiser, quoting from the Woodstock Sentinel, published on page two on August 5, 1858:

> *An important discovery has just been made in the Township of Enniskillen. A short time since, a party, in digging a well at the edge of the bed of Bitumen, struck upon a vein of oil, which combining with the earth forms the Bitumen.*

Some historians challenge Canada's claim to North America's first oil field, arguing that Pennsylvania's famous Drake Well was the continent's first. But there is evidence to support Williams, not least of which is that the Drake well did not come into production until August 28, 1859. The controversial point might be that Williams found oil above bedrock while "Colonel" Edwin Drake's well located oil within a bedrock reservoir.

History is not clear as to when Williams abandoned his Oil Springs refinery and transferred his operations to Hamilton. He was certainly operating there by 1860, however. Spectator advertisements offered coal oil for sale at 16 cents per gallon for quantities from 4000 US gallons (15,000 L) to 100,000 US gallons (380,000 L). By 1859 Williams owned 800 acres of land in Oil Springs. Williams reincorporated in 1860 as the Canadian Oil Company. His company produced oil, refined it, and marketed refined products. That mix of operations qualifies Canadian Oil Company as the world's first integrated oil company.

Exploration in the Lambton county backwoods quickened with the first flowing well in 1860; previous wells had relied on hand pumps. The first gusher erupted on January 16, 1862, when oil struck at 158 ft (48 m). For a week the oil gushed unchecked at levels reported as high as 3000 barrels per day, eventually coating the distant waters of Lake St Clair with a black film. There is historical controversy concerning whether it was John Shaw or

[1]A cairn in front of its administration building actually describes the university's origins two centuries ago from a fund created to launder the ill-gotten gains of an early 19th century war crime committed by the Royal Navy against a customs house in the U.S. state of Maine several months after Anglo-American hostilities of the War of 1812 had officially concluded.

another oil driller named Hugh Nixon Shaw who drilled this oil gusher; the newspaper article cited below identifies John Shaw.

News of the gusher spread quickly and was reported in the Hamilton Times four days later:

> *I have just time to mention that to-day at half past eleven o'clock, a.m., Mr. John Shaw, from Kingston, C. W., tapped a vein of oil in his well, at a depth of one hundred and fifty-eight feet in the rock, which filled the surface well, (forty-five feet to the rock) and the conductors [sic] in the course of fifteen minutes, and immediately commenced flowing. It will hardly be credited, but nevertheless such is the case, that the present enormous flow of oil cannot be estimated at less than two thousand barrels per day, (twenty-four hours), of pure oil, and the quantity increasing every hour. I saw three men in the course of one hour, fill fifty barrels from the flow of oil, which is running away in every direction; the flat presenting the appearance of a sea of oil. The excitement is intense, and hundreds are rushing from every quarter to see this extraordinary well.*

Historically, the ability of oil to flow freely has fascinated developers and at the same time ability of gas to leak and go out of control has intimidated them. Such fascination and intimidation continues today while nuclear electricity is considered to be benign while natural gas considered to be the source of global warming, all because it contains carbon—the very component nature needs for creating an organic product. Scientifically, however, the need for refining stems from the necessity of producing clean flame. Historically, Arabs were reportedly the first ones to use refined olive oil. They used exclusively natural chemicals in order to refine oil (Islam et al., 2010). However, such use of natural chemicals is nonexistent in modern day petroleum industry. When it comes to petroleum gas, it had been in use for millennia but only recent time "processing" of such gas has been introduced.

Natural gas seeps in Ontario County, New York were first reported in 1669 by the French explorer, M. de La Salle, and a French missionary, M. de Galinee, who were shown the springs by local Native Americans. This is the debut of natural gas industry in North America. Subsequently, William Hart, a local gunsmith, drilled the first commercial natural gas well in the United States in 1821 in Fredonia, Chautauqua County. He drilled a 27-foot deep well in an effort to get a larger flow of gas from a surface seepage of natural gas. This was the first well intentionally drilled to obtain natural gas. Hart built a simple gas meter and piped the natural gas to an innkeeper on the stagecoach route from Buffalo to Cleveland. Because there was no pipeline network in place, this gas was almost invariably used to light streets at night. However, in late 1800s, electric lamps were beginning to be used for lighting streets. This led to gas producers scrambling for alternate market. Shallow natural gas wells were soon drilled throughout the Chautauqua County shale belt. This natural gas was transported to businesses and street lights in Fredonia at the cost of US$.50 a year for each light (Website 1). In the mean time, in mid-1800s, Robert Bunsen invented the "Bunsen burner" that helped produce artificial flame by controlling air inflow in an open flame. This was significant because it helped producing intense heat and controlling the flame at the same time. This led ways to develop usage of natural gas for both domestic and commercial use.

The original Hart gas well produced until 1858 and supplied enough natural gas for a grist mill and for lighting in four shops. By the 1880s, natural gas was being piped to towns for lighting and heat, and to supply energy for the drilling of oil wells. Natural gas production from sandstone reservoirs in the Medina formation was discovered in 1883 in Erie County. Medina production was discovered in Chautauqua County in 1886. By the early years of the twentieth century, Medina production was established in Cattaraugus, Genesee, and Ontario counties.

Gas in commercial quantities was first produced from the Trenton limestone in Oswego County in 1889 and in Onondaga County in 1896. By the end of the nineteenth century, natural gas companies were developing longer intrastate pipelines and municipal natural gas distribution systems. The first gas storage facility in the United States was developed in 1916 in the depleted Zoar gas field south of Buffalo.

By the late 1920s, declining production in New York's shallow gas wells prompted gas companies to drill for deeper gas reservoirs in Allegany, Schuyler, and Steuben counties. The first commercial gas production from the Oriskany sandstone was established in 1930 in Schuyler County. By the 1940s, deeper gas discoveries could no longer keep pace with the decline in shallow gas supplies. Rapid depletion and overdrilling of deep gas pools prompted gas companies in western New York to sign long-term contracts to import gas from out of state.

It took the construction of pipelines to bring natural gas to new markets. Although one of the first lengthy pipelines was built in 1891—it was 120 miles long and carried gas from fields in central Indiana to Chicago — there were very few pipelines built until after World War II in the 1940s. Similar to all other developments in modern Europe, World War II brought about changes that led to numerous inventions and technological breakthroughs in the area of petroleum production and processing. Improvements in metals, welding techniques, and pipe making during the War made pipeline construction more economically attractive. After World War II, the nation began building its pipeline network. Throughout the 1950s and 1960s, thousands of miles of pipeline were constructed throughout the United States. Today, the US pipeline network, laid end-to-end, would stretch to the moon and back twice. The phenomenon of pipelining is of significance. Because of this, there has been tremendous surge in the corrosion control industry.

Onondaga reef fields were discovered by seismic prospecting in the late 1960s. Seven reef fields have been discovered to date in southern New York. Today, the Onondaga reef fields and many Oriskany fields are largely depleted and are being converted to gas storage fields. This state of depletion was achieved after a long production period and extensive hydraulic fracturing throughout 1970s and 1980s. These were considered to be tight gas sands. Recently, the same technology has made a comeback. The rapid development of New York's current Trenton-Black River gas play is made possible by technological advances in three-dimensional (3D) seismic imaging, horizontal drilling, and well completion. The surge in domestic oil and gas production through "fracking" emerges from technologies popularized in the 1970s. However, 3D seismic or multilateral drilling technology was not in place at the time. Figure 2.1 shows how natural gas production evolved in the state of New York throughout history.

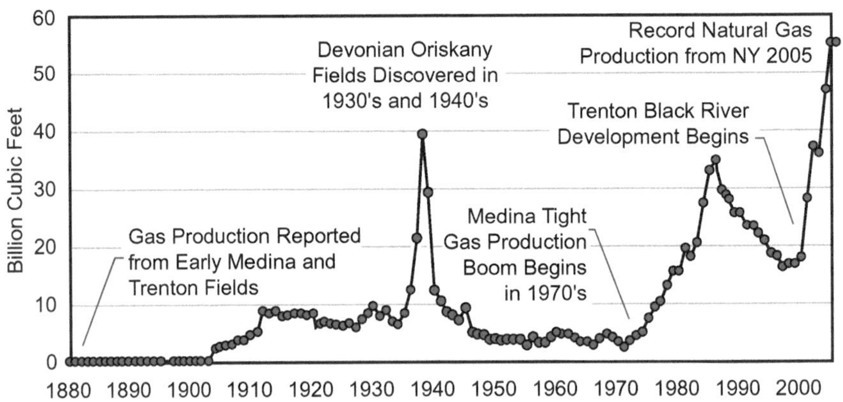

Figure 2.1 Natural gas production history in New York State. *From Website 1.*

In this figure, the first spike relates to discovery of Devonian shale. That spike led to a quick depletion. In early 1970s, production from "tight gas" formations led to another more sustained spike in gas recovery. During that period, extensive hydraulic was introduced as a means for increasing productivity. However, it was not considered to be a reservoir production enhancement scheme. In 2000, at the nadir of oil price, yet another spike took place in the state of New York. This related to the development of Trenton-Black River field. This gas production scheme would lead to record gas production in that state in 2005. This spike continued and led the way to producing domestic gas and oil from unconventional reservoirs in United States. Today, production from unconventional gas reservoirs has taken an unprecedented turn. In 2013, production from shale gas, tight gas, and coalbed methane (CBM) accounted for domestic production surpassing imports for the first time in 30 years. Shale gas, tight oil, or other unconventional resources are found in many of the states that had already produced from conventional sources. Figure 2.2 shows the locations of these unconventional formations.

The primary recovery techniques from these shale plays involve multilaterals and intense hydraulic fracturing, now known as "fracking." Two significant differences between fracking and old-fashioned hydraulic

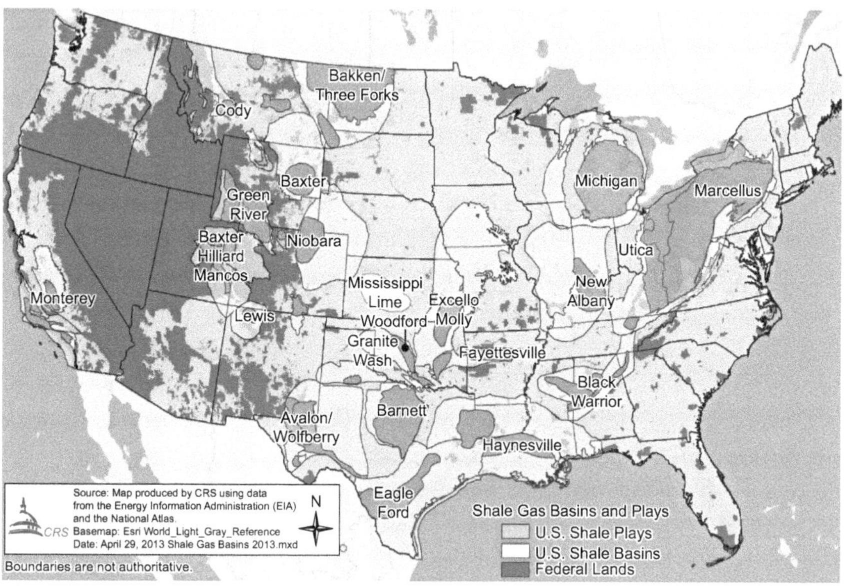

Figure 2.2 Locations of unconventional shale plays in lower 48 states. *From Ratner and Tiemann, 2014.*

Picture 2.1 Latest fracking techniques use artificial proppants (cylindrical shaped, magnified 25 times).

fracturing are (1) fracking uses multistage fractures with horizontal multilaterals and (2) fracking uses artificial sands as well fluid. One such example is shown in Picture 2.1. Figure 2.3 shows how these artificial proppants with cylindrical shape creates better fractures.

Figure 2.3 demonstrates that sands have the worst fracture efficiency, while the rod-shaped proppants have the highest efficiency. In this process, material cost of fracturing has skyrocketed and accounts for bulk of the fracturing scheme. The same can be said about fracturing fluid. It turns out water is not conducive to creating fractures in shale formations. In 1976, the US government started the Eastern Gas Shales Project, a set of dozens of public–private hydraulic fracturing pilot demonstration projects. During the same period, the Gas Research Institute, a gas industry research consortium, received approval for research and funding from the Federal Energy Regulatory Commission. That was the beginning of fracturing shale formations that gave boost in gas production throughout late 1970s and 1980s.

In 1997, based on earlier techniques used by Union Pacific Resources, now part of Anadarko Petroleum Corporation, Mitchell Energy, now part of Devon Energy, developed the hydraulic fracturing technique known as "slickwater fracturing" that involves adding chemicals to water thereby allowing increase to the fluid flow, which made the shale gas extraction economical. These chemicals are both expensive and toxic to the environment.

The fracturing fluid varies in composition depending on the type of fracturing used, the conditions of the specific well being fractured, and the

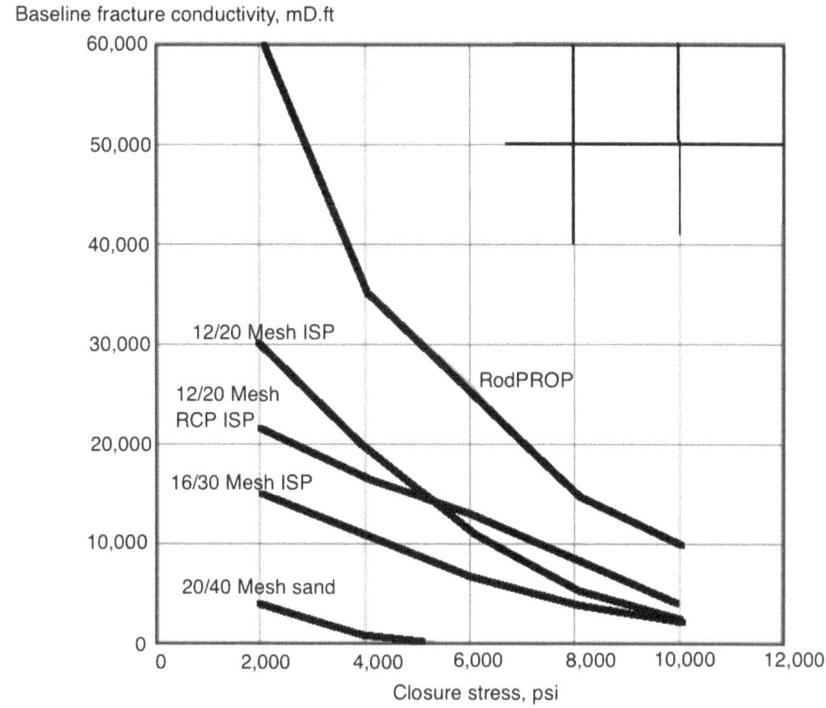

Figure 2.3 Comparison of long-term conductivity between rod-shaped proppant and conventional spherical proppant at 2 lbm/ft^2 at 250° F.

water characteristics. A typical fracture treatment uses between 3 and 12 additive chemicals. A typical fracturing operation involves the following chemicals.

Acids: hydrochloric acid (for carbonate cements) or acetic acid (for silicate cement) is used in the prefracturing stage for cleaning the perforations and initiating fissure in the near-wellbore rock.

Sodium chloride: delays breakdown of the gel polymer chains.

Polyacrylamide and other friction reducers: reduces turbulence (lower Reynold's number), while increasing proppant transport in the tubing or drill pipe.

Ethylene glycol: prevents formation of scale deposits in the pipe.

Borate salts: thermal stabilizers that maintain fluid viscosity under high temperature conditions.

Sodium and potassium carbonates: used for maintaining effectiveness of cross-linkers that stabilize the polymer.

Glutaraldehyde: used as disinfectant of the water to prevent bacterial growth and subsequent biodegradation of the fluid.

Guar gum and other water-soluble gelling agents: increases viscosity of the fracturing fluid to deliver more efficiently the proppant into the formation.

Citric acid: used for corrosion prevention as it is a milder form for corrosion inhibitors.

Isopropanol: increases the viscosity of the fracturing fluid.

The most common chemical used for hydraulic fracturing in the United States in 2005–2009 was methanol, while some other most widely used chemicals were isopropyl alcohol, 2-butoxyethanol, and ethylene glycol. New generation of chemicals include: Conventional linear gels (carboxymethyl cellulose, hydroxyethyl cellulose, carboxymethyl hydroxyethyl cellulose, hydroxypropyl cellulose, methyl hydroxyl ethyl cellulose), guar or its derivatives (hydroxypropyl guar, carboxymethyl hydroxypropyl guar), etc. These gels have higher viscosity at pH9 onwards and are used to carry proppants. After the fracturing job the pH is reduced to 3–4 so that the cross-links are broken and the gel is less viscous and can be pumped out. Organometallic cross-linked fluid zirconium, chromium, antimony, titanium salts are known to cross-link the guar-based gels. The cross-linking mechanism is not reversible. Aluminum phosphate–ester oil gels. Aluminum phosphate and ester oils are slurried to form a cross-linked gel. These are one of the first known gelling systems.

Hydraulic fracturing has been used for decades to stimulate increased production from existing oil or gas wells. This technique, along with other well stimulation techniques, has been regulated to varying degrees through state oil and gas codes. The detail and scope of applicable regulations vary across the states, and some states have regulated "well stimulation" broadly without addressing hydraulic "fracturing" explicitly. State regulators have noted that hydraulic fracturing operations are regulated through provisions that address various production activities, including requirements regarding well construction (e.g., casing and cementing), well stimulation (e.g., hydraulic fracturing), and well operation (e.g., pressure testing and blowout prevention). Nonetheless, state groundwater protection officials also have reported that development of shale gas and tight oil using high-volume hydraulic fracturing, in combination with directional drilling, has posed new challenges for the management and protection of water resources. Consequently, many of the major producing states have revised or are in the process of revising their oil and gas laws and regulations to respond to these advances in oil and natural gas production technologies and related changes in the industry.

Today, activism against fracking has taken an unprecedented twist. From California to New York, from Nova Scotia to British Columbia, people have lined up in almost a cultlike fervor to "ban fracking." This comes with a backdrop of months of intense debate, extensive reviews of the scientific literature, and input from citizens, industry, responsible environmentalists and academics, leading to several states writing legislature on the most far-reaching hydraulic fracturing law of modern time. Three things that did not happen are (1) scientific investigation pointing to the real short comings of fracking, (2) abatement in fracking operations in unconventional gas and oil reservoirs, and (3) research into how to turn fracking operations environmentally acceptable and sustainable. In the mean time, the United States is poised to continue to produce from unconventional reservoirs in an unprecedented pace.

Figure shows overall energy outlook of the United States. In terms of project, natural gas shows the highest growth in the 30-year projection. In fact, other than natural gas, only renewable and liquid biofuels show modest increase, while others drop or remain constant. Natural gas plays even more intense role when it comes to electrical power usage. Figure 2.5 shows the energy outlook of the United States. This figure shows natural gas is the most promising energy provider in coming decades. It indicates that US dry

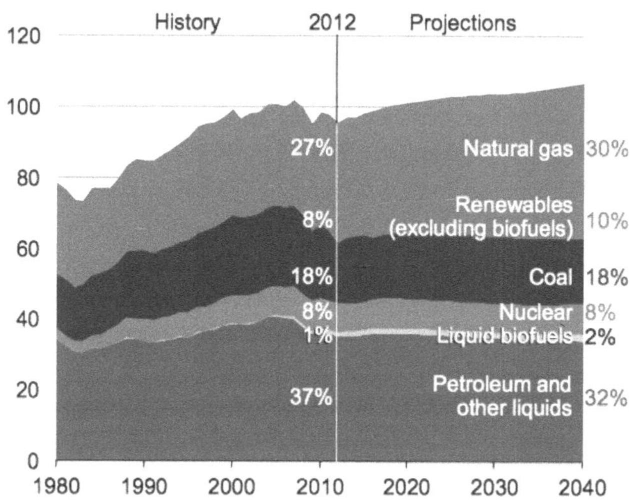

Figure 2.4 Overall energy outlook of the United States in quadrillion Btu. *From EIA Energy Outlook report, 2014.*

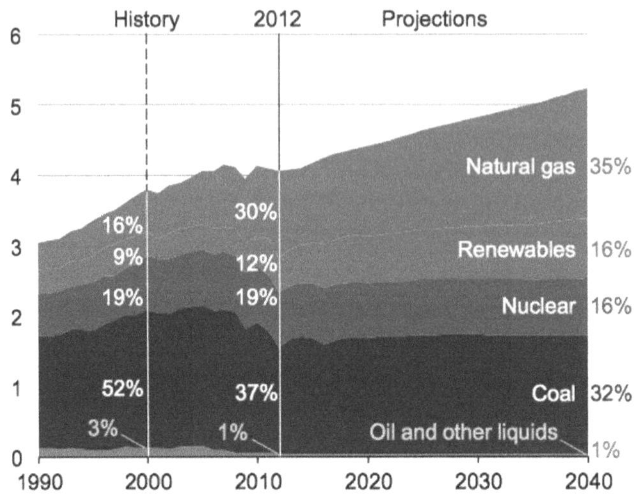

Figure 2.5 Past and projected electricity use in trillion Kwh. *From EIA Energy outlook report, 2014.*

natural gas production increases 1.3% per year throughout the reference case projection, outpacing domestic consumption by 2019. Higher volumes of shale gas production are central to higher total production volumes and a transition to net exports. As domestic supply has increased in recent years, natural gas prices have declined, making the United States a less attractive market for imported natural gas and more attractive for export.

Figure 2.6 shows total US natural gas production, consumption, and net imports in the reference case, 1990–2040 (trillion cubic feet, Tcf). This figure indicates that US net exports of natural gas grow to 3.6 Tcf in 2040. Most of the projected growth in US exports consists of pipeline exports to Mexico, which increase steadily as growing volumes of imported natural gas from the United States fill the widening gap between Mexico's production and consumption. Declining natural gas imports from Canada also contribute to the growth in US net exports. Net US imports of natural gas from Canada decline sharply from 2016 to 2022, then stabilize somewhat before dropping off again in the final years of the projection, as continued growth in domestic production mitigates the need for imports.

The share of coal in electric power generation falls over the projection period. Although coal is expected to continue its important role in the US electricity generation, there are many uncertainties that could affect future outcomes. The most important factor is the relationship between coal and natural gas prices and the potential for policies aimed at reducing

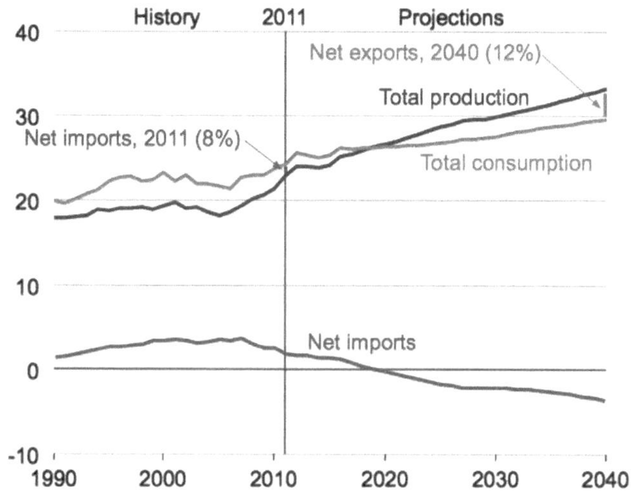

Figure 2.6 Natural gas production, consumption and import in trillion cubic feet. *From EIA Energy outlook report, 2013.*

greenhouse gas emissions. In 2012, natural gas prices were low enough for a few months for power companies to run natural gas-fired generation plants more economically than coal plants in many areas. During those months, coal and natural gas were nearly tied in providing the largest share of total electricity generation, something that had never happened before.

In the mean time, oil and gas production from unconventional sources is likely to be dominating the US domestic production. Figure 2.7 shows recovery projection from tight oil and shale gas formations. Note that shale gas production is projected to grow continuously, eclipsing tight oil in just over a decade. This potential of shale gas is credited to multilaterals and fracking techniques. However, other techniques should also be investigated and more research directed toward making fracking techniques environmentally sustainable.

This scenario plays out well with global energy outlook. Figure 2.8 shows the global energy outlook of the future along with the recent past. The International Energy Outlook 2013 (IEO2013) projects that world energy consumption will grow by 56% between 2010 and 2040. Total world energy use rises from 524 quadrillion British thermal units (Btu) in 2010 to 630 quadrillion Btu in 2020 and to 820 quadrillion Btu in 2040 (Figure 2.8). Much of the growth in energy consumption occurs in non-OECD (Organization for Economic Cooperation and Development), where demand is driven by strong, long-term economic growth. Energy

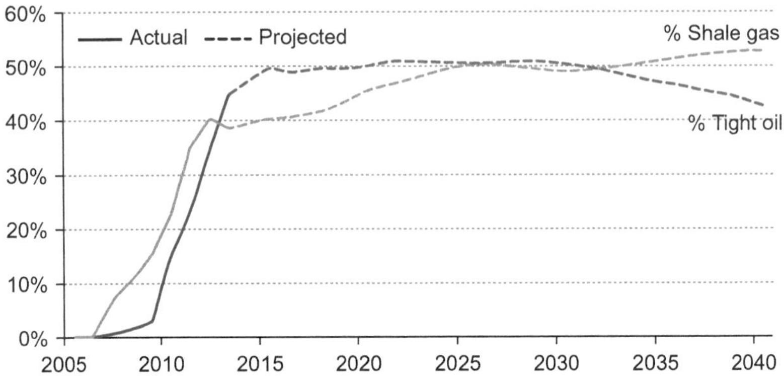

Figure 2.7 Percentage of U.S. oil and natural gas from tight oil and shale gas. *From EIA report, 2014.*

use in non-OECD countries increases by 90% as compared to 17% in OECD countries.

It is commonly perceived that renewable energy and nuclear power are the world's fastest growing energy sources, each increasing by 2.5% per year. However, fossil fuels continue to supply almost 80% of world energy use through 2040. Natural gas is the fastest growing fossil fuel in the

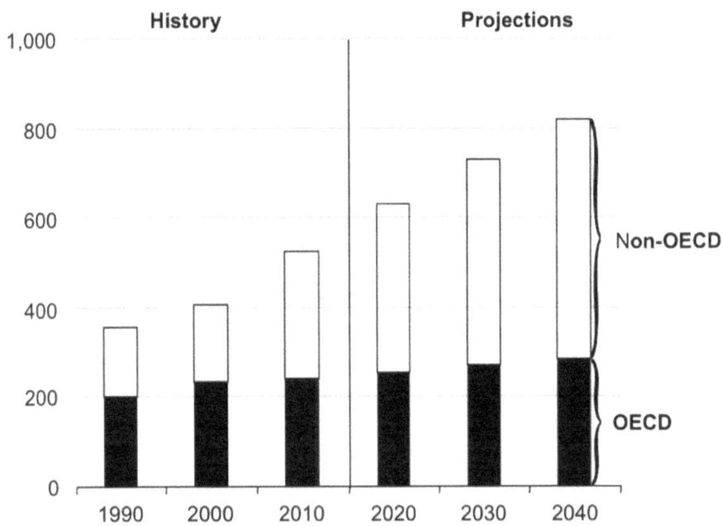

Figure 2.8 Global Energy consumption in the past and the future. *From EIA, 2013.*

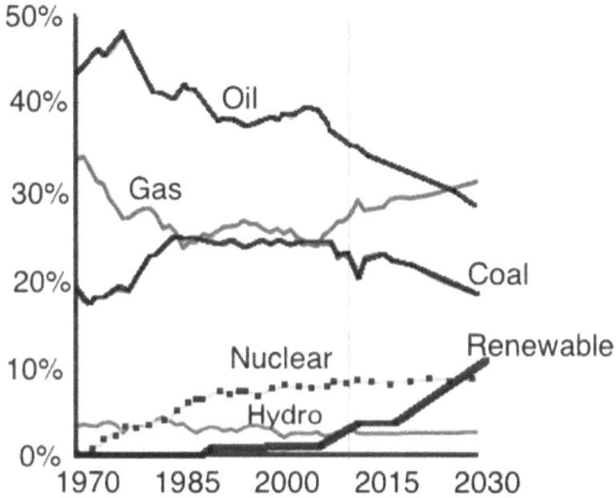

Figure 2.9 Shares of the US primary energy. *Courtesy BP.*

outlook. Global natural gas consumption increases by 1.7% per year. Increasing supplies of tight gas, shale gas, and CBM support growth in projected worldwide natural gas use. Coal use grows faster than petroleum and other liquid-fuel use until after 2030, mostly because of increases in China's consumption of coal and tepid growth in liquid demand attributed to slow growth in the OECD regions and high sustained oil prices.

Figure 2.9 shows the US shares of various energy sources. As far as the United States is concerned, all energy resources other than natural gas and renewable are in decline. Only renewable energy, such as wind, solar are predicted to be increasing in demand. However, Islam et al. (2010) have demonstrated that these "renewable" energy sources are neither renewable nor beneficial to the ecosystem. In addition, they are much more inefficient and toxic to the environment than the conventional fossil fuel burning. Even though "renewable" energy source is no threat to natural gas, one must note that renewable energy sources are subscribed and they are not renewable or sustainable, as noted by Chhetri and Islam (2008).

Figure 2.10 shows the breakdown of "renewable energy" outlook. Wind is the only energy source that is predicted to grow in future. The US oil and natural gas production is on the rise, primarily driven by resources from tight formations. The techniques developed to produce shale gas—directional drilling and hydraulic fracturing—have migrated to the oil sector. The United States is the third largest oil producer in the world, but also the fastest growing producer. The United States surpassed Russia in 2009 as the world's largest natural gas producer. Production from tight

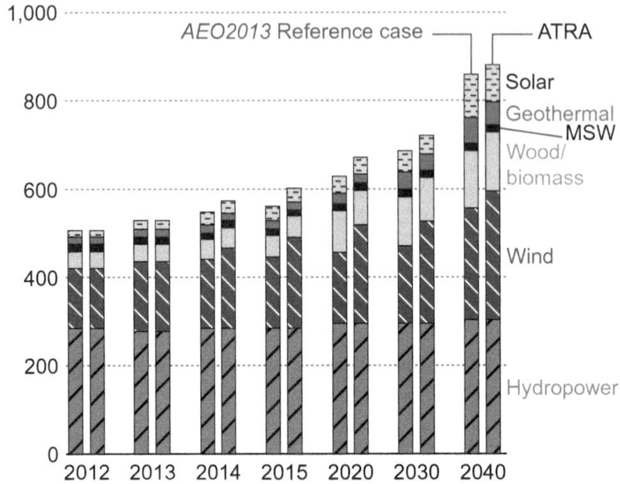

Figure 2.10 Renewable energy outlook.

formations is expected to make up a significant part of production of each commodity well into the future.

It is often suggested that unconventional oil and gas resources can pose environmental risks. These risks are purported to be due to contamination of water resources. For example, increased unconventional oil and gas production activity is thought to have placed a strain on water resources, and on wastewater treatment plants that were not designed to remove chemicals, salts, and other contaminants from hydraulic fracturing flow-back and produced water. The argument in favor of additional natural gas production is made by showing it as a "bridge" fuel that can provide more energy per unit of greenhouse gas produced than some alternatives (e.g., coal). While this second one is true, the science behind these conclusions is inaccurate. Hydraulic fracturing activities can be minimized, but more importantly, they can be rendered sustainable by removing artificial chemicals from the fracturing scheme. This aspect will be discussed in a later chapter of this book. In addition, natural gas itself is scientifically most environment-friendly natural resource and can be rendered environmentally sustainable by eliminating artificial chemicals during the processing cycle.

Nonetheless, the rapid expansion of oil and natural gas extraction using high-volume hydraulic fracturing has generated controversy in both rural and urban communities. Protests against "fracking" has become a routine phenomenon. Water quality issues have received the most attention, and of these, the potential risks associated with well stimulation by hydraulic fracturing have been at the forefront. Complaints of contaminated well water have emerged in various areas where unconventional oil and gas development is occurring, although regulators have not reported a direct connection between hydraulic fracturing of shale formations at depth and groundwater contamination. In shale formations, the vertical distance separating the target zone from usable aquifers generally is much greater than the length of the fractures induced during hydraulic fracturing. Thousands of feet of rock layers typically overlay the produced portion of shale, and these layers serve as barriers to flow. In these circumstances, geologists and state regulators generally view as remote the possibility of creating a fracture that could reach a potable aquifer. If the shallow portions of shale formations were developed, then the thickness of the overlying rocks would be less and the distance from the shale to potable aquifers would be shorter, posing more of a risk to groundwater. In contrast to shale, CBM basins often qualify as underground sources of drinking water. All these concerns can be mitigated by resorting to "green" fracturing fluid and natural materials that have coexisted with human activities for millennia.

State regulators have expressed more concern about the groundwater contamination risks associated with developing a natural gas or oil well (drilling through an overlying aquifer and casing, cementing, and completing the well), as opposed to hydraulic fracturing per se. The challenges of sealing off the groundwater and isolating it from possible contamination are common to the development of any oil or gas well, and are not unique to hydraulic fracturing. This special sensitivity is somewhat curious because natural gas is nature processed and can coexist with other natural systems, such as ground water. The confusion arises from the legislations that fail to distinguish between natural chemicals and artificial chemicals and all chemicals are painted with the same brush, regardless of their origins. In fact, natural gas and crude oil are considered to be the most toxic agent that needs to be removed from drilling mud (Khan and Islam, 2007). This reaction extends from oil to natural gas. Air pollution associated with unconventional oil and natural gas production has also raised public health concerns and has drawn regulatory scrutiny. Air pollutants can be released during various stages of oil and natural gas production, not just hydraulic fracturing. Emission sources include pad, road, and pipeline construction; well drilling and completion, and flowback activities; and natural gas processing, storage, and transmission equipment. Key pollutants include methane (the main component of natural gas and a potent greenhouse gas), volatile organic compounds (VOCs), nitrogen oxides, sulfur dioxide, particulate matter, and various hazardous air pollutants. According to Environmental Protection Agency (EPA), the oil and gas industry is a significant source of VOCs, which react with nitrogen oxides to form ozone (smog). EPA has identified hydraulically fractured gas wells during flowback as an additional source of VOC emissions in the natural gas industry.

The debate over the groundwater contamination risks associated with hydraulic fracturing operations has been fueled, in part, by the lack of scientific studies to assess more thoroughly the current practices and related complaints and uncertainties. To help address this issue, Congress has directed the EPA to conduct a study on the relationship between hydraulic fracturing and drinking water. The "hydraulic fracturing" debate also has been complicated by terminology. Many who express concern over the potential environmental impacts associated with hydraulic fracturing do not differentiate the well stimulation process of "fracking" from the full range of activities associated with unconventional oil and gas exploration and production. Figure 2.11 shows various regions of USA with their status vis-a-vis fracking practices.

2.3 PETROLEUM RESERVE AROUND THE WORLD

Scientifically, the word "reserve" implies technically and economically recoverable resource. However, technology as well as economic constraints both evolve continuously in addition to being intricately linked with sociopolitical aspects. Standardization of the term "reserve" has been a concern for some time.

Proven reserves are those reserves claimed to have a reasonable certainty (normally at least 90% confidence) of being recoverable under existing economic and political conditions, with existing technology. Industry specialists refer to this as P90 (i.e., having a 90% certainty of being produced). Proven reserves are also known in the industry as 1P (SPE, 2005). This "confidence" value arises from probability assignments that are themselves of spurious origin (Hossain and Islam, 2009). However, this denomination has been in place for sometime.

Proven reserves are further subdivided into "proven developed" (PD) and "proven undeveloped" (PUD) (Wright et al., 2008). PD reserves are reserves that can be produced with existing wells and perforations or from

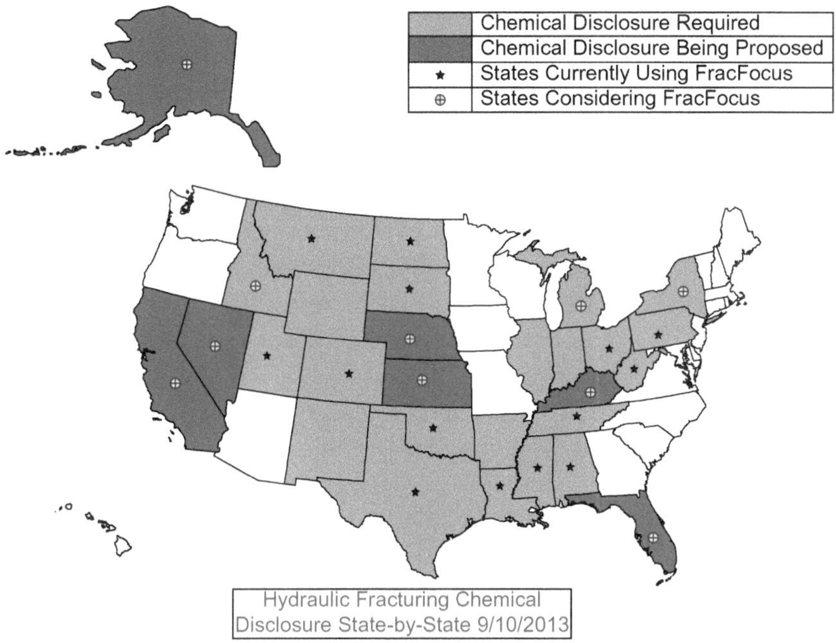

Figure 2.11 Hydraulic fracturing chemical disclosure by state. *From Ratner and Tiemann, 2014.*

additional reservoirs where minimal additional investment (operating expense) is required, keeping the infrastructure/capital cost to zero. PUD reserves require additional capital investment (e.g., infill drilling) to add to the oil and gas production.

Until December 2009 "1P" proven reserves were the only type the U.S. Securities and Exchange Commission (SEC) allowed oil companies to report to investors. Companies listed on US stock exchanges must substantiate their claims, but many governments and national oil companies do not disclose verifying data to support their claims. Since January 2010 the SEC now allows companies to also provide additional optional information declaring "2P" (both proven and probable) and "3P" (proven + probable + possible) provided the evaluation is verified by qualified third party consultants, though many companies choose to use 2P and 3P estimates only for internal purposes.

Unproven reserves are based on geological and/or engineering data similar to that used in estimates of proven reserves, but technical, contractual, or regulatory uncertainties preclude such reserves being classified as proven. Unproven reserves may be used internally by oil companies and government agencies for future planning purposes but are not routinely compiled. They are subclassified as probable and possible.

Probable reserves are attributed to known accumulations and claim a 50% confidence level of recovery. Industry specialists refer to them as "P50" (i.e., having a 50% certainty of being produced). These reserves are also referred to in the industry as "2P" (proven + probable).

Possible reserves are attributed to known accumulations that have a less likely chance of being recovered than probable reserves. This term is often used for reserves that are claimed to have at least a 10% certainty of being produced ("P10"). Reasons for classifying reserves as possible include varying interpretations of geology, reserves not producible at commercial rates, uncertainty due to reserve infill (seepage from adjacent areas), and projected reserves based on future recovery methods. They are referred to in the industry as "3P" (proven + probable + possible).

Figure 2.12 shows that world proved oil reserves at the end of 2012 reached 1668.9 billion barrels, sufficient to meet 52.9 years of global production. Even though an increase in official Iraqi reserves was the single largest addition, adding 6.9 billion barrels, subsequent data indicate that similar increase is also warranted due to current ability to access

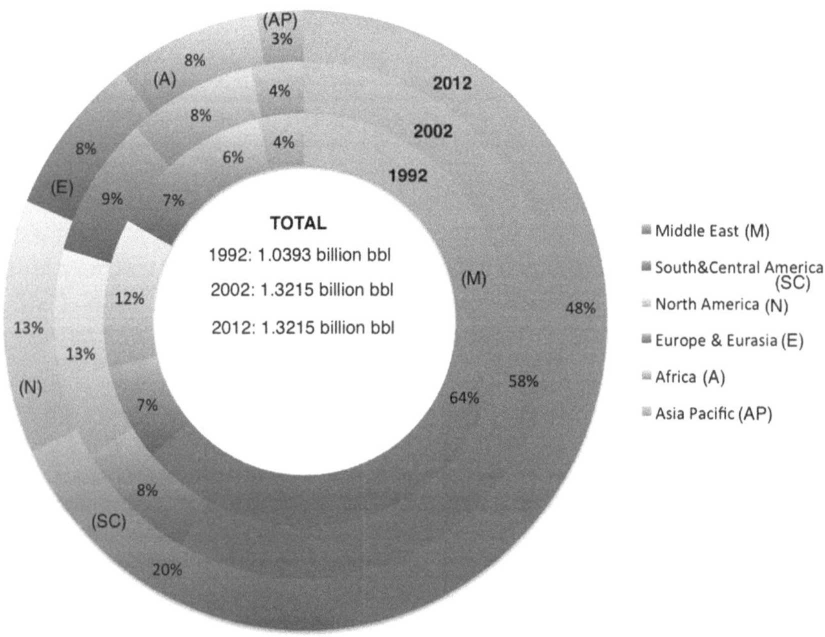

Figure 2.12 World proved reserve growth in last two decades. *Data from BP, 2013.*

unconventional oil and gas. Organization of the Petroleum Exporting Countries (OPEC) members continue to dominate, holding 72.6% of the global total. South and Central America continue to hold the highest reserve/production ratio. Global proved reserves have increased by 26%, or nearly 350 billion barrels, over the past decade. This is similar to the rise in reported in the previous decade.

Table 2.1 shows total oil reserve as well as reserve/production ratio of top oil producing countries. Each country is marked for its need for enhanced oil recovery (EOR). Note that the need does not imply suitability nor does it mean that other countries would not benefit from an EOR scheme. This is a crucial element as the status of "reserve" depends on the technology that can increase the recovery in the short term.

This table shows Venezuela's proved reserve has surpassed that of Saudi Arabia. This is a recent development and partly fueled by the addition of unconventional petroleum to the previous reserve. Few countries routinely

Table 2.1 Summary of Proven Reserve Data as of 2012

S. No.	Country	Reserves (10^9 bbl)	Production (10^6 bbl/day)	Reserve/ Production ratio (years)	Need for EOR
1	Venezuela	296.5	2.1	387	Low
2	Saudi Arabia	265.4	8.9	81	Medium
3	Canada	175	2.7	178	Low
4	Iran	151.2	4.1	101	Low
5	Iraq	143.1	2.4	163	Low
6	Kuwait	101.5	2.3	121	Low
7	United Arab Emirates	136.7	2.4	156	Low
8	Russia	80	10	22	High
9	Kazakhstan	49	1.5	55	High
10	Libya	47	1.7	76	Medium
11	Nigeria	37	2.5	41	High
12	Qatar	25.41	1.1	63	Medium
13	China	20.35	4.1	14	High
14	United States	26.8	7	10	High
15	Angola	13.5	1.9	19	High
16	Algeria	13.42	1.7	22	High
17	Brazil	13.2	2.1	17	High
	Total of top 17 reserves	1324	56.7	64	

EOR, Enhanced oil recovery.

update their proven reserves. In addition, due to lack of consistent criteria, some national oil companies have resorted to back calculate original oil in place from a proven reserve value that had little relevance to original in place and was decided by political events (e.g., quota in OPEC). This problem, along with others, prompted the Society of Petroleum Engineers (SPE) to appoint a committee, titled Oil and Gas Reserves Committee (OGRC) that compiled a study of reserve/resource classification systems published by the following eight international entities:

1. US Security and Exchange Commission (SEC-1978)
2. UK Statement of Recommended Practices (SORP-2001)
3. Canadian Security Administrators (CSA-2002)
4. Russian Ministry of Natural Resources (RF-2005)
5. China Petroleum Reserves Office (PRO-2005)
6. Norwegian Petroleum Directorate (NPD-2001)
7. United States Geological Survey (USGS-1980)
8. United Nations Framework Classification (UNFC-2004)

The overall structure of, and reserves definitions within, each system were compared to the 1997 SPE/WPC reserves definitions, the 2000 SPE/WPC/AAPG classification, the 2001 supplemental guidelines, and the 2004 glossary (hereafter referred to as the "SPE definitions"). The following conclusions were drawn (SPE, 2005):

- All systems define major resource categories that can be mapped directly to the SPE categories: undiscovered (prospective resources), discovered unrecoverable, discovered subcommercial (contingent resources), and discovered commercial (reserves).
- Most classifications recognize three deterministic scenarios with decreasing technical certainty: a low estimate, best estimate, and high estimate. It is generally accepted that the equivalent estimates on a cumulative probability distribution would be greater than or equal to P90, P50, and P10, respectively. For discovered and commercial volume estimates, the discrete (incremental) volumes within these bounds are generally referred to as proved, probable, and possible reserves. The Russian, UNFC, and USGS recognize similar certainty classes but use alternative terminology.
- The regulatory agencies typically define a subset of the total classification for disclosure to investors and further impose specific rules around technical and commercial certainty. The SEC guidance is the most restrictive while the Canadian and UK regulations allow disclosures more closely aligned with assessments used for internal resource management.

- The UNFC uniquely provides a high-level classification system that can be applied to all extractive industries including energy minerals (petroleum, coal, and uranium).

Based on the above conclusions, the following recommendations were made:

- Utilize a consistent set of criteria to segregate discovered from undiscovered without reference to ultimate commerciality. All such discovered volumes should be initially categorized as contingent resources.
- Estimates of recoverable quantities must clearly identify the development project(s) applied to a specific accumulation (reservoir) and its in-place hydrocarbons. The "project-reservoir" intersect becomes the resource entity for which an uncertainty distribution of recoverable quantities is defined. The project maturity/chance of reaching production status is used to segregate reserves from contingent resources.
- To maintain consistency, the same class confidence hurdles (P90/P50/P10) should be applied to estimates whether assessed using deterministic or probabilistic methods. Although the assessment should support either arithmetic summation or probabilistic aggregation, the guidelines should clearly identify that these certainty guidelines apply to the project-reservoir entity.
- From a business perspective, the inclusion of additional deterministic technical and commercial criteria for reserves classes (proved, probable, possible) or discrete estimates (1P, 2P, 3P) may have value in providing increased consistency in assessments. However, these should be provided as guidelines and not imbedded in the class definitions. The definitions should be broad enough to accommodate such criteria as imposed by regulatory agencies.
- Apply developed/undeveloped status to all reserve classes. Reserves that remain undeveloped beyond a reasonable period demonstrate lack of commitment and should be reclassified as contingent resources.
- The definitions should encompass all hydrocarbons whether conventional or unconventional (gas, liquid, or solid phases) irrespective of the extraction method and processing applied.
- The total system should provide for accounting of all components to support mass balance; that is, the sum of produced, recoverable, production/processing losses, and unrecoverable quantities should equal the estimated initially in-place hydrocarbons. The guidelines should provide the option, subject to regulatory rules, of including hydrocarbons to be consumed as fuel in production and processing as reserves and contingent resources.

- Documentation regarding reserves and resources is best presented in a more structured manner consisting of:
  Overall resource classification—chart and resource category definitions
  Reserves definitions—high level, principle-based
  Application guidelines—detailed guidance, subject to periodic revisions
  Application examples—illustrations of both common and exceptional issues

This work of SPE highlights the need of going back to original definition of petroleum resource as per geologic and geophysical data irrespective of the technology being used. This solves the problem of coupling technology to the definition of reserve, which is of particular relevance when it comes to unconventional oil and gas. The total estimated amount of oil and gas in a petroleum reservoir is called the original oil or gas in place. Even though only a fraction of this oil can be brought to the surface, the fraction of produced oil and gas depends on several factors, most important of which is the enhanced oil and gas recovery technique used. For instance, it is estimated that 12% of oil in place in the United States is recoverable through EOR. This amount is incremental to other technologies (e.g., primary, secondary). This number is estimated to be much higher for gas reservoirs. However, few gas reservoirs are actually subject to enhanced gas recovery scheme. For unconventional gas and oil, the problem arises from the fact that the entire volume of unconventional oil and gas is considered to be nonproducible with conventional definition of "proven reserve." As the prospect of unconventional oil and gas expands, the need to consider the entire oil and gas in place for future planning becomes a necessity (OPEC, 2011).

Even though Venezuela has exceeded Saudi Arabia in terms of oil (including gas condensate and liquid natural gas), it is generally recognized that Venezuelan deposits are more difficult to extract and are not readily available like Saudi Arabia. Figure 2.13 shows how proven reserve has evolved for some of the countries with greatest reserves. In recent years, Saudi Arabia and Canada both have held proved reserve at a constant value. Venezuela and Iran have made the most spectacular gains in proven reserve estimates. This gain is due to acceptance of internal data by OPEC and the rest of the world and it was not a matter of new technology of new discovery (Zatzman, 2012). There had to be some structural change in OPEC in order for it to recognize Venezuela's claim of greater reserve as legitimate. Of course, Iraq's reserve was updated due to uncertainties arising political unrests. A measure of the uncertainty about Iraq's oil reserves is indicated by widely differing estimates. The U.S. Department of Energy estimated in

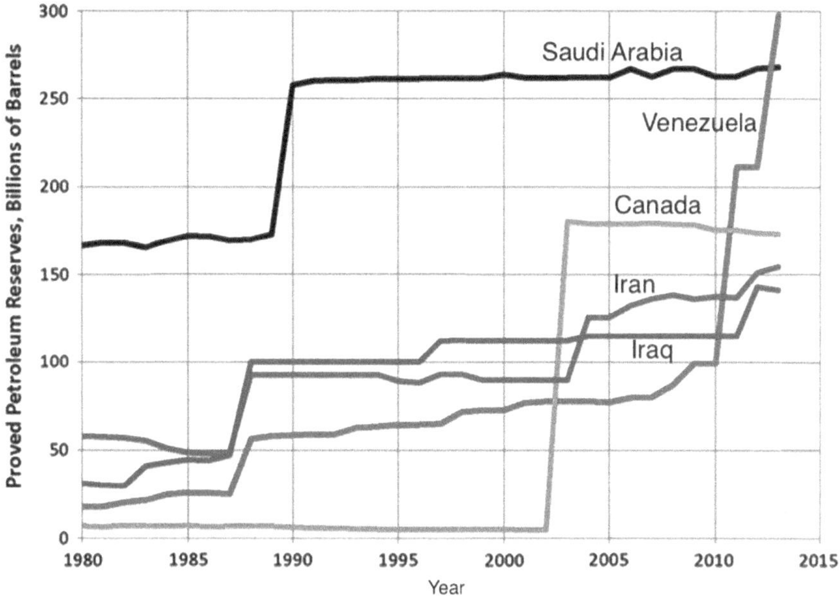

Figure 2.13 Proven reserve of top holders of petroleum reserve. *Data from USA EIA.*

2003 that Iraq had 112 billion barrels (17.8×10^9 m^3). In 1995, the USGS-estimated proven reserves were 78 billion bbl (12.4×10^9 m^3). Iraq's prewar deputy oil minister said that potential reserves might be 300 billion bbl (48×10^9 m^3). The source of the uncertainty is that due to decades of war and unrest, many of Iraq's oil wells are rundown and unkept. Repairs to the wells and oil facilities should make far more oil available economically from the same deposits. This intermingling of political, technical, and social issues is characteristic of oil business.

Even for the OECD countries, such as Canada, estimation of proven reserve is not without controversy. For instance, when Alberta's oil sand deposits of 174 billion barrels (28×10^9 m^3) were added as proven reserves by the Alberta Energy and Utilities Board, now known as the Energy Resources Conservation Board (ERCB), a great deal of skepticism was offered because oil sand is not "readily" usable as a petroleum resource.

While it is true that oil sand has to be mined, heated, or diluted with solvents to allow it to be produced and must be upgraded to lighter oil to be usable by refineries, it is still economically attractive, particularly in view of the vastness of the resource. Historically known as bituminous sands or sometimes as "tar sands," the deposits were exposed as major rivers cut through the oil-bearing formations to reveal the bitumen in the river banks.

In recent years technological breakthroughs have overcome the economical and technical difficulties of producing the oil sands, and by 2007 64% of Alberta's petroleum production of 1.86 million barrels per day (296,000 m^3/day) was from oil sands rather than conventional oil fields. The ERCB estimates that by 2017 oil sands production will make up 88% of Alberta's predicted oil production of 3.4 million barrels per day (540,000 m^3/day) (ERCB, 2008). Until now, OPEC does not recognize Canada's oil sand reserve as part of the "proved reserve."

2.4 WORLDWIDE GAS RESERVE

Some of the most dramatic energy developments of recent years have been in the realm of natural gas. Even before the "rush" of unconventional shale and tight gas, natural gas has been prominent in the post-Cold War era. In early 2000, the natural gas price remained low while huge quantities of unconventional gas entered commercial levels of production in the United States, changing the strategic picture for the United States by making it self-sufficient in natural gas for the foreseeable future. The benefit of such production boon transformed into US domestic production surpassing import in 2013. This would occur after some 16 years of opposite trend (Bloomberg, 2013). This phenomenon is somewhat different in Europe. Some 30% of European gas demand is met by Russian natural gas. Russia, even though second to USA, is the largest exporter.

Even before former USSR broke down, gas export from Russia had been declining. Even though such decline is often correlated with political events and US hegemony, the fact remains that the so-called "gas war" had to break out in order to restore values of natural gas to the level comparable to crude oil or petroleum liquids. Even though Gazprom was privatized in 2005, the Russian government has held a controlling share in Gazprom. The earliest sign of restoration of natural gas price to an equitable value was in place when on October 2, 2008, the Ukrainian Prime Minister Julia Timoshenko and the then Russian Prime Minister Vladimir Putin had agreed in a memorandum on the Ukraine raising the gas price to world market standards within the next three years. Previous to that, Russia had delivered gas to the Ukraine far below world market prices until the end of 2008, namely for US$ 180/1000 m^3. Gazprom itself had to import the gas that they delivered to the Ukraine, namely from Central Asian states—at a price of US$ 375/1000 m^3. This led to a billion dollar deficit for the Russian gas supplier.

At the end of 2008, the existing contract between the Russian Gazprom and the Ukrainian gas corporations expired—both gas corporations are under state control. A new contract about a new gas price in terms of the October 2008 memorandum and valid from January 1, 2009 on was prevented by the Ukraine, although Russia had made an offer to deliver the gas at a price of US$ 250/1000 m^3, which is less than the current world market price.

Thus Gazprom stopped its gas deliveries to the Ukraine on January 1, 2009. And this led to Ukraine unlawfully tapping the transit pipelines running through the Ukraine to other European states. Russia reacted by discontinuing the gas transfer across the Ukraine, completely.

Gazprom has an interesting background. Originally, Gazprom was an offshoot of the Soviet Gas Ministry, set up in 1965 when the USSR decided to develop the production and consumption of gas. In 1989, Gazprom became an "independent" company, in line with Khrushchev's original initiative to "modernize" USSR economy. Its first chairman was Viktor Chernomyrdin, a former prime minister of Russia under Boris Yeltsin and an ambassador to Ukraine (2001–2009). In August 1989, under the leadership of Chernomyrdin, the Ministry of Gas Industry was transformed itself into the State Gas Concern, Gazprom, which became the country's first state-corporate enterprise. Chernomyrdin was elected its first chairman. The company was still controlled by the state, but now the control was exercised through shares of stock, 100% of which were owned by the state. When the Soviet Union dissolved in late 1991, assets of the former Soviet state in the gas sector were transferred to newly created national companies such as Ukrgazprom and Turkmengazprom. Gazprom kept assets located in the territory of Russia and was able to secure a monopoly in the gas sector.

Gazprom's political influence increased markedly after the new Russian President Boris Yeltsin appointed the company's chairman Chernomyrdin as his Prime Minister in 1992. Rem Viakhirev took Chernomyrdin's place as Chairman both of the Board of Directors and of the Managing Committee. Gazprom was one of the backbones of the country's economy in 1990s, though the company underperformed during that decade. This was mainly because gas price was much lower than its equitable value. In the 2000s, Gazprom became the largest extractor of natural gas in the world and the largest Russian company. In 1993, the company was reorganized into a joint-stock company named RAO Gazprom, before becoming OAO Gazprom in 1998, the name it retains to this day.

In 2005, the Russian state became the company's majority shareholder with 50.01% of the share capital. At present Gazprom controls 25% of the world's gas reserve and 94% of Russia's natural gas. Gazprom owns the entire gas transport infrastructure in Russia—some 144,000 km of pipelines—and all of the country's compressor stations. It is the only company legally authorized to sell gas outside Russia. Gazprom employs around 300,000 people and represents 25% of Russia's total budgetary revenue.

This combination of energy and politics has made Gazprom an instrument of politico-social regulation in Russia. Russia is the world's second largest consumer of gas after the United States, although its economic power is almost 20 times less that of the United States. Russia therefore wastes its energy. Table 2.2 shows natural gas consumption of world's top 25 consumers.

The average price of 1000 cubic meters of gas for a Russian household has remained for a long time near US$ 15, although it did come close to US$ 40 in 2006. The low gas prices also reflect political interests: companies, which are not very competitive and major users of energy, can use them to lower their costs; low gas prices also prevent many Russian households from experiencing financial difficulties, which would happen if were prices to rise.

Soon after the first "gas war" the gas price rose to an unprecedented US$12/MMBtu (million Btu). However, the gas price has fluctuated ever since and hit a low value in mid-2012 in sync with "gas glut" due to unconventional gas production in the United States (Figure 2.14). This production of unconventional gas reverberated around the world and is synonymous with the culture of "fracking." In early 2014, gas price in the United States rose to a 6-year high mainly because of a severe cold spill. Even then, the real crisis began in late February of 2014.

It must be added that Russia has been trying for some time to gain access to the Ukrainian pipeline network in order to invest in the system of pipelines and guarantee an uninterrupted gas delivery to Europe. It has therefore suggested to establish an international consortium for the maintenance of that network. In 2014, the political crisis led to threat of invasion of Crimea; Gazprom declared it intends to start charging Ukraine around US$ 400/1000 m^3 for its gas, as opposed to the $270-odd it has been paying since Yanukovych spurned Brussels for Moscow.

In the mean time, since the first "gas war" of 2006, many European countries have made huge efforts to increase their gas storage capacity and stocks high. Some countries, such as Bulgaria, Slovakia, and Moldova,

Table 2.2 Natural Gas Consumption of Top 25 Consumers

Rank	Country	Natural gas consumption (cubic meters/year)	Date of information
1	United States	689,900,000,000	2011 est.
2	Russia	457,200,000,000	2012 est.
3	European Union	443,900,000,000	2010 est.
4	Iran	144,600,000,000	2010 est.
5	China	143,000,000,000	2011 est.
6	Japan	112,600,000,000	2011 est.
7	Saudi Arabia	99,230,000,000	2011 est.
8	Canada	82,480,000,000	2010 est.
9	United Kingdom	82,210,000,000	2011 est.
10	Germany	75,200,000,000	2010 est.
11	Italy	68,700,000,000	2010 est.
12	India	64,490,000,000	2010 est.
13	United Arab Emirates	60,540,000,000	2010 est.
14	Mexico	59,150,000,000	2011 est.
15	Ukraine	56,200,000,000	2010 est.
16	Netherlands	54,850,000,000	2010 est.
17	Korea, South	49,660,000,000	2011 est.
18	France	47,990,000,000	2010 est.
19	Uzbekistan	46,800,000,000	2012 est.
20	Egypt	46,170,000,000	2010 est.
21	Thailand	45,080,000,000	2010 est.
22	Argentina	43,290,000,000	2010 est.
23	Pakistan	42,900,000,000	2011 est.
24	Indonesia	39,560,000,000	2010 est.
25	Turkey	38,130,000,000	2010 est.

CIA Factbook.

which lack large storage capacity and depend heavily on gas supplies via Ukraine, would certainly suffer from any disruption in supplies. But Gas Infrastructure Europe, which represents the gas infrastructure industry, estimated that in late February European gas storage was 10% points higher than this time last year and about half full; the National Grid puts Britain's stocks at about 25% points above the average for that time of year.

Russia is Europe's biggest gas supplier, providing around a quarter of continental demand, which at current daily flows of 270 million cubic meters is worth almost US$ 100 million a day. Around a third of Russia's gas is exported through Ukraine (Figure 2.15).

In this energy politics scenario, natural gas from unconventional sources looms large. Access to such reservoirs means longstanding gas exporters,

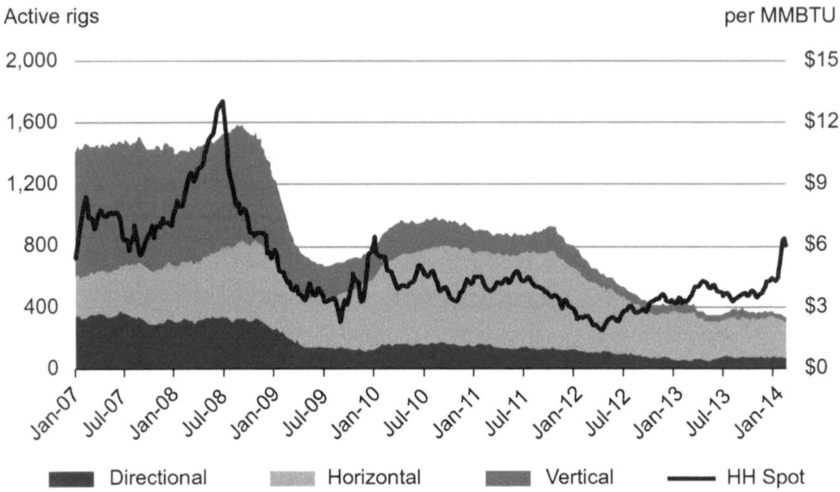

Figure 2.14 History of gas price and active rigs. *From EIA.*

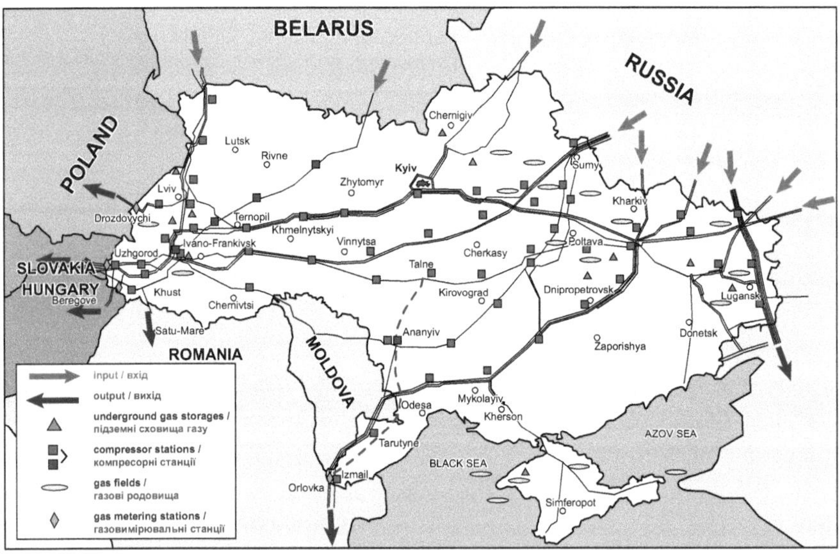

Figure 2.15 One-third of Russia's supply to Europe go through Ukraine.

such as Russia, now have to fight for their share of the market. Previously such competition was offered only from producers such as Qatar that would export to Europe in the form of liquefied natural gas. The unconventional reserves will also compensate for any "glut" created through North Sea gas that has captured a good share of the European market. The recent improvement in fracking technologies has opened up eastern Europe to major oil companies such as Chevron, who have been very active in western Ukraine, Poland, and Romania over the last two years, signing agreements to commence drilling operations in these countries. Fracking technology has been proven to be efficient in numerous cases in the United States. However, this is not the only technology that can make unconventional resources appealing. Emerging technologies as well as technologies well established in other areas of petroleum engineering will make much higher amount of natural gas available with minimum additional cost of production and operations. This contribution is markedly superior both in environmental and economic term than offshore reserves that are currently considered to be the alternative to "political gas."

Table 2.3 shows the yearly natural gas production from various countries, as supplied by CIA Factbook.

The natural gas production should be tallied against natural gas export in order to determine the overall dependence or dominance of a country in natural gas trade. Table 2.4 shows natural gas export by various countries. Note how the ranking changes from producers to exporters.

Of significance is how the United States has grown both in its production as well as export of natural gas. Figure 2.16 shows how exponential growth has taken place in US export of natural gas.

During the same period, considerable decline in natural gas import has taken place. This is evident from Figure 2.17. Coinciding with the first "gas war" that saw a price hike of natural gas worldwide, the United States has successfully decreased its net import of natural gas. This trend continues even when unusually cold winter in United States (simultaneous with unusually warm winter in Europe) led to record consumption of natural gas in 2013.

Figure 2.18 shows the natural gas export history of Russia. Interestingly Russia and USA share opposite profiles in terms of net export. Soon after the gas crisis of 2006, Russia's export declined sharply whereas that of United States increased, while its imports decreased. The United States exports continue to increase while imports continue to decline right through the financial crisis of 2008. This surplus can be explained with the emergence of "fracking" technology that unlocked tremendous amount of unconventional gas, particularly that from shale gas resources. Figure 2.19 shows shares of natural gas from various sources in the United States.

Table 2.3 Natural Gas Production from Various Countries

Rank	Country	Natural gas production (cubic meters/year)	Date of information
1	United States	681,400,000,000	2012 est.
2	Russia	669,700,000,000	2012 est.
3	European Union	164,600,000,000	2012 est.
4	Iran	162,600,000,000	2011 est.
5	Canada	143,100,000,000	2012 est.
6	Qatar	133,200,000,000	2011 est.
7	Norway	114,700,000,000	2012 est.
8	China	107,200,000,000	2012 est.
9	Saudi Arabia	103,200,000,000	2012 est.
10	Algeria	82,760,000,000	2011 est.
11	Netherlands	80,780,000,000	2012 est.
12	Indonesia	76,250,000,000	2011 est.
13	Uzbekistan	62,900,000,000	2012 est.
14	Malaysia	61,730,000,000	2011 est.
15	Egypt	61,260,000,000	2011 est.
16	Turkmenistan	59,500,000,000	2011 est.
17	Mexico	53,960,000,000	2012 est.
18	United Arab Emirates	52,310,000,000	2011 est.
19	Bolivia	48,970,000,000	2012 est.
20	Australia	48,240,000,000	2012 est.
21	United Kingdom	40,990,000,000	2012 est.
22	Trinidad and Tobago	40,600,000,000	2011 est.
23	India	40,380,000,000	2012 est.
24	Pakistan	39,150,000,000	2011 est.
25	Argentina	38,770,000,000	2011 est.
26	Thailand	36,990,000,000	2011 est.
27	Oman	35,940,000,000	2012 est.
28	Peru	32,400,000,000	2012 est.
29	Nigeria	31,360,000,000	2011 est.
30	Venezuela	25,280,000,000	2012 est.
31	Kazakhstan	20,200,000,000	2011 est.
32	Bangladesh	20,110,000,000	2011 est.
33	Ukraine	19,800,000,000	2011 est.
34	Azerbaijan	17,860,000,000	2011 est.
35	Brazil	17,030,000,000	2012 est.
36	Kuwait	13,530,000,000	2011 est.
37	Bahrain	12,620,000,000	2011 est.
38	Brunei	12,440,000,000	2011 est.
39	Burma	11,910,000,000	2011 est.
40	Colombia	10,950,000,000	2011 est.
41	Romania	10,610,000,000	2011 est.

Table 2.3 Natural Gas Production from Various Countries—cont'd

Rank	Country	Natural gas production (cubic meters/year)	Date of information
42	Yemen	9,620,000,000	2011 est.
43	Vietnam	9,300,000,000	2012 est.
44	Germany	9,000,000,000	2012 est.
45	Syria	7,870,000,000	2011 est.
46	Libya	7,855,000,000	2011 est.
47	Italy	7,800,000,000	2012 est.
48	Equatorial Guinea	6,880,000,000	2011 est.
49	Israel	6,860,000,000	2011 est.
50	Denmark	6,412,000,000	2012 est.
51	Poland	6,193,000,000	2012 est.
52	Portugal	4,904,000,000	2012 est.
53	New Zealand	4,590,000,000	2012 est.
54	Philippines	3,910,000,000	2012 est.
55	Mozambique	3,820,000,000	2011 est.
56	Japan	3,273,000,000	2012 est.
57	Hungary	2,462,000,000	2012 est.
58	Tunisia	1,930,000,000	2011 est.
59	Austria	1,906,000,000	2012 est.
60	Croatia	1,850,000,000	2012 est.
61	Cote d'Ivoire	1,500,000,000	2011 est.
62	South Africa	1,280,000,000	2011 est.
63	Chile	1,144,000,000	2012 est.
64	Cuba	1,030,000,000	2012 est.
65	DR Congo	946,000,000	2012 est.
66	Iraq	880,000,000	2011 est.
67	Tanzania	860,000,000	2011 est.
68	Angola	752,000,000	2011 est.
69	Turkey	632,000,000	2012 est.
70	Serbia	557,000,000	2012 est.
71	France	508,000,000	2012 est.
72	Korea, South	424,900,000	2012 est.
73	Bulgaria	410,000,000	2011 est.
74	Ireland	373,000,000	2012 est.
75	Taiwan	330,200,000	2011 est.
76	Ecuador	240,000,000	2011 est.
77	Jordan	230,000,000	2011 est.
78	Belarus	220,000,000	2011 est.
79	Czech Republic	200,000,000	2012 est.
80	Cameroon	150,000,000	2011 est.
81	Afghanistan	140,000,000	2011 est.

Continued

Table 2.3 Natural Gas Production from Various Countries—cont'd

Rank	Country	Natural gas production (cubic meters/year)	Date of information
82	Slovakia	105,000,000	2012 est.
83	Papua New Guinea	100,000,000	2011 est.
84	Gabon	70,000,000	2011 est.
85	Spain	61,000,000	2012 est.
86	Morocco	60,000,000	2011 est.
87	Ghana	50,000,000	2010 est.
88	Tajikistan	40,000,000	2012 est.
89	Barbados	20,000,000	2010 est.
90	Senegal	20,000,000	2011 est.
91	Kyrgyzstan	10,000,000	2011 est.
92	Albania	10,000,000	2011 est.
93	Georgia	9,151,000	2012 est.
94	Greece	6,000,000	2011 est.
95	Slovenia	2,000,000	2011 est.

It can be seen from Figure 2.19 that shale gas is the only one that has been rising continuously during the period in question. The U.S. Energy Information Association (EIA) projection shows that such rise will continue for next few decades. What is of importance here is that tight gas as well as CBM resources are predicted to hold steady. However, new technologies or more sustainable techniques can unlock significant amount from these reservoirs that have quite high original gas in place. That would change the dynamics of natural gas market in the United States as well as the world. Russia is estimated to have 5–20 trillion cubic meters (Tcm) of natural gas in unconventional reservoirs. However, these reservoirs remain untapped mainly because Russia's conventional reserve is so vast that they do not see prudence in creating gas glut by exploiting unconventional resources.

Canada offers an interesting perspective to global natural gas market. Figure 2.20 shows Canadian production and consumption history over the years 2000–2011. In most part Canadian production and net export has remained steady throughout this period. During the 2008-financial crisis and its aftermath, there was some decline in consumption of natural gas, which resulted in decline in overall production of natural gas but this did not affect the net export. Canada does have an estimated 1000 Tcf of unconventional gas, but most of it remains undeveloped. However, this scenario is changing as several "fracking" projects are underway.

Table 2.4 Natural Gas Export by Various Countries

Rank	Country	Natural gas export (cubic meters/year)	Date of information
1	Russia	196,000,000,000	2012 est.
2	Qatar	113,700,000,000	2011 est.
3	Norway	107,300,000,000	2012 est.
4	European Union	93,750,000,000	2010 est.
5	Canada	88,290,000,000	2012 est.
6	Netherlands	74,060,000,000	2012 est.
7	Algeria	52,020,000,000	2011 est.
8	Turkmenistan	46,100,000,000	2011 est.
9	United States	45,840,000,000	2012 est.
10	Slovakia	45,430,000,000	2012 est.
11	Bolivia	40,280,000,000	2012 est.
12	Indonesia	38,670,000,000	2011 est.
13	Austria	34,750,000,000	2012 est.
14	Malaysia	33,100,000,000	2011 est.
15	Australia	30,270,000,000	2012 est.
16	Nigeria	25,960,000,000	2011 est.
17	Czech Republic	25,230,000,000	2012 est.
18	Poland	25,010,000,000	2012 est.
19	Belgium	21,180,000,000	2012 est.
20	Germany	18,170,000,000	2012 est.
21	Trinidad and Tobago	17,640,000,000	2011 est.
22	Uzbekistan	13,400,000,000	2012 est.
23	United Kingdom	11,970,000,000	2012 est.
24	Oman	10,930,000,000	2011 est.
25	Egypt	10,510,000,000	2011 est.
26	Kazakhstan	9,700,000,000	2011 est.
27	Brunei	9,420,000,000	2011 est.
28	Iran	9,050,000,000	2011 est.
29	Yemen	8,750,000,000	2011 est.
30	Peru	8,730,000,000	2012 est.
31	Burma	8,570,000,000	2011 est.
32	Switzerland	8,494,000,000	2012 est.
33	France	5,994,000,000	2012 est.
34	Azerbaijan	5,550,000,000	2011 est.
35	Equatorial Guinea	5,260,000,000	2011 est.
36	United Arab Emirates	5,180,000,000	2011 est.
37	Spain	4,414,000,000	2012 est.
38	Libya	3,666,000,000	2011 est.
39	Mozambique	3,300,000,000	2011 est.
40	China	3,195,000,000	2011 est.

Continued

Table 2.4 Natural Gas Export by Various Countries—cont'd

Rank	Country	Natural gas export (cubic meters/year)	Date of information
41	Denmark	2,983,000,000	2012 est.
42	Hungary	2,837,000,000	2012 est.
43	Ukraine	2,600,000,000	2010 est.
44	Colombia	2,110,000,000	2011 est.
45	Slovenia	1,181,000,000	2012 est.
46	Turkey	600,000,000	2012 est.
47	Brazil	400,000,000	2012 est.
48	Italy	324,000,000	2012 est.
49	Croatia	222,000,000	2012 est.
50	Argentina	200,000,000	2011 est.
51	DR Congo	39,000,000	2012 est.
52	Mexico	11,000,000	2012 est.

From CIA Factbook.

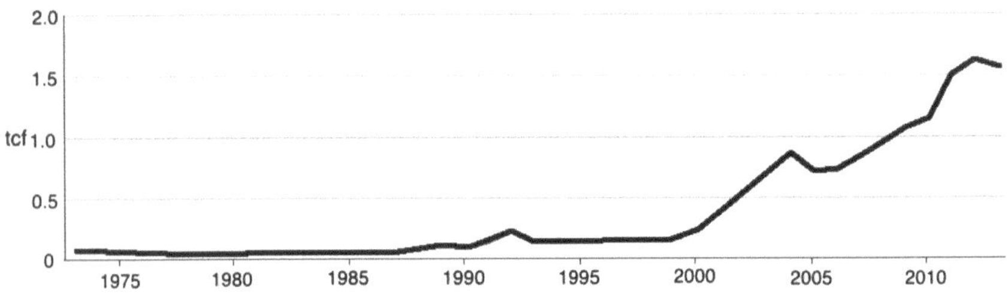

Figure 2.16 US export of natural gas.

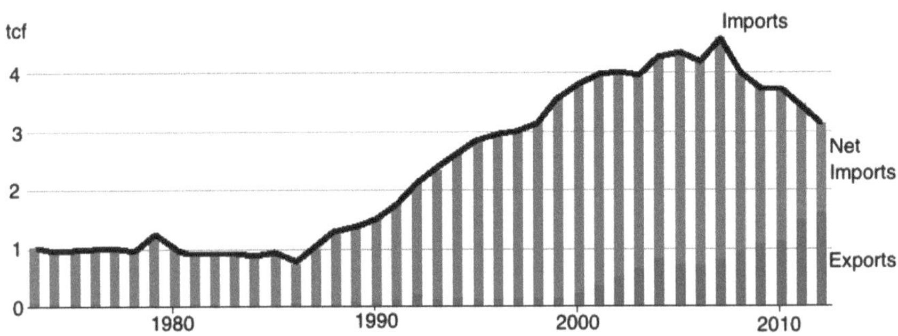

Figure 2.17 US export and import of natural gas. *Data from EIA.*

Of the above, a significant natural gas production comes from unconventional sources. For instance, see Figure 2.21 that shows the gas production from shale gas.

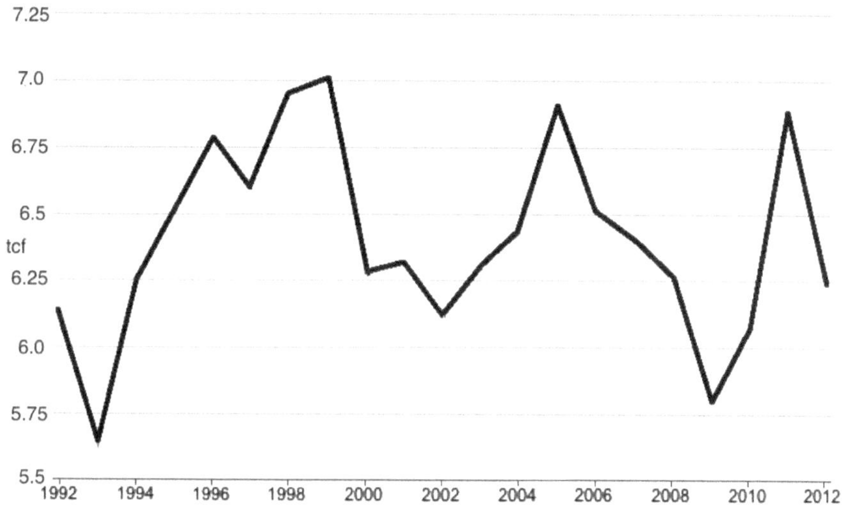

Figure 2.18 Net export of Russia. *Data from EIA.*

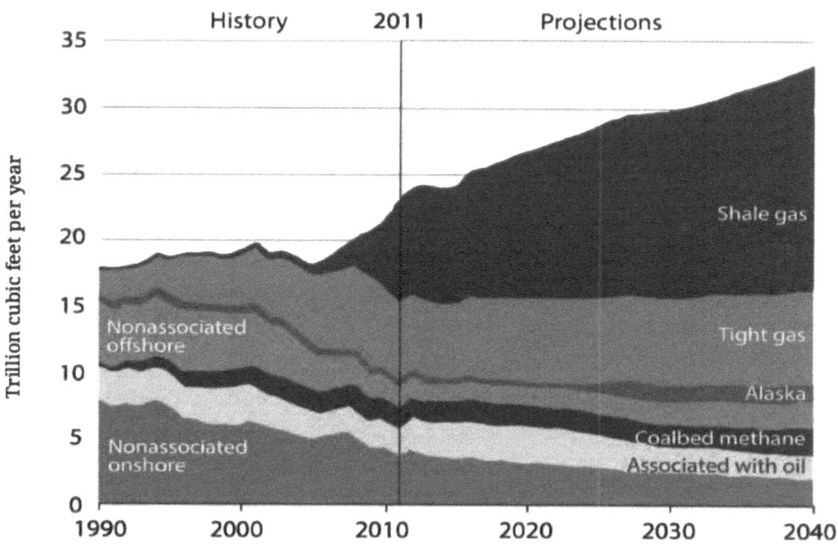

Figure 2.19 Shares of various natural gas sources in the United States. *Data from EIA.*

The term "proven gas reserve" is not free of controversy, even though natural gas is free from the constraints associated to liquid petroleum recovery. Based on data from BP, at the end of 2009 proven gas reserves were dominated by three countries: Iran, Russia, and Qatar, which together held nearly half the world's proven reserves.

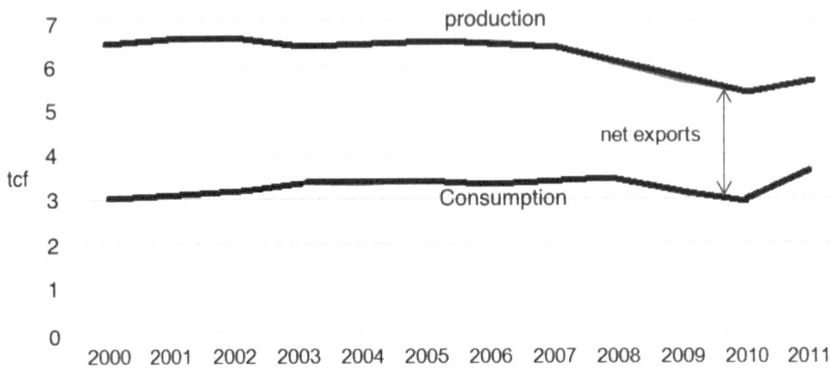

Figure 2.20 Import and export of Canada. *Data from EIA.*

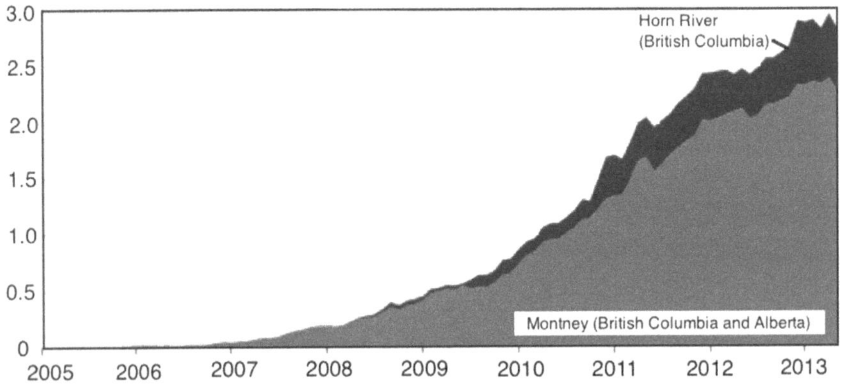

Figure 2.21 Shale gas production (billion cubic feet per day) from Canada. *Data from Canadian National Energy Board.*

There is some disagreement on which country has the largest proven gas reserves. Sources that list Russia as having by far the largest proven reserves include the US CIA (47.6 Tcm), the Oil and Gas Journal (47.8 Tcm), the U.S. EIA (47.8 Tcm), and OPEC (48.7 Tcm). However, BP credits Russia with only 32.9 Tcm, which would place it in second place, slightly behind Iran (33.1–33.8 Tcm, depending on the source). CIA Factbook gives the following list (Table 2.5) of proven gas reserve of countries that are known to have gas.

BP on the other hand gives the following list (Table 2.6) of reserves.

Figure 2.22 shows historical listing of proved natural gas reserves for top five countries. During last 30 years, Iran has been adding to its proved gas reserve steadily. This is typical of an underdeveloped natural resource.

Table 2.5 Proven Gas Reserve of Different Countries as per CIA Factbook

Rank	Country	Proven gas reserve (Cubic meter)	Date of information
1	Russia	47,800,000,000,000	January 1, 2013 est.
2	Iran	33,610,000,000,000	January 1, 2013 est.
3	Qatar	25,200,000,000,000	January 1, 2013 est.
4	United States	9,459,000,000,000	January 1, 2012 est.
5	Saudi Arabia	8,150,000,000,000	January 1, 2013 est.
6	Turkmenistan	7,504,000,000,000	January 1, 2013 est.
7	United Arab Emirates	6,089,000,000,000	January 1, 2013 est.
8	Venezuela	5,524,000,000,000	January 1, 2013 est.
9	Nigeria	5,153,000,000,000	January 1, 2013 est.
10	Algeria	4,504,000,000,000	January 1, 2013 est.
11	Iraq	3,158,000,000,000	January 1, 2013 est.
12	China	3,100,000,000,000	January 1, 2013 est.
13	Indonesia	3,069,000,000,000	January 1, 2013 est.
14	Kazakhstan	2,407,000,000,000	January 1, 2013 est.
15	Malaysia	2,350,000,000,000	January 1, 2013 est.
16	Egypt	2,186,000,000,000	January 1, 2013 est.
17	Norway	2,070,000,000,000	January 1, 2013 est.
18	European Union	1,955,000,000,000	January 1, 2012 est.
19	Canada	1,930,000,000,000	January 1, 2013 est.
20	Uzbekistan	1,841,000,000,000	January 1, 2013 est.
21	Kuwait	1,798,000,000,000	January 1, 2013 est.
22	Libya	1,547,000,000,000	January 1, 2013 est.
23	India	1,241,000,000,000	January 1, 2013 est.
24	Netherlands	1,230,000,000,000	January 1, 2013 est.
25	Australia	1,219,000,000,000	January 1, 2013 est.
26	Ukraine	1,104,000,000,000	January 1, 2013 est.
27	Chad	999,500,000,000	January 1, 2012 est.
28	Azerbaijan	991,100,000,000	January 1, 2013 est.
29	Oman	849,500,000,000	January 1, 2013 est.
30	Vietnam	699,400,000,000	January 1, 2013 est.
31	Pakistan	679,600,000,000	January 1, 2013 est.
32	Mexico	487,700,000,000	January 1, 2013 est.
33	Yemen	478,500,000,000	January 1, 2013 est.
34	Brazil	395,500,000,000	January 1, 2013 est.
35	Brunei	390,800,000,000	January 1, 2013 est.
36	Trinidad and Tobago	375,400,000,000	January 1, 2013 est.
37	Angola	366,000,000,000	January 1, 2013 est.
38	Peru	359,600,000,000	January 1, 2013 est.
39	Argentina	332,500,000,000	January 1, 2013 est.
40	Thailand	284,900,000,000	January 1, 2013 est.

Continued

Table 2.5 Proven Gas Reserve of Different Countries as per CIA Factbook—cont'd

Rank	Country	Proven gas reserve (Cubic meter)	Date of information
41	Burma	283,200,000,000	January 1, 2013 est.
42	Bolivia	281,500,000,000	January 1, 2013 est.
43	Israel	268,500,000,000	January 1, 2013 est.
44	United Kingdom	246,000,000,000	January 1, 2013 est.
45	Syria	240,700,000,000	January 1, 2013 est.
46	Timor-Leste	200,000,000,000	January 1, 2006 est.
47	Bangladesh	183,700,000,000	January 1, 2013 est.
48	Colombia	169,900,000,000	January 1, 2013 est.
49	Papua New Guinea	155,300,000,000	January 1, 2013 est.
50	Cameroon	135,100,000,000	January 1, 2013 est.
51	Mozambique	127,400,000,000	January 1, 2013 est.
52	Germany	125,000,000,000	January 1, 2013 est.
53	Romania	105,500,000,000	January 1, 2013 est.
54	Philippines	98,540,000,000	January 1, 2013 est.
55	Chile	97,970,000,000	January 1, 2013 est.
56	Bahrain	92,030,000,000	January 1, 2013 est.
57	Poland	92,000,000,000	January 1, 2013 est.
58	DR Congo	90,610,000,000	January 1, 2013 est.
59	Cuba	70,790,000,000	January 1, 2013 est.
60	Tunisia	65,130,000,000	January 1, 2013 est.
61	South Sudan	63,710,000,000	January 1, 2013 est.
62	Italy	62,350,000,000	January 1, 2013 est.
63	Namibia	62,290,000,000	January 1, 2013 est.
64	Rwanda	56,630,000,000	January 1, 2013 est.
65	Afghanistan	49,550,000,000	January 1, 2013 est.
66	Serbia	48,140,000,000	January 1, 2013 est.
67	Denmark	42,980,000,000	January 1, 2013 est.
68	Equatorial Guinea	36,810,000,000	January 1, 2013 est.
69	New Zealand	29,420,000,000	January 1, 2013 est.
70	Gabon	28,320,000,000	January 1, 2013 est.
71	Mauritania	28,320,000,000	January 1, 2013 est.
72	Cote d'Ivoire	28,320,000,000	January 1, 2013 est.
73	Ethiopia	24,920,000,000	January 1, 2013 est.
74	Croatia	24,920,000,000	January 1, 2013 est.
75	Ghana	22,650,000,000	January 1, 2013 est.
76	Sudan	21,240,000,000	January 1, 2013 est.
77	Japan	20,900,000,000	January 1, 2013 est.
78	South Africa	16,000,000,000	January 1, 2012 est.
79	Uganda	14,160,000,000	January 1, 2013 est.
80	Slovakia	14,160,000,000	January 1, 2013 est.
81	Austria	10,820,000,000	January 1, 2013 est.
82	France	10,700,000,000	January 1, 2013 est.

Table 2.5 Proven Gas Reserve of Different Countries as per CIA Factbook—cont'd

Rank	Country	Proven gas reserve (Cubic meter)	Date of information
83	Ireland	9,911,000,000	January 1, 2013 est.
84	Georgia	8,495,000,000	January 1, 2013 est.
85	Hungary	8,098,000,000	January 1, 2013 est.
86	Ecuador	6,994,000,000	January 1, 2013 est.
87	Tanzania	6,513,000,000	January 1, 2013 est.
88	Taiwan	6,229,000,000	January 1, 2013 est.
89	Turkey	6,173,000,000	January 1, 2013 est.
90	Jordan	6,031,000,000	January 1, 2013 est.
91	Kyrgyzstan	5,663,000,000	January 1, 2013 est.
92	Tajikistan	5,663,000,000	January 1, 2013 est.
93	Bulgaria	5,663,000,000	January 1, 2013 est.
94	Somalia	5,663,000,000	January 1, 2013 est.
95	Korea, South	5,269,000,000	January 1, 2013 est.
96	Czech Republic	3,964,000,000	January 1, 2013 est.
97	Guatemala	2,960,000,000	January 1, 2006 est.
98	Belarus	2,832,000,000	January 1, 2013 est.
99	Spain	2,548,000,000	January 1, 2013 est.
100	Madagascar	2,010,000,000	January 1, 2012 est.
101	Morocco	1,444,000,000	January 1, 2013 est.
102	Benin	1,133,000,000	January 1, 2013 est.
103	DR Congo	991,100,000	January 1, 2013 est.
104	Greece	991,100,000	January 1, 2013 est.
105	Albania	849,500,000	January 1, 2013 est.
106	Barbados	113,300,000	January 1, 2013 est.

The political side of it has been discussed in the context of proved oil reserve. Qatar had similar increase but it was halted in 2004. The proven reserve of Saudi Arabia has been increasing steadily since 1980. For the United States, there has been a period of decline in proven gas reserve. However, shortly after 2000, the gas reserve has been increasing mainly due to the addition of unconventional gas.

2.5 UNCONVENTIONAL GAS AROUND THE WORLD

The notion that conventional gas and oil is miniscule compared to unconventional gas reserve is decades old. With it comes the notion that it is more challenging to produce unconventional petroleum resource. In addition, at least for petroleum oil the notion that unconventional resource

Table 2.6 World Proven Gas Reserve in Recent Decades as per BP

	At end 1992	At end 2002	At end 2011	At end 2012			
Natural Gas: Proven reserves	**Tcm**	**Tcm**	**Tcm**	**Tcf**	**Tcm**	**Share of total**	**R/P ratio**
United States	4.7	5.3	8.8	300.0	8.5	4.5%	12.5
Canada	2.7	1.7	2.0	70.0	2.0	1.1%	12.7
Mexico	2.0	0.4	0.4	12.7	0.4	0.2%	6.2
Total North America	**9.3**	**7.4**	**11.2**	**382.7**	**10.8**	**5.8%**	**12.1**
Argentina	0.5	0.7	0.3	11.3	0.3	0.2%	8.5
Bolivia	0.1	0.8	0.3	11.2	0.3	0.2%	17.0
Brazil	0.1	0.2	0.5	16.0	0.5	0.2%	26.0
Colombia	0.2	0.1	0.2	5.5	0.2	0.1%	12.9
Peru	0.3	0.2	0.4	12.7	0.4	0.2%	27.9
Trinidad & Tobago	0.2	0.6	0.4	13.3	0.4	0.2%	8.9
Venezuela	3.7	4.2	5.5	196.4	5.6	3.0%	*
Other South & Central America	0.2	0.1	0.1	2.0	0.1	◆	15.5
Total South & Central America	**5.4**	**7.0**	**7.5**	**268.3**	**7.6**	**4.1%**	**42.8**
Azerbaijan	n/a	0.9	0.9	31.5	0.9	0.5%	57.1
Denmark	0.1	0.1	0.0	1.3	0.0	◆	5.9
Germany	0.2	0.2	0.1	2.0	0.1	◆	6.1
Italy	0.3	0.2	0.1	1.9	0.1	◆	7.0
Kazakhstan	n/a	1.3	1.3	45.7	1.3	0.7%	65.6
Netherlands	1.7	1.4	1.0	36.7	1.0	0.6%	16.3
Norway	1.4	2.1	2.1	73.8	2.1	1.1%	18.2
Poland	0.2	0.1	0.1	4.2	0.1	0.1%	28.3
Romania	0.5	0.3	0.1	3.6	0.1	0.1%	9.3

Russian Federation	n/a	29.8	32.9	1162.5	32.9	17.6%	55.6
Turkmenistan	n/a	2.3	17.5	618.1	17.5	9.3%	★
Ukraine	n/a	0.7	0.7	22.7	0.6	0.3%	34.6
United Kingdom	0.6	1.0	0.2	8.7	0.2	0.1%	6.0
Uzbekistan	n/a	1.2	1.1	39.7	1.1	0.6%	19.7
Other Europe & Eurasia	34.7	0.4	0.3	10.1	0.3	0.2%	29.2
Total Europe & Eurasia	**39.6**	**42.1**	**58.4**	**2062.5**	**58.4**	**31.2%**	**56.4**
Bahrain	0.2	0.1	0.2	7.0	0.2	0.1%	14.0
Iran	20.7	26.7	33.6	1187.3	33.6	18.0%	★
Iraq	3.1	3.2	3.6	126.7	3.6	1.9%	★
Kuwait	1.5	1.6	1.8	63.0	1.8	1.0%	★
Oman	0.2	0.9	0.9	33.5	0.9	0.5%	32.8
Qatar	6.7	25.8	25.0	885.1	25.1	13.4%	★
Saudi Arabia	5.2	6.6	8.2	290.8	8.2	4.4%	80.1
Syria	0.2	0.3	0.3	10.1	0.3	0.2%	37.5
United Arab Emirates	5.8	6.1	6.1	215.1	6.1	3.3%	★
Yemen	0.4	0.5	0.5	16.9	0.5	0.3%	63.1
Other Middle East	0.0	0.1	0.2	7.3	0.2	0.1%	78.0
Total Middle East	**44.0**	**71.8**	**80.4**	**2842.9**	**80.5**	**43.0%**	★
Algeria	3.7	4.5	4.5	159.1	4.5	2.4%	55.3
Egypt	0.4	1.7	2.2	72.0	2.0	1.1%	33.5
Libya	1.3	1.5	1.5	54.6	1.5	0.8%	★
Nigeria	3.7	5.0	5.2	182.0	5.2	2.8%	★
Other Africa	0.8	1.1	1.3	44.3	1.3	0.7%	68.1
Total Africa	**9.9**	**13.8**	**14.7**	**512.0**	**14.5**	**7.7%**	**67.1**

Continued

Table 2.6 World Proven Gas Reserve in Recent Decades as per BP—cont'd

	At end 1992	At end 2002	At end 2011	At end 2012			
Natural Gas: Proven reserves	**Tcm**	**Tcm**	**Tcm**	**Tcf**	**Tcm**	**Share of total**	**R/P ratio**
Australia	1.0	2.5	3.8	132.8	3.8	2.0%	76.6
Bangladesh	0.3	0.3	0.3	6.5	0.2	0.1%	8.4
Brunei	0.4	0.3	0.3	10.2	0.3	0.2%	22.9
China	1.4	1.3	3.1	109.3	3.1	1.7%	28.9
India	0.7	0.8	1.3	47.0	1.3	0.7%	33.1
Indonesia	1.8	2.6	3.0	103.3	2.9	1.6%	41.2
Malaysia	1.7	2.5	1.2	46.8	1.3	0.7%	20.3
Myanmar	0.3	0.4	0.2	7.8	0.2	0.1%	17.4
Pakistan	0.6	0.8	0.7	22.7	0.6	0.3%	15.5
Papua New Guinea	0.4	0.4	0.4	15.6	0.4	0.2%	⋆
Thailand	0.2	0.4	0.3	10.1	0.3	0.2%	6.9
Vietnam	0.1	0.2	0.6	21.8	0.6	0.3%	65.6
Other Asia Pacific	0.3	0.4	0.3	11.8	0.3	0.2%	18.6
Total Asia Pacific	**9.4**	**13.0**	**15.5**	**545.6**	**15.5**	**8.2%**	**31.5**
Total World	**117.6**	**154.9**	**187.8**	**6614.1**	**187.3**	**100.0%**	**55.7**

R/P ratio, Reserve/production ratio.
⋆more than 100 years, diamond: less than 0.05%

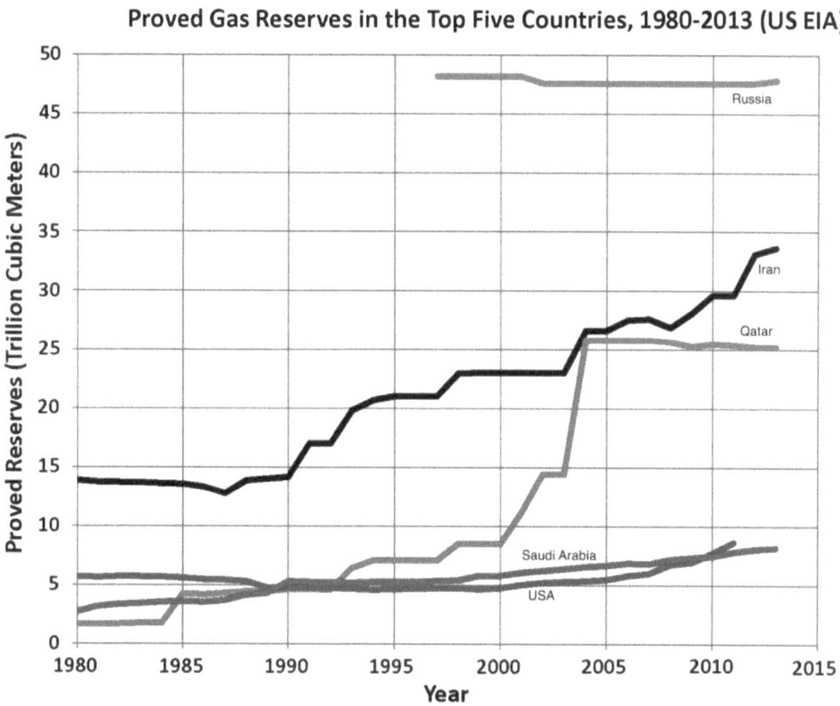

Figure 2.22 History of proven natural gas reserve for top five countries.

is more challenging to process is prevalent. This notion is false. With the renewed awareness of the environmental sustainability it is becoming clear that unconventional resources offer more opportunities to produce environment-friendly products than conventional resources. Figure 2.23 shows the pyramid of both oil and gas resources.

On the oil side, the quality of oil is considered to be declining as the American Petroleum Institute gravity declines. This correlation is related to the processing required for crude oil to be ready for conversion into usable energy, which is related to heating value. Heating value is typically increased by refining crude oil upon addition of artificial chemicals that are principally responsible for global warming (Chhetri and Islam, 2008; Islam et al., 2010). In addition, the process is inefficient resulting harmful products to the environment. Figure 2.24 shows the trend in efficiency, environmental benefit, and real value with the production cost of refined crude. This figure shows clearly there is great advantage in using petroleum products in their natural state. This is the case for unconventional oil. For instance, shale oil burns naturally. The color of flames (left image of Picture

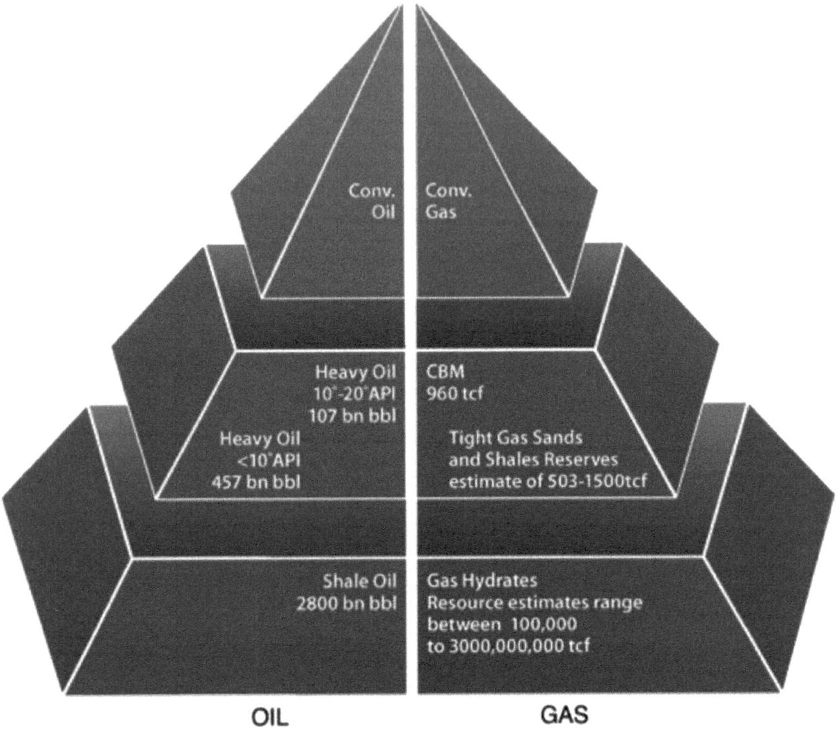

Figure 2.23 The volume of petroleum resources increases as one moves from conventional to unconventional.

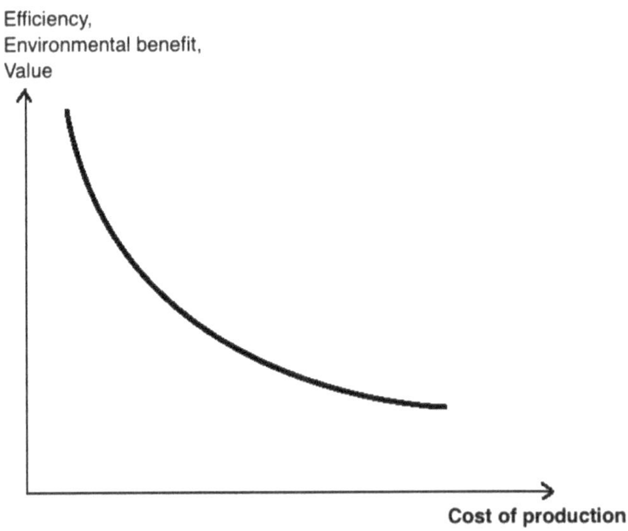

Figure 2.24 Cost of production increases as efficiency, environmental benefits, and real value of crude oil declines. *Modified from Islam et al., 2010.*

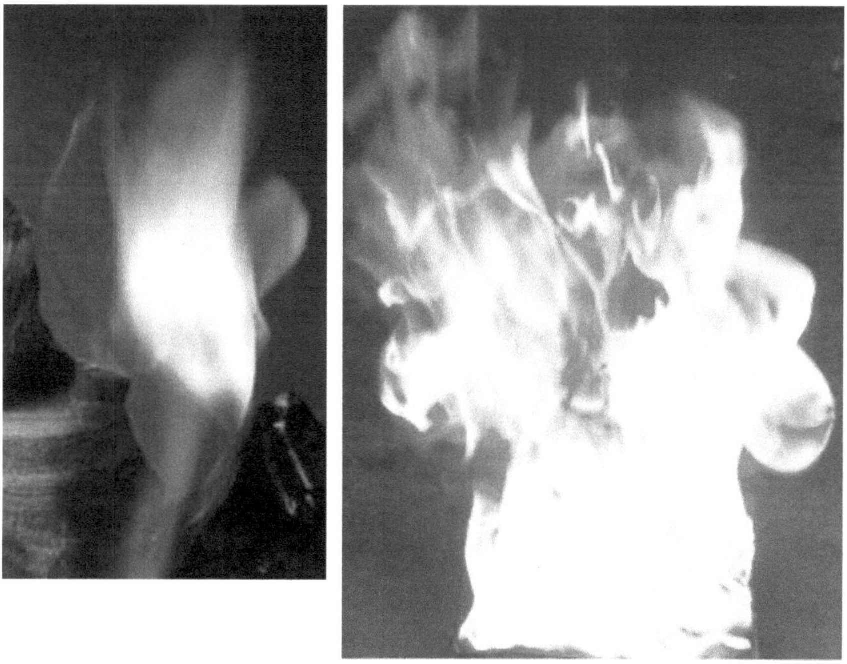

Picture 2.2 Images of burning crude oil from shale oil (left) and refined oil (right).

2.2) indicates that crude oil produced from shale oil does not need further processing. The right image of Picture 2.2 emerges from burning gasoline and has similar colors to those of the left.

In addition, crude oil from shale oil is "cleaner" than other forms of crude oil because of the fact that it is relatively low in tar content as well as sand particles. Another crucial aspect is the fact that sulfur content or other toxic elements of crude oil have no correlation with unconventional or conventional sources. Also, heavier oils do not have more of these toxic elements and are not in need of refinement to be usable.

Lighter crudes are considered to be easier and less expensive to produce only because modern engineering uses refined version of the crude oil and all refining technologies are specially designed to handle light crude oil. If sustainable refining techniques are used, lighter or conventional oil offers no particular advantage over unconventional one and yet the volume and ease of production of unconventional are greater in unconventional resources.

For natural gas, the quality of gas actually improves with unconventional resources. For oil, the lighter the oil the more toxic it is considered. Whereas for gas, more readily available resources are less toxic. For instance, biogas is least toxic, whereas it is most plentiful. As can be seen in Figure 2.23, as one transits from conventional gas to CBM to tight gas and shale gas all the way to hydrates, one encounters more readily combustible natural resources. In fact, CBM burns so readily that coal mine safety primarily revolves around combustion of methane gas. Processing of gas does not involve making it more combustible, it rather involves the removal of components that do not add heating value or create safety concerns (e.g., water, CO_2, H_2S).

Figure 2.25 shows how the volume of resources goes up as one moves from conventional to unconventional resources. In this process the quality of gas also increases. For instance, hydrate has the purest form of methane and can be burnt directly with little or no safety concern. At the same time, the volume of natural gas in hydrate is very large. The concentration of "sour" gas components also decreases with abundance of the resources. Such trend can be explained by the processing time of a particular resource. There is a continuity in nature that dictates that the natural processing

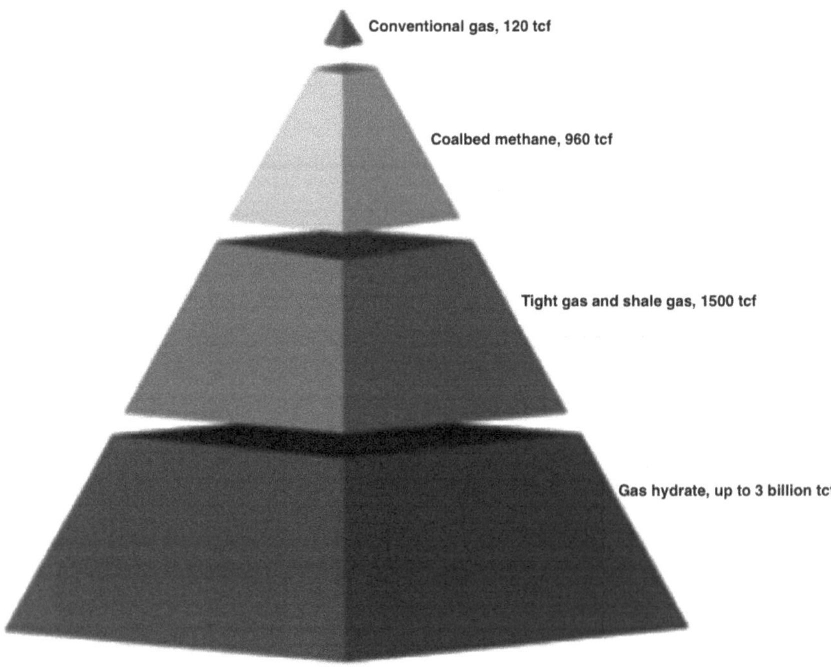

Figure 2.25 Current estimate of conventional and unconventional gas reserve.

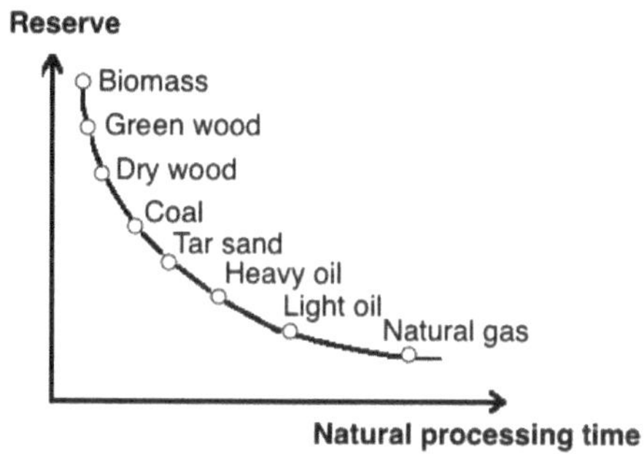

Figure 2.26 Abundance of natural resources as a function of time.

increases both the value and global efficiency of energy sources (Chhetri and Islam, 2008). Figure 2.26 depicts the nature of volume of natural resources as a function of processing time.

In this picture, "natural gas" relates to petroleum products in conventional sense. This figure shows natural gas, in general, is most suitable for clean energy generation. Within unconventional gas sources, there exists another correlation between reserve volume and processing time. In general, the processing time for various energy sources is not a well-understood science. Scientists are still grappling with the origin of earth or universe; some discovering only recently that water was and remains the matrix component for all matter (Pearson et al., 2014). Figure 2.27 shows how natural evolution on earth involved a distinctly different departure point not previously recognized. Pearson et al. (2014) observed a "rough diamond" found along a shallow riverbed in Brazil that unlocked the evidence that a vast "wet zone" deep inside the earth that could hold as much water as all the world's oceans put together. This discovery is important for two reasons. Water and carbon are both essential for living organisms. They also mark the beginning and end of a life cycle. All natural energy sources have carbon or require carbon to transform energy in usable form (e.g., photosynthesis).

With this information and uncertainty in the backdrop, let us discuss geologic age of unconventional resources. CBM resides within a "coal zone," as discrete coal seams and/or packages with several thin and thick seams interbedded with noncoaly rock layers or beds. Coal zones are found in strata ranging in age from Late Jurassic (approximately 145 million years old) to Tertiary (approximately 65 million years old). Shale gas and tight gas can have older age depending on the age of the rock. At present, gas

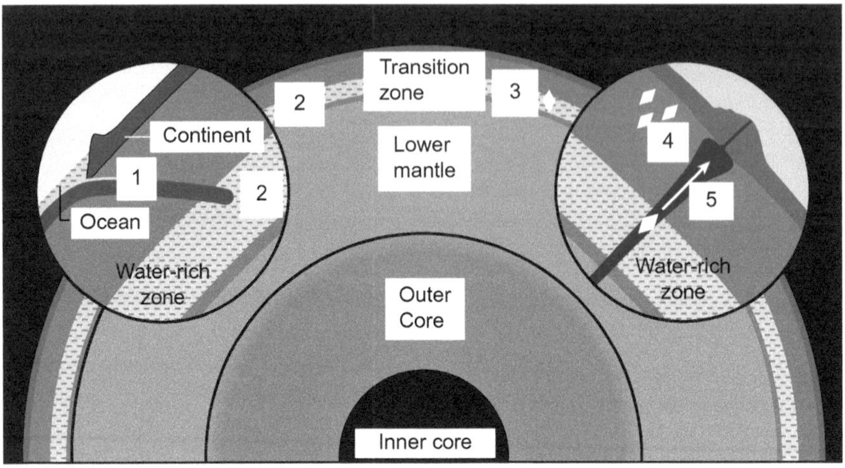

Figure 2.27 Water plays a more significant role in material production than previously anticipated. *(Modified from the Guardian, March 12, 2014).*

hydrates are known to enclose the freshest methane. Methane hydrates, icelike compounds in which methane is held in crystalline cages formed by water molecules, are widespread in areas of permafrost such as the Arctic and in sediments on the continental margins. However, gas hydrates are not confined to the surface of the earth. They can exist deep in the ocean and some can enclose much older methane. Either way, gas hydrates provide one with the cleanest gas that can be burnt readily.

2.5.1 Shale Gas Reserve

Figure 2.28 shows the locations of shale gas reserves around the world. This distribution has to be explained in terms of current status of exploration as well as interest in expanding unconventional resources. Only a few of these locations are also rich in conventional gas reserves. In addition, this ranking is different from the ranking in conventional natural gas reserves. Table 2.7 shows the ranking of top 10 countries in terms of shale gas reserve.

Even though, shale gas is distributed around the globe, North America dominates shale gas development and production. According to EIA estimate, the world has 7299 Tcf of technically recoverable shale gas resources, adding approximately 47% to the 15,500 Tcf of technically recoverable conventional gas resources. This number is much larger if one considers the fact that the technologies for shale gas production are evolving and several of the most promising techniques are not commercially applied to date.

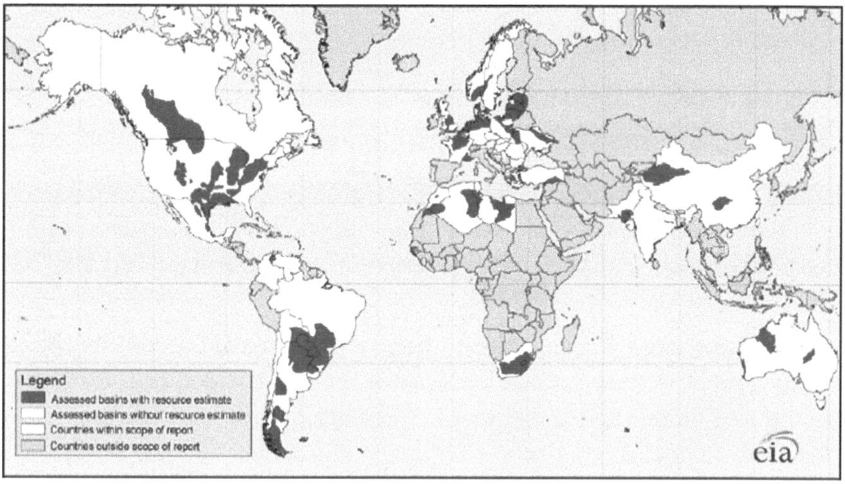

Figure 2.28 Locations of major shale gas reservoirs.

World energy outlook 2012 projects in their new policies scenario that Europe will produce 7.7 Tcf of gas in 2035 and import 16 Tcf (68% of demand), mainly from Russia, Africa, and the Middle East.

2.5.2 CBM Gas Reserve

CBM reserve estimates are not readily amenable to calculations. The difficulty arises from the fact that only a fraction of coal reserve is counted in estimating CBM reserves. Figure 2.29 shows how the CBM reserve estimates have fluctuated in the United States.

Table 2.7 Shale Gas Reserve for Top 10 Countries and Their Rankings in Terms of Conventional Gas Reserve

Countries	Shale gas ranking	Conventional gas ranking	Shale gas reserve (Tcf)
China	1	12	1115
Argentina	2	39	802
Algeria	3	10	707
U.S.A.	4	4	665
Canada	5	19	573
Mexico	6	32	545
Australia	7	25	437
South Africa	8	78	390
Russia	9	1	285
Brazil	10	34	245

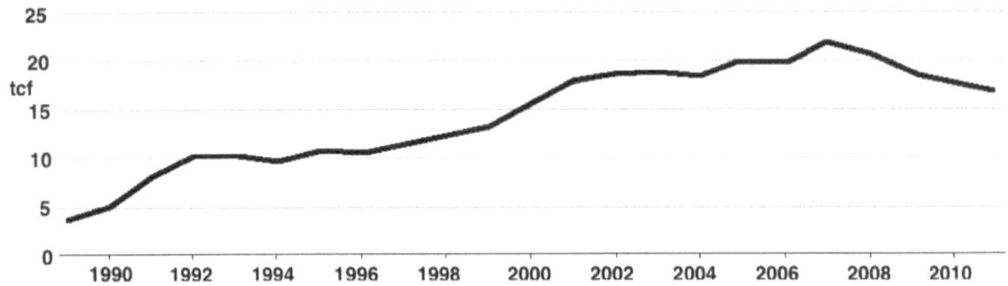

Figure 2.29 History of coalbed methane reserve in the United States. *Data from EIA.*

CBM is related to coal reserve. Coal remains a crucial contributor to energy supply in many countries. Coal is the most widespread fossil fuel around the world, and more than 75 countries have coal deposits. The current share of coal in global power generation is over 40% (WEO, 2013). These coal reserves are directly related to CBM. However, not all coal reserve owners develop their CBM reserve. Table 2.8 shows the coal reserve for countries with greatest reserve.

This reserve connects to following major distribution of CBM around the world (Table 2.9).

The above ranking shows that Canada and UK follow a different ranking than the ranking accorded by coal reserve. Such is the case because coal reserves in these countries have declined whereas CBM estimates have been accelerated. In fact, the search for and production of CBM follows a different trend. Figure 2.30 shows various spots that have active CBM projects or have dealt with CBM production.

Table 2.8 Coal Reserve in Countries with Most Reserve

Country	Reserves (Mt) 2011	Reserves (Mt) 1993	Production (Mt) 2011	Production (Mt) 1993	2011 R/P years
United States of America	237,295	168,391	1092	858	>100
Russian Federation	157,010	168,700	327	304	>100
China	114,500	80,150	3384	1150	34
Australia	76,400	63,658	398	224	>100
India	60,600	48,963	516	263	>100
Rest of World	245,725	501,748	1805	1675	>100
Global total	891,530	1,031,610	7520	4474	>100

R/P = reserve/yearly production
From WEO, 2013.

Table 2.9 CBM Reserve in Major Reserve Containing Countries

	Coalbed methane (Tcf)	Ranking
Russia	1730	1
United States	711 (lower 48) + 1037 Alaska	2
China	1307	3
Australia	1037	4
Canada	699	5
United Kingdom	102	6
India	71	7
Ukraine	42	8
Kazakhstan	23	9

2.5.3 Tight Gas Reserve

Tight gas reserves are usually linked to currently producible oil and gas reservoirs. The United States is the leader of tight gas production. Not surprisingly, the United States has developed extensive data base on tight gas reserve. Figure 2.31 shows history and projection of tight sand gas in the United States, along with other unconventional reserves.

The estimates of gas production, reserves, and potential from the tight gas basins in the United States are compatible with the concept of the resource triangle. Figure 2.32 illustrates the tight gas resource base estimates from the

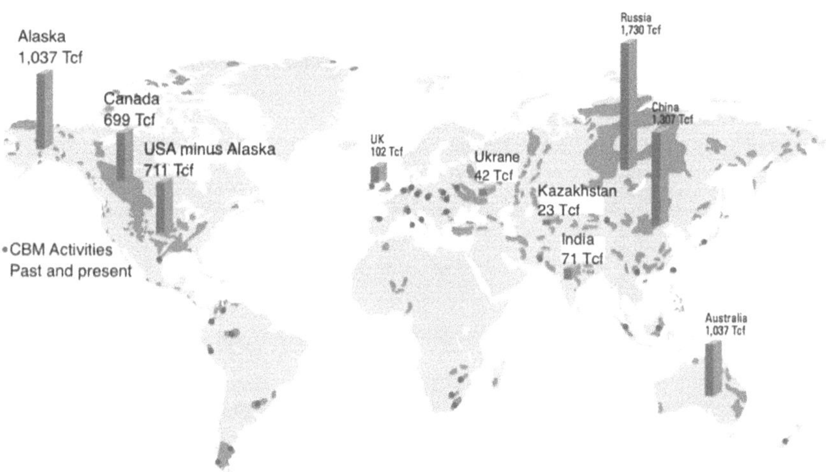

Figure 2.30 Coalbed methane distribution around the world (red spots show past or present activities). For interpretation of the references to color in this figure legend, the reader is referred to the online version of this book.

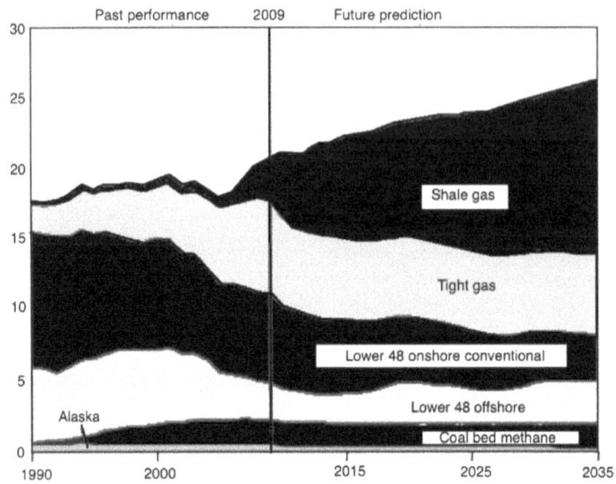

Figure 2.31 Recovery and projection of various unconventional gas resources. *(Redrawn from BP, 2013).*

Gas Technology Institute (GTI). The gas produced throughout the year 2000 from tight gas reservoirs is estimated to be 58 Tcf. Proven reserves in tight gas reservoirs are 34 Tcf. Thus, the sum of produced gas plus proven reserves adds up to 92 Tcf. GTI estimates the volume of technically recoverable gas from known US tight gas accumulations at 185 Tcf. The term "technically recoverable" means that the gas is known to exist; the technology is available to drill, complete, stimulate, and produce this gas; but the gas cannot be booked as reserves until the wells are drilled and the reservoirs are developed. The next category in Figure 2.32 is called undiscovered, which represents the GTI estimate of gas that is likely to be discovered in known tight gas basins. Finally, the largest category is called resources. This

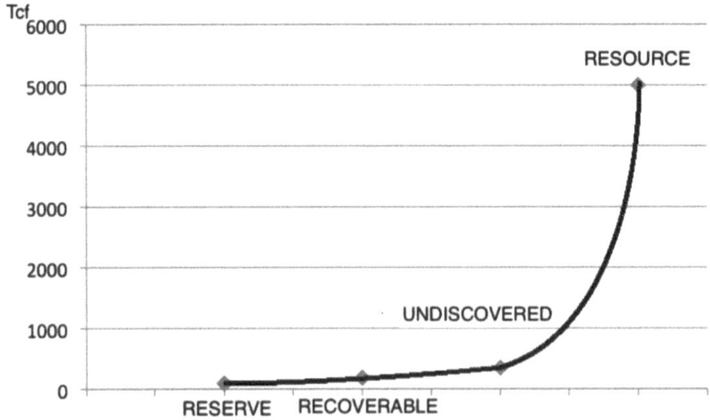

Figure 2.32 Resource triangle for tight gas in the United States.

value represents the gas in place in the US tight gas basins. Substantial improvements in technology or changes in the gas market are required before the gas in the resources category can be produced economically.

Figure 2.32 highlights the need for improving technologies in order to access much larger reserve. It is to be noted, almost all large conventional oil reservoirs will have tight gas or oil attached to them. However, they are not counted for political and economical reasons. Figure 2.33 shows how unconventional resources will make an impact on some countries whereas others, such as Saudi Arabia, would remain unaffected. For the United States, China, and Canada, unconventional gas production will become more prominent than conventional production, whereas countries, such as Iran, Qatar, Algeria, Saudi Arabia and Turkmenistan will have no recovery from unconventional resources (Figure 2.34).

In United States, it is expected that shale gas production, will grow by 113% from 2011 to 2040. Tight gas, on the other hand, is expected to grow by 24% (Figure 2.35). Such prediction is pessimistic and does not consider the full potential of tight gas.

2.5.4 Gas Hydrate

It is well known that gas hydrates as a source of natural gas have tremendous potentials. It is so vast that the estimates of global reserves are only sketchy, but range from 2800 trillion to 8 billion trillion m^3 of natural gas. This is several times higher than global reserves of 440 Tcm of conventional gas. While gas hydrate burns readily (Picture 2.3), exploitation of hydrate reserve is considered to be a difficult task.

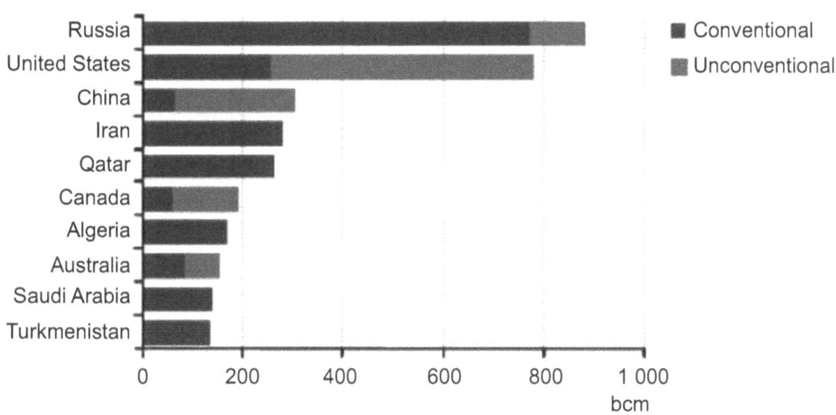

Figure 2.33 Projection of gas production in 2035. *From IEA Energy outlook, 2011.*

Figure 2.34 Distribution of various unconventional gas reserves around the world.

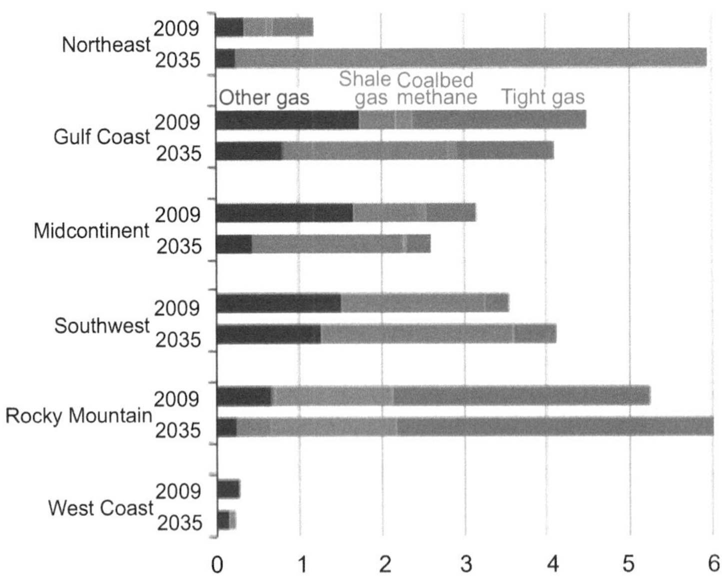

Figure 2.35 Past and future production of gas from unconventional resources. *From EIA, 2013.*

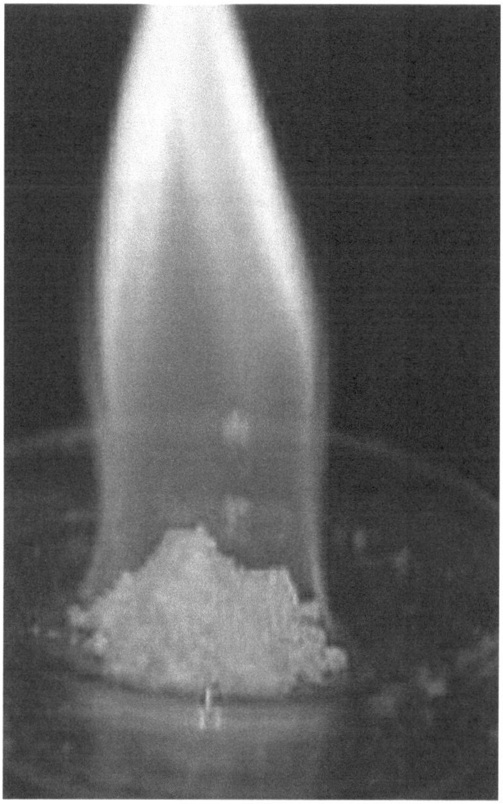

Picture 2.3 Hydrate burns readily without any safety or environmental hazard.

EIA global estimates place the gas volume resident in oceanic natural gas hydrate deposits in the range of 30,000 to 49,100,000 Tcf, and in continental natural gas hydrate deposits in the range of 5000 to 12,000,000 Tcf. Comparatively, current worldwide natural gas resources are about 13,000 Tcf and natural gas reserves are about 5000 Tcf. The current mean (expected value) estimate of domestic natural gas hydrates in place is 320,222 Tcf. In comparison, as of 1997 the mean estimate of all untapped technically recoverable US natural gas resources was 1301 Tcf, US proven natural gas reserves were 167 Tcf, and annual US natural gas consumption was about 22 Tcf.

Large volumes of natural gas hydrates are known to exist in both onshore and offshore Alaska, offshore the States of Washington, Oregon, California, New Jersey, North Carolina, and South Carolina, and in the deep Gulf of Mexico. Most of the volume is expected to be in Federal jurisdiction offshore waters, although 519 Tcf of hydrated gas in place was assessed for onshore Alaska—more than three times the 1997 level of US proven natural gas reserves. The USGS assessment indicates that the North Slope of Alaska may host about 85 Tcf of undiscovered technically recoverable gas hydrate resources (Figure 2.36). According to the report, technically recoverable gas hydrate resources could range from a low of 25 Tcf to as much as 158 Tcf on the North Slope. Total US consumption of natural gas in 2007 was slightly more than 23 Tcf. Of the mean estimate of 85 Tcf of technically recoverable gas hydrates on the North Slope, 56% is located on federally managed lands, 39% on lands and offshore waters managed by the state of Alaska, and the remainder on Native lands. The total area comprised by the USGS assessment is 55,894 square miles, and extends from the National Petroleum Reserve in the west to the Arctic National Wildlife Refuge in the east (Figure 2.36). The area extends north from the Brooks Range to the state–federal offshore boundary three miles north of the Alaska coastline. Gas hydrates might also be found outside the assessment area.

Global estimates by the committee for gas estimates reported methane in gas hydrate deposits to be in the range of 3.1×10^{15} to 7600×10^{15} m^3 for oceanic sediments and from 0.014×10^{15} to 34×10^{15} m^3 for polar regions (Figure 2.37) (Max, 2003). Boswell and Collett (2011) puts this

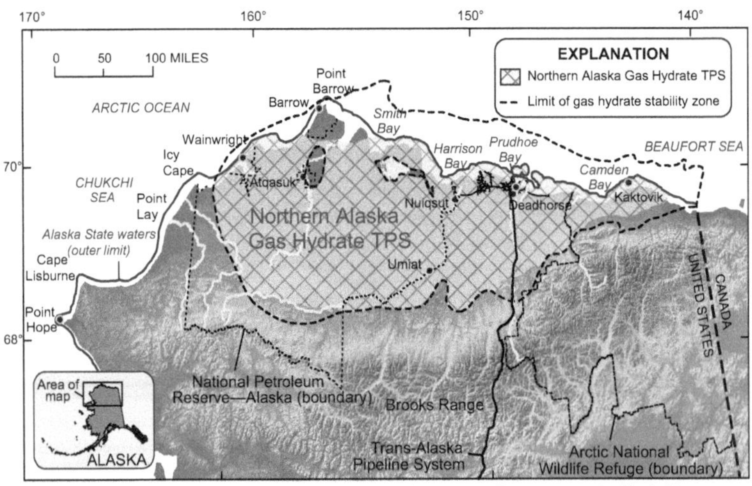

Source: USGS Fact Sheet 2008-3073, Assessment of Gas Hydrate Resources on the North Slope, Alaska, 2008, at http://pubs.usgs.gov/fs/2008/3073/.

Figure 2.36 Gas hydrate deposits of Alaska.

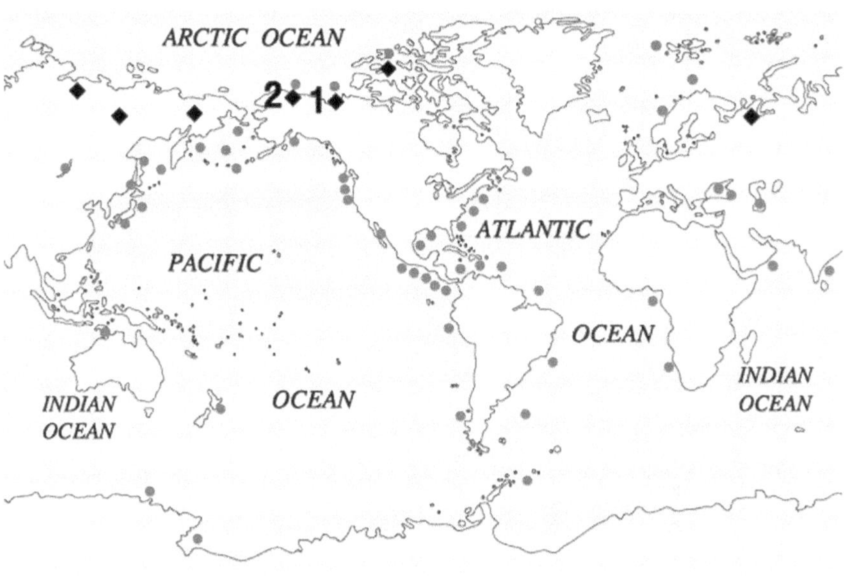

Figure 2.37 Known and inferred natural gas hydrate occurrences in marine (red circles) and permafrost (black diamonds) environments. *(Figure from USGS http://pubs.usgs.gov/fs/fs021-01/fs021-01.pdf).*

reserve to a volume of 1–120 × 10^{15} m^3 of methane trapped within global reserve. In near future, hydrates can alter the energy demography for the world. Such efforts are in progress in India (Sain, 2012), Japan (Pfeifer, 2014), and others. This has the potential of creating another energy revolution the likes of which did not occur in last 100+ years. Following is an estimate of currently known hydrate reserve for some of the leading countries.

USA-318,000 Tcf
Alaska North Slope-590 Tcf
Japan-1765 Tcf
India-4307 Tcf
Canada-1550–28,600 Tcf

2.6 SUMMARY AND CONCLUSIONS

A scientific approach to looking at natural gas resources of the world reveals that future of energy lies within exploiting unconventional gas. Unlike

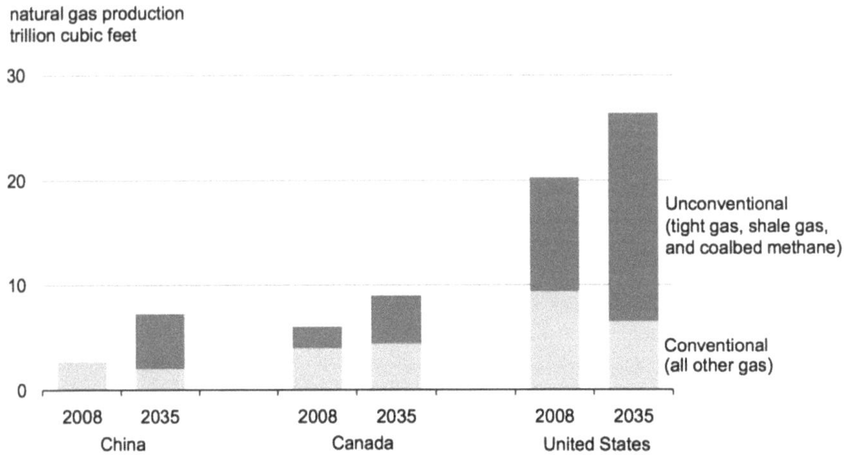

Figure 2.38 Future trends in some of the major future user of unconventional gas. *From EIA report, 2013.*

what has been promoted in recent decades, unconventional gas is more likely to generate truly "clean" energy. Thankfully this fact is being recognized by future players. Figure 2.38 shows how there will be major

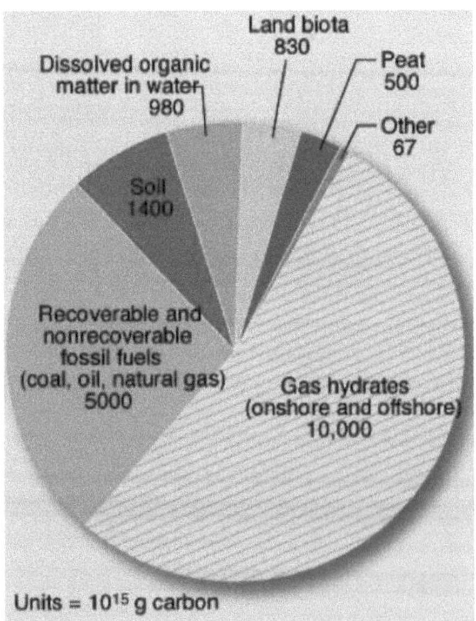

Figure 2.39 Gas hydrates that are the largest global sink for organic carbon offer the greatest prospect for the future of energy.

Table 2.10 Shale Gas Potential for Various Regions around the World

Continent	Shale gas potential (billion cubic feet)	Percentage of total area examined
North America	1931	29%
Asia	1389	21%
South America	1225	18%
Europe	639	10%
Africa	1042	16%
Australia	396	6%

shift in terms of usage of unconventional gas. This trend will shape the future of energy.

In even longer term future, gas hydrates offer the greatest promise of meeting energy needs. Gas hydrates are the most abundant of unconventional gas resources and they are the largest global sink for organic carbon (Figure 2.39).

At present, over 30 countries are actively involved at least in exploration of unconventional gas. For even the most active countries, this assessment of unconventional reserve is at its nascent state. For instance, for shale gas, Table 2.10 shows how different continents have addressed very limited percentage of the total possible gas in place. These numbers are lower for CBM, and much lower for gas hydrates.

CHAPTER 3

Important Features of Unconventional Gas

3.1 SUMMARY AND INTRODUCTION

Unconventional gas reservoirs are loosely defined as those that cannot be produced with conventional techniques. It turns out that the volume of gas available increases exponentially as one moves from conventional gas to coalbed methane (CBM), then to tight sand and shale gas, and then finally to gas hydrate deposits. For instance, gas hydrate volume is hundreds of times greater in methane volume than other sources of methane. At the same time, the quality as well as environmental integrity is also higher with more abundant gas resources. In this chapter the most important features of gas are presented. They deal with both source and the process of gas formation. As such, geological background is presented. This is followed with the process that is involved in forming natural gas through various physicochemical processes. At the end, it is demonstrated that unconventional gas is also the "cleanest" form of energy available today. This finding demystifies the global warming debate and puts all petroleum resources in its proper perspective.

U.S. Bureau of Mines and U.S. Geological Survey (1980) produced the following figure (Figure 3.1) that shows the dynamic relationship between two most important factors, namely, economics and geology. The overall movement of petroleum resources is to the right as accumulations are discovered and upward as development and production ensue. The degree of uncertainty as to the existence of resources decreases to the right in the diagram. The degree of economic viability decreases downward and also implies a decreasing certainty of technologic recoverability. The two boundaries related to conventional– unconventional and recoverable–unrecoverable are flux due to changing economic viability over time, the role of sustainability, and social constraints. Significant changes in the cost/price relationship or fundamental changes in technologic capabilities can shift these boundaries, causing modifications in perceptions and the practical meaning of the definitions. Thus, uncertainties in economic and technologic

Unconventional Gas Reservoirs
ISBN 978-0-12-800390-9
http://dx.doi.org/10.1016/B978-0-12-800390-9.00003-7

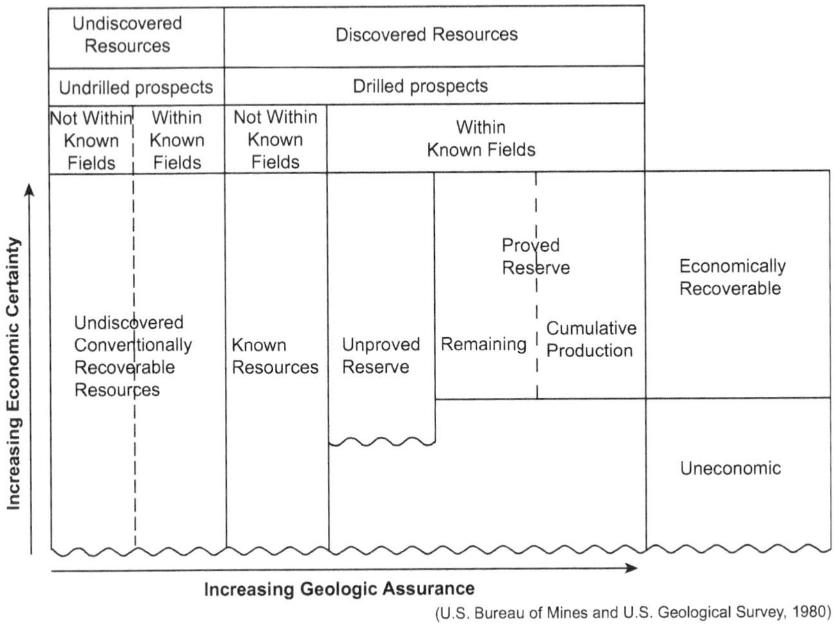

Figure 3.1 Interplay between economics and geological assurance.

conditions contribute to the substantial uncertainties in the resource assessment.

As perceptive Lewis Weeks (1958), in considering this issue, wrote:

"While research adds to our proved reserves by developing new ways to find and produce oil, it is a field of activity whose advances are impossible to predict. This is because they depend to a large degree on such important, intangible human resources as initiative and ingenuity."

"... man's mind is his most valuable asset— a 'natural resource' of unlimited potential— and the key to an abundant supply of fuel in the future."

It is to be recognized that technological development is linked with economics; this aspect will be discussed later. In order to assure cohesion with existing literature, the following conventional definitions are introduced.

Conventionally recoverable: Producible by natural pressure, pumping, or secondary recovery methods such as gas or water injection.

Marginal probability of hydrocarbons: An estimate, expressed as a decimal fraction, of the chance that an oil or natural gas accumulation exists in the area under consideration. The area under consideration is typically a geologic entity, such as a pool, prospect, play, basin, or province; or a large geographic area such as a planning area or region. It is a matter of probability before drilling but it becomes a certainty for a drilled well but a matter of probability with greater certainty for other regions.

Resources: Concentrations in the earth's crust of naturally occurring liquid or gaseous hydrocarbons that can conceivably be discovered and recovered. Normal use encompasses both discovered and undiscovered resources.

Recoverable resources: The volume of hydrocarbons that is potentially recoverable, regardless of the size, accessibility, recovery technique, or economics of the postulated accumulations.

Conventionally recoverable resources: The volume of hydrocarbons that may be produced from a wellbore as a consequence of natural pressure, artificial lift, pressure maintenance (gas or water injection), or other secondary recovery methods. They do not include quantities of hydrocarbon resources that could be recovered by enhanced recovery techniques, gas in geopressured brines, natural gas hydrates (clathrates), or oil and gas that may be present in insufficient quantities or quality (low-permeability "tight" reservoirs) to be produced via conventional recovery techniques.

Remaining conventionally recoverable resources: The volume of conventionally recoverable resources that has not yet been produced and includes remaining proved reserves, unproved reserves, reserves appreciation, and undiscovered conventionally recoverable resources.

Economically recoverable resources: The volume of conventionally recoverable resources that is potentially recoverable at a profit after considering the costs of production and the product prices.

Undiscovered resources: Resources postulated, on the basis of geologic knowledge and theory, to exist outside of known fields or accumulations. Included also are resources from undiscovered pools within known fields to the extent that they occur within separate plays.

Undiscovered conventionally recoverable resources: Resources in undiscovered accumulations analogous to those in existing fields producible with current recovery technology and efficiency, but without any consideration of economic viability. These accumulations are of sufficient size and quality to be amenable to conventional primary and secondary recovery techniques. Undiscovered conventionally recoverable resources are primarily located outside of known fields.

Undiscovered economically recoverable resources: The portion of the undiscovered conventionally recoverable resources that is economically recoverable under imposed economic and technologic conditions.

Reserves: The quantities of hydrocarbon resources, which are anticipated to be recovered from known accumulations from a given date forward. All reserve estimates involve some degree of uncertainty.

Proved reserves: The quantities of hydrocarbons, which can be estimated with reasonable certainty to be commercially recoverable from known accumulations and under current economic conditions, operating methods, and government regulations. Current economic conditions include prices and costs prevailing at the time of the estimate. Estimates of proved reserves equal cumulative production plus remaining proved reserves and do not include reserves appreciation.

Remaining proved reserves: The quantities of proved reserves currently estimated to be recoverable. Estimates of remaining proved reserves equal proved reserves minus cumulative production.

Unproved reserves: Reserve estimates based on geologic and engineering information similar to that used in developing estimates of proved reserves, but technical, contractual, economic, or regulatory uncertainty precludes such reserves being classified as proved.

Reserves appreciation: The observed incremental increase through time in the estimates of reserves (proved and unproved (P and U)) of an oil and/or gas field. It is that part of the known resources over and above proved and unproved reserves that will be added to existing fields through extension, revision, improved recovery, and the addition of new reservoirs. Also referred to as reserves growth or field growth.

Total reserves: All hydrocarbon resources within known fields that can be profitably produced using current technology under existing economic conditions. Estimates of total reserves equal cumulative production plus remaining proved reserves plus unproved reserves plus reserves appreciation.

Total endowment: All conventionally recoverable hydrocarbon resources of an area. Estimates of total endowment equal undiscovered conventionally recoverable resources plus cumulative production plus remaining proved reserves plus unproved reserves plus reserves appreciation.

3.2 OVERVIEW OF UNCONVENTIONAL GAS RESERVOIRS

Law and Curtis (2002) defined unconventional gas accumulations as follows:

"Conventional gas resources are buoyancy-driven deposits, occurring as discrete accumulations in structural and/or stratigraphic traps, whereas unconventional gas resources are generally not buoyancy- driven accumulations. They are regionally pervasive accumulations, most commonly independent of structural and stratigraphic traps."

The regionally pervasive nature of unconventional oil and gas accumulations gives rise to very large resource volumes in most cases, and hence their significance is critical to assess. Natural distribution of energy sources is such that the volume of the reserve capacity increases as one moves away from conventional toward unconventional hydrocarbon resources.

Cander (2010) addressed the geological understanding of unconventional plays, noting:

"...the term 'resource play' implies to some that subsurface risks are either minimized or irreducible. As well, the term 'unconventional gas' connotes that little is to be gained from application of conventional principles of basin evolution and petroleum generation, migration, and entrapment. Under these circumstances, the value of regional geologic understanding of an entire basin prior to acreage capture can be overlooked."

According to Cander (2010), it is important to understand a basin from basement to surface in assessing potential for unconventional hydrocarbon resources. This also bears relevance in terms of technologies required to unlock the gas.

3.2.1 The Resource Triangle

Figure 3.2 shows the schematic of the resource triangle concept that has been widely utilized for exploration as well as resource assessment of various

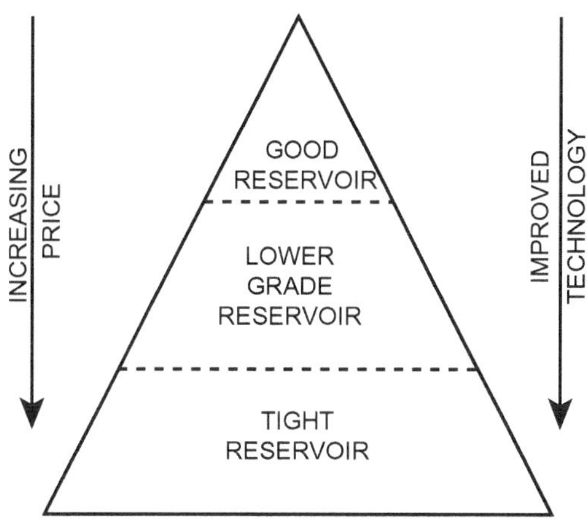

Figure 3.2 The resource triangle. *After Masters, 1979.*

commodities. Masters (1979) popularized the concept for gas reservoirs. This model has been applied in exploration and resource assessment for a number of different commodities and was popularized in the assessment of gas resources by Masters (1979). This triangle concept states that of the total resource base of a particular commodity, only a small proportion—near the apex of the triangle—is contained in high-quality reservoirs or deposits. Much greater resource volumes occur in poorer quality deposits, lower in the body of the triangle. For gas and oil, it is recognized that only a very small proportion of the resource has migrated into and can be produced from high-quality "conventional" reservoirs—those that the industry has been drilling throughout most of its history. Much larger hydrocarbon volumes occur in the "unconventional" reservoirs.

It is generally understood that the relative magnitude of this growth is proportionally larger, the younger the field. This continuity is supported by the sustainability argument discussed in the previous chapter that showed biomass, which is the youngest form of carbon-energy source, is also the most abundant. However, when unconventional reserves are included a different factor enters the interplay. This is due to the types of reservoirs that enclose older form of energy but were previously excluded due to the lack of proper technology and/or favorable economic factors. Overall, the following factors play a role (Lore et al., 1999):

- areal extension of existing reservoirs,
- discovery of new reservoirs,
- increases in reserve estimates in existing reservoirs as production experience is gained,
- improved recovery technologies,
- increases in prices and/or reductions in costs, which reflect the influences of market economics and technology (revisions),
- field expansion via mergers with newer fields (extensions),
- systematic assessment bias toward conservatism, which typically exists in initial estimates of field sizes (revisions), and
- reporting practices with respect to proved reserves.

This triangle model also assumes that the technology has to be improved for accessing "lower quality" reservoirs. This assumption implies that the price of an energy source (or commodity) will increase. Even though this model is widely accepted in the petroleum industry, only recently Zatzman and Islam (2007) have challenged the validity of the model. Zatzman (2012a, 2012b) has subsequently demonstrated that these models are in fact misleading with worse implication for unconventional resources. This point should be discussed to some extent.

Traditionally, economic evaluations are based on cost per unit output, which is only suitable for determining short-term and tangible outlooks. A comprehensive economic evaluation of any system should include long-term considerations that are only captured through intangible elements. An evaluation that incorporates both the tangible and intangible elements may be considered truly comprehensive. An engineering decision support system that follows such an evaluation process will focus on the long term, even as it tests and selects ingenious solutions that are suitable for tangible and short-term applications. By focusing on the long term, the sustainability criterion is fulfilled, thereby eliminating long-term negative consequences of a short-term remedy. This chapter proposes a guideline of economic evaluation that will truly identify the best process among different processes for both short-term and long-term applications. The triangle model assumes that the technology developed strictly for short-term gains are indeed sustainable. This assumption is incorrect (Khan and Islam, 2012).

In most cases, the comparison of different processes is based on economic evaluation. In a conventional economic analysis, there is no room for distinguishing between an energy source that is nonrenewable and one that is renewable. In the information age, it has become clear that such a "fit for all" analysis technique is not appropriate for meeting the energy needs of the future (Zatzman and Islam, 2007). In the information age, it has become both necessary and possible to custom design a specific engineering application in order to ensure long-term sustainability. One can no longer count on the future in order to ensure that the short-term needs are fulfilled. Because of the sustainability crisis, conventional economics and accounting theories have lost their effectiveness. For instance, according to the supply-and-demand theory, the cost of products from limited resources will increase continuously with the increase of demand and depletion of the resources. This scenario holds the most severe consequences for any resource that is the driving force of civilization and is very limited in supply. With this mode, if current practices of energy production and utilization continue, there will be a huge shortage of energy in the near future. This is evident from the increase in gas prices within the last two decades, the recent tripling of oil prices that sparked worldwide financial crisis, and the most recent sharp drop in oil prices that made the financial crisis worse. Yet, the real value of crude oil did not change, just like the real value of soil or any other natural commodity. This is the reason Zatzman and Islam (2007) characterized this economic system as being based on perception and not knowledge.

The conventional economic theories can be directly linked to disinformation regarding the intangible–tangible nexus. The theories of modern economics taught today, all start from something called "the theory of marginal utility." The English economics writer William Stanley Jevons (1870) first developed this theory in the 1870s, and it was furthered in the works of Carl Menger (1871), Leon Walras (1874), and Alfred Marshall (1890). Its underlying thesis, which became the basis of an elaborate theory known as neoclassical economics, is that endogenous "choices" about price operate entirely according to personal choice or desire for access to, and use of, some good or service, with the last unit of demand determining the price "at the margin." All of this takes place without reference to any exogenous conditions such as the monopolized character of production, the cartelized character of international trade, or the role of imperial dictates, rivalries, and/or wars in suppressing or further distorting the operation of supply and demand. Consumption takes place without reference to how commodities were produced in the first place. How, indeed, could that which has not yet been produced supposedly be distributed and consumed? "At the margin," reply the neoclassical economists. But one only has to ask the question, "whence the originating intention to consume or to produce?"

The specific role asserted for the theory of marginal utility in this arrangement reduces the domain of concern for "the margin," so as to simplify handling the relevant variables. However, it is precisely in this process that the counterfeiting job is carried out, as all exogenous conditions beyond "the margin" are simply removed from the domain of consideration, including any condition (such as intention) that may play some role in defining the boundary of, and/or the conditions at, "the margin."

The mathematical assumptions fundamental to the basic theory of marginal utility are another source of serious misdirection. Here, the still unaddressed question of intention becomes layered in further opacity. In order to model the individual's "choice" behaviors in economic reality, an untestable assumption that is completely subjective is used to measure what happens on a societal scale. That untestable assumption asserts that individuals' behaviors consist of maximizing personal pleasure and minimizing personal pain. This assumption is untestable because it assumes that society is composed of the individual multiplied uniformly and homogeneously over and over, or, in other words, that the individual exists, but society as a real category in its own right does not exist when it comes to economic analysis or decisions. To assert the existence of each would be true, but to hitch one (society) as subordinate to and derivative of the other (the individual) is false. In addition to this fundamental difficulty, this procedure also transfers the short-term perspective of the individual to society as a whole. However, since the death of any individual or individuals obviously cancels their personal term without shortening the term of society's existence, such a procedure is inherently and patently absurd.

Jevons, in particular, declared economic behavior to be nothing more or less than the materialization in social form of this allegedly universal and thoroughly selfish principle. What the worker does to avoid starvation is, thereby, equated with what the business owner does to get another few pennies of profit out of the powerless public. Jevons even suggested a mathematical model that justified his standpoint, arguing that the discrete choices of millions of economic actors may be approximated meaningfully or usefully by continuous-type mathematical functions. He combined this with the notion of processes that would eventually reach steady-state conditions, adapting to economics a mode of analysis that researchers in thermodynamics had pioneered widely by the middle third of the nineteenth century.

If the spurious assumption is removed, a new picture emerges from Figure 3.2. For instance, using truly sustainable technologies will make the downward arrows reversed. Indeed, Islam et al. (2010; 2012) and Khan and Islam (2012) have demonstrated that proper selection of technologies can align efficiency with sustainability, i.e., environmental integrity and economic benefits increase as one applies a sustainable technology to vaster resources. In this process, the qualities of gas, including its heating value and

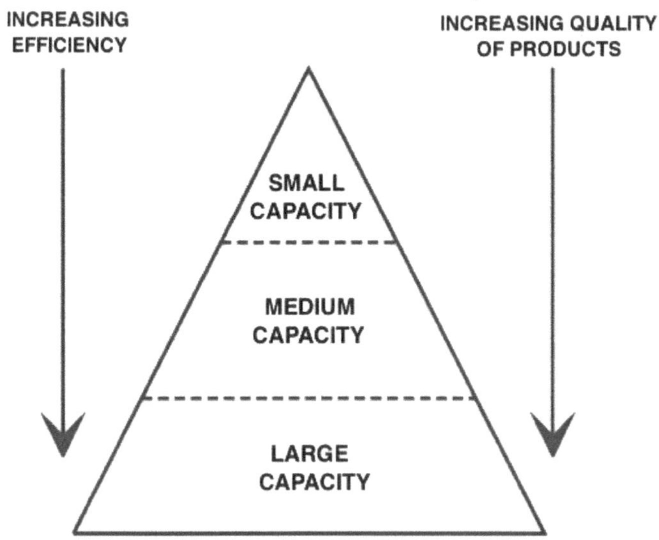

Figure 3.3 Paradigm shift in unconventional gas development.

environmental integrity, are improved. Figure 3.3 shows this paradigm shift in unconventional gas.

In the incubation era of unconventional gas (2000–2008), the industry saw large gas price increases and tremendous advances in drilling and completions practices, such as extended-reach horizontal wells and multiple staged hydraulic fracture ("fracking") stimulations. Unconventional gas production thus increased dramatically in the United States and to a lesser extent in Canada. Since 2008, however, world economic issues and the glut of new gas production on the North American market have depressed gas prices in North America. Industry is now turning more attention to unconventional plays that can produce liquid hydrocarbon resources from the lower part of the resource triangle. Such measures are not necessary if custom designed techniques are used rather than using blanket technologies for all resources. For instance, gas hydrate does not require any gas processing and burns easily. In addition, it is as abundant at the earth surface as tar sand. Why should this process be more expensive than natural gas that is produced from 10,000 ft below surface and contain high amount of toxic components?

By convention, several types of gas reservoirs are recognized, that are listed below:

- Coalbeds, hosting CBM, also known as coal seam gas;
- Tight reservoirs, both clastics and carbonates;
- shale reservoirs; and
- gas hydrates.

Shallow biogenic gas reservoirs have been considered by some as a distinct unconventional gas resource type (e.g., Shurr and Ridgley, 2002), but are now generally included with shale gas and tight gas resources.

3.3 SPECIAL FEATURES OF UNCONVENTIONAL GAS RESERVOIRS

The most commonly considered unconventional gas is shale gas, whereas the least heard unconventional resource is gas hydrate. Following is a brief description of these reservoirs.

3.3.1 Coalbed Methane

CBM is, as the name implies, a natural gas hosted in seams or beds of coal. Bustin and Clarkson (1998) described it as:

"Coalbed methane, unlike conventional gas resources, is unique in that gas is retained in a number of ways including: (1) adsorbed molecules within micropores (<2 nm in diameter); (2) trapped gas within matrix porosity; (3) free gas (gas in excess of that which can be adsorbed) in cleats and fractures; and (4) as a solute in ground water within coal fractures."

Coalbeds are generally self-sourcing reservoirs—they contain gas evolved either biogenically or thermogenically from organic material within the coals themselves. In their original form, this gas contains all vital components of life form, such as, oxygen, nitrogen, water vapor, carbon dioxide, carbon monoxide, methane, ozone, nitrogen dioxide, nitric acid, ammonia and ammonium ions, nitrous oxide, sulfur dioxide, hydrogen sulfide, carbonyl sulfide, dimethyl sulfide, and a complex array of nonmethane hydrocarbons. It is interesting to note that only nitrogen and oxygen are not "greenhouse gases." These biogenic gases are not accounted for global warming vulnerability when emitted from biomass. However, they are considered to be the primary reason for global warming when emitted through 'petroleum activities'. CBM is distinct from a typical sandstone or other conventional gas reservoir, as the methane is stored within the coal by a process called absorption and adsorption. The methane is in a near-liquid state, lining the inside of pores within the coal (called the matrix). The open fractures in the coal (called the cleats) can also contain free gas or can be saturated with water.

Absorption refers to the process by which one substance, such as a solid or liquid, takes up another substance, such as a liquid or gas, through minute pores or spaces between its molecules. The absorption capacity of the absorber depends on the equilibrium concentrations between gaseous phase and the liquid phase. For diluted concentrations in many gases and in a wide interval of concentrations, the equilibrium relation is given by Henry's Law, which quantifies the gas absorption capacity in the fluid (Cengel and Boles, 2008). The most effective gas absorbing unit would ensure complete contact between the gas and the solvent, in such a way that diffusion occurs at the interphase. This is the case with CBM. For instance, with water, methane absorption efficiency is around 97%. It also simultaneously removes H_2S when H_2S is less than 300 cm^3 and is adjustable to changing pressure or temperature.

In addition, adsorption also plays a big role in CBM. Adsorption refers to the process by which molecules of a substance, such as a gas or a liquid, collect on the surface of a solid. It differs from absorption, in which a fluid permeates or is dissolved by a liquid or solid (Tondeur and Teng, 2008). This process can be physical (reversible) or chemical (irreversible). In physical adsorption processes, gas molecules adhere to the surface of the solid adsorbent as a result of the molecule's attraction force (Van der Walls Forces).

Chemical adsorption involves a chemical reaction. Coalbed is known to be the best of adsorbers. For instance, coalbed acts as a highly efficient methane adsorber (95–98% CH_4). It is also excellent for capturing H_2S at relatively low energy. The microstructure of coalbed is such that it acts as a natural filter. It has a high surface area ranging from 0.68 m^2/g to 9.7 m^2/g.

Furthermore, condensation plays a role in CBM. It is the process of converting a gas into a liquid by reducing temperature and/or increasing pressure. Condensation occurs when partial pressure of the substance in the gas is lower than the vapor pressure of the pure substance at a given temperature. As a coalbed is exposed to high pressure, condensation leads to compaction of the gas resource, making CBM to be commercially available within coalbed.

Finally, natural membrane separation plays a role in purifying methane. CBM as well as water films serve as a selective barrier between two phases and remain impermeable to specific particles, molecules, or substances when exposed to the action of a driving force. The driving force is the pressure difference between both sides of the membrane. Gas permeability through a membrane is a function of the solubility and diffusivity of the gas into the material of the membrane. Membranes are expensive and their separation efficiencies are low (Ramirez, 2007).

This gas is distinctly different from gas from other sources. The term refers to methane adsorbed into the solid matrix of the coal. It is called "sweet gas" because of its lack of hydrogen sulfide. The presence of this gas is well known from its occurrence in underground coal mining, where it presents a serious safety risk. CBM is distinct from a typical sandstone or other conventional gas reservoir, as the methane is stored within the coal by a process called adsorption. The methane is in a near-liquid state, lining the inside of pores within the coal (called the matrix). The open fractures in the coal (called the cleats) can also contain free gas or can be saturated with water.

Most facies models for coal-bearing strata show peat accumulating close to active clastic depositional environments, such as floodplains of meandering rivers, coastal mires, behind beach barriers or in interdistributary bays of deltas. In this, muds and slits are introduced during frequent floods, storm surges, or high tides (McCabe, 1988). Thick, low ash peat accumulations are most commonly formed in areas that were removed from active clastic deposition for long periods of time. Coal seams have complex internal stratigraphies that lead to considerable variation in plant types; early alteration of organic material and climate changes are important factors during peat accumulation. Variations in vertical profile of coal seams are observation based on maceral

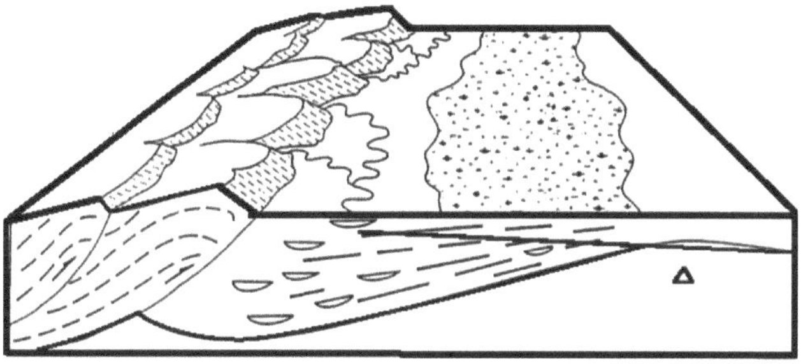

Figure 3.4 Schematic showing tectonic controls of coal deposition during the early tertiary in the Alberta basin. *Richardson et al., 1988.*

distribution and palynological studies (Hacquetard and Donaldson, 1969). Although some coals and adjacent clastics may have been deposited simultaneously, most common cases show that the contact between the two lighologies represents a considerable time gap in deposition. Peats commonly overlie much older sediments representing a 100,000 year of more hiatus (Figure 3.4).

In some cases, thick accumulations of peat sit directly on Pleistocene beach ridges in areas unaffected by storms and tidal surges. These mires develop extensive, low ash peats unhindered by clastic sources. Mires can be classified into three types: raised, low-lying, and floating. Low-lying mires form in areas of poor drainage. Floating mires develop over lakes until the lake infills with sediment or peat and can later develop into larger low-lying mires. If climate conditions permit, a raised mire can develop. Raised mires occur in areas where annual precipitation is greater than annual evaporation, which occur in maritime environments. There has to be a balance between preservation and deterioration of organic matters. For instance, most tropical rainforests are not sites of peat accumulation as the organic matter deteriorates rapidly.

The alluvial plain environment is characterized by widespread mires and very long gradients. For many cases, associated sediments are predominantly fluvial in origin. A moderately low ash content and laterally persistent and relatively think nature of many of these coal seams would indicate the development of widespread mires isolated from clastic deposition for long periods of time.

The key parameters that control gas resource quantities and producibility are

- thermal maturity—sorption capacity generally increases with maturity;
- maceral composition—sorption capacity increases with vitrinite content, and decreases with inertite content and mineral matter;

- reservoir pressure—generally increases with depth;
- coal thickness;
- fracture (cleat) density;
- permeability, which is a function of burial history, burial depth, and in situ stresses; and
- hydrologic setting, which dictates how much water is present in coal seams (Ayers, 2002).

CBM is formed in confined coalbed aquifers through complex bio-geophysical processes and remains trapped by water pressure. Coalbed natural gas-produced water can have high concentrations of soluble salts (Rice et al., 2000). The above mentioned factors play significant roles in forming and transporting of CBM. Figure 3.5 shows how the process acts like a water filter. CBM has a biogenic root. The biogenic gas falls into two distinct systems that have different attributes. Early-generation systems have blanketlike geometries, and gas generation begins soon after deposition of the source rock. Late-generation systems have ringlike geometries, and long time intervals separate deposition of reservoir and source rocks from gas generation. For both types of systems, the gas is dominantly methane and is associated with a water system. In certain cases, biogenic gas will be transformed through thermogenic processes. When these fine-grained sediments are buried by deposition of later, overlying sediments, the increasing heat and pressure resulting from burial turns the soft sediments into hard rock strata. If further burial ensues, then temperatures continue to increase. When temperatures of the organic-rich matrix exceed 120 °C the organic remains begin to be "cooked" and more natural gas is formed.

Figure 3.6 shows coalbed matrix that embeds gas surrounding the coal bound by water and rock. Note that the light blue represents gas that is held

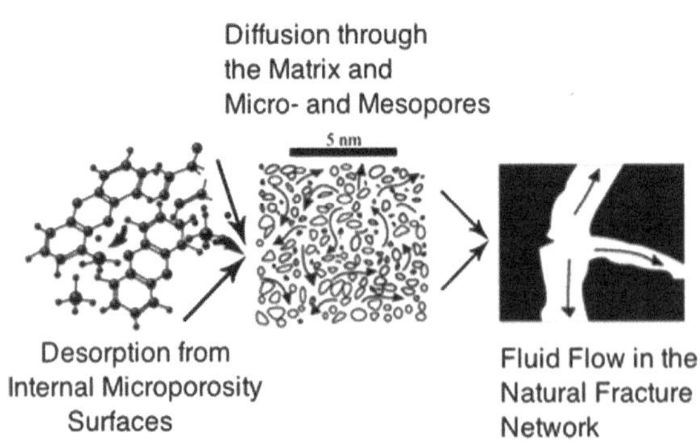

Figure 3.5 Microscopic process of CBM transport.

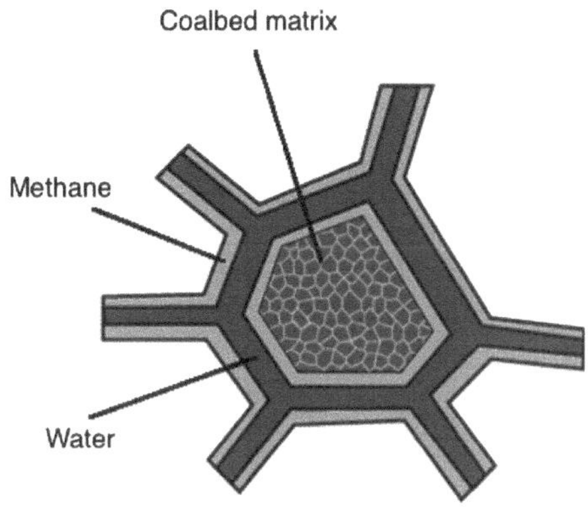

Figure 3.6 Schematic of CBM surrounded by water.

in contact with the coal through high water pressure. Water remains in cleats and fractures. This schematic explains how CBM is one of the cleanest site of methane. Both coal and water are most potent contaminant separating agents. Both gaseous and liquid contaminants can be separated in this process.

Cleats are fractures that usually occur in two sets that are, in most instances, mutually perpendicular and also perpendicular to bedding. Although pre-1990 geologic literature and current mining usage distinguishes these sets on the basis of factors such as "prominence" that are difficult to quantify, abutting relations between cleats generally show that one set predates the other. Picture 3.1 shows the existence of cleats in Alberta outcrops. Through-going cleats formed first and are referred to as face cleats; cleats that end at intersections with through-going cleats formed

Picture 3.1 Coal outcrop from Utah. *USGS file photo.*

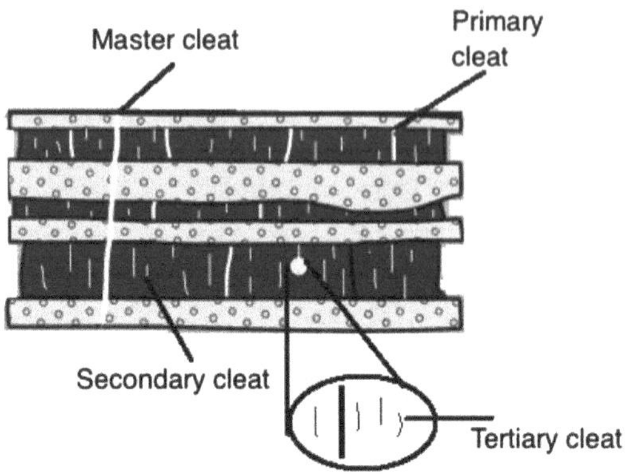

Figure 3.7 The existence of various cleats within coalbed.

later and are called butt cleats (Laubach and Tremain, 1991; Kulander and Dean, 1993). These fracture sets, and partings along bedding planes, impart a blocky character to coal.

Coal cleat is important in enhancing CBM potential, as it is these interconnected fracture networks (permeability) that allow fluids (water and gas) to move through the coals and into a wellbore for production. The widths or openings of cleats are generally wider near the surface, but with increasing depth the weight of the overlying rock compresses, or closes the cleats, reducing their width and connectivity (reducing reservoir permeability). Cleats may be enhanced in areas of faulting, fracturing or where coal seams may be draped over sandstone channels. CBM companies are always looking for areas of enhanced permeability to improve their exploration success potential. Figure 3.7 shows the existence master cleat, primary, secondary, and tertiary cleat within the cola matrix.

3.3.1.1 Coal Rank, Depth of Cover, and CBM Potential

One of the most important features of coalbed is that as the depth increases the gas content is increased whereas the fracture openings are restricted. This leads to the existence of optima in terms of depth of a coalbed.

3.3.2 Tight Gas

Unconventional tight oil and gas resources are generally found in basin-centered systems, defined by Law (2002) as:

"...regionally pervasive accumulations that are (hydrocarbon) saturated, abnormally pressured, commonly lack a downdip water contact, and have low-permeability reservoirs."

In general, the term "hydrocarbon saturated" refers to oil or gas that occupies most of the reservoir pore volume (generally >75%), and that one or both of these phases flow from the formation; however, there is almost always some residual water saturation. "Abnormal pressures" indicate that the hydrocarbon phase is not connected to a regional aquifer—pressures may be relatively high or low compared to a normal hydrostatic gradient (and may be both in different regions of a given basin). "Low permeability" is a term that has been used in a variety of ways, but the most commonly accepted definition is that "tight" gas reservoirs have maximum in situ permeabilities of <0.1 mD, with the implication that natural or artificial fracture stimulation is required for economic production. This definition of Law (2002) excludes many reservoirs that have high water content. Also, the definition in terms of low permeability has been accepted worldwide, while keeping open the arbitrary nature of the cutoff point of 0.1 mD. Often, this "low permeability" is linked with need to fracture a formation in order to make the gas flow economically. Others have defined tight gas reservoirs as being the ones that cannot be produced at economic flow rates or recover economic volumes of gas unless the well is stimulated by a large hydraulic fracture treatment and/or produced using horizontal wellbores (Holditch, 2007, 2006). This definition is also applied to CBM, shale gas, and tight carbonate reservoirs. There is no typical tight gas reservoir, it can be: deep or shallow, high pressure or low pressure, high temperature or low temperature, homogeneous or naturally fractured (heterogeneous), single or multilayered, high transient decline rates, comingled production; require fracturing jobs and/or horizontal well. In other words, tight gas is gas that is "trapped" in a very tight formation underground, stored within low-porosity and low-permeability rock formations. In general, it is commonly understood, there are only two options for recovering this gas. They are fracturing and drilling horizontal wells.

Source rocks for basin-centered accumulations are commonly found interbedded within or downdip of the tight reservoir section. Hydrocarbons thus charge the system directly, without migration through and exposure to regional aquifer systems (Welte et al., 1984; Law, 2002). Strong aquitards, such as thick, continuous marine shales, also assist in isolating basin-centered accumulations from contact with regional aquifers.

A play is defined as a group of reservoirs genetically related by depositional origin, structural style or trap type, and nature of source rocks or seals (White and Gehman, 1979; White, 1980). A typical tight play consists of grouped reservoirs (termed pools or accumulations) within individual fields that produce from the same chronozone and depositional sequence. These pools often vertically stacked within fields (Figure 3.8).

A key feature in the current exploitation of most basin-centered tight gas systems is the presence of a relatively thick (tens to hundreds of meters) column of hydrocarbon-saturated rock. Gas resources in place per unit area (often expressed as Bcf per section—billion cubic feet per one-mile-square section of land) are thus high—up to tens of Bcf per section. Alternatively, a tight gas or oil reservoir may be thinner, but very laterally extensive—allowing highly-repeatable drilling strategies over large areas. Using appropriate drilling and completions strategies, each well can access an economic reserves volume, despite low formation permeabilities.

A typical depositional model is shown in Figure 3.9. This particular example has three depositional styles (retrogradational, aggradational, and progradational) and one depositonal facies (fan). The retrogradational style, characterized by thick shale sections and thin sandstone beds, represents major or widespread transgressive events. The lower part of the retrogradational section commonly contains thin sandstone units that are products of reworking of the top of the underlying shallow-water sandstones. Within the retrogradational package are thinner packages of sandstone that typically comprise upward-coarsening progradational parasequences. When stacked, the thin progradational parasequences form a back-stepping architecture, reflecting the increasing amount of accommodation space and the retreat of depositional environments during relative sea-level rise.

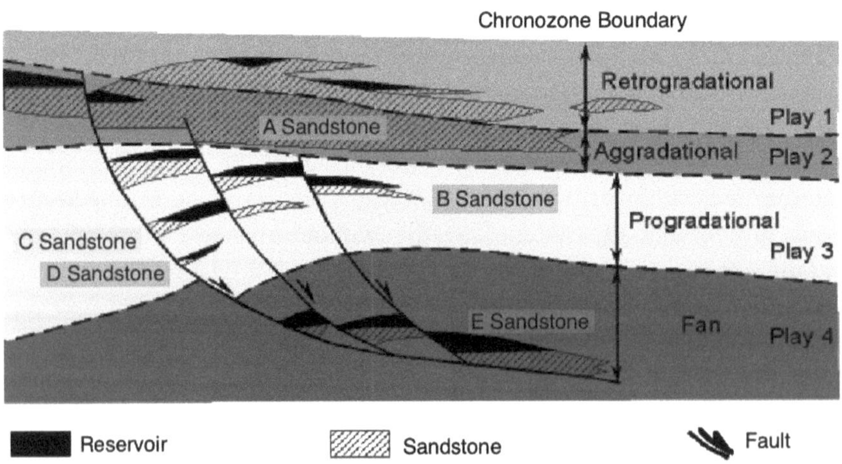

Figure 3.8 Stacked pools. *From Lore et al., 1999.*

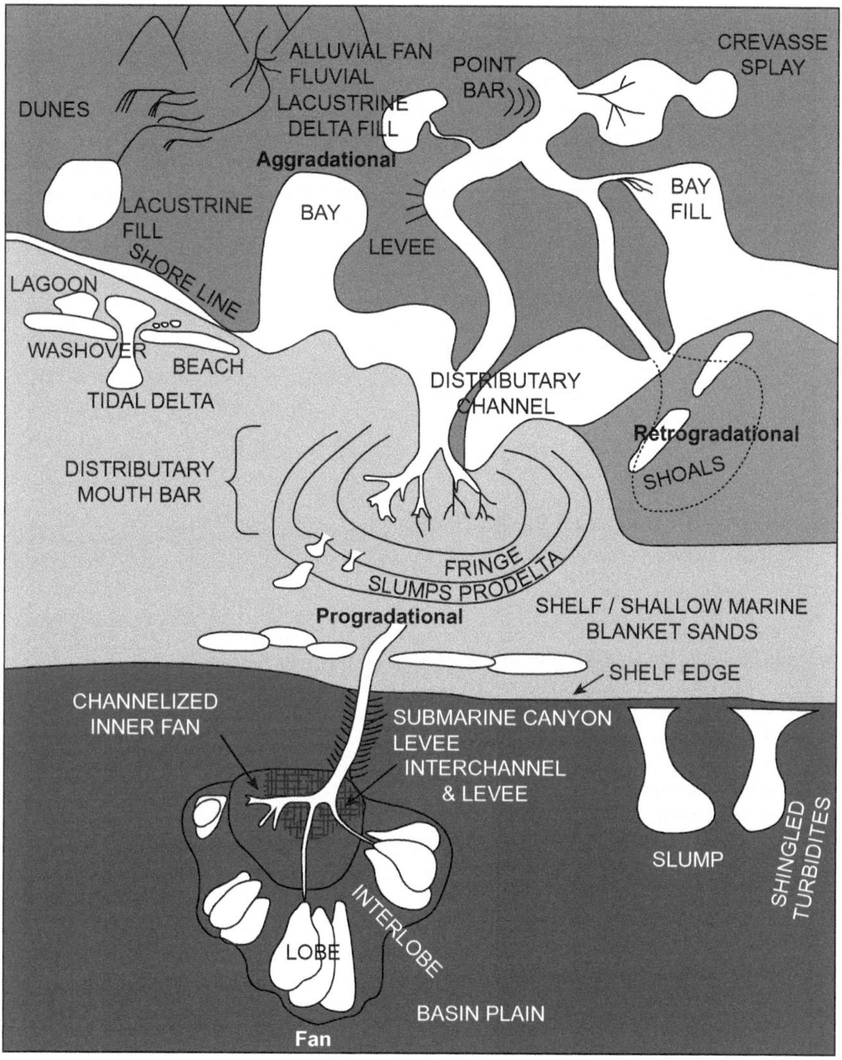

Figure 3.9 Model for deltaic deposition. *From Lore et al., 1999.*

The aggradational style comprises thick sandstone beds separated by thin shale units. Depositional environments represented by aggradational sediments include fluvial–streamplain, bay–lagoon, barrier island, coastal strandplain, and marine shelf (Morton et al., 1988). Fluvial and strandplain depositional environments dominate the aggradational depositional style.

The progradational style is characterized by deeper water shale at the base, along with thin sandstone units that grade upward into dominantly shallow marine deltaic and shoreline sandstones that are topped by thin shale interbeds. A broad spectrum of paralic depositional environments, including deltaic, shoreline, strandplain, barrier bar, shelf, and coastal plain, are subsumed under the progradational style. Deltaic depositional environments are dominant. Progradational architecture is constructed of thinner packages of dominantly progradational parasequence sets. Minor or local retrogradational events are typically interspersed within the overall progradational style.

The fan facies is a sandstone-rich, deepwater environment, characterized by a variable pattern of sandstone-body thickness (including thick to thin and blocky to upward-fining sandstones), sharp-based channel-fill sandstones, and serrated, thin to thick sandstones interbedded with thick shale units. Fan environments are characteristically overlain by hundreds of feet of deepwater shale.

The continental environment has the following systems:

Eolian systems

Lacustrine systems

Fluvial systems

Terrigenous fan (alluvial fan and fan-deltaic systems)

The shoreline environment has the following systems:

Delta systems

Barrier-strandplain systems

Lagoon, bay, estuarine, and tidal-flat systems

Marine environments has the following systems:

Continental and intracratonic shelf systems

Continental and intracratonic slope and basinal systems

These features show up in spontaneous potential data, as shown in Figure 3.10. This figure also shows characteristic features of each layer.

For Gulf of Mexico, Lore et al. (2001) described some of the important features of various styles.

The main characteristic of retrogradational deposit is the existence of upward-coarsening and upward fining of thin sandstone and upward-thinning packages of sandstone. The depositional environments are back-stepping assemblage of shoreline, deltaic, and nearshore environments that culminate in open-shelf mud-rich settings. They are typically capped by a flooding surface coincident with a chronozone boundary.

The aggradational style shows thick, blocky stacked sandstone. The environment shows the existence of vertically stacked upper alluvial plain, valley-fill, fluvial crannel, overbank, upper-delta plain, sand-rich deposits.

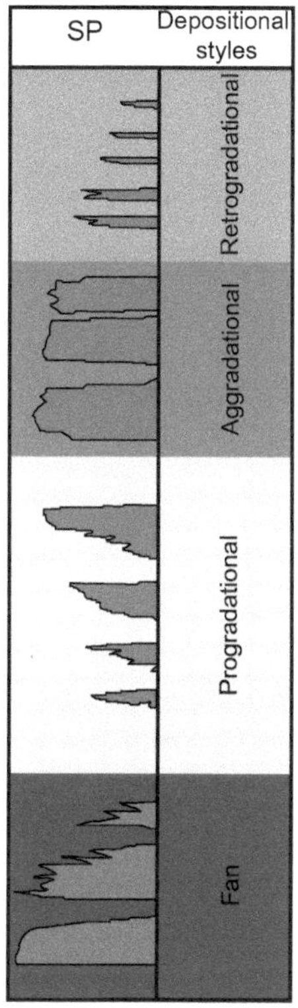

Figure 3.10 Electric characteristics of various depositional styles. *From Lore et al., 1999.*

The progradational layer is characterized by thin to thick, upward-coarsening sandstone and sandstone packages. The depositional environment is regressive assemblage of environments grading from relatively deepwater mud-rich distal deltaic environments that grade upward to relatively shallow-water paralic and sand-rich deltaic and shoreline environments. Typical overlying a chronozone boundary in proxima position are fan systems in distal position.

The fan style is characterized with serrated, thin to thick sandstone packages, thick shale at top, upward fining, blocky at base with singular or stacked settings. The environment has upper-slop to abyssal-plain environment comprising channel fill, levees, and overbank sands deposited in a relatively sand-rich deepwater environment (Table 3.1).

The classic example of a basin-centered gas system is the San Juan Basin of the Western United States. Masters (1979) demonstrated the widespread occurrence of gas in low-permeability Cretaceous reservoirs in the center of the basin, downdip from regional aquifers around the basin margins

Table 3.1 Summarizes the Clastic Sedimentary Environment

Environment	Agent of transportation deposition	Sediments
Alluvial	Rivers	Sand, gravel, mud
Lake	Lake currents, waves	Sand, mud
Desert	Wind	Sand, dust
Glacial	Ice	Sand, gravel, mud
Delta	River, waves, tides	Sand, mud
Beach	Waves, tides	Sand, gravel
Shallow shelf	Waves, tides	Sand, mud
Deep sea	Ocean currents, settling	Sand, mud

(Figure 3.11). Hayes (2005) reviewed some of the key advances throughout this time in recognizing the importance and characteristics of tight gas resources.

Law (2002) formalized the definition of basin-centered gas, and tabulated numerous examples of basin-centered gas accumulations in the United States and internationally. Moslow (2005) summarized characteristics of Western U.S. Rocky Mountain tight gas reservoirs as follows:

- Thick (tens to hundreds of meters), laterally extensive sandstones;
- Low porosities (8–12%) and permeabilities (<0.1 mD);
- Generally high sand/shale ratios;
- Extensive natural fracturing, arising from the complex tectonic history of most U.S. Rocky Mountain basins; and
- Abnormal reservoir pressures (predominantly overpressured).

Masters (1979, 1984) proposed a somewhat different model for regionally extensive low-permeability gas reservoirs along the western flank of the Western Canada Sedimentary Basin within a basin-centered gas system, calling it the Deep Basin. This model is shown in Figure 3.12.

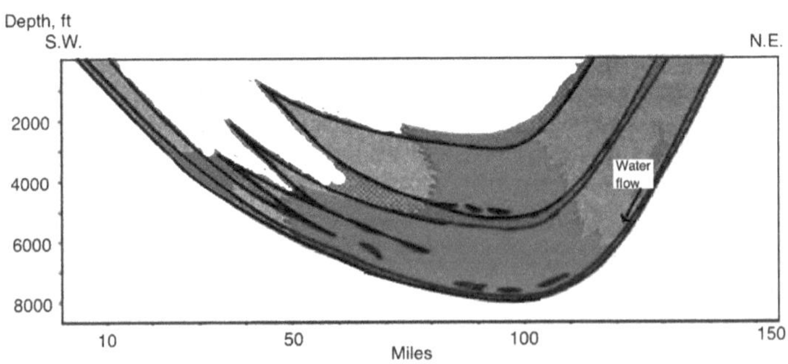

Figure 3.11 Schematic cross section of San Juan Basin, showing basin-centered gas system (in deep grey) *From Masters (1979).*

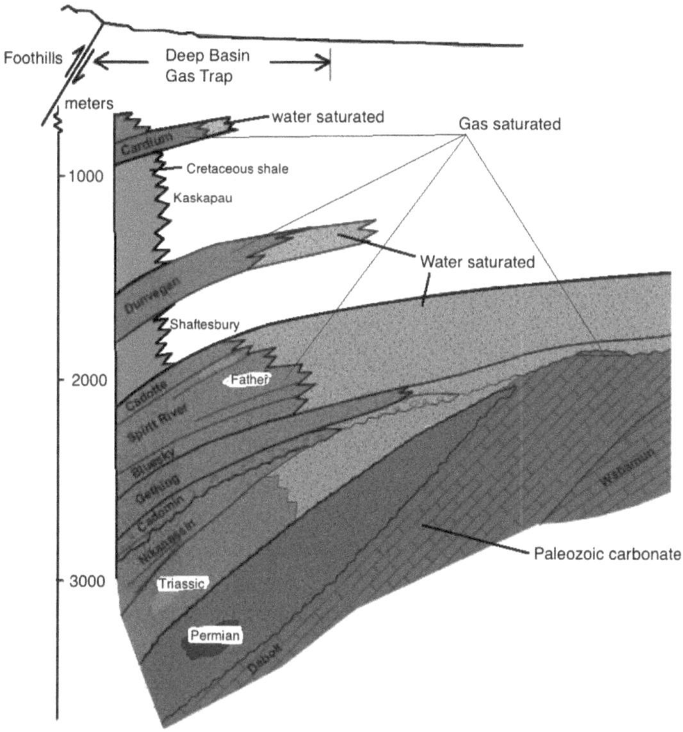

Figure 3.12 Schematic cross section of Western Canada Deep Basin, showing continuous gas-saturated reservoirs downdip from regional aquifers. *(Redrawn from Hayes and Archibald, 2012).*

The system is sourced from coaly strata interbedded within the tight gas section and from regional marine source rocks, occurring downsection and downdip in the gas window. Thick marine shales cap Deep Basin reservoirs and isolate them from shallow aquifers and meteoric waters. While many geologists were initially reluctant to accept the Deep Basin concept, the success of Canadian Hunter Exploration in discovering widespread sub-normally pressured gas pools, not connected to updip regional aquifers, provided evidence of the existence of Deep Basin. However, these early discoveries tapped into "stratigraphic sweet spots" within the Deep Basin—relatively small bodies of coarse clastic strata featuring conventional reservoir quality—such as shoreline-deposited conglomerates in the Cadotte and Falher members (Masters, 1984). In this, enormous amount of gas and liquid resources in the truly low-permeability rocks were not being accessed.

Deep Basin tight gas reservoirs host huge gas resources in Alberta and British Columbia. Initial estimates show the following numbers:

- Gas in place, Alberta Deep Basin, Townships 45–75: 430 Tcf (Hayes et al., 2009);
- Gas in place, northeast British Colombian Deep Basin: 111–260 Tcf (Hayes, 2003; Hayes and Hayes, 2004); and
- Gas in place, Montney Deep Basin: corporate report and analyst figures ranging from tens to hundreds of Bcf per section, for a total play potential of hundreds of Tcf.

Hayes (2010), in an initial scoping assessment of tight gas potential in the Northwest Territories, concluded that tight gas potential may exist in specific Cambrian depocenters in the Colville Hills area, in the Devonian Imperial Formation of Peel Plain, and in Cretaceous, Triassic, and Devonian Jean Marie tight gas plays mapped northward from Alberta and British Colombia into southern Northwest Territories. Elsewhere in Canada, tight gas potential has generally not been assessed systematically—in part because it is difficult to establish the existence of all the components of a tight gas petroleum system where drilling and testing are sparse. Documentation of complete petroleum systems and conventional petroleum plays in Paleozoic basins of eastern Canada by Lavoie et al. (2009) and Dietrich et al. (2011) suggests that numerous tight gas and oil plays may exist there.

3.3.3 Shale Gas Reservoirs

Shale gas been in the core of recent domestic gas production revolution in the United States. It is expected that this superflux of gas would reverse the expectation that the United States remain a substantial gas importer, and has contributed to reduced gas prices, and enhanced industrial competitiveness. Overall, it is predicted that shale gas would exhibit the fastest growth in production within the United States, making an example for the rest of the world to follow.

Curtis (2002) defined shale reservoirs as:

> *fine-grained, clay- and organic carbon-rich rocks, [which] are both gas source and reservoir rock components of the petroleum system…Gas is of thermogenic or biogenic origin and stored as sorbed hydrocarbons, as free gas in fracture and intergranular porosity, and as gas dissolved in kerogen and bitumen.*

Similar to most nomenclature, the term "shale" is not scientific because gas can be present in many other forms of rocks that have similar properties.

This was pointed out by Rokosh et al. (2009) that stated: gas can be present "not only in shale, but also a wide spectrum of lithology and texture from mudstone to siltstone and fine-grained sandstone, any of which may be of siliceous or carbonate composition."

Hamblin (2006) also noted the compositional variability of "gas shales," and defined them more broadly, in terms of unconventional accumulations:

> *These are unconventional, basin-centered, self-sourced, continuous-type gas accumulations where the total gas charge is represented by the sum of free gas and adsorbed gas...In effect, these shale gas plays represent discrete, self-enclosed petroleum systems which do not rely on hydrocarbon expulsion/migration/trapping because the premise is that the hydrocarbon stays in the original source rock; if they were well-connected to conventional plays, then they wouldn't provide a new play at all.*

Practically all caprocks will have similar properties. The common practice is to use certain cutoff point in terms of porosity or saturation or permeability in conventional reservoir analysis. However, this practice loses meaning when it comes to unconventional gas and oil. Vast majority of caprock will have certain amount of gas that can be considered part of the unconventional reserve.

Shale gas reservoirs have the following features:

- Rock volume—the entire thickness and areal extent of the shale has the potential of producing gas as long as special techniques are used;
- Total organic carbon content will be mainly distributed as adsorbed to the rock surface and along the fractures and fissures;
- Mature organic materials that are closest to the source rock;
- Mineralogical composition of the shale and related rocks is such that the rock is brittle;
- Natural fractures and fissures are present; and
- High water saturation.

The above features are all relevant and exploitable as long as the right technology is employed. To date, horizontal wells and fracking remain the only two techniques vastly employed in the petroleum engineering industry. However, there are no optimal values for these parameters, but the characteristics of each shale reservoir dictate the most appropriate drilling, completion, and production strategies. As each new shale play unfolds, operators must undertake a period of experimentation in order to refine these strategies. Several workers, such as Curtis (2002), have tabulated selected parameters for various shale plays (Table 3.2).

Table 3.2 Geological, Geochemical, and Reservoir Parameters for Five Shale Gas Systems in the United States

Property	Antrim	Ohio	New Albany	Barnett	Lewis
Depth (ft)	600–2400	2000–5000	600–4900	6500–8500	3000–6000
Gross thickness (ft)	160	300–1000	100–400	200–300	500–1900
Net thickness (ft)	70–120	30–100	50–100	50–200	200–300
Bottom-hole temperature (°F)	75	100	80–105	200	130–170
TOC (%)	0.3–24	0–4.7	1–25	4.50	0.45–2.5
Vitrinite reflectance (% R_0)	0.4–0.6	0.4–1.3	0.4–1.0	1.0–1.3	1.6–1.88
Total porosity (%)	9	4.7	10–14	4–5	3–55
Gas-filled porosity (%)	4	2.0	5	2.5	1–3.5
Water-filled porosity (%)	4	2.5–3.0	4–8	1.9	1–2
Permeability thickness [Kh (mD-ft)]	1–5000	0.15–50	NA	0.01–2	6–400
Gas content (scf/ton)	40–100	60–100	40–80	300–350	15–45
Adsorbed gas (%)	70	50	40–60	20	60–85
Reservoir pressure (psi)	400	500–2000	300–600	3000–4000	1000–1500
Pressure gradient (psi/ft)	0.35	0.15–0.40	0.43	0.43–0.44	0.20–0.25
Well costs ($1000)	180–250	200–300	125–150	450–600	250–300
Completion costs ($1000)	25–50	25–50	25	100–150	100–300
Water production (b/day)	5–500	0	5–500	0	0
Gas production (mdt/day)	40–500	30–500	10–50	100–1000	100–200
Well spacing (ac)	40–160	40–160	80	80–160	80–320
Recovery factor (%)	20–60	10–20	10–20	8–15	5–15
Gas in place (Bcf/section)	6–15	5–10	7–10	30–40	8–50
Reserves (mmcf/well)	200–1200	150–600	150–600	500–1500	600–2000
Historic production area basis for data	Otsego Country, Michigan	Pike County, Kentucky	Harrison County, Indiana	Wise County, Texas	San Juan & Rio Arriba Counties, New Mexico

ac = acre

From Curtis, 2002.

Historically, shale gas development in the United States has been synonymous Devonian shales. As early as 1821, these formations have been set in production in the Appalachian Basin (e.g., Curtis, 2002). Production from these shales has been important locally since that time. In the past, a relatively small number of shale reservoirs, primarily Devonian and Mississippian organic-rich systems such as the Antrim, Ohio, New Albany, and Barnett shales, were considered viable. Horizontal drilling technologies developed in 1980s and 1990s unlocked much greater resources of shale gas. This followed improvements in fracking technologies, including multistage fracturing, giving shale gas an additional boost. For instance, Mitchell Energy and others developed previously unexploited Barnett shale in the Fort Worth Basin of Texas, initially with horizontal wells, then with multistage fracturing. By 2000, the Barnett Newark East Field was the largest producing gas field in Texas, with multi-Tcf reserve potential (Montgomery et al., 2005). Since 2006, shale drilling and production has exploded in the United States, supported by continued advances in technology and rising gas prices. Shale gas plays are producing in numerous basins across the country (Figure 3.13), although most of the activity is focused on a few plays such as the Barnett,

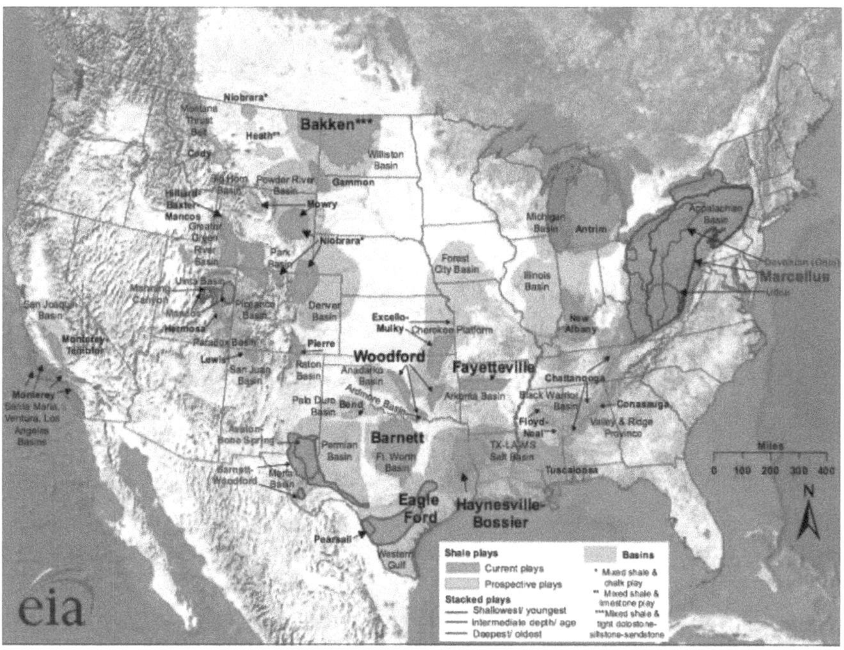

Figure 3.13 Shale gas basins and play trends, the continental United States. *From U.S. Energy Information Administration, 2011.*

Haynesville, Fayetteville, Woodford, and Marcellus. Huge efforts are being made to better understand shale reservoirs (e.g., Montgomery et al., 2005; Bowker, 2007; Jarvie et al., 2007; Pollastro, 2007; Pollastro et al., 2007).

Canada got into shale gas business much later than the United States. Only after 2000s that Candian shale plays have been systematically exploited. Hamblin (2006) reviewed 50 shale units in seven major regions of Canada to produce a geographic inventory of significant shale potential across the country (Figure 3.14–3.17). He concluded that many of these shales have interesting geological characteristics that make them worthy of further investigation. Hamblin (2006) judged the following shale gas units as worthy of concerted geological examination:

- Middle to Upper Ordovician of Appalachian Mountains, St. Lawrence Platform, and Anticosti Island (Utica, Collingwood, Blue Mountain);
- Upper Devonian of St. Lawrence Platform and Michigan Basin (Marcellus, Kettle Point);

Figure 3.14 Shale plays in Atlantic Canada, Quebec, Ontario, and Hudson margin. *From Hamblin, 2006.*

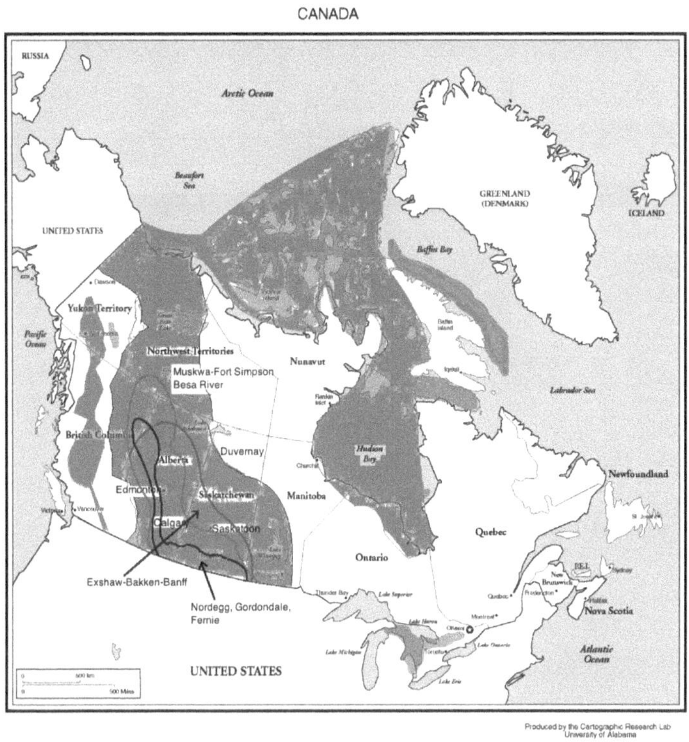

Figure 3.15 Shale gas play areas, western Canada Paleozoic and Mesozoic passive margin platform. *From Hamblin, 2006.*

- Middle and Upper Devonian of western Alberta, northeastern BC, Liard Basin, and Mackenzie Corridor (Horn River, Muskwa, Imperial, Fort Simpson, Besa River);
- Lower to Middle Triassic of northwestern Alberta and northeastern BC (Montney, Doig);
- Jurassic of western Alberta and northeastern BC (Nordegg, Gordondale, Fernie); and
- Middle-Upper Cretaceous of Alberta, Saskatchewan, and Manitoba (Colorado).

Figure 3.18 shows all North American shale play areas. While broadly distributed, North American shale gas basins generally follow a trend of thrust belts and a Mississippian/Devonian shale fairway from Western Canada and into the Western, Southern, and Eastern United States. The Laramide Thrust Belt bounds the Horn River and Montney Play in British Columbia and Alberta, as well as many of the Western US shale gas fields,

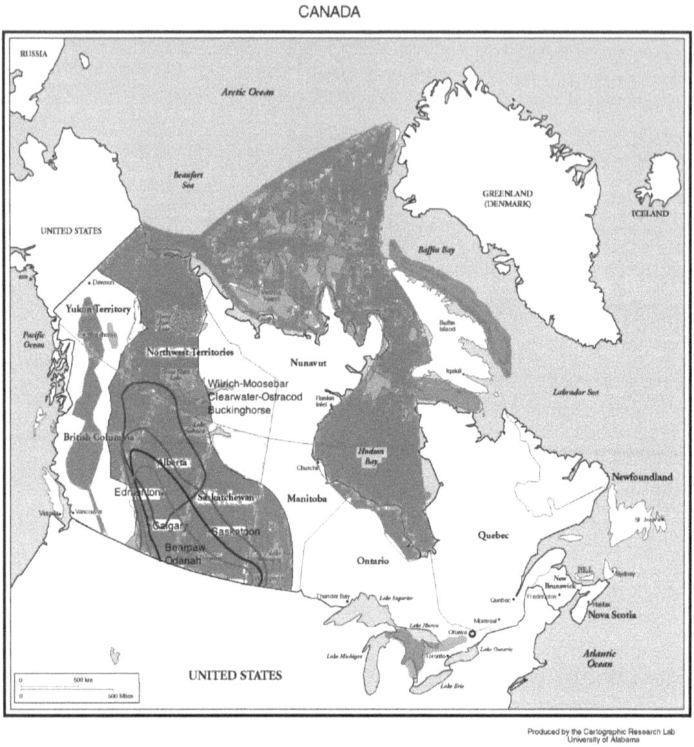

Figure 3.16 Shale gas play areas, western Canada Mesozoic foreland basin. *From Hamblin, 2006.*

including Jonah and Pinedale. Starting in South Texas, the Ouchita Thrust Belt bounds Southern US gas basins, including Eagle Ford, Barnett, Woodford, Haynesville and Fayetteville. Finally, the merger of the Ouchita and Appalachian Thrust Belts define the broad extents of the Marcellus shale gas basin.

North American shale gas reservoirs currently rank as 6 of the largest 22 global gas fields (Figure 3.19), based upon estimated recoverable reserves, with average recovery factors of about 20%. This recovery factor can be increased significantly if new techniques for recovery are used. Enhanced gas recovery schemes, not yet commercially applied, offers great promises for the future. In the mean time, 3D seismic, microseismic, Formation MicroImager/Formation MicroScanner and other measurements, are pivotal in unlocking North American gas supplies for the decades ahead.

Tables 3.3 and 3.4 shows some of the main features of North American shale plays. Note how reservoir properties vary in such a way that

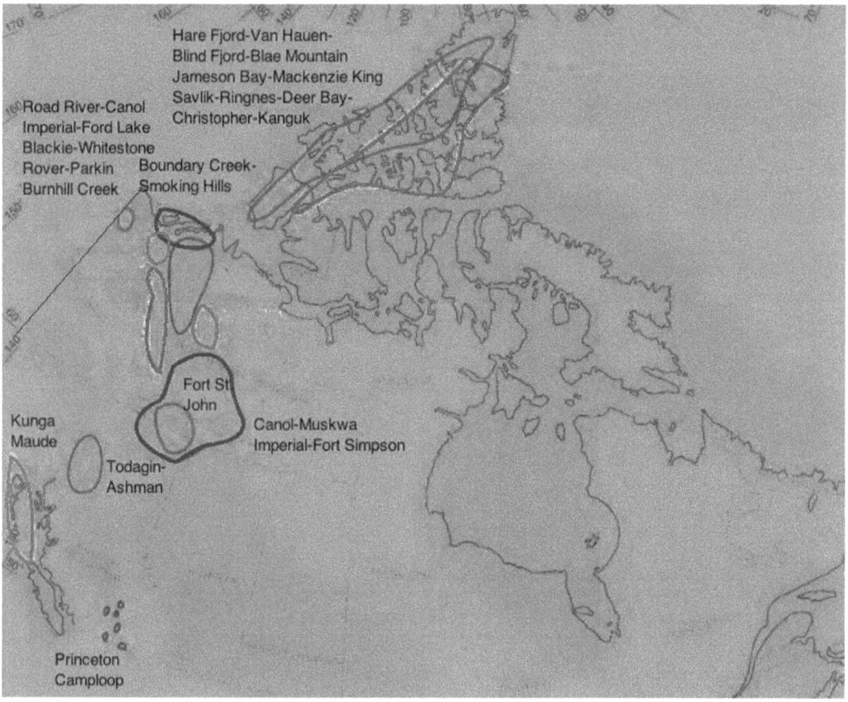

Figure 3.17 Shale gas play areas, Northwest Territories, Arctic Islands, and intermontane basins of BC. *From Hamblin, 2006.*

Figure 3.18 North American shale gas basins.

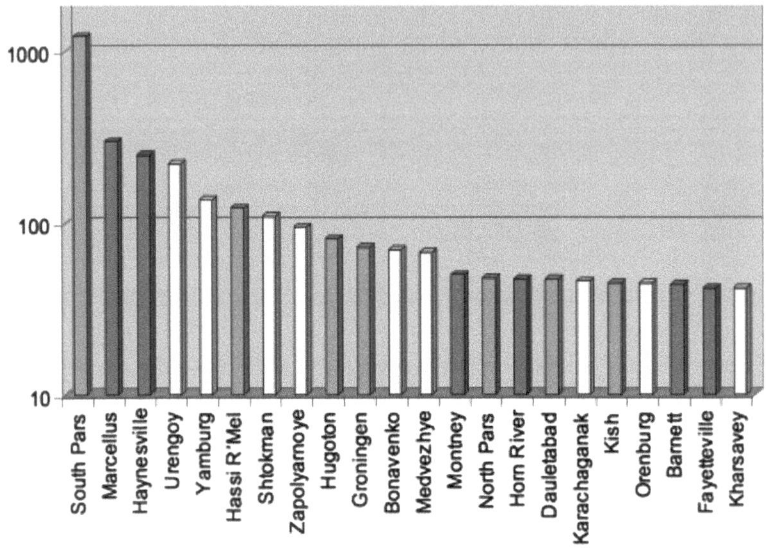

Figure 3.19 World's largest proven gas reservoirs.

production features are not amenable to Darcian correlation. Further complication arises from the variability in fracturing as well as development of horizontal wells. Surprisingly, after nearly 30 years of development and over 10,000 wells, wellbore lengths and completions parameters in the Barnett Shale of Texas can vary by factors of two or more—pointing to the challenge and non-uniqueness of production optimization (Website 4).

At present, the overall effort has been to identify shale gas "sweet spot" and avoid "dead zones." Accordingly, multilaterals as well fracturing techniques are designed to maximize access to "sweet spots." Following is a description of several shale plays of the United States and Canada along with their characteristic features.

Barnett Shale—Fort Worth Basin: These reservoirs are most suitable for production through horizontal wells with multistage completions. The Mississippian Barnett Shale averages 300 ft in thickness, at an average depth of about 7500 ft, with average porosities of 6% and low permeabilities of 250nD. Average well production reaches 2.65 Bcf, with initial production rates of up to 13 mmcf/day. Seismic data is generally of moderate quality, but is essential for tracking the key bounding horizons and identifying karst "collapse chimneys" that can cause severe decline in individual well production. This is shown in Figure 3.20. In addition, vitrinite reflectance averages 2% in the Barnett, but drops down in the northern basin.

Table 3.3 Shale Gas Reservoir Properties

Shale Play	Barnett	Haynesville	Marcellus	Horn River	Eagle Ford
Age	Mississippian	Upper Jurassic	Middle Devonian	Upper Devonian	Cretaceous
MYA	340	170	385	370	100
Total area size (sq mi)	50,000	9000	95,000	5000	1350
Total gas (Tcf)	327	717	1500	370	N/A
Gas in place (Bcf/sq mi)	150	175	200	250	200
Producible gas (Tcf)	50	251	356	47	N/A
Depth (ft)	7,500	12,000	7000	8800	11,500
Thickness (ft)	300	225	350	450	250
Horizontal well cost ($M)	2.8	7.0	3.5	7.0	4.8
Average EUR (Bcf)	2.65	6.5	3.75	N/A	5.5
Pressure (psi)	4000	8500	4000	N/A	5200
Temperature (°F)	200	340	N/A	160	N/A
R_o (%)	2	2.2	1.3	2.5	1.5
TOC (%)	4.5	3	3.3	3	4.5
Porosity (%)	6	10	8	3	11
Matrix permeability (nD)	250	658	1000	230	1100
Pressure gradient (psi/ft)	0.53	0.95	0.4	0.6	0.65

From Website 4.

Table 3.4 Canadian Shale Reservoirs

Property	Horn River	Montney	Colorado	Utica	Horton Bluff
Depth (m)	2500–3000	1700–4000	300	500–3300	1120–2000+
Thickness (m)	150	Up to 300	17–350	90–300	150+
Gas-filled porosity (%)	3.2–6.2	1.0–6.0	<10	2.2–3.7	2
TOC (%)	0.5–6.0	1–7	0.5–12	0.3–2.25	10
Maturity (R_o)[a]	2.2–2.8	0.8–2.5	Biogenic	1.1–4	1.53–2.03
Silica (%)	45–65	20–60	Sand and silt	5–25	38
Calcite or dolomite (%)	0–14	Up to 20%	–	30–70	significant
Clay (%)	20–60	<30	High	8–40	42
Free gas (%)	66	64–80	–	50–65	–
Adsorbed gas (%)	34	20–36	–	35–50	–
CO_2 (%)	12	1	–	<1	5
GIP/section (million m^3)[b]	1700–9000+	230–4500	623–1800	710–5950	2000–1700+
GIP/section (Bcf)[b]	60–318+	8–160	22–62	25–210	72.4–600+
Play area GIP (billion m^3)[b]	4,100–17,000	2300–20000	>2800	>3400	>3700
Play area GIP (Tcf)[b]	144–600+	80–700	>100	>120	>130
Horizontal well cost, including fracking (million $Cdn)	7–10	5–8	0.35 (vertical only)	5–9	Unknown

[a]R (vitrinte reflectance) is a measure of the thermal maturity of organic matter, values above 1.2 are in the gas window.
[b]GIP, gas in place; recoverable gas may be on the order of 20%.
From National Energy Board, 2009.

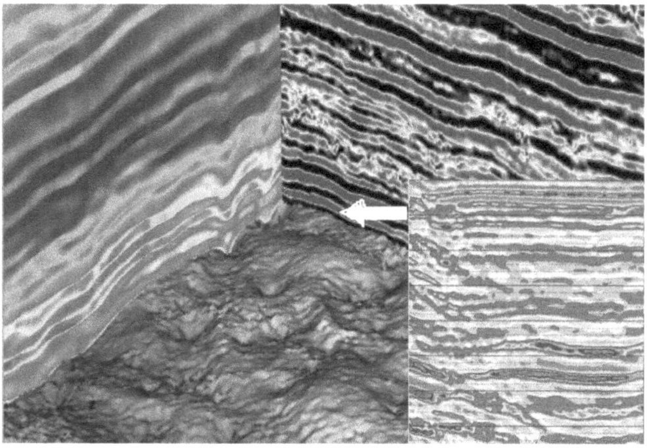

Figure 3.20 Barnett seismic, inversion and surface showing karst collapse chimneys. *(Modified from Devon Energy, website 4).*

Marcellus Shale—Appalachia: This is the largest of the United States shale gas plays and covers nearly 100,000 sq mi in Pennsylvania, Virginia, and New York state. Comparable to the Barnett in depth (7000 ft), thickness (350 ft) and porosity (8%)—permeabilities reaching 1 μD are favorable conditions for as high as 3.75 Bcf/day gas production. Such high productivity is attributed to faults, abundance of fractures, fissures, and variable shale and carbon lithology. Figure 3.21 shows a typical cross section.

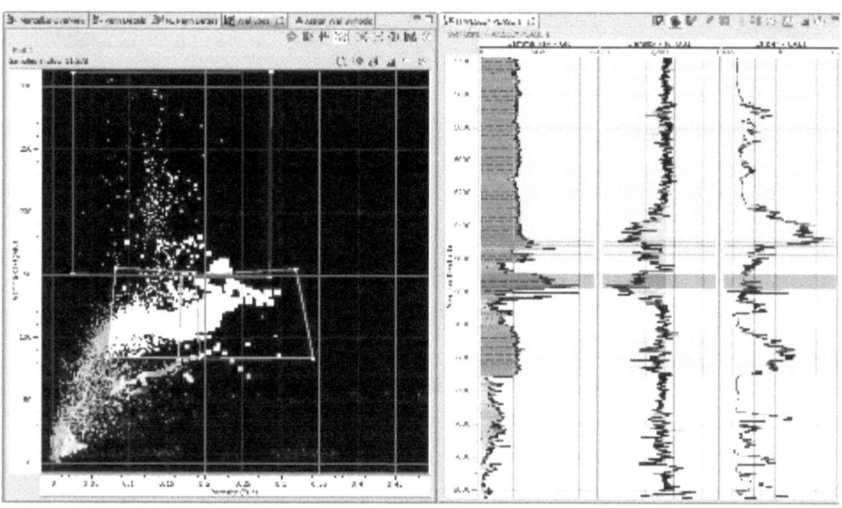

Figure 3.21 Marcellus lithology variations.

Horn River Basin—North-Eastern British Columbia: The unique feature of the Horn River is that it contains multiple prospective shales including the Carboniferous–Devonian Muskwa, Otter Park, Klua, and Evie formations. Prospectivity of these formations vary across the Horn River Basin, leading to alternating or combined completions across the play. The Horn River is a reasonable facsimile of the Barnett, with similar porosities and permeabilities and well depths of about 8800 ft and an average pay zone thickness of 450 ft.

Eagle Ford Shale—South Texas: One of the newest shale gas plays is the Cretaceous Eagle Ford, located in South Texas. It has porosities of 11% and permeability in excess of 1 μD, making it a good candidate for gas recovery. This play is not well studied but initial seismic and microseismic data are encouraging.

In general, Gas shales are source rocks that have not released all of their generated hydrocarbons. In fact, source rocks that are "tight" or "inefficient" at expelling hydrocarbons may be the best prospects for shale gas potential. In this, more than any other factor is the surrounding environment that makes the migration of gas out of the source rock as shale is a reservoir, source rock as well as a trap for natural gas. Most gas is in the adsorbed form, in which gas forms mono- or multilayer. In order to unlock the adsorbed gas, there needs to be a change in either pressure or temperature or both. To date, the main technique for extraction of natural gas from shale gas formations has been through depressurization, which can be achieved by increasing the surface area (horizontal well drilling) and/or by increasing permeability near the wellbore through fracturing.

Unlike tight gas reservoirs, desorption in shale gas would involve separation of gas phase from the bitumen. Recall that shale gas is generated by the following mechanisms:

- primary thermogenic degradation of organic matter;
- secondary thermogenic cracking of oil;
- biogenic degradation of organic matter.

Mono- or multilayer is formed between adsorbed gas and rock in pores, whereas gas remains in diffused/absorbed forms within bitumen and oil. Free gas in shale gas can remain only in fractures and fissures. A higher amount of free gas results in high production rate in the beginning of the production history. As this gas is exhausted, the production rate declines. Any increase in production rate has to be invoked by pressure or thermal change in the reservoir.

In general, gas shales have much lower permeability than sandstone, limestone, or dolomite. Its effective permeability is several orders of magnitude less than 0.1 mD, the cutoff point for "unconventional" reservoirs.

With the exception of a few profusely fractured shale gas reservoirs (e.g., Antrim shale in the Michigan basin of the United States), they are in need of artificial stimulation, such as fracturing.

3.3.4 Gas Hydrate Methane

Gas hydrates are unlike any other form of unconventional gas reservoirs. The process of hydrate formation is unique as it involves the trapping of methane molecules directly within water body. Even though biogenic in its root, methane gas is purified during the process of hydrate formation.

Gas hydrate and its potential for generating energy-giving methane is a topic of research that is only a few decades old. Gas hydrates are not like ice, do not melt; they feel more like Styrofoam. Gas hydrates can burn pure methane safely, even inside of house or a meeting room.

Research on methane hydrate has increased in the last few years, particularly in countries such as Japan that have few native energy resources. As scientists around the world learn more about this material, new concerns surface. For example, ocean-based oil-drilling operations sometimes encounter methane hydrate deposits. As a drill spins through the hydrate, the process can cause it to dissociate. The freed gas may explode, causing the drilling crew to lose control of the well. This concern is similar to CBM that posed hazards to miners long before assuming its potential as an energy source.

Another concern is that unstable hydrate layers could give way beneath oil platforms or, on a larger scale, even cause tsunamis. While research in hydrate has many facets, most research has focused on understanding how hydrates form, and little on how hydrate can be used to unlock huge amount of clean methane. Recent work of the Livermore-USGS team shed new light on hydrate formation. They attempted an entirely new procedure. They mixed sieved granular water ice and cold, pressurized methane gas in a constant–volume reaction vessel and slowly heated it. Warming started at a temperature of 250 K with a pressure of about 25 MPa. The reaction between methane and ice started near the normal melting point of ice at this pressure (271 K) and continued until virtually all of the water ice had reacted with methane, forming methane hydrate. The team studied the resulting material by X-ray diffraction and found pure methane hydrate with no more than trace amounts of water. As a result, they developed low-porosity, cohesive samples with a uniformly fine grain size and random crystallographic grain orientation. This laboratory synthesis is useful for shedding light on actual process of natural hydrate formation. It is also remarkable that the study showed that the ice did not liquefy as it should have when its melting temperature was reached and surpassed. In fact, methane hydrate was formed over a period of 7 or 8 h, with the temperatures inside the reaction vessel reaching 290 K before the last of the ice was consumed.

In other experiments conducted with neon, no hydrate was formed. Under otherwise identical experimental conditions, the ice melted as it should, showing reversibility of the process. Other experiments replaced the methane with both gaseous and liquid carbon dioxide, which formed a hydrate. Here the superheating phenomenon reappeared, indicating that it is not unique to methane hydrate. It appears that formation of hydrate and the thermal hysteresis observed are related to water. Water forms such hydrates with methane and carbon dioxide but does not form with neon or other inert gases that are not directly involved with organic matter.

Another surprising finding was when the team studied the formation of dissociation of gas hydrates. After the samples were created, the pressure was reduced to 0.1 MPa, the pressure at sea level. They did this in two ways: by slow cooling and depressurization and by rapid depressurization at a range of temperatures. The compound decomposed to ice and gas as expected in all experiments except those that involved rapid depressurization at temperatures from 240 to 270 K. In these experiments, the team found yet another surprise. Even after the pressure drop, the methane hydrate was "preserved" as a compound for as long as 25 h before it decomposed. This behavior may have implications for future exploitation of the material. Preserving the mixed hydrates may be possible at an easily accessible temperature, just a few degrees below ice's melting temperature.

In another series of experiments, the team is looking at the strength of gas hydrate samples in various temperature and pressure scenarios. Results of these experiments may indicate the possible effects that stresses from gravity, tectonic activity, or human disturbance might have on gas hydrate deposits. Thus far, the team has found that water ice and methane hydrate have about the same strength at very low temperatures of 180 K and below. But the hydrate is much stronger than ice at temperatures of 240 K and above. The most recent data indicate that methane hydrate is several times stronger than ice, although not as strong as rocks.

Figure 3.22 shows how methane seeps through a marine environment and forms hydrates. In both marine and arctic environment, hydrates can form. Before forming solid hydrates, the system is purged of salinity as well as contaminants. The hydrate may mix with solid depositions, but does not

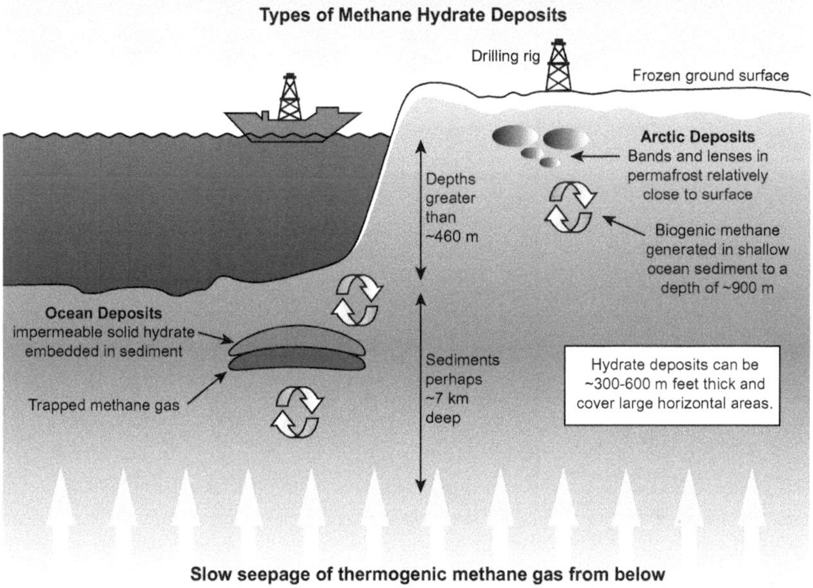

Figure 3.22 Hydrate formation in natural environment.

lose its potential as an energy retainer. This is evidenced in the phase diagram in Figure 3.23.

Figure 3.23 shows the phase diagram of free methane gas and methane hydrate for a pure watered and pure methane system. The addition of NaCl to water shifts the curve to the left. Adding CO_2, H_2S, C_2H_6, C_3H_8 to methane shifts the boundary to the right, thereby increasing the area of the hydrate stability field. It means, natural purification of methane before hydrate is formed.

Methane hydrate is a cage-like lattice of ice inside which are trapped molecules of methane, the chief constituent of natural gas. If methane hydrate is either warmed or depressurized, it will produce natural gas that is pure and ready to burn. When brought to the earth's surface, one cubic meter of gas hydrate releases 164 m^3 of natural gas. Hydrate deposits may be several hundred meters thick and generally occur in two types of settings: under Arctic permafrost, and beneath the ocean floor. Methane that forms hydrate can be both biogenic, created by biological activity in sediments, and thermogenic, created by geological processes deeper within the earth. Hydrates contain the most naturally purified methane available on earth. The processing involves water, thereby making the final product most suitable for human energy consumption.

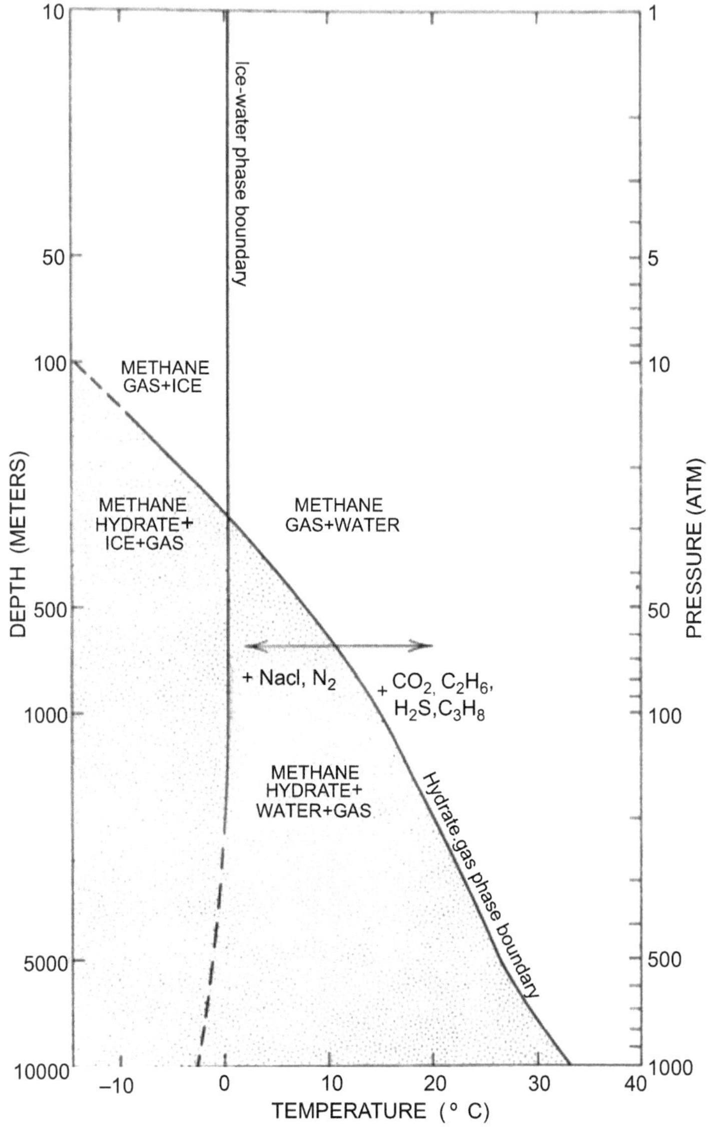

Figure 3.23 Phase diagram of brine and methane. *From Kvenvolden, 1993.*

During gas hydrate formation, water molecule crystallize into cubic lattice structures (Figure 3.24).

Figure 3.24 shows the structure of gas hydrates. The rigid cages in the figure are composed of hydrogen-bonded water molecules and each cage contains a methane molecule. During the formation process, hydrate

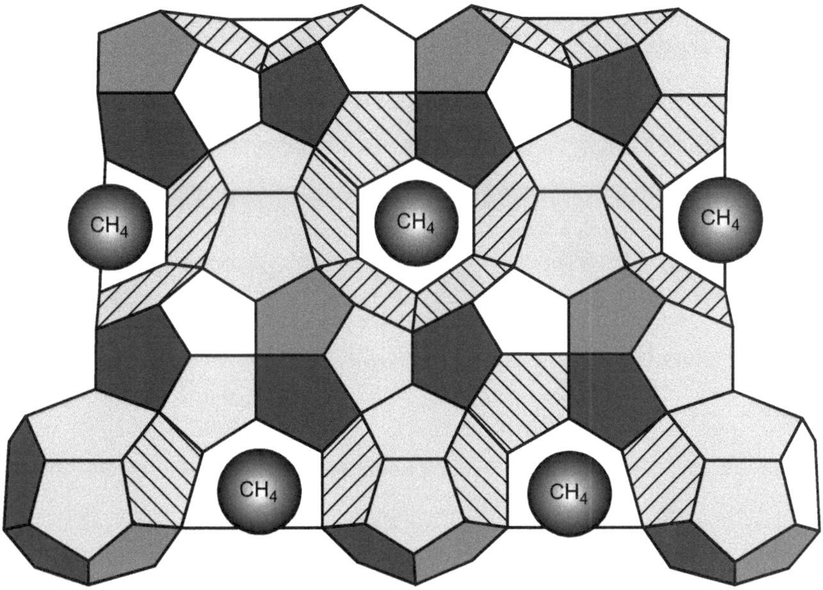

Figure 3.24 Molecular structure of hydrate.

crystals expel salt ions from the crystal structure. Measurements of chlorine in gas hydrate samples indicate far less concentration of salt than sea water (Kvenvolden and Kastner, 1990). The salt concentration of around 2% represents the salinity level of normal drinking water in contrast to some 20% concentration in sea water.

The U.S. Department of Energy methane hydrate program aims to develop the tools and technologies to allow environmentally safe methane production from arctic and domestic offshore hydrates. This research is in its nascent state and is discussed below in order to familiarize the reader of the unknown nature of hydrate. The program includes research and development in the following areas:

Production feasibility: Methane hydrates occur in large quantities beneath the permafrost and offshore, on and below the seafloor. DOE R&D is focused on determining the potential and environmental implications of production of natural gas from hydrates. Figure 3.25 shows the phase diagram for the two types of hydrates.

DOE's methane hydrate program is focused on developing the technology and knowledge that will allow industry to commercialize methane hydrate production. DOE's field programs are focused on the Alaska North Slope and offshore Gulf of Mexico as both contain prolific known petroleum systems

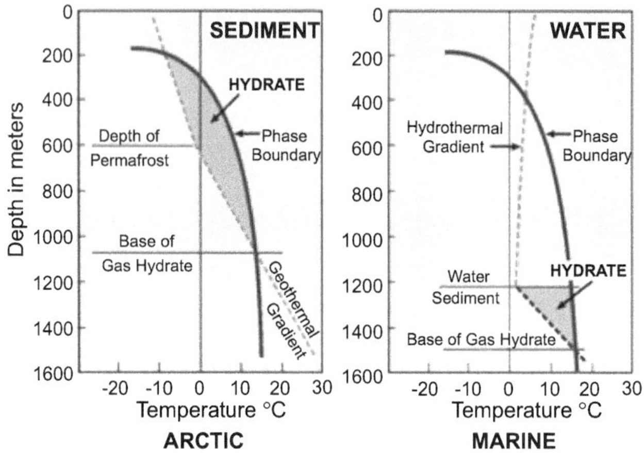

Figure 3.25 Phase diagram of hydrate under arctic and marine environments.

conducive to the formation of gas hydrates, as well as a wealth of information and infrastructure developed during decades of industry hydrocarbon exploration. The well-delineated gas hydrate occurrences on the Alaska North Slope provide relatively low-cost opportunities to conduct extended duration field production trials. The Gulf of Mexico provides a laboratory for testing gas hydrate exploration technology and providing an initial confirmation of the scale of marine gas hydrate occurrence in the United States.

Research and modeling: DOE is studying innovative ways to predict the location and concentration of subsurface methane hydrate before drilling. DOE is also conducting studies to understand the physical properties of gas hydrate-bearing strata and to model this understanding at reservoir scale to predict future behavior and production. Research is focused on understanding the physical and chemical nature of gas hydrate-bearing sediments.

Gas hydrate characterization in the Gulf of Mexico, Scripps Institute of Oceanography: Investigates the feasibility of using marine electromagnetic surveying as a tool for characterizing and quantifying the occurrence of hydrate in the seafloor section.

Electrical resistivity investigation of gas hydrate distribution in Mississippi Canyon Block, Baylor University: Will evaluate the direct-current electrical resistivity method for remotely detecting and characterizing the concentration of gas hydrates in the deep marine environment.

Gas hydrate research in deep sea sediments, Naval Research Lab: Develops and tests a bottom-mounted seismic source for mapping gas hydrates in marine environments.

Detection and production of methane hydrate, Rice University: Investigates the local and regional variations in methane hydrate deposits, where differences in in situ concentrations are relevant to the importance of gas hydrate as a resource, a geohazard, and a factor in the carbon cycle.

Heat flow and gas hydrates on the Continental Margin of India, University of Oregon: Investigates the relationship of residual heat flow anomalies to fluid flow and gas hydrate distribution in the subsurface.

Mechanisms leading to coexistence of gas and hydrate in ocean sediments, University of Texas at Austin: Understands the manner in which methane is transported within the hydrate stability zone and consequently, the growth behavior of methane hydrates at both the grain scale and bed scale.

Methane recovery from hydrate-bearing sediments, Georgia Institute of Technology: Develops observational and experimental data relative to the basic mechanisms at work in a methane hydrate reservoir that is under production.

Natural Gas hydrates in permafrost and marine settings—resources, properties, and environmental issues, U.S. Geological Survey: assist in characterizing and modeling arctic and marine methane hydrates, and participate in design and execution of field studies.

Modeling activities are centered on a multinational effort to optimize computer codes for simulating methane hydrate reservoir behavior. The code comparison study compares the results of simulations of field and laboratory data sets by several commercial and public methane hydrate reservoir simulators. This allows for improvement in the models and also generates a repository of simulations of hydrate formation and dissociation behavior. Research at DOE national labs provides laboratory data on methane hydrate formation and dissociation that is used to enhance the models.

Laboratory studies in support of characterization of recoverable resources from methane hydrate deposits, Lawrence Berkeley National Lab: Studies in support of reservoir simulation include hydrologic measurements, combined geomechanical/geophysical measurements, and synthetic hydrate formation studies.

Formation and dissociation of methane hydrates, National Energy Technology Lab: Validates results from reservoir simulations relative to hydrate re-formation during gas production, gas migration within hydrate-bearing sediments, and carbon dioxide–methane exchange within sediments.

Climate change: DOE is studying the role of methane hydrate formation and dissociation in the global carbon cycle. Another aspect of this research is incorporating gas hydrate science into climate models to understand the relationship between global warming and methane hydrates.

Methane hydrates store huge volumes of methane formed by the bacterial decay of organic matter or leaked from underlying oil and natural gas deposits. The active formation of methane hydrates in the shallow crust prevents methane, a greenhouse gas, from entering the atmosphere. On the other hand, warming of arctic sediments or ocean waters has the potential to cause methane hydrate to dissociate, releasing methane into the deepwater sediments, the ocean or atmosphere. DOE is conducting research to understand the mechanisms and volumes involved in these little-studied processes.

DOE environmental and climate change research projects related to Arctic methane hydrate deposits include:

Characterization of methane degradation and methane-degrading microbes in Alaska coastal waters, University of Delaware: Studies methane degradation and methane-degrading microbes in order to improve predictive models of methane fluxes in the Arctic.

Source characterization and temporal variation of methane seepage from thermokarst lakes on the Alaska North Slope in Response to Arctic climate change, University of Alaska Fairbanks and U.S. Geological Survey, Woods Hole, Massachusetts: Characterize the source, magnitude, and temporal variability of methane seepage from thermokarst lakes in the Alaskan North Slope, and to assess the vulnerability of these areas to ongoing and future arctic climate change.

Interrelation of global climate and the response of oceanic hydrate accumulations, Lawrence Berkeley National Lab and Los Alamos National Lab and U.S. Geological Survey's Study of Beaufort Sea Hydrate DOE environmental and climate change research projects related to marine or global methane hydrate deposits include:

Gulf of Mexico gas hydrates seafloor observatory project, University of Mississippi Center for Marine Resources and Environmental Technology: The Gulf of Mexico-Hydrate Research Consortium (GOM-HRC) is developing a seafloor station to monitor hydrates in situ.

Integrating natural gas hydrates in the global carbon cycle, University of Chicago: Develops a two-dimensional, basin-scale model for the deep sediment biosphere with methane dynamics to understand the distribution of hydrates on the seafloor and their vulnerability to warming of the deep ocean, thus integrating methane hydrates into the global carbon cycle.

Remote sensing and sea-truth measurements of methane flux to the atmosphere, Texas A&M University, Corpus Christi: Improves estimates of methane flux from submarine seeps and associated gas hydrate deposits on continental margins by compiling a remote sensing inventory of active gas and oil vents.

Assessing the efficacy of the aerobic methanotropic biofilter in methane hydrate environments, University of California, Santa Barbara: Assesses the efficacy of aerobic methanotrophy in preventing the escape of methane from marine, hydrate-bearing reservoirs to the atmosphere and ultimately to better define the role of aerobic methanotrophy in the global carbon cycle.

Interrelation of global climate and the response of oceanic hydrate accumulations, Lawrence Berkeley National Lab and Los Alamos National Lab: Investigates the effect of rising water temperatures on the stability of oceanic hydrate accumulations, and estimate the global quantity of hydrate-originating carbon that could reach the upper atmosphere and its impact on global climate.

Natural gas hydrates in permafrost and marine settings—resources, properties, and environmental issues, U.S. Geological Survey: Studies the links between Late Pleistocene to contemporary climate change and the state of the gas hydrate reservoir in high-latitude regions, specifically the US Arctic.

International collaboration: International collaboration continues to be a vital part of the program since gas hydrates represent research challenges and resource potential that are important on a global scale.

In 1982 the multinational Deep Sea Drilling Program recovered the first subsea substantial methane hydrate deposits, which spurred methane hydrate research in the United States and other countries. The successor programs, the Ocean Drilling Program (ODP) and the Integrated Ocean Drilling Program (IODP) sampled hydrate deposits off Oregon (ODP 204, 2002) and in the Cascadia Margin off Vancouver Island, Canada (ODP 146, 1992 and IODP 311, 2005). In the Atlantic Ocean off the United States, ODP Leg 146 sampled hydrate deposits on the Blake Ridge and Carolina Rise in 1995.

International cooperation helps scientists in the United States and other countries expand their understanding of the variety of naturally occurring methane hydrate deposits and develop expertise in sampling and measurement techniques. Although methane hydrate deposits are inferred to exist on all continental shelves, only a few sites have been drilled and logged or cored, which makes this information valuable to all scientists. DOE is participating as a steering committee member in a global assessment of methane hydrates in the natural environment, led by the United Nations

Environment Programme. Information on the assessment is available at http://www.methanegashydrates.org.

Because of the broad level of international cooperation, information is available on research programs and sampling expeditions for Canada, India, China, Japan, and Korea. Additional information about specific international methane hydrate research expeditions is available in numerous articles in the newsletter, Fire in the Ice.

Overall, gas hydrate methane is an excellent source of natural gas. Most commonly found hydrates (e.g. Gulf of Mexico) is mostly microbial with others being thermogenic. In these cases, 99% of gas is methane. Gas hydrates associated with mud volcanoes in the Caspian Sea contain hydrocarbon gases with 59–96% methane having carbon isotopic compositions of 44.857.3%. These are essentially thermogenic methane.

3.4 GLOBAL WARMING AND NATURAL GAS

Previous sections confirm that natural gas from unconventional sources are the most naturally processed energy source. Often, natural gas is considered to be "lesser evil." This results from years of campaign that is devoid of scientific premises. Aggressive development of petroleum resources has been curtailed by the recent "awareness" of environmental consequences of petroleum production and utilization. "Petroleum production and utilization is the cause for environmental disaster" is the predominant theme today (Picture 3.2).

Picture 3.2 "Petroleum is dirty, anything else is clean" is the most dominant environmental theme today.

Yet, petroleum fluid is the 2nd most abundant fluid (second only to water) on earth and presents the most potent purely naturally processed energy source. Also, petroleum contains the most amount of carbon, the essential ingredient of life on earth.

Despite the bad name, oil and gas continue to be the most dominant energy provider of the world. For many decades, oil production from all regions have been increasing steadily. World oil production increased by 1.9 million bbl/day in 2012, more than double the growth of global consumption. US output grew by 1 million bbl/day, the largest increase in the world and in the country's history. Interestingly, global oil consumption grew at a less steep pace and Europe and North America marked decline in oil consumption over the past years. Figure 3.26 shows the history of production from various regions.

Figure 3.27 shows future projections of accelerated production from OPEC countries. OPEC production in the reference case grows from 35 million bbl/day in 2011 to 48 million bbl/day in 2040. Although the

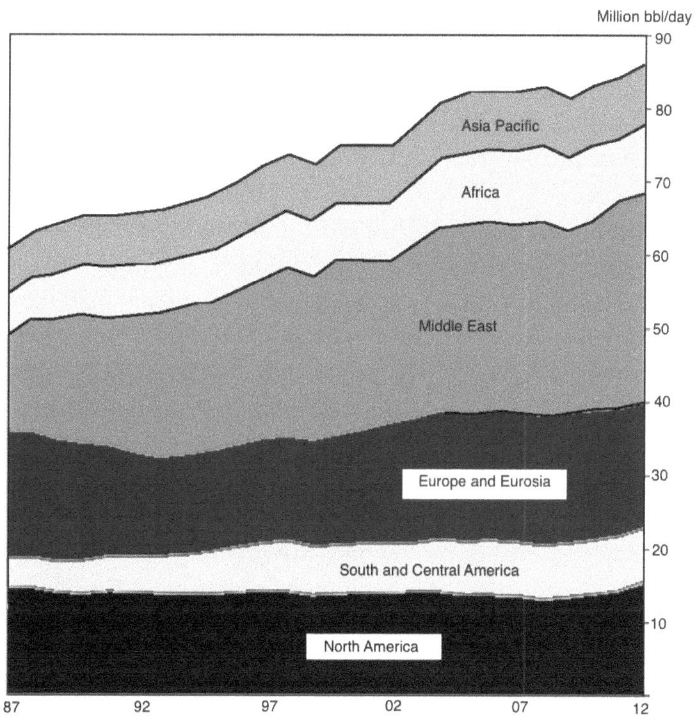

Figure 3.26 Oil productions (million bbl/day) for various regions. *Redrawn from Courtesy BP Outlook, 2013.*

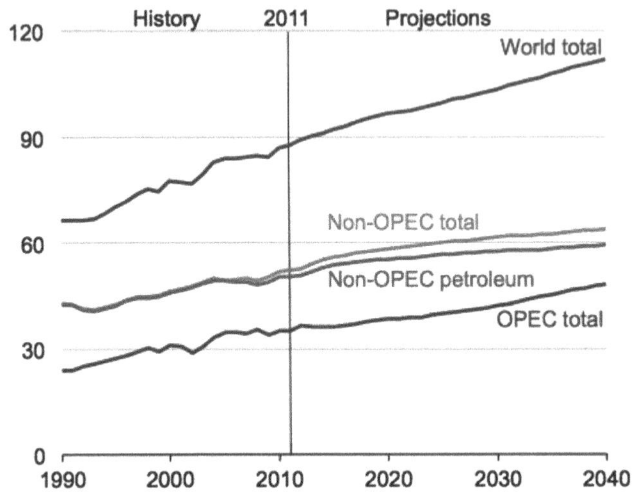

Figure 3.27 Future prediction of global oil production. *EIA, 2013 annual report.*

OPEC resource base is sufficient to support much higher production levels, the OPEC countries have an incentive to restrict production in order to support higher prices and sustain revenues in the long term. The role of OPEC will become more intense after 2020. The reference case assumes that OPEC will maintain a cohesive policy of limiting supply growth, rather than maximizing total annual revenues. The reference case also assumes that no geopolitical events will cause prolonged supply shocks in the OPEC countries that could further limit production growth. The uncertainty in the oil pricing due to political instability cannot be ruled out and is likely to be a major factor that cannot be predicted. Figure 3.28 shows how previous oil market has been affected by political events.

The growth in non-OPEC production results primarily from the development of new fields and the application of new technologies, such as enhanced oil recovery, horizontal drilling, and hydraulic fracturing, which increase recovery rates from existing fields. The average cost per barrel of non-OPEC oil production rises as production volumes increase, and the rising costs dampen further production growth.

Non-OPEC production of other liquids grows from 1.8 million bbl/day in 2011 to 4.6 million bbl/day in 2040, as Brent crude oil prices remain sufficiently high to make other liquids production technologies economically feasible. Non-OPEC liquids production in the reference case totals 58 million bbl/day in 2020, 61 million bbl/day in 2030, and 64 million bbl/day in 2040.

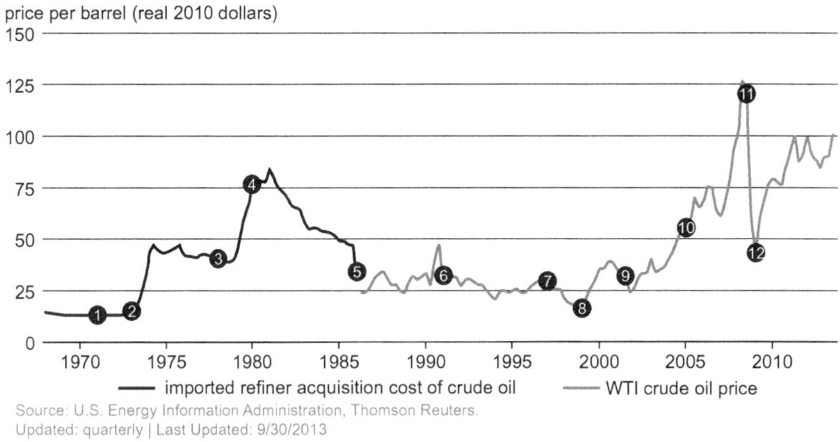

1: US spare capacity exhausted
2: Arab Oil Embargo
3: Iranian Revolution
4: Iran-Iraq War
5: Saudis abandon swing producer role
6: Iraq invades Kuwait
7: Asian financial crisis
8: OPEC cuts production targets 1.7 mmbpd
9: 9-11 attacks
10: Low spare capacity
11: Global financial collapse
12: OPEC cuts production targets 4.2 mmbpd

Figure 3.28 Oil price and world political events.

The apparent "vilification" of petroleum activities did not change the prediction that petroleum resources will continue to dominate the world energy scenario for many decades to come. However, at the same time, few have countered the environmental arguments with proper science. Instead, there has been contradictory evidence and argument shifting blames on everything but the process that "refines" petroleum. The debate has taken an unprecedented acrimony between supporters and detractors of petroleum industry. The topic is global warming and the two camps have been entrenched in their respective positions with utmost rigidity. These positions being: (1) There is no global warming, (2) Carbon is the enemy. Figure 3.29 shows the weather map of June, 2013 that reported record breaking temperature in Alaska.

This event prompted detractors to quickly condemn all forms of petroleum production and attacking petroleum supporters as "flat earth theoretician."

Within a few months, another report came up stating that Antarctic has an unprecedented build up of ice, prompting supporters to dismiss global warming as a "myth." Figure 3.30 shows the buildup of ice in the Antarctic as reported in the Fall of 2013. Figure 3.31 shows the temporal extent of the ice buildup.

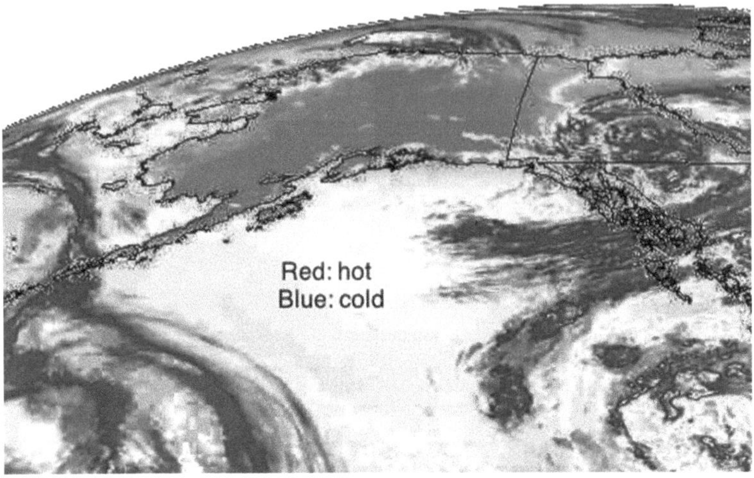

Figure 3.29 June 18, 2013 saw record breaking heat in Alaska.

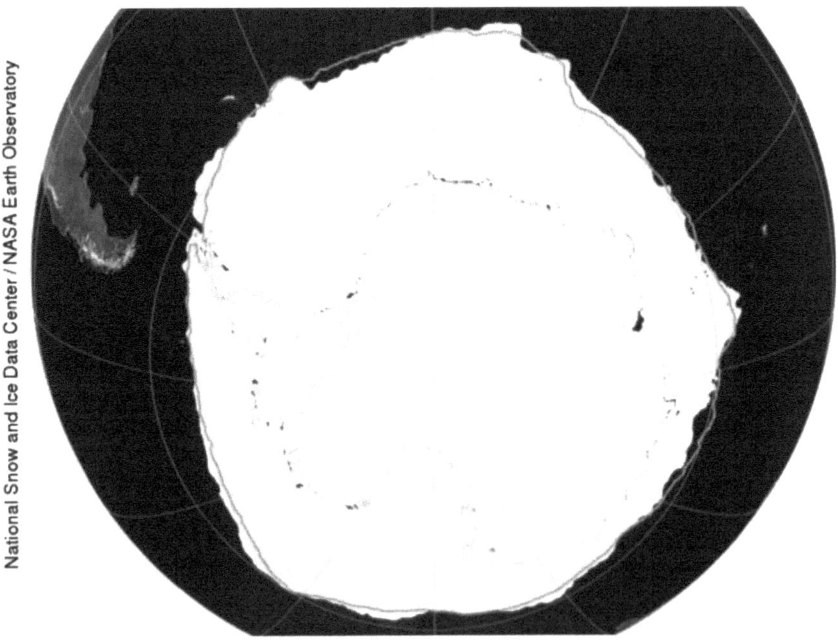

Antarctic sea ice extent on September 22 compared to 1981-2010 median depicted by orange curve (NSIDC)

Figure 3.30 September 22, 2013 picture shows unprecedented ice buildup in the Antarctic.

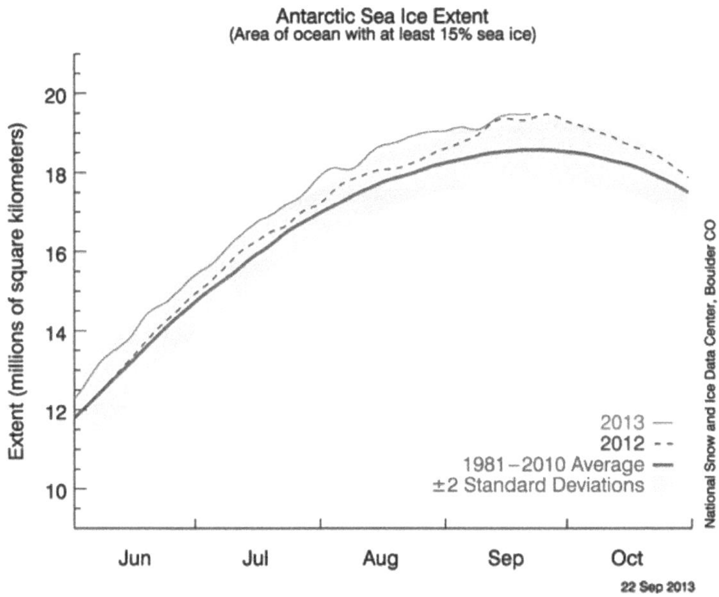

Figure 3.31 Recent report shows unprecedented buildup of ice in the Antarctic.

This fueled the notion that Sorokhtin et al. (2007) were correct in predicting we are actually in the midst of a global cooling. Fluctuations in CO_2 leads to fluctuations in global temperatures (Figure 3.32). This figure is used to claim that we are in fact in a cooling cycle if we consider natural phenomena in the geological time scale. These supporters argue that ice and soil cores the earth has gone through many up and downs. We have surpassed 400 ppm of CO_2 now which we have many times before. They claim that oxygen levels were extremely high during the ages of dinosaurs. Some scientists point to the Sun for the increase in temperatures, which they say is happening to all the

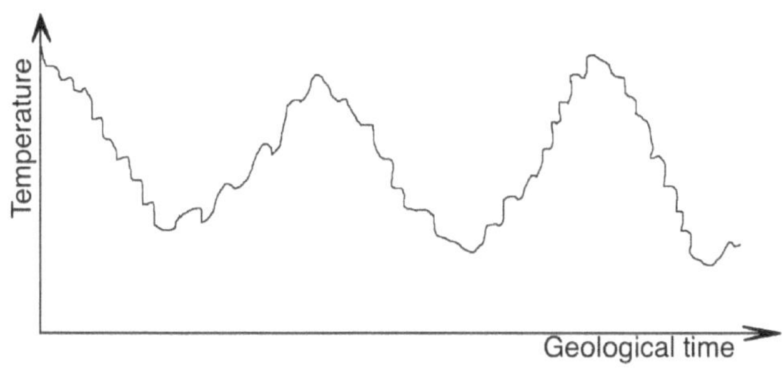

Figure 3.32 In geological scale, the current epoch is actually in a cooking cycle.

inner orbit planets. The Sun is hitting maxima now but the hope is that it will stabilize in time before the earth becomes in habitable. All these come from the premise that the Sun, Earth, etc., are all part of a random process. What is rarely talked about is that such "natural phenomena" have been severely altered because of the emergence of artificial materials. When Freon was just introduced, it was said, there is no way it can make a dent in the atmosphere, the emission being so insignificant. After all, they argued, nature is too vast to be impacted by human activities. At present it is said, we only burn 100 million bbl of oil, a peanut compared to the total CO_2 emission by the entire biomass of this planet. New science does not account for the theory that these artificially generated CO_2 are to the earth what cancer causing chemicals are to the human body.

The opponents of petroleum activities claim that because CO_2 is the source of global warming, any process that uses carbon is harmful and must be eliminated. This fateful conclusion that "Carbon is the enemy" is deconstructed by asking the pointed question: "Are you saying CO_2 that is the only source of carbohydrate is a problem while nuclear is not?" In order to remove the paradox, one must invoke a paradigm shift in the first premise. It has been done by Islam et al. (2010) in the following sequence.

1. Nature is perfect. This is the major premise. It means only natural is sustainable and readily accepted by nature.
2. In order to posit a minor premise, it is asked: What is natural? The answer is put as a minor premise.
3. Then, an axis is formed with the major premise as the source and minor premise as the other point of the axis. That created a unique set of scaling laws.
4. With this set of scaling laws (or standards), it can be concluded that
   a. CO_2 is not the problem;
   b. It is the artificial (not naturally occurring) additives that are the problem. These polluted CO_2 are the source of global warming.

If natural, excess CO_2 would increase vegetation, thereby bringing down global temperature and increasing O_2 supply to the ecosystem.

This is captured in the Figure 3.33 that shows in exaggerated scale what happened ever since artificial chemicals have been in use for refining crude oil.

The millions of barrels are burnt everyday that emit large amount of artificial CO_2 that cannot be absorbed by the ecosystem. This is the type of CO_2 that contributes to the rapid increase in CO_2 in the atmosphere, despite that fact that natural carbon dioxide in the atmosphere is decreasing,

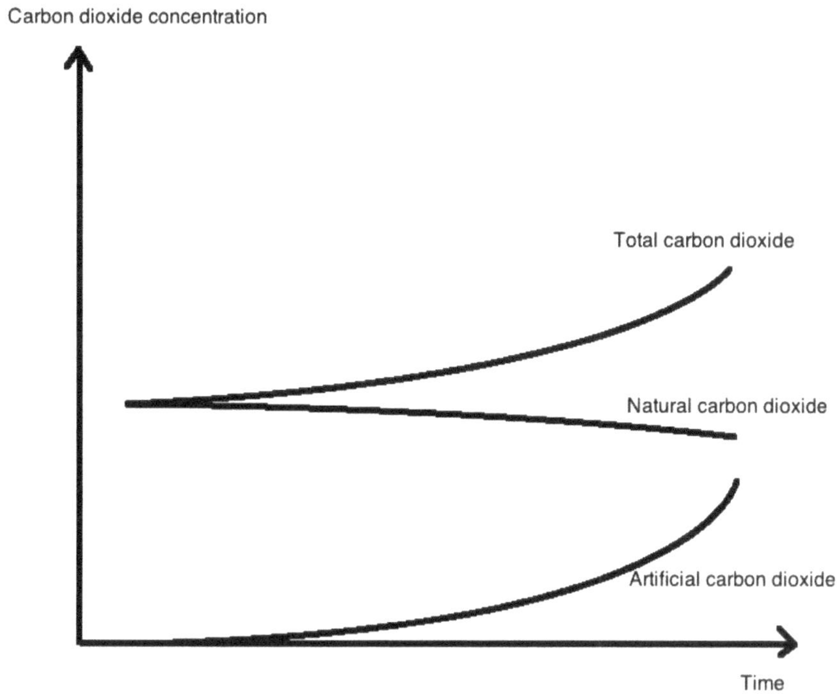

Figure 3.33 Schematic of temporal change in various types of carbon dioxide.

albeit very slightly. Note that the overall rise of the CO_2 content or the global temperature is conflated by most supporters as "natural phenomena" that one should be able to observe as an extension of the variation in geological time. The inability to distinguish between natural and artificial (polluted with additives) CO_2 has made it impossible to pin point the cause of global warming. It also turns out that the same additives are all carcinogenic, thereby creating strong correlation between the usage of petroleum products with devastating impact on human health.

As for the controversy regarding the buildup of extra ice in the Antarctic region is explained as an effect of global warming. The following depiction (Figure 3.34) by Argonne National laboratory explains such a buildup in terms of global ocean circulation between deep, colder water and warmer, surface water, strongly influencing regional climates around the world (NASA, 2004). It was in fact predicted that the entire North America and Western Europe would plunge into ice age. This gave a significant fighting weapon to the detractors of petroleum activities, while the other group avoided giving reference to it until the report of massive build up of ice in

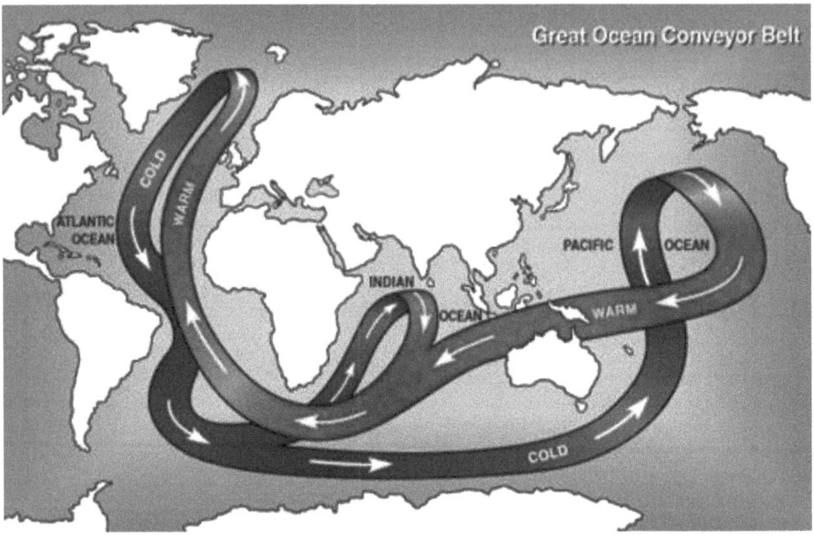

Figure 3.34 As early as 2004, NASA raised the alarm that global warming can lead to more ice buildup in the Antarctic.

the Antarctic was reported (NASA, 2013a). For them, it is a clear evidence that global warming is a hoax.

In 2012, NASA raised alarm regarding north pole ice being sixth lowest in recorded history (NASA, 2012). In 2013, NASA recorded highest accumulation of ice in the Antarctic (NASA, 2013).

However, another report of NASA (2013b) showed loss of ice in the North pole. This report was not used by the supporters of petroleum activities.

While the diminishing north pole ice can be linked to "warming," the simultaneous rise in water temperature and build up of record level of ice in Antarctic as well as record breaking high temperature in Alaska created the so-called "Arctic sea ice paradox." (Figure 3.35) Similar to numerous other paradoxes that have crippled New Science, this paradox cannot be removed unless contradictions at the fundamental levels are removed. Once again, one is faced with distinguishing between types of ice that emerge from different sources. For instance, continental ice is different from sea ice. Continental ice is formed by the gradual deposition, buildup, and compaction of snow, resulting in ice that is hundreds to thousands of meters thick. This ice stores and releases freshwater that can affect the ecosystem for thousands of years. On the other hand, sea ice is a thin layer (not more than 2 m) of ice that forms on the surface of the ocean when the atmospheric temperature is below the freezing point. Once ice is formed, it acts as an insulator and the

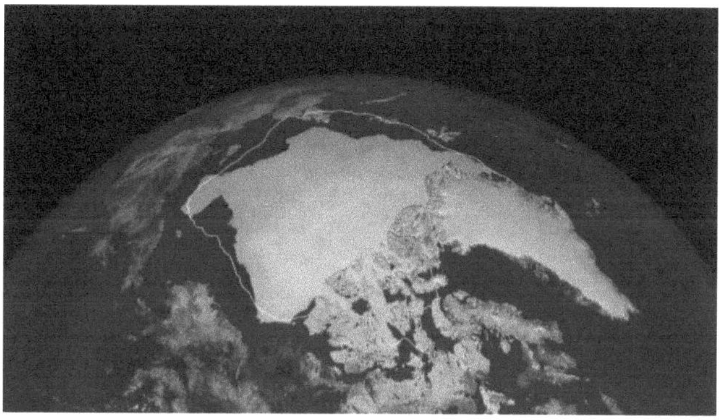

Figure 3.35 The diminishing ice in the North pole has given detractors weapon for an argument to curtail petroleum activities.

sea temperature is conserved. This is necessary for the aquatic life to be protected. The sea ice is mobile and transitory. It can revert to water as the prevailing conditions change. From there sea ice can move with the winds and currents, continuing to grow both by freezing and through collisions (between the floes that make up the ice cover). When the atmosphere, and/or ocean is suitably warm again, such as in spring or if the sea ice has moved sufficiently towards the equator, then the sea ice melts again. In the Antarctic, approximately 80% of the sea ice area grows each autumn to winter and decays each spring to summer. This annual expansion–contraction from about 4 to 19 million square kilometers is one of the greatest seasonal changes on the earth's surface.

In addition, difficulties arise from the false estimation of volume from the area that can be easily monitored. However, monitoring of depth is far from accurate when it comes to sea ice.

In 2013 winter, the Antarctic sea ice coverage was reported to be 3.6% above the winter average calculated from 1981 to 2010, calling for fresh debate on global warming.

This continues a trend that is weakly positive and remains in stark contrast to the decline in Arctic summer sea ice extent (2013 was 18% below the mean from 1981 to 2010). Nobody being able to conduct an accurate heat balance, it becomes anyone's guess as to what conclusion should be drawn from these observations. In this the 2004 NASA model for hot and cold stream circulation only explains part of the observations. Furthermore, any density variation of sea ice, along with spatial variation can throw the heat balance calculations. If the global water temperature

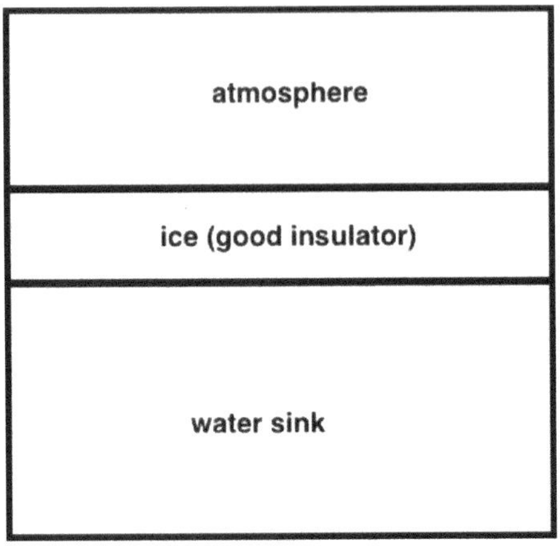

Figure 3.36 Water represents a gigantic heat sink that can store enough heat to stabilize the global temperature in the long term.

does go up, it would correspond to retention of more heat in a typical three-layer model as shown in Figure 3.36. Such change in water temperature will affect both the density and areal extent of sea ice.

Other factors involve ocean/ice interactions, compositional change in atmosphere, wind velocity, all of which will have possible contradictory impacts on the sea ice. Overall, water is an excellent heat sink. Warm water means a great amount of heat loss, which can translate into a huge amount of ice formation. Ice being a great insulator, the atmosphere is not affected by the heating phenomena.

3.5 DISTINGUISHING FEATURES OF UNCONVENTIONAL RESERVOIRS

Even though different forms of unconventional reservoirs are lumped under the title "Unconventional reservoirs," they have distinguishing features that separate them from each other. On the other hand, they all share certain features that are different from those of conventional reservoirs. In general, it is understood that the differences between conventional and nonconventional lie within the following:

- low-permeability structure itself: this is the most important feature of unconventional reservoirs. Even though the permeability value cut off point is set at 0.1 mD arbitrarily, the fact remains that the matrix permeability of unconventional reservoirs is less than 0.1 mD. Because modern day petroleum engineering is based on Darcy's equation that puts permeability as the most important parameter governing flow rate, the demarcation based on permeability is understandable.

- response to overburden stress: the peculiar response to overburden gives unique characteristics of unconventional reservoirs, resulting in low permeability, devoid of fractures or the presence of random fracture network (e.g. coalbed).
- impact of the low-permeability structure on effective permeability relationships under conditions of multiphase saturation: this feature results in the presence of "permeability jail"—a range of saturation for which none of the fluid flows.
- entrapment of gas in both adsorbed and adsorbed conditions in the reservoir.

The volume of unconventional gas is much more than the volume of conventional gas. Unlike commonly held perception, this additional volume also represents higher quality of natural gas. The most abundant gas resource, hydrate represents the highest quality gas. CBM also shows equally high quality. Even though tight gas and shale gas composition is relatively 'impure', they are naturally processed and if sustainable techniques are used to purify offer the best type of natural energy source. Deconstruction of commonly believed accounting theories showed that the relationship between technical feasibility and economics is a fictitious one. With proper science, the cost of developing larger resources is minimized. A discussion on scientific cause of global warming reveals that natural gas from unconventional resources can help reverse global warming if artificial chemicals are avoided during the gas processing phase.

CHAPTER 4

Future Potentials of Unconventional Gas: Challenges and Opportunities

4.1 INTRODUCTION

The United States has led the world in unconventional gas production. Natural gas demand in the United States is expected to increase from 25 Tcf/year currently to 30–34 Tcf/year by the year 2025. Conventional gas sources (sandstone reservoirs) will not be able to satisfy this demand and thus unconventional gas sources (tight gas, shale gas, and coalbed methane (CBM)) are expected to be a major component of this production. In the long term, methane from gas hydrates will add to the unconventional production of natural gas. At present, gas hydrates are only a research theme and there is no existing project on production of natural gas from gas hydrates. In terms of technological development, only horizontal well and fracturing have been used to produce natural gas from unconventional reservoirs. While these technologies will continue to produce more natural gas from unconventional resources, more resources will be added to the "proven" reserve as well as newer technologies (or older technology from enhanced oil and gas recovery schemes) will determine the truly long-term future of unconventional gas. This chapter discusses enhanced gas recovery (EGR) potential of unconventional gas and then presents the best way to increase recovery of unconventional gas. EGR has been a fleeting concept mainly because conventional gas reservoirs have very high recovery factor, thereby making EGR techniques redundant. In addition, the prevalent concept in the petroleum industry is that any enhanced recovery scheme requires large investment, which cannot be justified considering low gas prices. This scenario changes for unconventional reservoirs that has much bigger prize embedded in large formations and have the potential of multifold increase in recovery factors over conventional techniques of horizontal drilling and fracking.

Unconventional Gas Reservoirs
ISBN 978-0-12-800390-9
http://dx.doi.org/10.1016/B978-0-12-800390-9.00004-9

4.2 LESSONS LEARNT FROM ENHANCED OIL RECOVERY

This section is an overview of enhanced oil recovery (EOR) schemes. Even though the focus is oil, the discussion helps the readership identify technologies that are proper for gas recovery from unconventional sources.

Primary recovery is the oil recovery by natural drive mechanisms. Such natural drives may be through solution gas expansion, water drive, gas-cap drive, or simply gravity drainage. Secondary recovery is known to be the oil recovery technique in which gas or water is injected in order to maintain the reservoir pressure. Tertiary recovery is any oil recovery scheme, conventionally applied after secondary recovery. However, for over two decades, there is a tendency to use the term *enhanced oil recovery* in order to define a wide range of recovery processes. The EOR is an oil recovery scheme that uses the injection of fluids, not normally present in the reservoir. For instance, chemical injection, steam injection, in situ combustion (ISC), or even microbial enhanced recovery will be considered as EOR. This definition, while encompasses many recovery schemes beyond the scope of *tertiary recovery* (recovery scheme that follows secondary recovery), does not include techniques such as electromagnetic heating even when this could be an effective technique for increasing oil production from a reservoir. However, recent publications acknowledge electromagnetic heating as an EOR process. A proper definition of EOR should, therefore, be any oil recovery technique that improves oil recovery from a reservoir beyond primary recovery. While fluid injection may be required for some techniques, energy dissipation may be sufficient in some cases. This definition may seem to be too broad because it does not exclude waterflood or pressure maintenance gas injection from the definition of EOR. In fact, many waterflood and gas injection schemes are indeed displacement-type recovery processes and should be called an EOR scheme. A purely pressure maintenance scheme is usually well defined and no confusion as to its distinction from EOR schemes exists.

The word "EOR" fell out of grace shortly after tax incentive for EOR schemes were repealed in 1980s. This saw the sudden drop of EOR projects in the United States that peaked in 1986.

Table 4.1 shows total oil reserve as well as reserve/production ratio of top oil producing countries. Each country is marked for its need for EOR. Note that the need does not imply suitability nor does it mean that other countries would not benefit from an EOR scheme.

Table 4.1 Summary of Proven Reserve Data as of 2012

—	Country	Reserves (10^9 bbl)	Production (10^6 bbl/day)	Reserve/production ratio (years)	Need for EOR
1	Venezuela	296.5	2.1	387	Low
2	Saudi Arabia	265.4	8.9	81	Medium
3	Canada	175	2.7	178	Low
4	Iran	151.2	4.1	101	Low
5	Iraq	143.1	2.4	163	Low
6	Kuwait	101.5	2.3	121	Low
7	United Arab Emirates	136.7	2.4	156	Low
8	Russia	80	10	22	High
9	Kazakhstan	49	1.5	55	High
10	Libya	47	1.7	76	Medium
11	Nigeria	37	2.5	41	High
12	Qatar	25.41	1.1	63	Medium
13	China	20.35	4.1	14	High
14	United States	26.8	7	10	High
15	Angola	13.5	1.9	19	High
16	Algeria	13.42	1.7	22	High
17	Brazil	13.2	2.1	17	High
	Total of top 17 reserves	1324	56.7	64	

More recently, environmental concerns have become part and parcel of EOR activities. In part this is fueled by renewed awareness of the environmental impact of unsustainable engineering practices, even though it is perceived as a greenhouse gas emission problem.

Energy production and use are considered major causes of greenhouse gas emissions. The emission of greenhouse gases, particularly CO_2, is of great concern today. Even though CO_2 is considered one of the major greenhouse gases, production of natural CO_2 is essential for maintaining life on earth. Note that all CO_2 is not the same and plants do not accept all types of CO_2 for photosynthesis. There is a clear difference between old CO_2 from fossil fuels and new CO_2 produced from renewable biofuels (Dietze, 1997; Islam et al., 2010). The CO_2 generated from burning fossil fuel is an old and contaminated one. Because various toxic chemicals and catalysts are used for oil and natural gas refining/processing, the danger of generating CO_2 with higher isotopes cannot be ignored. Hence, it is clear that CO_2 itself is not a culprit for global warming, but the industrial CO_2

that is contaminated with catalysts and chemicals likely becomes heavier with higher isotopes and plants cannot accept this CO_2. Plants always accept the lighter portion of CO_2 from the atmosphere (Bice, 2001). Thus, CO_2 has to be distinguished between natural and industrial CO_2, based on the source from which it is emitted and the pathway that the fuel follows from the source to combustion. Islam et al. (2010) showed that even though the total CO_2 is increasing in the atmosphere, the natural CO_2 has been decreasing since the industrial revolution. They further argued that industrial CO_2 is responsible for global warming. Thus, generalizing CO_2 as a precursor for global warming is an absurd concept and is not valid. See discussion below for more details.

In absence of solidly founded scientific investigations, world governments have at least agreed that the greenhouse emission is a concern and have been under pressure to ratify Kyoto Agreement. Even though the Kyoto protocol as well as Copenhagen agreement (or lack of it) have not been implemented, they managed to extract more funding for greenhouse mitigation projects than EOR projects. For instance, in 2011, the petroleum sector in the United States invested US$ 71.1 billion, whereas US$ 73.7 billion came from the private sector other than the petroleum industry. The rest (23%) that amounts to US$ 43.4 billion came from the government that tagged the amount to greenhouse gas mitigation projects. It turns out the 39% from the "other industry" also invested in these "environmental" projects.

Most recent U.S. Department of Energy (DoE)/Energy Information Association (EIA) report (2013) shows how climate change concerns and specifically greenhouse gas emissions have been instrumental in determining future energy outlook and regulatory policies. Regulators and investment companies have been pushing energy companies to invest in technologies that are less greenhouse gas intensive. Federal government grants are often linked to environmental aspects of petroleum engineering. Even within the petroleum industry, the focus has shifted toward mitigation of greenhouse gases. This provides a unique opportunity for coupling EOR projects that can bring in double dividends, namely environmental integrity and financial boon.

Figure 4.1 shows CO_2 emissions per sector. On average, energy-related CO_2 emissions in the reference case decline by 0.2% per year from 2005 to 2040, as compared with an average increase of 0.9% per year from 1980 to 2005. Reasons for the decline include: an expected slow and extended recovery from the recession of 2007–2009; growing use of renewable technologies and fuels; automobile efficiency improvements; slower growth

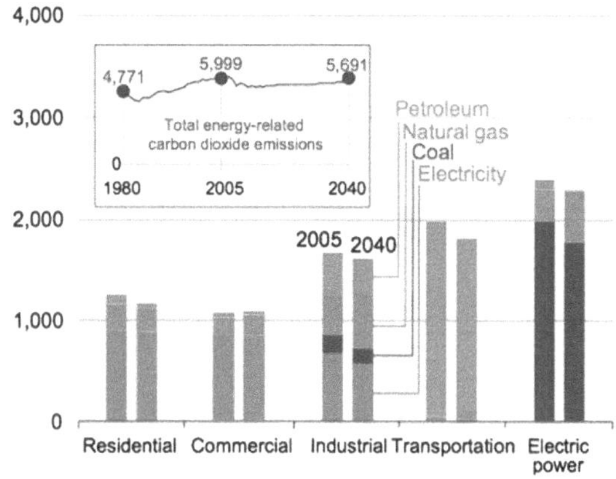

Figure 4.1 Past performance and future projections of greenhouse gases by sector (million metric tons).

in electricity demand; and more use of natural gas, which is less carbon intensive than other fossil fuels. In the reference case, energy-related CO_2 emissions in 2020 are 9.1% below their 2005 level. Energy-related CO_2 emissions total 5691 million metric tons in 2040, or 308 million metric tons (5.1%) below their 2005 level.

As indicated earlier, the reserve/production ratio of oil is declining around the world, some hitting critical need for EOR. The exception is the middle eastern region that continues to produce under par, as evident from Figure 4.2. World proved oil reserves at the end of 2012 reached 1668.9 billion bbl. This is sufficient to meet 60 years of global production, without tapping into additional sources. Note that additional sources include heavier or nonconventional resources and new discoveries. Global proved reserves have increased by 26%, or nearly 350 billion bbl, over the past decade. This trend is likely to continue.

Of significance is the fact that there is much more nonconventional petroleum reserve than the convention "proven" reserve. This point is made in Figure 4.3. Even though it is generally assumed that more abundant resources are "dirtier," hence in need of processing that can render the resource economically unattractive, sustainable recovery techniques can be developed that are more efficient for these resources and also economically attractive and environmentally appealing (Islam et al., 2010). In addition, natural gas quality is little affected by the environment. For instance, gas hydrate that represents the most abundant source of natural gas is actually far cleaner than less abundant resources.

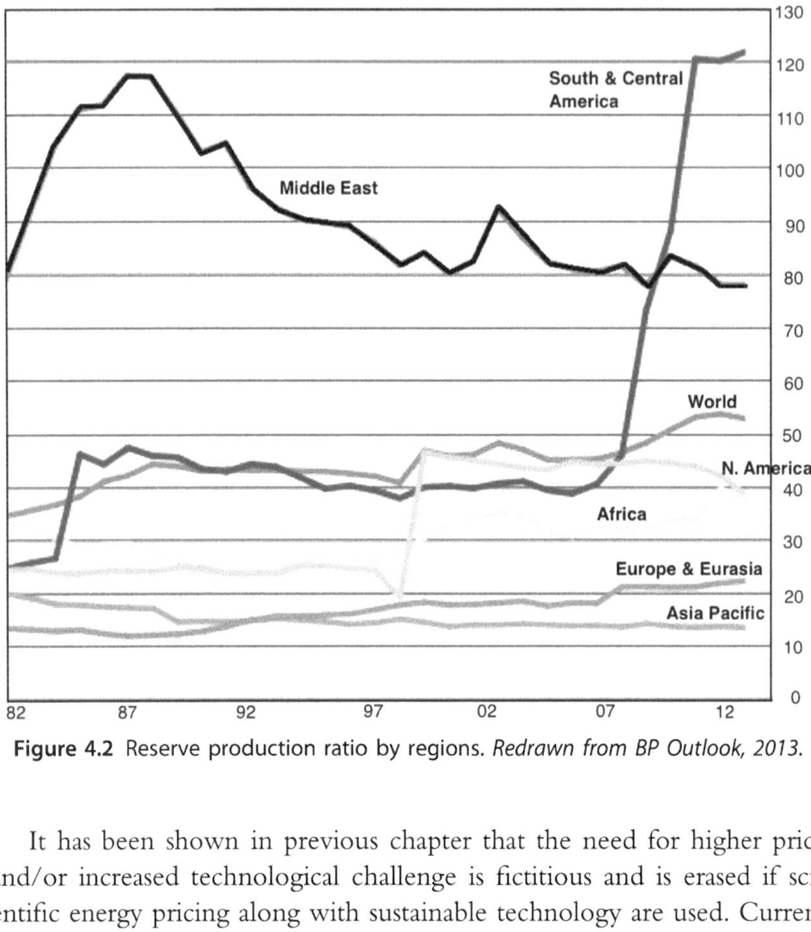

Figure 4.2 Reserve production ratio by regions. *Redrawn from BP Outlook, 2013.*

It has been shown in previous chapter that the need for higher price and/or increased technological challenge is fictitious and is erased if scientific energy pricing along with sustainable technology are used. Current investment strategy has fueled this misconception.

In terms of oil industry, the main focus has been in nonconventional petroleum extraction. For instance, Figure 4.4 shows major investments in oil sands in Canada. In 2013, 7299 Tcf of shale gas 345 billion bbl of shale/tight oil has been added. In the United States, the focus has been on unconventional oil and gas. A government report published in summer of 2013 revealed "U.S. domestic crude oil production exceeded imports last week for the first time in 16 years. Output was 32,000 bbl a day higher than imports in the seven days ended May 31, according to weekly data today from the Energy Information Administration, the Energy Department's statistical arm" (Bloomberg, 2013). Production had been lower than international purchases since January 1997. This surge in oil is attributed to the influx of horizontal drilling and hydraulic fracturing (popularly known as "fracking").

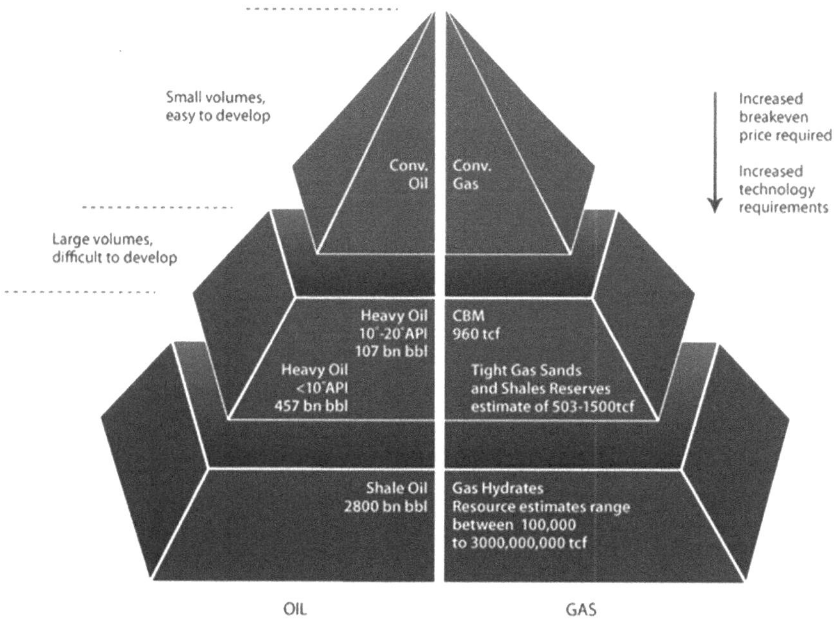

Figure 4.3 There is a lot more oil and gas reserve than the "proven" reserve.

For over 20 years, horizontal drilling has been the most common drilling technique in the United States. However, the unlocking of tight formations, including shale, has become the most important reason for the surge. Large schemes of fracking have been implemented in the states of North Dakota,

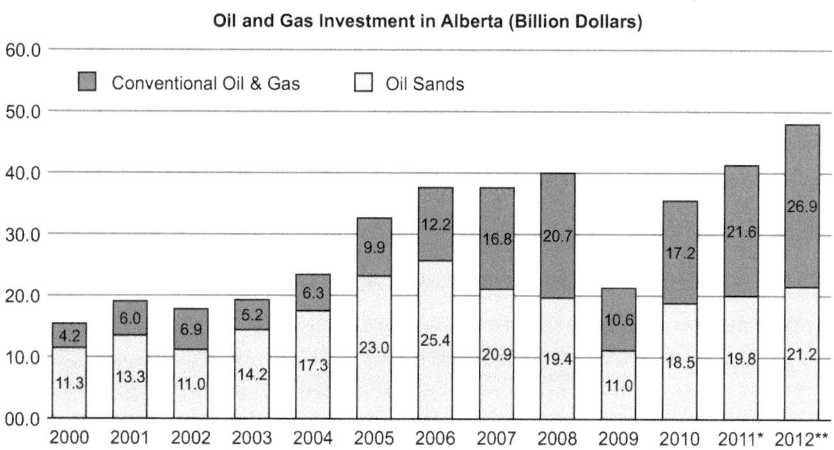

Figure 4.4 Major investment in oil sands in Canada. *2011 data are preliminary actuals, **2012 data are intentions. *Statistics Canada, public and private investment.*

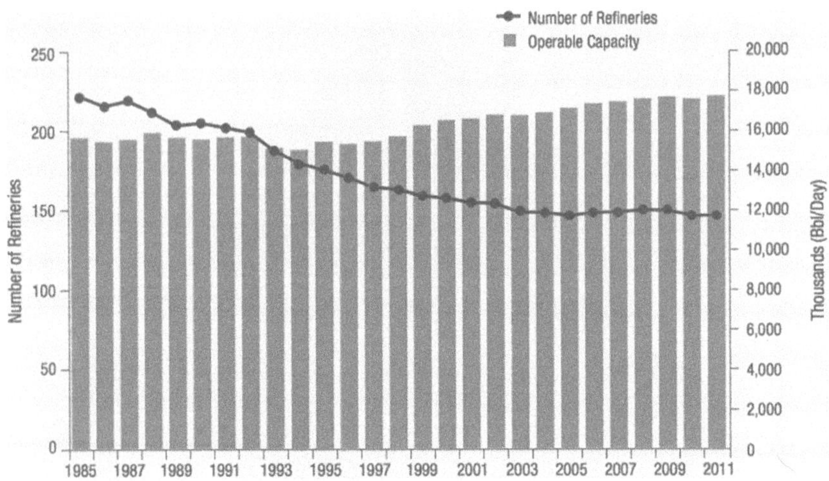

Figure 4.5 Last few decades have seen an increase in efficiency of refineries.

Oklahoma, and Texas. According to the EIA data, the surge in oil and gas production helped the United States meet 88% of its own energy needs in February, the highest monthly rate since April 1986. Crude inventories climbed to the highest level in 82 years in the week ended May 24, 2013.

This has been accentuated with an increasing efficiency in refining. Figure 4.5 shows how refining capacity has grown despite declining number of refineries.

4.2.1 Need for EOR

There are primarily three reasons given for increasing oil recovery:

1. Primary recovery techniques leave behind more than half of the original oil in place. This is a tremendous reserve to forego.
2. Increased drilling activities do not increase new discoveries of petroleum reserve. While this has been replaced with new technological opportunities (e.g., fracking technology creating oil and gas reserve in unconventional reserve), the argument is made to justify EOR.
3. Environmental concern of CO_2 emission. Ever since signing of Kyoto Agreement, US government has led the movement of CO_2 sequestration, thereby increasing oil recovery.

From the beginning of oil recovery, scientists have been puzzled by the huge amount of oil leftover following primary recovery. Naturally occurring drive mechanisms recover anything from 0% to 70% of the oil in place. In most cases, recovery declines rapidly as viscosity of oil increases.

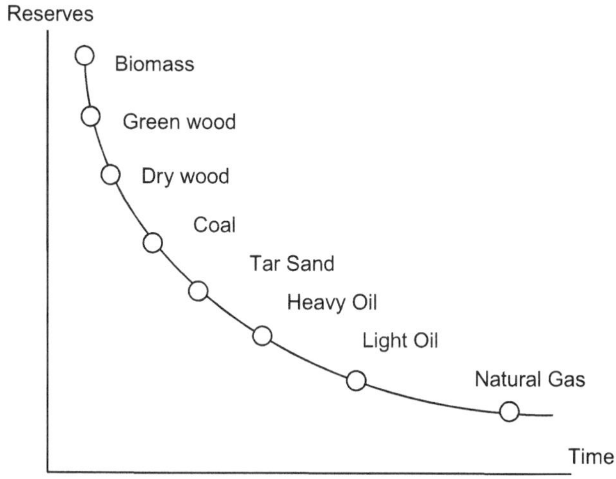

Figure 4.6 As natural processing time increases so does reserve of natural resources. *From Chhetri and Islam, 2008.*

For instance, primary recovery is less than 5% when oil viscosity exceeds 100,000 cp. This is not to say that heavy oil recovery was the primary incentive for EOR, even though most EOR projects in the United States, Canada, and Venezuela involve heavy oil recovery. The primary incentive for EOR is the fact that a typical light oil reservoir would have more than 50% of the original oil in place leftover, while a small investment can recover over 70% of the oil in place. For heavy oil, the room for improvement is much higher. Even though theoretically there is much more recovery potential of heavier energy sources all the way up to biomass (Figure 4.6), the current recovery techniques are geared toward light oil. This figure shows that natural gas is the most efficient with the most environmental integrity. This argument has been sharpened in the previous chapter that breaks down natural gas further into various forms of unconventional reservoirs. Within petroleum itself, the "proven reserve" is miniscule compared to the overall potential, as depicted in Figure 4.7.

Of course, if one includes solar energy that would be the highest reserve possible. Because all energy source utilization techniques are equipped with processing light oil as a reference, the primary focus of EOR has been light oil. In early 2000, US tertiary recovery was estimated to be 12% (Figure 4.8). This number has held steady until the huge surge in unconventional recovery of oil and gas that increased the oil and gas production by 40% in 2013. It is difficult to characterize unconventional

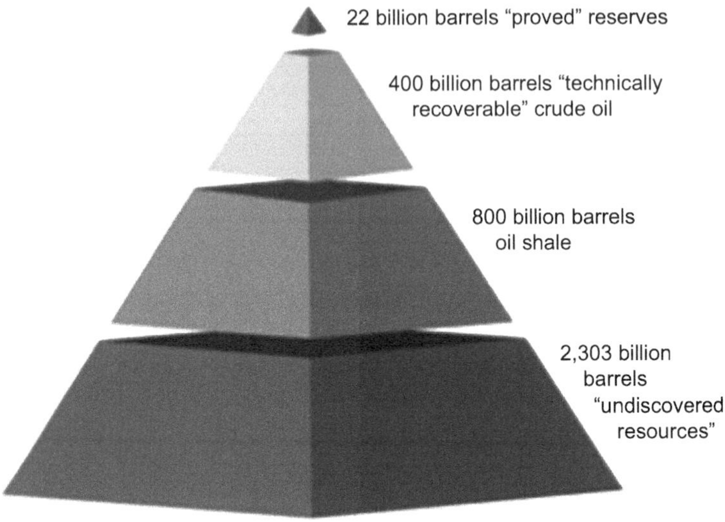

Figure 4.7 "Proven" reserve is miniscule compared to total potential of oil.

recovery under a known category of oil and gas production. In any event, the knowledge of EOR is invaluable for developing any form of petroleum production scheme.

One of the most important criteria for selection of an EOR scheme is the reserve/production ratio. This ratio is low for the United States. Consequently, the United States has been the leader in implementing EOR techniques. Of the total recoverable oil in the United States, 12% lends itself to EOR (Figure 4.8). Further clarification must be made about the term "recoverable" oil. Some of the OPEC countries calculate this number by multiplying total petroleum in place with the recovery factor, which has little scientific merit. Other countries, than OPEC, calculate initial oil in place by dividing "recoverable reserve" by the recovery factor, which is often low and without scientific justification. A recent survey shows little is known about how numbers such as "recovery factor," "recoverable reserve," etc., come to exist. However, it is commonly accepted that the "recoverable reserve" that is published worldwide and is used as the basis for determining OPEC quota for production is the most accurate starting point. This itself creates confusion in the scientific community that is vastly unfamiliar with how those "recoverable reserve" numbers are calculated. As for the pie chart in Figure 4.8, it concerns the contribution of various recovery techniques in the United States. It turns out that the recovery factor in the United States is much higher. For instance, many miscible floods as well as

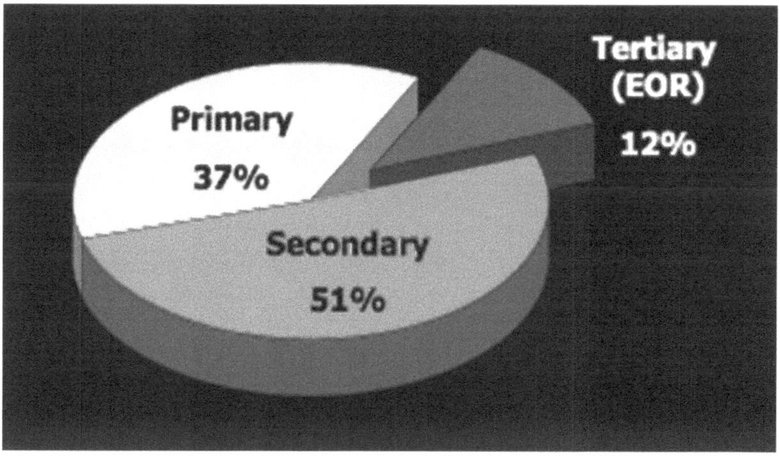

Figure 4.8 Until 2012, tertiary recovery in the United States is estimated to be 12% of recoverable oil in place.

steamfloods have recovery factors in the vicinity of 85% and 70%, respectively. In order to avoid confusion as to what this pie chart should represent in countries that have yet to start an EOR scheme, it is best to determine primary recovery potential and expect 30% of that oil during EOR schemes for light oil reservoirs. For heavy oil reservoirs, this percentage of recovery is much higher, mainly because the primary oil recovery factor is very low. In addition, many heavy oil formations do not "see" primary recovery as is the case for most heavy oil reservoirs in the United States. Some Canadian heavy oil reservoirs do produce in primary mode but with an extremely low recovery factor (less than 10% of initial oil in place).

With regards to EOR technology development and implementation, the United States has been the world leader in implementing EOR. Figure 4.9 shows that the United States is ahead both in time of implementation and magnitude of EOR recovery in the world scale. This trend continues today despite the new found domestic resources in unconventional oil and gas and shale oil megaproject of Canada. Such leadership emerges from the US superiority in related areas of new drilling and well technologies, intelligent reservoir management and control, advanced reservoir monitoring techniques, and the application of different enhancements of primary and secondary recovery processes. However, the present paper presents a comprehensive review of EOR status and opportunities to increase oil recoveries and final recovery factors in reservoirs ranging from extra-heavy oil to gas condensate.

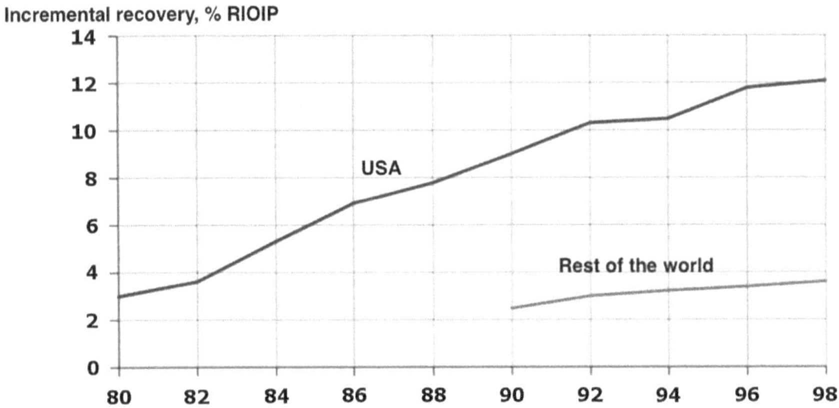

Figure 4.9 Growth of EOR in the United States as compared to the rest of the world.

It is well known that EOR projects have been strongly influenced by economics and crude oil prices. The initiation of EOR projects depends on the preparedness and willingness of investors to manage EOR risk and economic exposure and the availability of more attractive investment options. For an economic scheme to be successful in the long term, the technology cannot be expensive beyond the rate of return. In addition, the incremental recovery has to be substantially more than recovery with status quo. In this regard, the recovery/reserve ratio is important and is the most important criterion for implementation of EOR. The following Table shows the ratio of reserve over production rate. When recovery potential is estimated, the actual recovery with EOR can be calculated. These numbers are listed in Table 4.2. It is to be noted that sustainable EOR is the most economic and environmentally appealing option. Unless such scheme is implemented, infill drilling is the most economic scheme. This is especially true for reservoirs with high reserve/production ratios.

It is difficult to comment on the validity of the "reserve" numbers. A recent survey shows that there is no standard for this reserve or the definition of "proven" reserve is so varied and subjective that there is a need for a comprehensive study of this subject. It seems clear that politics plays a significant role in claiming "proven reserve." While the developing countries in general and OPEC countries in particular are cited for politicizing the reserve numbers, history tells us there is a systematic lack of transparency in both numbers and the process involved in determining these numbers. Resolution of this problem is beyond the scope of this document.

Table 4.2 Reserve/Recovery Ratio for Different Countries

Rank	Country	Reserves (10^9 bbl)	Reserve/production ratio (years)	EOR reserve (10^9 bbl)	EOR suitability with existing technology	EOR suitability with sustainable technology
1	Venezuela	296.5	387	44.5	Low	High
2	Saudi Arabia	265.4	81	39.8	Medium	High
3	Canada	175	178	26.25	Low	Medium
4	Iran	151.2	101	22.7	Low	Medium
5	Iraq	143.1	163	21.5	Low	Medium
6	Kuwait	101.5	121	15.2	Low	Medium
7	United Arab Emirates	136.7	156	20.5	Low	Medium
8	Russia	80	22	12	High	High
9	Kazakhstan	49	55	7.35	High	High
10	Libya	47	76	7.0	Medium	High
11	Nigeria	37	41	5.5	High	Medium
12	Qatar	25.41	63	3.8	Medium	Medium
13	China	20.35	14	3.1	High	High
14	United States	26.8	10	4.0	High	High
15	Angola	13.5	19	2.0	High	High
16	Algeria	13.42	22	2.0	High	High
17	Brazil	13.2	17	2.0	High	High

The need for EOR comes from the declining nature of worldwide reserve. Practically all countries have reached peak recovery rate as evidenced from Figure 4.10. This is the case despite the fact that new discoveries continue to grow. This is not to subscribe to the theory of peak oil because this graph does not consider increase in reserve due to the improvement of recovery technologies and the addition of unconventional oil and gas. This unconventional reserve is the principal reason that the United States has seen some 40% increase in domestic oil and gas recovery in the year 2013. Figure 4.11 shows how US reserve in oil and gas have progressively declined since 1970 only to pick up in 2011 onward.

There is a scientific group that believes that the above graph is misleading. As can be seen in Table 4.2, recovery alone cannot be an evidence of declining reserve because the recovery/reserve ratio varies largely among different countries. Figure 4.12 lends credibility to this statement.

Figure 4.12 also shows how countries with the exception of Venezuela have added no new reserve in the last decade. Despite this, there have been claims that major OPEC countries have inflated their reserves in order to gain more share in the competitive world market. This scenario is a pessimistic one because other countries do not actively look for or necessarily declare new reserves or reserves that have become "recoverable"

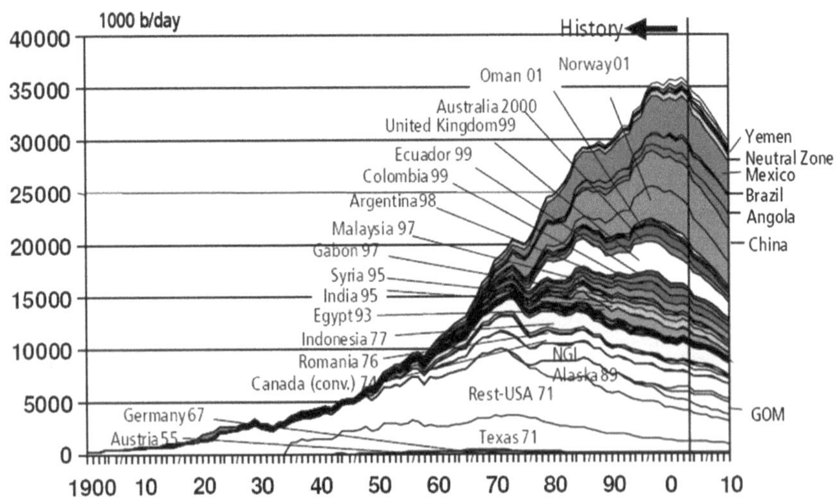

Figure 4.10 Recovery rates decline around the world. *Picture from http://en.wikipedia.org/wiki/Hubbert_peak_theory.*

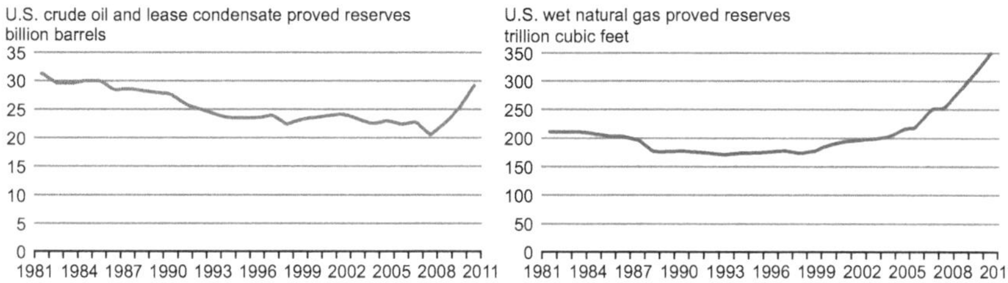

Figure 4.11 Declared reserve for various countries.

because of technological improvements. The most remarkable case here is Saudi Arabia. World's 25% of recoverable reserve is in Saudi Arabia and until now the entire recovery process is through primary. Because the recovery/reserve ratio is still fairly high, EOR suitability of Saudi fields is a question mark. While it is considered as low risk to develop Saudi reservoirs for secondary recovery because of the low-cost of implementation of waterflood schemes, the benefit of implementing suitable EOR schemes directly after primary remain very high, at least in theory. Also, it is of importance to note that Saudi Arabia has significant amount of tar and other heavy oil deposits that are ignored in their reserve estimates. However, considering latest technological breakthroughs in tar sand and heavy oils,

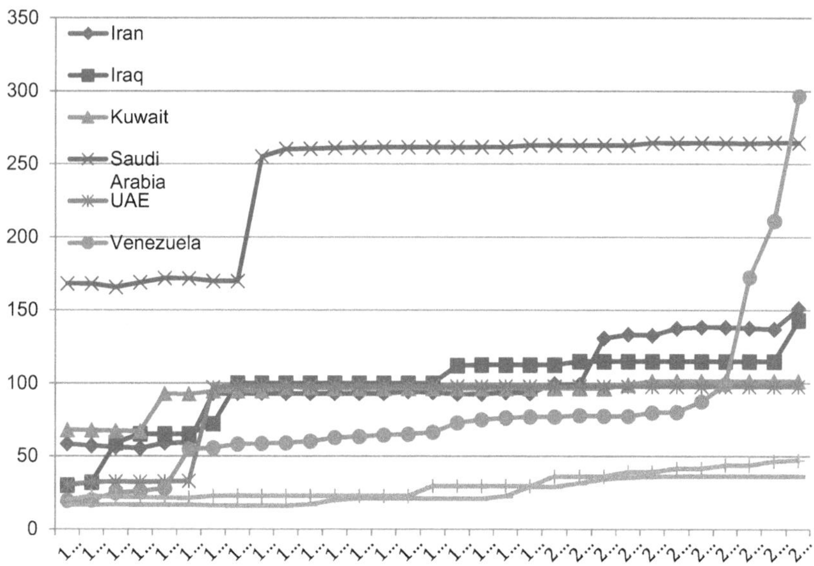

Figure 4.12 USA reserve variation in recent history. *U.S. Energy Information Administration, Form EIA-23, Annual Survey of Domestic Reserves.*

due to mega projects in Canada, Saudi Arabian heavy oil reserves, every well, can become very prominent in the world scale. Developments in the next most important case is that of Venezuela. Venezuela has the highest reserve/production ratio in the world. With EOR implementations, it has the capacity to double the daily output or total recoverable reserve.

In the United States, total oil reserve has been declining since 1970 only to peak up in 2011. Wet natural gas reserve, on the other hand has been declining since 1981 but started to increase shortly after 1999. Figure 4.11 shows the US reserve as a function of time. The decline in reserve was reversed shortly after the 2008 financial crisis. Such rise in reserve is attributed to horizontal drilling and hydraulic fracturing in shale and other "tight" (very low permeability) formations. Because the oil price increased sharply during that period, more drilling was performed. This added to the proven reserve of the United States. US proven reserves of natural gas began growing sharply in the mid-2000s as operators adopted expanded horizontal drilling programs and applied new hydraulic fracturing techniques in shale formations. Starting with 2009, similar horizontal drilling programs were applied in several of the nation's tight oil formations—reserves additions from tight oil plays have reversed the long-term trend of generally declining proved US oil and lease condensate reserves (Figure 4.11). Proven reserves of crude oil and lease condensate increased in each of the five largest crude oil and lease condensate areas (Texas, the Gulf of Mexico federal offshore, Alaska, California, and North Dakota) in 2011. Of these, Texas had the largest increase by a large margin, about 1.8 billion bbl (46% of the net increase), resulting mostly from ongoing development in the Permian and Western Gulf Basins in the western and south-central portions of the state. North Dakota reported the second largest increase, 771 million bbl (20% of the net increase), driven by development activity in the Williston Basin. Collectively, North Dakota and Texas accounted for two-thirds of the net increase in total US proven oil reserves in 2011.

Proven wet natural gas reserves increased in each of the five largest natural gas producing states (Texas, Wyoming, Louisiana, Oklahoma, and Pennsylvania) in 2011. Pennsylvania's proven natural gas reserves, which more than doubled in 2010, rose an additional 90% in 2011, contributing 41% of the overall US increase. Combined, Texas and Pennsylvania added 73% of the net increase in US proven wet natural gas reserves. Expanding shale gas developments in these and other areas, particularly the Pennsylvania and West Virginia portions of the Marcellus formation in the Appalachian Basin, drove overall increases.

In terms of technical recoverability, both oil and gas reserves changed over the last decade. Figure 4.13 shows how technical recoverability has changed for both oil and gas reserves in the United States. Even with reduced aggressive research, technological developments in various aspects of petroleum engineering made it possible to upgrade the reserve estimates (Figure 4.13). This decline has been accompanied with increasing sulfur content of US crude. Figure 4.14 shows general trends in sulfur content of crude oil in USA.

Figure 4.15 shows American Petroleum Institute (API) gravity decline in US crude oil. Together Figures 4.14 and 4.15 show that the overall quality of US crude oil is declining. Figure 4.16 shows both API gravity and sulfur content of crude oil from around the world. Light and sweet crude oil is the most desirable. However, any change of the quality of the crude implies both economic and technological drain on the crude oil. Light and sweet grades are desirable because they can be processed with far less sophisticated and energy-intensive processes/refineries. The figure shows selected crude types from around the world with their corresponding sulfur content and density characteristics. One particular advantage of certain EOR techniques is the in situ upgrading of in situ oil. While no data is available on the quality of oil recovered with EOR as compared to the same without EOR, it is

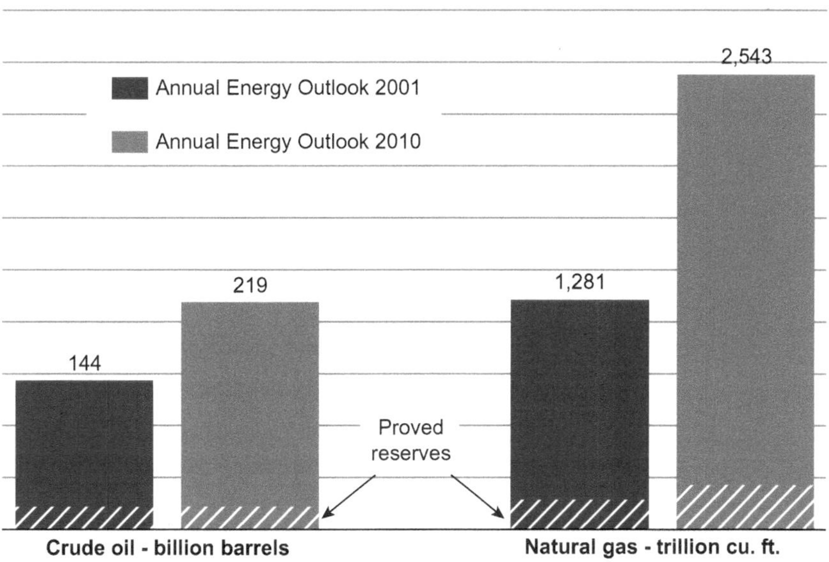

Figure 4.13 Technically recoverable oil and gas reserve in the United States. *Energy Information Administration, based on 1999 and 2008 USGS assessments.*

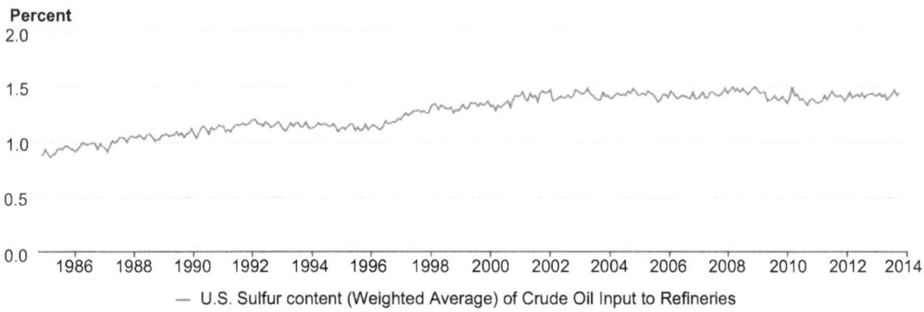

Figure 4.14 Sulfur content of the US crude over last few decades. *U.S. Energy Information Administration.*

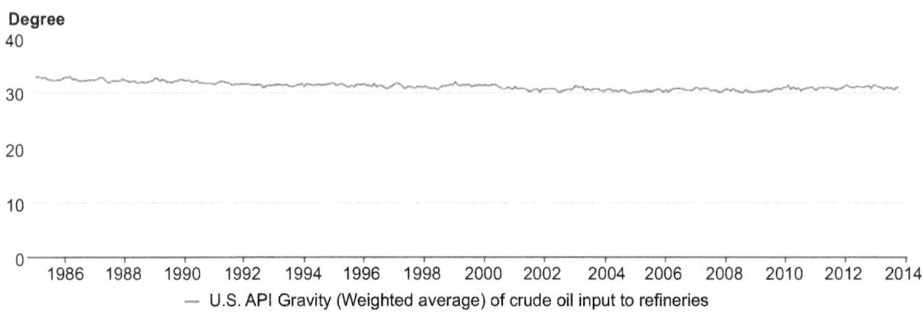

Figure 4.15 Declining API gravity of US crude oil. *U.S. Energy Information Administration.*

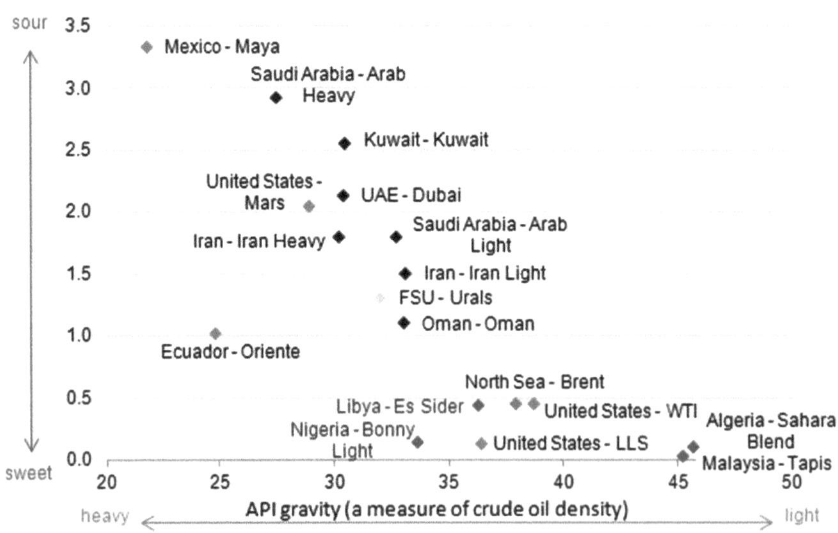

Figure 4.16 Worldwide crude oil quality. *U.S. Energy Information Administration, based on Energy Intelligence Group–International Crude Oil Market.*

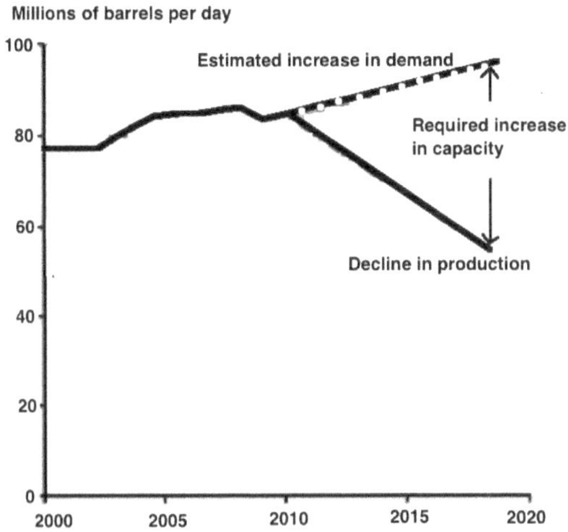

Figure 4.17 The need for EOR is evident in production and oil quality decline. *Redrawn from Total.*

reasonable to assume that in situ upgrading would improve the quality of produced oil.

The selected crude oils in the Figure 4.16 show the "sweetness" of various crude oils from around the world. These grades were selected for the recurrent and recently updated EIA report, "The Availability and Price of Petroleum and Petroleum Products Produced in Countries other than Iran."

Figure 4.17 demonstrates the need for EOR. EOR involves making up for the loss of natural production cycle in order to meet the growing need of petroleum. However, for reservoirs with high reserve/production ratio, it is most cost-effective to infill drill. By carefully selecting infill drilling sites, the recovery factor can be increased even with primary production mode. For reservoirs that have seen significant rise in water cut during primary production, one should consider local improvement of mobility ratio by adding chemicals, such as polymer. However, increasing EOR performance with polymer is not recommended because polymer slugs do not travel in the reservoir beyond a few meters. The economics of EOR changes drastically if waste gas, produced gas or locally available gas, is used for EOR injection. Figure 4.18 presents a qualitative comparison among various modes of EOR. This figure that is modified from Zatzman and Islam (2007) shows how local fluid injection gives higher return in investment than conventional turn key projects, even though the return is lower at early stages of EOR. Using local fluid requires more investment in

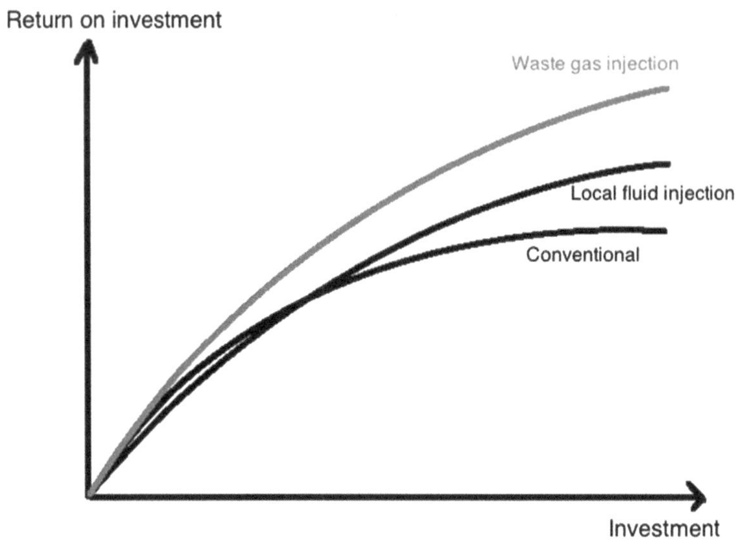

Figure 4.18 For the same investment, return is much different depending on type of fluid injected.

infrastructure than turn key projects but the investment pays off quickly and much higher return is posted at later stages. Local fluids may be produced hydrocarbon gas, locally available CO_2, or other gas/fluid available in and around the reservoir. Waste gas, on the other hand, shows higher return throughout the duration of the project. Waste gas may include produced hydrocarbon gas that is normally flared, flue gas, sour gas, or any others that are considered to be liability to the producer.

Figure 4.19 shows drilling activities in the United States over the last decade. This represents enhanced level of drilling throughout to match with the production boost during the same period. Note that the reserve/recovery ratio in the United States is quite low. Such intense drilling activities would represent far greater output in high ratio reservoirs. As stated earlier, infill drilling can increase both production rate and recoverable reserve for cases for which reserve/recovery ratio is low, as is in most OPEC countries.

4.2.2 State of the Art of EOR

Most EOR techniques are based on oil viscosity reduction and/or improvement of mobility ratio by increasing the displacement phase viscosity or by reducing oil viscosity and/or the interfacial tension (IFT)

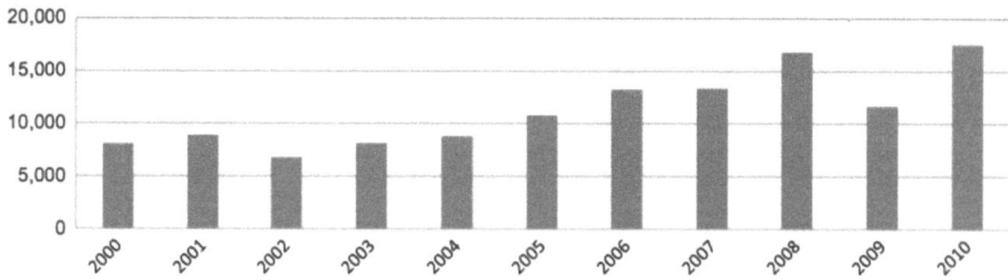

Figure 4.19 Drilling activities in the United States for various years.

between injected fluid and oil. They can be categorized into two broad types: thermal and chemical recovery processes. This follows the natural cleaning technique of hot water wash with soap. It turns out the most potent cleaning agents of petroleum (second most abundant fluid) in nature are (1) Water (most abundant fluid), (2) Clayey material (most abundant solid material on earth), (3) Wood ash (solid products of oxidation of organic products), (4) Carbon dioxide (gaseous product of oxidation of organic products). It turns out that organic products are the second most abundant solid on earth and oxygen is the most abundant gas. Hot water naturally offers the most effective cleaning product as long as the heating is done with the most abundant energy source, viz. solar energy. Finally, one should note heating through combustion of organic energy source (e.g., petroleum, wood) is the second most efficient heating mechanism.

Even though thermal recovery would include several methods in addition to steamflooding, steamflooding remains by far the most widely successful thermal EOR technique. Recently, it has been realized that mobility control can be realized by using surfactant with steam when steam/foam is generated. Ever since this realization, most emerging technologies in steamflooding involve some kind of surfactant application. EOR in light oil reservoirs have mainly focused on surfactant and/or polymer injection. Hundreds of patents have been issued on different forms of surfactant and polymer injection (in form of surfactant-water flood, micellar flood, surfactant/polymer enhanced waterflooding, etc.). Even though the chemical EOR has been recently marked as too expensive ever since the drop in oil prices in 1982, surfactants continue to play an important role in virtually all forms of successful EOR, be it in form of foam (mobility control in gas injection), steam/foam (mobility control in steamflooding), micellar, alkaline/polymer flooding or others.

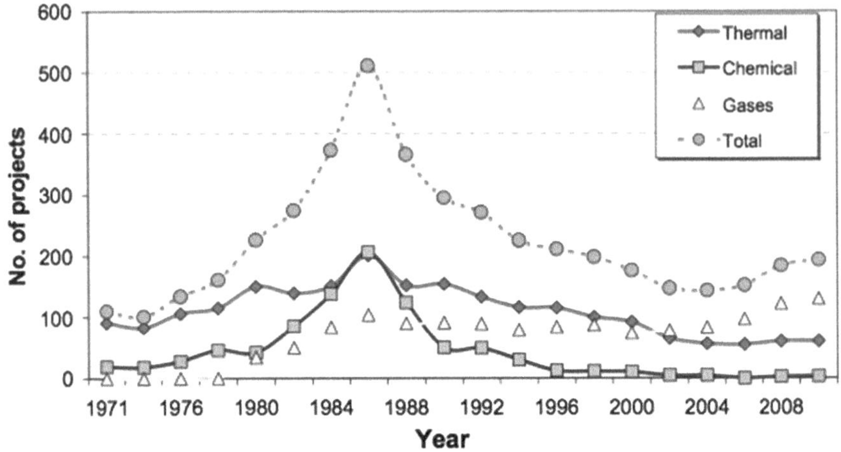

Figure 4.20 EOR projects over the year. *From Alvarado and Manrique (2010).*

Figure 4.20 shows the growth of different EOR schemes in last 40 years. This figure draws a particularly grim picture of the contribution of chemical flooding schemes. One must realize that Figures 4.20 and 4.21 do not consider the chemical applications in thermal and other EOR schemes and the scenario would be likely to change if they were considered.

Recent reports show that the contribution of EOR in total oil production has increased steadily throughout last two decades, despite fluctuating oil prices (Figure 4.20). Also, the number of EOR projects has steadily declined since the peak in 1986. This shows more efficiency of the EOR projects, indicating a trend that efficiency is the focus of the future EOR schemes. This trend is likely to continue even if the oil price continues to rise. Oil industries seem to be convinced that they can no longer afford to experiment with EOR schemes that do not show immediate improvement in oil production.

In using the United States as a reference, one must be cautious about the context in the United States. In United States, tax benefits for EOR projects were repealed in mid-1980s during Reagan era. This followed the sudden decline of the number of projects that were declared as "EOR." In absence of tax breaks, there was no benefit of declaring a project as EOR. Of particular consequence was the "chemical/polymer" processes. There has been nil contribution of these techniques over the last two decades. With the exception of China, no other country developed any commercial EOR project using chemical methods. There have been several pilot tests involving chemical methods in the North Sea, but the results have been

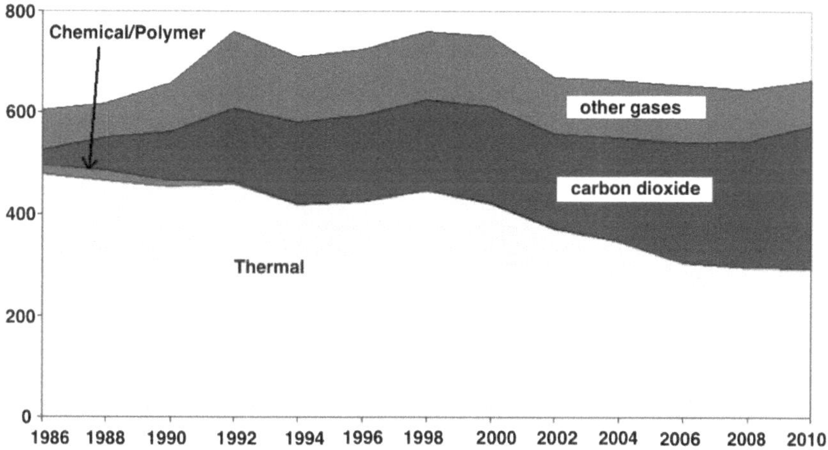

Figure 4.21 Contribution (1000 bbl/day) of various EOR projects.

dismal. During the slump period, the production continued to rise, albeit with a small slope. The rise subsided after 2000. Note that this is the year oil price was unusually low. The original prediction was \$200/bblin 2000. Instead, it was 10 times that number. In that decade, oil price was so low it was hard to justify investments in aggressive EOR techniques. The notion of sustainable EOR that gives one multiple dividend and is environmentally appealing was only a theoretical concept at that time. During that period recoveries with gas, thermal, and carbon dioxide schemes started to decline keeping pace with decreasing number of EOR projects. The number of thermal projects decreased because main fields in California began to reach maturity and the use of expensive mobility control agents with thermal EOR did not bear fruit. While this is theoretically demonstratable, the petroleum industry had to experiment with it before shutting down many thermal projects. For the same reason chemical EOR practically disappeared. By contrast, carbon dioxide projects continued to be in operation and after 2004, the number actually increased. Even though the number of projects with "other gases" was also increased, CO_2 projects showed marked increase in oil recovery. In fact, CO_2 projects continued to increase with the new incentive related to green house gas emission, as discussed in earlier sections. Since 2002, EOR gas injection projects outnumbered thermal projects for the first time in the last three decades. However, thermal projects have shown a slight increase since 2004 due to the increase of High pressure air injection (HPAI) projects in light oil reservoirs. This technique originally perfected with heavy oil and tar sand (through ISC projects) has the potential of increasing oil recovery from light oil reservoirs to a great extent. The technique is simple and cost-effective.

Chemical EOR methods that were highly successful in laboratory tests have failed miserably in field trials. Once again, this was theoretically expected but the industry could not anticipate in absence of scaled model studies or even scaling laws that capture chemical flooding effectively. Only two projects in chemical injection were reported in 2008. However, there are consorted efforts in the United States as well as the rest of the world to promote chemical flooding, particularly those involving mobility control agents. The focus now is not to create miscible fronts with micelle etc. Instead, new genre of chemical flooding schemes involve the introduction of new polymers and surface active agents. This reattachment to chemical flooding is reminiscent of 1970s and 1980s research in which hundreds of patents "proved" that chemical flooding would be effective in the field but not a single project became cost-effective or even technically successful, despite enjoying tax benefits in the United States for such projects. These techniques are not likely to produce positive results, as will be discussed in the case studies section.

4.2.3 Carbon Dioxide Injection

The most significant development in terms of EOR has been in CO_2 projects. Figure 4.22 shows various US basins with increased recovery throughout the last decade. It is this time that there has been a global effort to link CO_2 to global warming. The use of CO_2 provides one with double dividends. Based on this principle, numerous CO_2 projects have surfaced. While theoretically, any CO_2 project is both effective and environment friendly, a CO_2 project cannot be sustainable unless proper process is followed. This aspect will be considered in a later section.

The second most important considerations in CO_2 floods is the fact that it is considered to be inexpensive, at least in the United States (US$ 1–2/Mscf). In addition, the United States has existing network of pipelines that can be readily used for distribution of CO_2. There is one significant case study in Weyburn field of Canada, for which an entire pipeline was created in order to dispatch CO_2 from United States to Canada. This CO_2 was deemed most cost-effective than Canadian CO_2 that would have to be extracted from local coal-fired power plants. The project received US $1 billion in government grants and more in tax rebates and flagged as the most important CO_2 sequestration project of the time. This project was

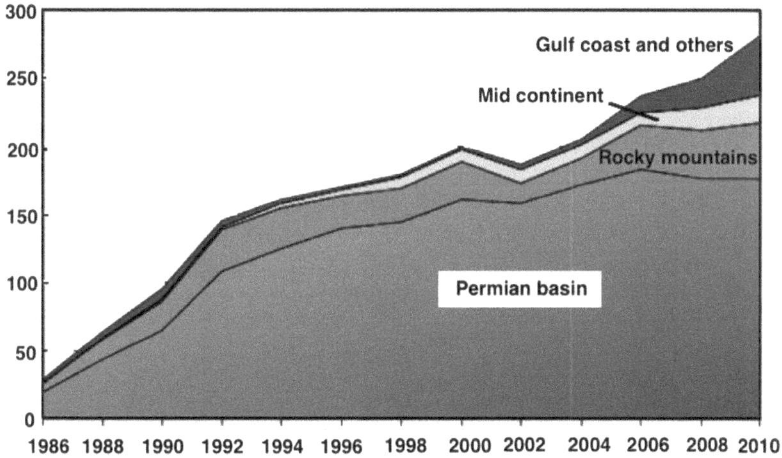

Figure 4.22 Evolution of oil production (1000 bbl/day) of EOR projects in the United States. *From Oil & Gas Journal EOR Surveys 1976–2010, based on Oil and Gas Journal, 2010.*

"profitable" only because of the government grant and some ten-fold increase in oil price. This will be discussed in later sections.

It is also important to note that the CO_2 pipeline system in the United States was built in a 30-year (1975–2005) time span when oil prices and tax incentives were sufficiently attractive to ensure security of supply as main drivers. These are viable only because of government interference in name of climate change, funding, and investment. Figure 4.23 shows evolution of CO_2 projects in the United States and average crude oil prices for the last 30 years. This figure is extracted from Alvarado and Manrique (2010). They used oil prices of the refiner average domestic crude oil acquisition cost reported by the Energy Information Administration (EIA). For reference purposes, crude oil price used in Figure 4.23 was arbitrarily selected for every month of June except for year 2010 (oil price as of March 2010). These CO_2 projects led to significant recovery (Figure 4.24).

Although it can be concluded that CO_2 EOR ("from natural sources") is a proven technology with oil prices less than US $20/bbl, this EOR method represents a specific opportunity in the United States and not necessarily can be extrapolated to all producing basins in the world. This conclusion is based on the selection criteria listed in Table 4.3. This cannot be generalized to other countries, where different economic, environmental, and technical conditions prevail. From sustainability point of view, there must be questions that should be asked in proper sequence. For

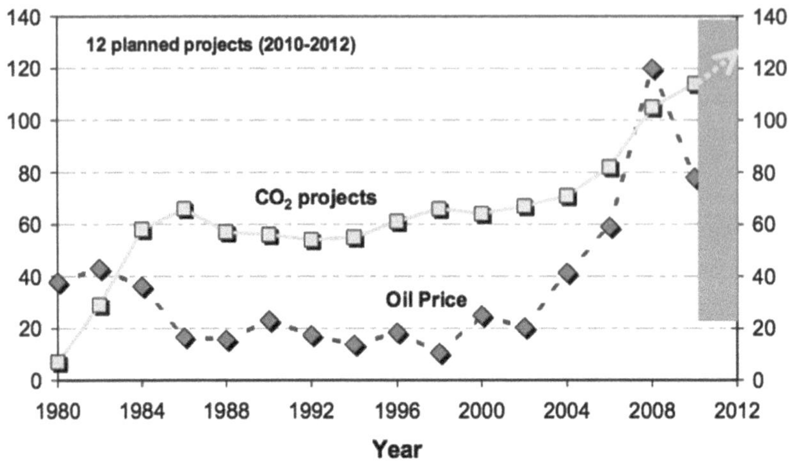

Figure 4.23 Evolution of CO_2 projects and oil prices in the United States. *From Oil & Gas Journal EOR Surveys 1980–2010 and U.S. EIA 2010. From Alvarado and Manrique (2010).*

instance, if the technological feasibility question is asked before availability of the carbon dioxide, the answer would be irrelevant at best. Similarly, the presence of an existing infrastructure for both CO_2 purification and distribution can alter the decision tree. Finally, recent findings indicate that CO_2 is quite effective in recovering heavy oil. In fact, with the new

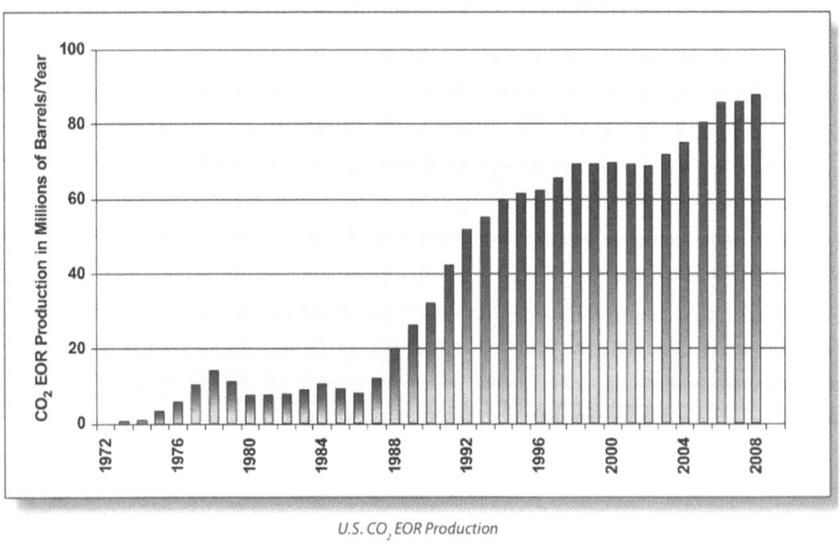

Figure 4.24 CO_2-EOR recovery in the United States throughout history.

Table 4.3 Screening Criteria for CO_2 Projects as Used in the United States

Depth (ft)	<9800 and >2000
Temperature (°F)	<250, but not critical
Pressure (psi)	>1200 to 1500
Permeability (mD)	>1 to 5
Oil gravity (°API)	>27 to 30
Viscosity (cp)	≤10 to 12
Residual oil saturation after waterflood, fraction of pore space	>0.25 to 0.30

incentive of CO_2 sequestration, heavy oil reservoirs offer the greatest potential for CO_2 injection.

Figure 4.25 shows the strategy developed by the government of Alberta. This program shows equal importance to conventional and heavy oil formations. Scaled model studies show that heavy oil recovery with CO_2 can lead to 70% of the oil in place. This is tremendous considering the fact that primary recovery of heavy oil is less than 5% and similar recovery factor with steamflooding would require significant cost increase while having bigger footprint on the environment.

Figure 4.26 shows the importance given to CO_2-EGR. The use of CO_2 injection in EOR is a mature well practice technology. Enhancing gas recovery through the injection of CO_2 however is yet to be tested in the

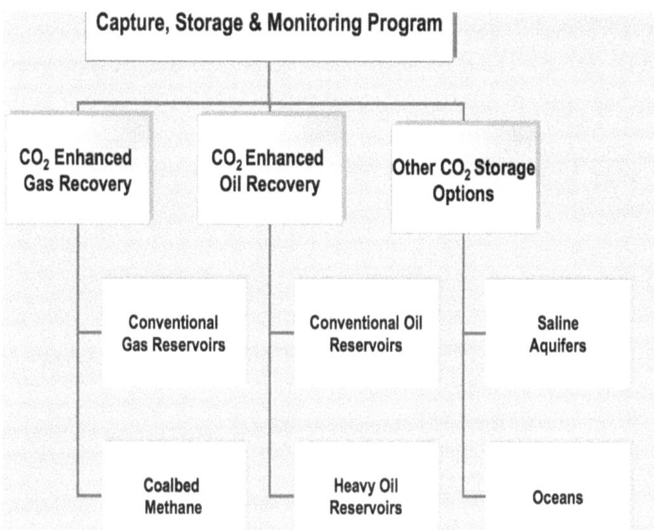

Figure 4.25 Alberta government strategy.

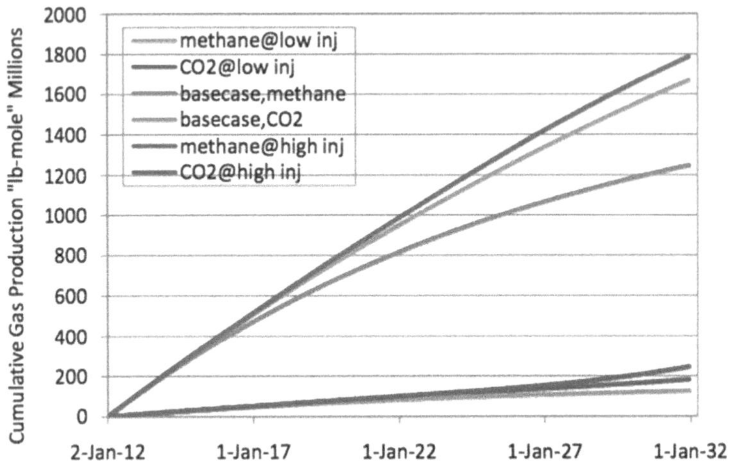

Figure 4.26 Natural gas production with CO_2 injection schemes. *From Khan et al., 2012.*

field (Hussen et al., 2012). Numerous simulation studies ever since the early work of Islam and Chakma (1990) have appeared to support high recovery of gas and heavier components from a gas reservoir along with high capacity of CO_2 sequestration. Although there are some published simulation studies that have been carried out to comprehend by which process CO_2 sequestration in a depleted gas reservoir could lead to EGR, none of these studies have ever attempted to manifest the effect of mixing (CO_2–CH_4) on the recovery process prior to depleted reservoir. These studies were mainly aimed to reduce greenhouse gas emission in the atmosphere and sequestrating in a depleted gas reservoir or in an aquifer. In the year 2005, a project by Gas de France Production, The Netherlands was in progress to assess the feasibility of CO_2 injection prior to depletion of the gas reservoir (K12-B) for EGR and storage. However, since then no follow up results have been published on the final gain in reserve recovery (van der Meer, 2005).

Generally, high natural gas recovery factors along concerns with degrading of the natural gas resource through mixing of the natural gas and CO_2 have led to very little interest shown in CO_2-EGR (Clemens, 2002). In terms of sequestration, natural gas reservoirs can be a perfect place for carbon dioxide storage by direct carbon dioxide injection. This is because of the ability of such reservoirs to permeate gas during production and their proven integrity to seal the gas against future escape (Oldenburg et al., 2001).

However, displacement of natural gas by injection of CO_2 at supper critical state has not been studied extensively and not well understood (Mamora and Seo, 2002). Despite of the fact that CO_2 and natural gas are mixable, their physical properties such as viscosity, density, and solubility are potentially favorable for reservoir repressurization without extensive mixing. This phenomenon of gas–gas mixing can be controlled by controlling the operating parameters.

The injected CO_2 in geological formations undergo geochemical interactions, such as structural, stratigraphic and hydrodynamic trapping. The injected CO_2 is trapped either in the form of physical trapping as a separate phase or as a chemical trapping where it reacts with other minerals present in the geological formation (International Energy Agency, 2010). As time passes, CO_2 becomes immobilized in the geological formation as a function of given long time scales. This is known as geological sequestration. Oldenburg (2003) simulated CO_2 as a storage gas. The results suggested that CO_2 injection as a supercritical fluid allows more CO_2 storage as the pressure increases due to its high compressibility factor. Thus, an expansion of the compressed gas is expected due to changes in pressure and temperature. As a result, there will be a point when gas production no longer is economically feasible.

In terms of economics, not unsurprisingly, Gaspar et al. (2005) claimed the major obstacle for applying CO_2-EGR is the high costs involved in the process of CO_2 capture and storage. The experience from oil recovery schemes indicate that the economics look quite different when purity in injected CO_2 is not sought. It turns out that the purity does not need to be high, and naturally available CO_2 or even flue gas would accomplish the same outcome. It is in line with pressure maintenance schemes in oil reservoirs. This option that would make CO_2 injection appealing without tax incentive as claimed by IEA (2010).

Khan et al. (2012) conducted economic feasibility study of carbon dioxide into a natural gas reservoir and found the scheme economically attractive because of EGR. Figure 4.26 shows results of CO_2 injection at high and low injection rates. Natural gas production is the highest for CO_2 injection at high rate. It is because the mixing is the greatest under high injection rates. However, one should note that this study used a stable displacement front. This is a reasonable assumption because CO_2 is more viscous and denser than natural gas. Such results are not expected in oil reservoirs. In terms of overall gas injection for EGR, there are 50 projects in North America that employs sour gas injection for treatment of natural gas and produced CO_2 has been injected in Dutch sector of North Sea for years (K-12B gas reservoir).

Based on the CO_2 capture, utilization, and sequestration strategy, government of Alberta has drafted a comprehensive scheme as shown in Figure 4.27. "CO_2 Backbone" is a network or manifold of pipelines that can be used for transporting CO_2 from emission hubs as well as taking CO_2 to customer sites. The idea is to create an infrastructure based on the "CO_2 culture." Because CO_2 is ultimately a valuable commodity, it is suggested that industrial complexes, including pharmaceutical industry, be developed along the backbone. This is a powerful template for developing a comprehensive carbon dioxide based EOR technique.

Figure 4.28 shows locations for various CO_2 sequestration projects around the world. These projects are in support of greenhouse gas mitigation.

4.2.4 Thermal Methods

By far the most important EOR scheme in the United States and the world has been the thermal EOR. It has been in operation in heavy oil formations for over five decades. Obviously, the advantage gained by exponential

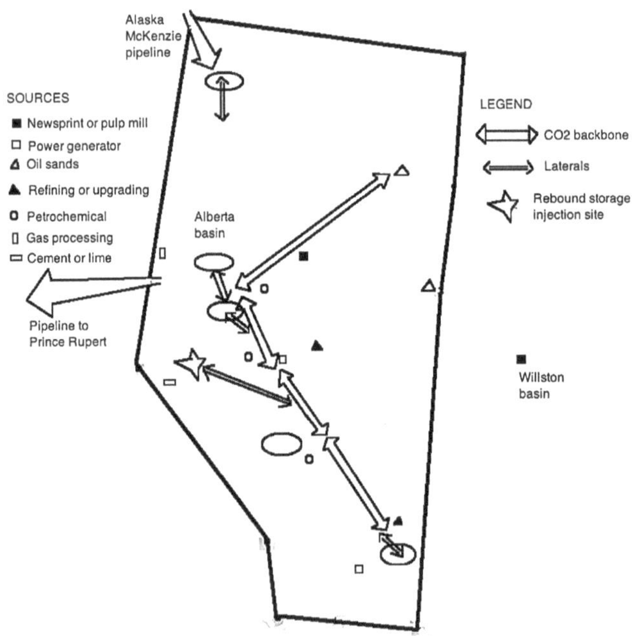

Figure 4.27 Alberta's plan to implement comprehensive Carbon management scheme. *From Islam, 2014.*

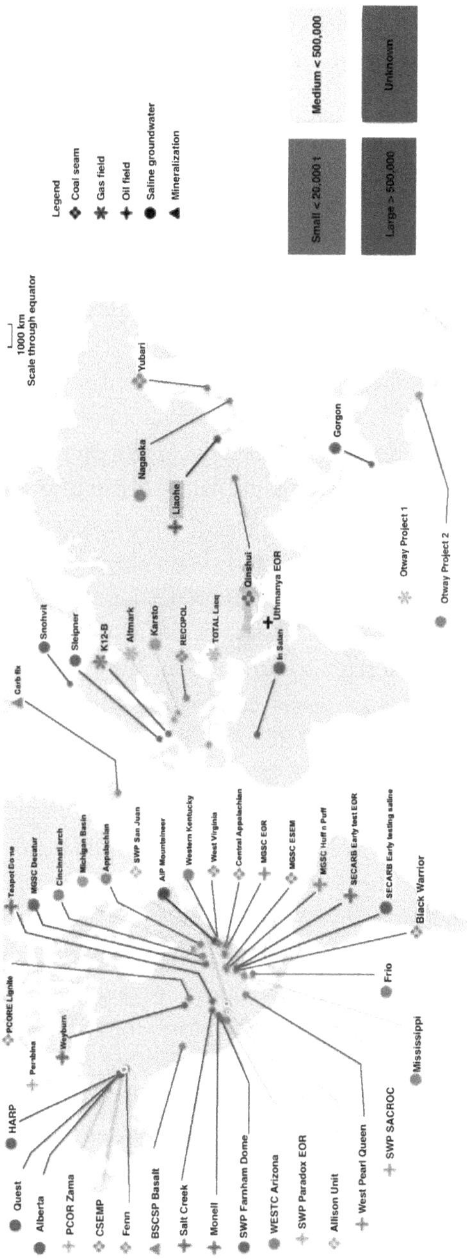

Figure 4.28 CO_2 sequestration demonstration projects around the world.

decrease in oil viscosity due to linear increase in temperature has been the focal point of all thermal EOR schemes. Among thermal methods, steam injection has been the most dominant EOR scheme. Simplicity of the scheme and the unique latent heat properties of water are the major reasons why oil industry has been active in steamflooding. Besides, primary oil recovery being practically impossible, huge heavy oil reserves are left as target of the steam injection scheme.

Just as water is the best cleaner, thereby, making waterdrive the most common drive for oil production, steam injection is the most common technique for heavy oil reservoirs. Table 4.4 lists the contrasting features of water and oil. This list makes it clear why the use of water is the most effective technique for oil recovery.

Steam injection process involves conversion of scale-free water to high-quality steam of about 232 °C temperature and at a pressure higher than the corresponding saturation pressure. Generally, using direct fired heaters the water is converted to steam. Using insulated distribution lines, steam is transported to various injection wells. Steam injection can be done by two different methods: steam stimulation and steam displacement.

In the stimulation method a predetermined volume of steam is injection into well and the well is shut in to allow to stimulate the wellbore area. After a few days of shut in the well starts to production. If necessary the stimulation process repeat again. In the steam displacement process, continuous injection of steam, usually apply at lower rates. The steam is injected in place as to distance and direction form production wells. Steam injection is a highly sophisticated process and it requires extensive engineering and analytical inputs.

It is estimated that there are 85–110 billion bbl of heavy oil reserve in the United States. Since the 1960s, steam has become the predominant EOR method for these high-viscosity heavy oil reservoirs worldwide. However, factors such as steam channeling, gravity segregation, and reservoir heterogeneity often result in poor contact of the heavy oil formation by the injected steam, leading to low recoveries. One method of conformance control that has received considerable attention is the use of surfactant foams that reduce steam mobility. Numerous laboratory and technically successful field studies have been reported.

Figure 4.29 shows the production of Syncrude and bitumen in Alberta. Syncrude and bitumen represent the two extremes of the viscosity spectrum. Syncrude is synthesized from natural gas, whereas bitumen represents the heaviest (and most viscous) components of petroleum. Note that both products grew exponentially in the last few decades, ever since

Table 4.4 Contrasting Features of Water and Petroleum

Water	Petroleum
Source of all organic matter	End product of all organic matter
Most abundant fluid on earth	Second most abundant fluid on earth
Oxygen 85.84; sulfur 0.091	Carbon, 83–87%
Hydrogen 10.82; calcium 0.04	Hydrogen, 10–14%
Chloride 1.94; potassium 0.04	Nitrogen, 0.1–2%
Sodium 1.08; bromine 0.0067	Oxygen, 0.05–1.5%
Magnesium 0.1292; carbon 0.0028	Sulfur, 0.05–6.0%
	Metals, < 0.1%
Mostly homogeneous	Hydrocarbon (15–60%), napthenes (30–60%), aromatics (3–30%), with asphaltics making up the remainder.
Reactivity of water towards metals. Alkali metals react with water readily. Contact of cesium metal with water causes immediate explosion, and the reactions become slower for potassium, sodium, and lithium. Reaction with barium, strontium, calcium are less well known, but they do react readily.	Nonreactive toward metal.
Nonmetals like Cl_2 and Si react with water $Cl_{2(g)} + H_2O_{(l)} \rightarrow HCl_{(aq)} + HOCl_{(aq)}$ $Si_{(s)} + 2H_2O_{(g)} \rightarrow SiO_{2(s)} + 2H_{2(g)}$ Some nonmetallic oxides react with water to form acids. These oxides are referred to as acid anhydrides.	Reaction with nonmetals is faster
High cohesion	Low cohesion
Unusually high surface tension; susceptible to thin film	Unusually low surface tension
Adhesive to inorganic	Adhesive to organic
Unusually high specific heat	Unusually low specific heat

Continued

Table 4.4 Contrasting Features of Water and Petroleum—cont'd

Water	Petroleum
Unusually high heat of vaporization	Unusually low heat of vaporization
Has a parabolic relationship between temperature and density	Has a monotonous relationship between temperature and density
Unusually high latent heat of vaporization and freezing	Unusually low latent heat of vaporization and freezing
Versatile solvent	Very poor solvent
Unusually high dielectric constants	Unusually low dielectric constants
Has the ability to form colloidal solutions	Destabilizes colloids
Can form hydrogen bridges with other molecules, giving it the ability to transport minerals, carbon dioxide, and oxygen	Poor ability to transport oxygen and carbon dioxide
Unusually high melting point and boiling point	Unusually low melting point and boiling point
Unusually poor conductor of heat	Unusually good conductor of heat
Unusually high osmotic pressure	Unusually low osmotic pressure
Nonlinear viscosity pressure and temperature relationship (extreme nonlinearity at nanoscale; Hussain and Islam, 2010)	Mild non-linearity in viscosity pressure and temperature relationship
Enables carbon dioxide to attach to carbonate	Absorbs carbon dioxide from carbonate
Allows unusually high sound travel	Allows unusually slow sound travel
Large bandwidth microwave signals propagating in dispersive media can result in pulses decaying according to a nonexponential law (Pieraccini et al., 2009)	Faster than usual movement of microwave
Unusually high confinement of X-ray movement (Davis, 2005)	Unusually high facilitation of X-ray movement.

From Hutchinson, 1957; Attwood, 1949; Handbook of Chemistry and Physics, 1981.

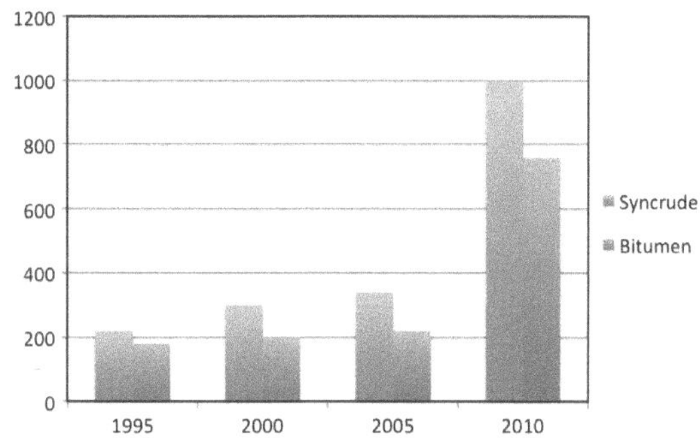

Figure 4.29 Synthetic crude and bitumen production from Canada's tar sand. *NRCan, 1998.*

implementation in 1980s. While the economics of these products have been reported to be attractive, often the government contribution in building the infrastructure has been overlooked or not included in the analysis. Without significant government involvement, these projects would not be implemented, particularly during the time when oil price was in the range of US $10/bbl. With the increase in oil price, these schemes have become attractive and mega projects are being implemented in bitumen extraction and processing. The future of Syncrude has a somewhat conflicted scenarios. Gas price has increased making it more comparable with oil than previous years. In addition, Alberta has suffered from lack of enough natural gas to meet local needs. This is due to the fact that the population of the province has increased manifold in a country that has seen almost zero growth in population over the same period.

For a long time, steam has been used as the driving fluid in heavy oil reservoirs. The steam injection scheme has been very popular because of its simplicity. However, steam injection leads to an unfavorable mobility ratio in most applications. Besides, gravity lay over is a problem with most reservoirs with little or no dip. Injected steam, because of its low density, rises to the top of the reservoir and tends to form a channel beneath the cap rock to the production well. Early steam breakthrough can occur at producing wells owing to override, channeling, unfavorable, and viscous fingering mobility ratio, resulting in low oil recovery efficiency. Because of high steam mobility, there is little pressure differential between injector and producer once steam breakthrough occurs. The majority of subsequently injected steam follows this established path of least resistance and the process efficiency is impaired. Injecting surfactants to generate foam in situ can reduce steam mobility and improve the volumetric sweep efficiency in oil reservoirs. There have been many examples of increased oil production in Californian heavy oil reservoirs when steam r foam was used.

In the last three decades, there have been many attempts to improve steam injection efficiency by the use of additives. Among many additives tried, the aqueous surfactant solution appears to be the most promising one. The objective of such surfactant injection is either to increase the pressure gradient across the region of interest by generation of foam or to use the surface active properties of the surfactant to reduce the oil–water IFT and to alter the relative permeability curve. Following is a list of research areas in this topic:

(1) *Surfactant Selection Criteria for Steamflooding.* In selecting surfactants for application in thermal recovery, two criteria are set, namely, the resistance of surfactants to hydrolytic degradation and to thermal degradation. It is a common practice to study surfactants at elevated temperatures exterior to porous media. The foam tube test is the most commonly applied technique for determining foam stability exterior to porous media (DeVries, 1958). Some studies found foam stability outside of porous media to be an important indicator of potential mobility reduction within porous media (Doscher and Hammershaimb, 1981). In other studies, however, no such correlation was found (Dilgren et al., 1982). It is likely that the tube test represents foam behavior in very large pore throats and may not represent foam stability in a confined case as in a real porous medium. This observation has been further confirmed by Zhong et al. (1999).

Handy et al. (1982) indicated that thermal stability is a critical factor in the choice of a foaming agent for thermal EOR processes. It has been demonstrated through many studies that foam can be used for flow diversion in a steamflood process. Recently, Djabbarah et al. (1990) reported thermal stability of several surfactants. Despite many disjointed efforts, a comprehensive selection criterion applicable to steamflooding has not been developed yet (Zhong et al., 1999).

2. *Microscopic Behavior of Surfactant Steamflooding.* It is important to understand microscopic behavior of a system before a field application can be recommended. In steamflooding research, little effort has been spent in studying microscopic behavior and extending that observation to the scaled-up version. Several theories have been proposed to try to explain surface phenomena for a surfactant-steam system (Ransohoff and Radke, 1988; Falls et al., 1988, 1989; Hirasaki and Lawson, 1985). However, very little agreement among researchers exist and fundamental questions, such as

the role of gas rate on apparent viscosity of foam, mechanism of bubble generation, effect of surfactant concentration, or the effect of temperature on foam flow cannot be answered without some degree of ambiguity.

3. *Role of Residual Oil on Foam.* This fundamental aspect of the steam/ foam process has not been addressed properly. Most papers on the topic claim to offer different solutions. One possible way to address this process is to conduct research on the microphysical aspect of the process (Zhong and Islam, 1995).

When heat is combined with water, producing steam becomes an effective displacement tool for additional heavy oil recovery. The decrease in heavy oil viscosity being log with increase in temperature, any heating unlocks tremendous amount of oil from the porous medium. Figure 4.30 shows general trend in viscosity versus temperature. Note that the temperature scale is linear whereas the viscosity scale is logarithmic. It translates into a sharp decline in viscosity for moderate increase in temperature. Darcy's law being linear, such decrease in viscosity leads to immediate flow rate increase. In addition, the larger change in viscosity takes place in the lower temperature region and the sharpest decline in higher viscosity oils.

Also affected by temperature is the IFT. This alteration in interfacial difference comes from the fact that surface tensions of water and various

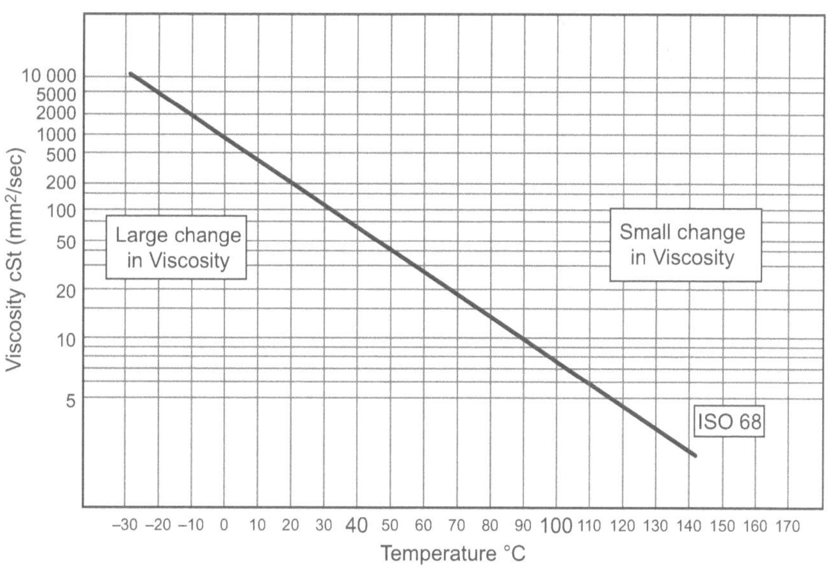

Figure 4.30 Change in viscosity for change in temperature.

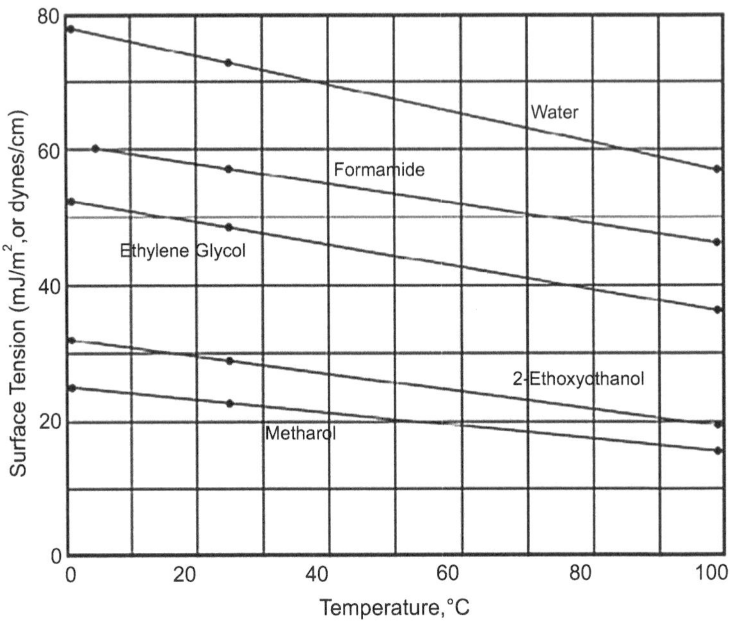

Figure 4.31 Surface tension changes with temperature with different slopes for different chemicals.

petroleum fluids are affected differently, even though each of them varies linearly. Figure 4.31 shows how surface tension varies for various liquids.

Figure 4.32 provides a summary of the measured endpoint residual oil saturations to waterflood as a function of temperature for certain Canadian bitumen samples.

Figure 4.32 shows two different regimes exist as a function of temperature. At the lower temperature range (below 100 °C), there is a rapid decline in oil saturation. In the range of 120–200 °C, the decrease rate is subsided. However, at higher range of temperature (beyond 200 °C), the saturation declines rapidly once again.

It is well documented that residual oil saturation tends to reduce at constant temperature by steamflooding in comparison to conventional waterflooding at the same temperature condition. This is believed to be due to turbulence effects associated with the vaporization of pellicular films of water underlying trapped bitumen as well as possible changes in IFT and wettability during the steam displacement process. Also active is the steam distillation factor that can improve the efficiency of oil recovery with steamflooding. Overall, steamflooding represents optimum cleaning of oil.

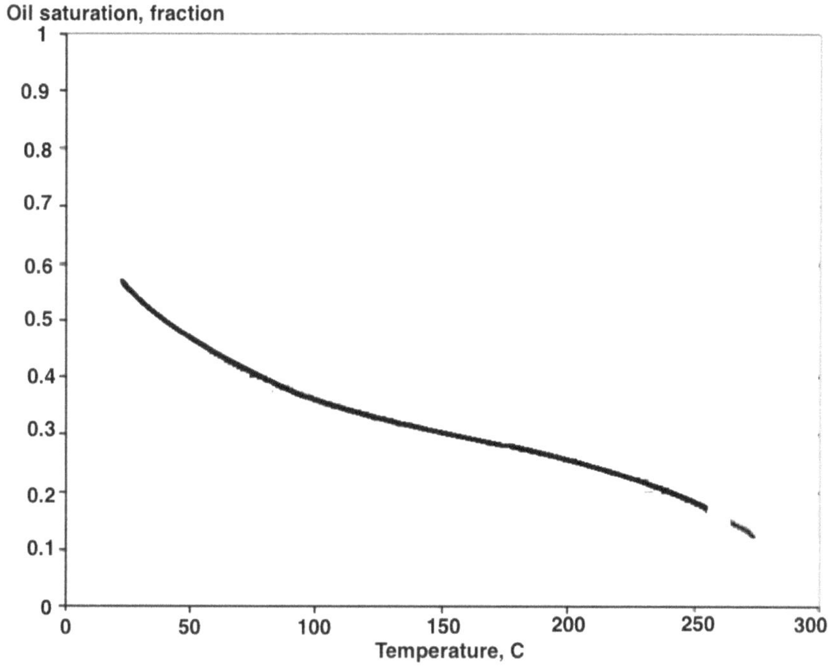

Figure 4.32 Residual oil saturation as a function of temperature. *Modified from Bennion et al., 2006.*

Figure 4.33 illustrates the trend of pre- and post-steamflood residual oil saturation as a function of steamflood temperature. This figure demonstrates the superiority of steam over hot water injection at the same temperature.

Cyclic steam injection (Huff & Puff), steamflooding and Steam-Assisted Gravity Drainage (SAGD) have been the most widely used recovery methods of heavy and extra-heavy oil production in sandstone reservoirs during last decades. Thermal EOR projects have been concentrated mostly in Canada, former Soviet Union, the United States and Venezuela, and Brazil. Recently, China has made good progress in thermal EOR. Steam injection began approximately five decades ago. Mene Grande and Tia Juana field in Venezuela, and Yorba Linda and Kern River fields in California are good examples of steam injection projects over four decades. They are considered to be some of the most successful EOR projects of all time. The lessons learned have been immense. However, little of that knowledge has been transferred to conventional light oil recovery processes. There is a one-way disconnection between EOR in heavy oil and

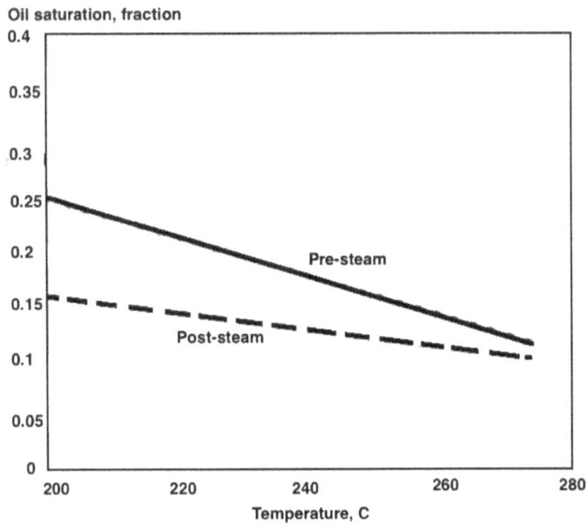

Figure 4.33 Residual oil reduction with temperature for pre- and post-steamflood. *Modified from Bennion et al., 2006.*

EOR in light oil. It is so because a great deal of the knowledge from light oil recovery has been transferred to recently developed heavy oil processes, such as steamfloods in the Crude E Field in Trinidad, Schoonebeek oil field in Netherlands and Alto do Rodrigues in Brazil. In addition, heavy oil recovery processes such as VAPEX has used light oil solvent flooding technologies, developed in the 1960s and 1970s. Ironically, "mistakes" of light oil recovery, particularly when it applies to much discredited chemical flooding, have filtered through heavy oil recovery schemes. If it was not for the subsidy of the government and the tax credits offered to stimulate heavy oil and tar sand recovery, these projects would not be viable.

Capitalizing on the success of steamfloods, numerous "improvements" have been suggested for steam-related recovery techniques. They include the use of solvents, gases, chemical additives, and foam in an attempt to control the mobility of the displacement front. Laboratory results shows great recovery potentials of these "novel" techniques. However, similar to chemical flood schemes, field experimentation with this mobility control chemicals have failed to produce satisfactory results. This failure is mainly due to the fact that (1) any use of solvent is deemed uneconomical, (2) it is impossible to control mobility with chemicals beyond a few feet from the wellbore, (3) original steam flood of cyclic steam injection produce significant amount of heavy oil, leaving behind little room for improvement. One example is the LASER (for Liquid Addition to Steam for Enhancing

Recovery) process, which consists of the injection of C5+ liquids as a steam additive in cyclic steam injection processes. Although the LASER process was tested at pilot scale in Cold Lake, Canada, the process has not been expanded at a commercial scale. As stated in the previous paragraph, commercial viability of these projects is nil because of inherent issues.

Steam injection has also been tested in medium and light oil reservoirs being crude oil distillation and thermal expansion the main recovery mechanisms in these types of reservoirs. Because light oil reservoirs are often fractured that pose a scenario different from conventionally homogenous formations of heavy oil, considerations must be made in designing steamflood in light oil reservoirs. To be remembered also that light oil reservoirs are already hot and the temperature range from which the maximum decrease in viscosity occurs does not apply to light oil reservoirs. Any heat in the formation will expand the rock/fluid system in such a way that the displacement front is altered. Steam in light oil reservoirs will distillate the crude oil, creating in situ refining. The precipitation of heavier component and ensuing adsorption on the rock surface can change the rock wettability that may favor the oil production. Steam injection in light oil formations does hold promises but has rarely been investigated with rigor.

On the other extreme of the oil viscosity spectrum, SAGD has been employed in recovering tar sand in Canada. This process is particularly suitable for unconsolidated reservoirs with high vertical permeability and has become standard in many fields of Canada. Even though SAGD pilot tests have been reported in China, the United States, and Venezuela, commercial applications of this EOR process have been reported in Canada only and more specifically those implemented in McMurray Formation, Athabasca (e.g., Hanginstone, Foster Creek, Christina Lake, and Firebag, among others). These projects were all subsided by the government of Alberta that spent practically all extra revenues of additional income due to oil boom in the province on these and similar landmark projects. From technological perspective, these projects have been successful. However, their economics have been good only because of the new surge in oil prices. Some argue that they were attractive even when oil price was US$ 12/bbl. These calculations do not account for government subsidies and the tax breaks. A more realistic estimate is the oil price has to be at least US$ 20/bbl for these projects to be economically viable.

Figure 4.34 shows reservoir depths, average horizontal permeability and formation of several SAGD (pilot and large scale) projects, as documented

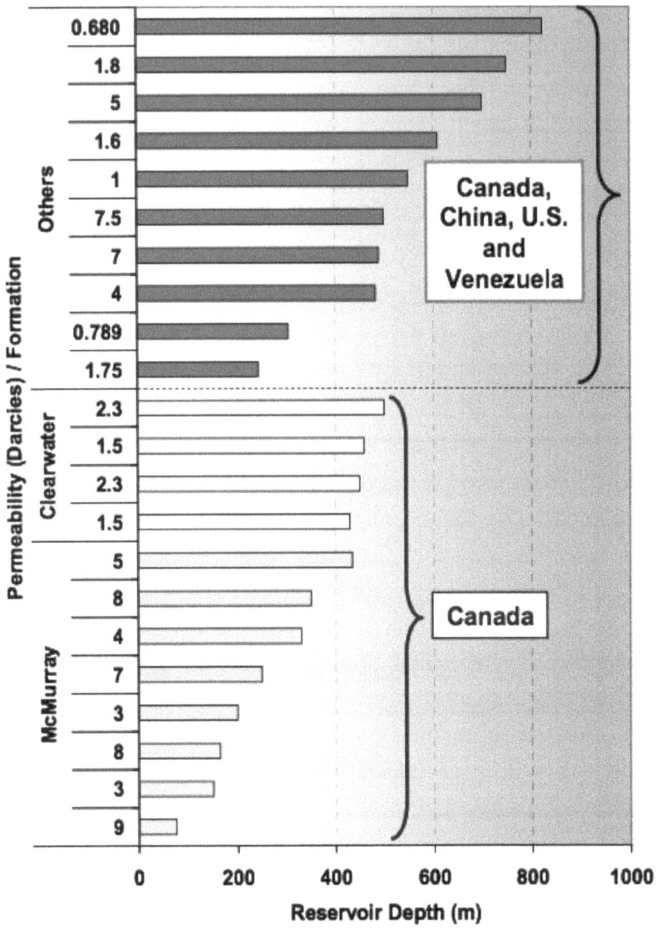

Figure 4.34 Average permeability for various formation and their depth. *From Alvarado and Manrique (2010).*

in the literature. Among these projects, only those developed in McMurray Formation (blue bars of Figure 4.34) operate commercially. SAGD projects tested in Clearwater formation in Cold Lake, Canada (yellow bars of Figure 4.34) have proven to be uneconomic.

Commercial SAGD projects in McMurray formation validate the importance of the geology and reservoir characteristics for this EOR method, findings that have been reported by Rottenfusser and Ranger (2004), Putnam and Christensen (2004), and Jimenez (2008), among others. For any formation beyond 400-m depth, the nature of vertical permeability is such that the horizontal extent of the steamflood becomes more dominant leading to loss of steam in non-extractable zones. With such loss, economics of the system cannot be attractive. From technical perspective, there is a need to study the lateral extents of SAGD wells so that the fact that vertical permeability is lower than horizontal permeability can be used to the benefit of the project.

With the current level oil prices, it is anticipated that the SAGD processes will continue to expand, mainly in Athabasca's McMurray formation. More research and pilot projects should be done for implementation of SAGD to formations that are deeper than 400 m or have low vertical permeability.

Alternatives to SAGD have been proposed. As stated earlier, most alternatives involve "improvements" with chemicals that are meant to reduce mobility of the displacement phase and/or increase extraction of the oil through mixing with solvents (e.g. VAPEX, SW-SAGD, ES-SAGD). In addition, the well configuration or number of wells is also changed for some applications. As examples, one can cite X-SAGD, Fast SAGD, and single well SAGD or SW-SAGD. Well configuration should be designed based on individual formations and typically one should not adhere to a rigid set of well configurations. The use of chemicals, on the other hand, is unlikely to yield positive results because of inherent technical flaws. In addition, they are not sustainable from both economic and environmental aspects.

It is recognized that ISC is the second most important thermal recovery method. Even though, ISC has been applied in tar sand and extra-heavy oil formations, evidence has surfaced that tells us that it is most applicable to medium heavy or even light oil formations. This new evidence explains why most of the ISC pilot projects have yielded inconclusive or failed pilot results. It is increasingly being clear that heavy oil and tar sands are wrong candidates for ISC. In the last decade, ongoing ISC projects in heavy oil reservoirs such as Battrum Field in Canada, Suplacu de Barcu, Romania, Balol, Bechraji, Lanwa and Santhal in India, and Bellevue in the United States demonstrate that a much better candidate for ISC is medium heavy oil formation.

It is worth noting also that hot air injection is the first EOR scheme known to the modern petroleum industry. It is not well publicized because it was not implemented by design. The injection of air leads to ISC and every oil reservoir is a potential candidate of this application. It turns out that recently popularized HPAI is only an offshoot of the original hot air injection concept. The successful application of air injection projects in light oil reservoirs like West Hackberry in the United States demonstrate

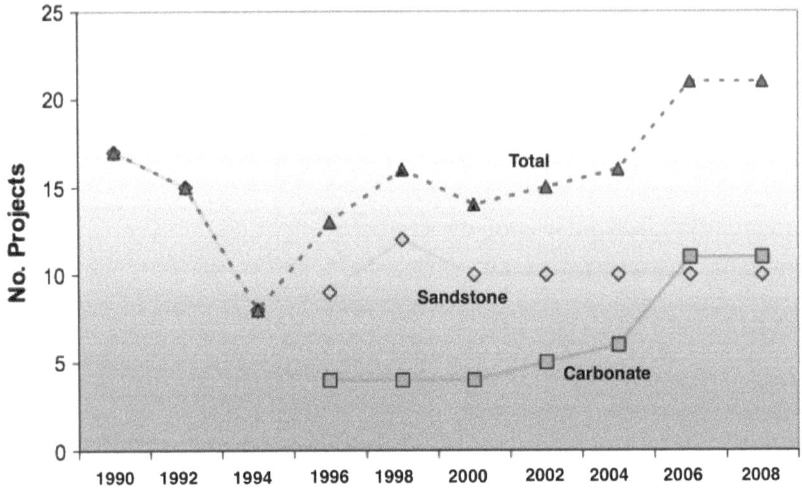

Figure 4.35 Trends in ISC and HPAI. *From Moritis, 2008. From Alvarado and Manrique (2010).*

that this recovery process is a viable EOR strategy for high dipping angle reservoirs combined with double displacement strategies. Since 2000, the number of ISC projects has been steady with 10 projects in sandstone formations, whereas the number of HPAI projects in US light oil reservoirs has shown an important increase during the same period (Figure 4.35). These HPAI projects have been implemented in carbonate formations exclusively.

ISC in light oil reservoirs does not need to be at high pressure. In fact, simple air injection can lead to the onset of the ISC, making the drive turn into an effective recovery technique. There have been several reports on such applications, such as the one reported by Duiveman et al. (2005) and Hongmin et al. (2008) on air injection projects in Handil Field, Indonesia and Hu 12 Block, Zhong Yuan Field in China, respectively. Although Handil Field HPAI pilot (0.5–1 cp oil) reported injectivity problems due to lack of reservoir communication in the pilot area, the results were reported as encouraging. Injectivity problem in this field is most likely due to reasons other than oil viscosity. During injection of air, low temperature oxidation may occur leading to precipitation of plugging agents that would not generally occur under original field conditions. In addition, air injection that does not necessarily have a fire front at the leading edge can lead to intense viscous fingering making the process extremely inefficient. This may result in low recovery. However, such instability problem cannot be alleviated with the use of chemicals, mainly because chemicals do not travel beyond a few meters in the formation. In addition, use of chemicals is inherently uneconomic and can render the process environmentally unsustainable.

In an attempt to improve air injection, foam assisted water alternating air was used in a pilot project in China (Hongmin et al., 2008). This reservoir has an oil viscosity of 3.9 cp. The results were reported to be encouraging but it is difficult to determine how much of the result can be assigned to improvements with foam. It is likely that the presence of foam did not alter the mechanism of water injection alternated with air. An improvement is expected when one combines these two techniques. Other examples can be given from Rio Preto West onshore Brazil reported by Moritis (2008) and studies reported by Hughes and Sarma (2006), Sarma and Das (2009) and Teramoto et al. (2005), and Onishi et al. (2007) evaluating technical feasibilities and potential of HPAI in Australia and Japan, respectively. All these suggest both technical feasibility and future potential of HPAI in light oil formations.

Other alternatives to ISC has been proposed as well. One alternative involves "Toe-to-heel air injection" or "THAI". It is an integrated reservoir–horizontal wells process, which uses air injection to propagate a combustion front from the toe-position to the heel of the horizontal producer. Figure 4.36 is a schematic representation of the basic features of the process. This process is meant to minimize gravity override.

The stability of the THAI process depends on two key factors: (1) a high temperature burning zone, which is more advanced in the top part of the oil layer, exhibiting controlled (stable) gas override behavior, and (2) deposition of coke, or heavy residue, inside the horizontal producer. The coke that is deposited inside the horizontal producer acts as a gas seal.

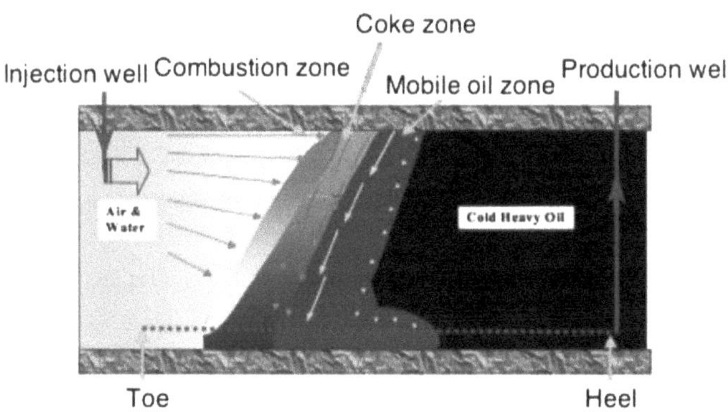

Figure 4.36 Schematic of THAI process (from Greaves and Xia, 2004).

THAI is a new, more advanced variant of the conventional ISC process, which operates as a short distance, as opposed to long-distance displacement process. This is equivalent to SAGD version of steamflooding. Due to the well arrangement used in THAI, the mobilized oil ahead of the combustion front only travels a short distance (down) to the exposed section of the horizontal producer. Since THAI operates at much higher temperatures than SAGD, it can achieve significant in situ upgrading, and thereby maximize oil recovery. THAI is currently the subject of a pilot development at Christina Lake, Canada.

Controlled atmospheric pressure resin infusion (CAPRI) is the catalytic extension of the THAI process, incorporating an annular layer of catalyst, emplaced on the outside of the perforated horizontal producer well, along its whole length. The reaction conditions created ahead of the combustion front, prior to reactants passing down through the mobile oil zone to contact the catalyst, are established by the THAI process. Further upgrading of the produced oil is achieved by catalytic conversion, as the mobilized oil passes through the catalyst layer. This process is the first step toward downhole refining. While this system works well in laboratory, field application of refining technique is economically unattractive. This would not be the case if (1) expensive catalysts are replaced with natural, yet effective catalysts; (2) the produced fluid is considered to be upgraded, thereby being assigned a higher grade at the refinery; (3) custom-designed well placement for each application, depending on formation and fluid characteristics.

Figure 4.37 shows some of the test results using THAI and CAPRI. The test was combination of dry and wet THAI/CAPRI test, in which water was injected together with the injected air (CAPRI), as tracer during the second wet combustion period. A stable, high temperature combustion front (500–600 °C) was propagated along the horizontal producer, during the dry and wet combustion periods. The excellent sweep of the combustion front, in a "toe-to-heel" manner, achieved a high oil recovery, at 87% original oil in place (OOIP). The figure shows the variation of the API gravity and viscosity for samples of the produced oil collected during the experiment. Before the combustion front reached the catalyst layer, the degree of thermal upgrading of the produced oil was only about two API points. This is very low, compared to a normal THAI test on Wolf Lake heavy oil. This is because no clay was added into the sand pack for this particular trial. The effect of catalyst on the produced oil is clearly evident in Figure 4.37. The API gravity of the produced oil

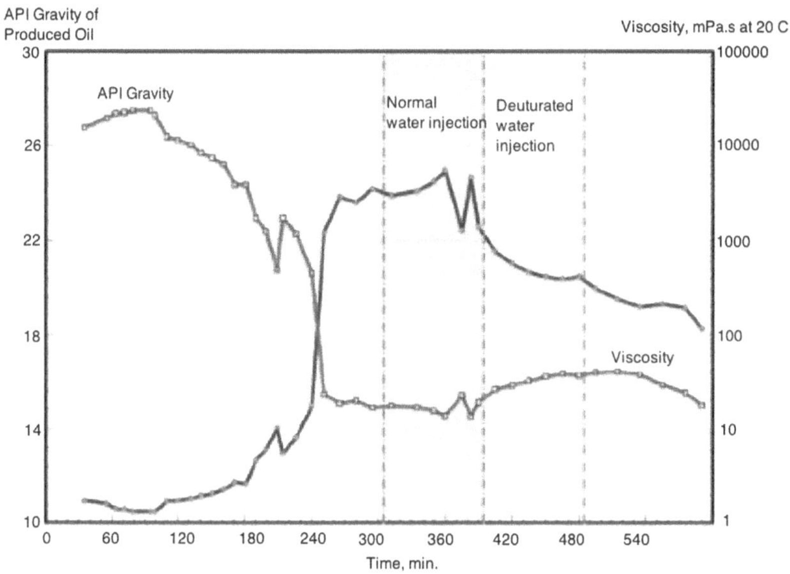

Figure 4.37 Upgrading with THAI and CAPRI (from Greaves and Xia, 2004).

jumped from an API value of 14, up to 24, during the period of 240–400 min. As the combustion front approached the catalyst section along the horizontal producer, mobilized oil, already partially upgraded in the THAI process, underwent further catalytic conversion reactions. During the second wet combustion mode, with water injection, the upgrading trend was reduced slightly from 22 °API to 20 °API. The viscosity of the produced oil achieved by CAPRI was 10–40 mPas, down from the original 24,400 mPas for Wolf Lake crude oil.

In order to seek increased efficiency, other forms of thermal EOR have been proposed. For instance, downhole steam generation (Eson, 1982; Donaldson, 1997), electric heating (Sierra et al., 2001) or electromagnetic heating (Islam et al., 1991; Das, 2008), and microwave (Hascakir et al., 2008) technologies. Some of these technologies were touted decades ago in the form of electrical heating. Later on, great promises were made with electromagnetic heating. A company, called Electromagnetic Oil Recovery was formed in 1980s in Calgary. This company implemented a number of field applications of the electromagnetic heating technology. However, none produced satisfactory result and the company went bankrupt. This technique promised to eliminate one of the bigger technical problems with conventional steam injection in regions where permafrost exists. Several options have been tried for the use of electricity to heat oil reservoirs. These methods can be classified according to the mechanism of thermal dissipation that dominates the recovery process (Pizzaro and Trevisan, 1990). They range from dielectric heating with high frequency range to radio frequency in the microwave range. Wadadar and Islam (1994) had investigated the possibility of using electromagnetic heating with horizontal wells. Islam and Chilingar (1995) reported the results of a series of numerical simulation tests based on the method originally proposed by Islam and Chakma (1990). They showed that by coupling electromagnetic heating with other EOR schemes, one can increase the recovery of heavy oil or tar sand significantly. To date, however, not a single field project has been reported to be successful on the use of this technology.

4.2.5 Chemical Methods

Chemical EOR methods lived their best times in the 1980s, most of them in sandstone reservoirs. These methods typically promise alteration of rock and/or fluid properties so that irreducible oil saturation is decreased. The total of active projects using chemical peaked in 1986 with polymer flooding as the most important chemical EOR method. However, since 1990s, oil production from chemical EOR methods has been negligible around the world except for China (Han et al., 1999; Delamaide et al., 1994; Wang et al., 2002). Nevertheless, chemical flooding has been shown to be sensitive to volatility of oil markets despite recent advances (e.g., low surfactant concentrations) and lower costs of chemical additives. Technically polymer alone does not increase oil recovery efficiency. In addition, polymer does not propagate within the formation beyond a few meters, mainly because polymer adsorption rate is very high and if polymer concentration is increased, one runs into serious injectivity problem. Polymer seems to be effective when combined with surfactants and sacrificial chemicals. However, cost of such process remains prohibitively high.

Polymer flooding has been tried for a long time and is considered to be a mature technology and still the most important EOR chemical method in sandstone reservoirs based on the review of full-field case histories. Even though all forms of chemical methods have been practically abandoned in the United States, according to the EOR survey presented by Moritis in 2008, there are ongoing pilots or large-scale polymer floods in Argentina (El Tordillo Field), Canada (Pelican Lake), China with approximately 20 projects (e.g., Daqing, Gudao, Gudong, and Karamay fields, among others), and India (Jhalora Field). It is important to mention that a commercial polymer flood was developed in North Burbank during the 1980s, demonstrating that this EOR method may still have potential to increase oil recovery in mature basins (i.e. mature floods with movable and/or bypassed oil). North Burbank reinitiated polymer flooding on a 19-well pattern in December. Other reported polymer flooding projects include Brazilian Carmopolis, Buracica, and Canto do Amaro fields. India also reports a polymer flood in Sanand Field. Oman documented a polymer flood pilot developed in Marmul Field and almost 20 years later a large-scale application is under way (Moritis, 2008). Additionally, Argentina (El Tordillo Field), Brazil (Voador offshore Field), Canada (Horsefly Lake Field) and Germany (Bochstedt Field) announced plans to implement polymer flood projects.

Colloidal dispersion gels (CDGs) and BrightWater® also represent novel polymer-based technologies that are currently under evaluation at field scale. These chemicals are meant for mobility control of a displacement drive. By injecting these mobility control agents, reservoir heterogeneities are homogenized, thereby avoiding channeling or fluid loss in unproductive areas. Documented CDGs projects include Daqing Field in China, El Tordillo and Loma Alta Sur fields in Argentina, and in multiple US oilfields. Regarding BrightWater®, at the present time Milne Point in Alaska is the only field application discussed or documented in the public domain. While it is expected that the number of CDGs and BrightWater® field applications will increase in the near future based on recent field and laboratory studies underway, none of them is expected to give positive results. If the oil price continues to be high, producers would be satisfied with the investment, irrespective of the actual benefit of the recovery scheme.

While polymer flooding has been the most applied EOR chemical method in sandstone reservoirs, the injection of alkali, surfactant, alkali-polymer (AP), surfactant-polymer (SP) and Alkaline Surfactant-Polymer (ASP) have been tested in a limited number of fields. In this application, alkali plays the role of a sacrificial agent. New genre of surfactants have been developed that reduce the dynamic IFT to a very low number (Islam and Farouq Ali, 1990; Taylor et al., 1990). These surfactants are highly unstable extremely toxic, and exuberantly costly.

As mentioned earlier, micellar polymer flooding had been the second most used EOR chemical method in light and medium crude oil reservoirs until the early 1990s. Although this recovery method was considered a promising EOR process since the 1970s, the high concentrations and cost of surfactants and co-surfactants, combined with the low oil prices during mid-1980s limited its use. The development of the ASP technology since mid-1980s and the development of the surfactant chemistry have brought up a renewed attention for chemical floods in recent years, especially to boost oil production in mature and waterflooded fields.

Several EOR chemical methods, other than polymer flood, have been extensively documented in the literature during the last two decades. However, at the present time Daqing Field represents one of the largest, if not the largest, ASP flood implemented as of today. ASP flooding has been studied and tested in Daqing for more than 15 years though several pilots of different scales. According to the EOR survey presented by Moritis in 2008, there are ongoing ASP pilots in Delaware Childers Field (Oklahoma) and planned ASP floods in Lawrence Field (Illinois) and Nowata Field (Oklahoma), and SP floods in Midland Farm Unit, Texas (Grayburg Carbonate Formation) and in Minas Field, Indonesia. The surging number of the chemical projects, however, does not tell the full story. There are many other unreported cases that are either in the plan or have been implemented but have not been reported because of lack of success. Fundamentally, ASP or any other chemical EOR technique is economically unattractive and environmentally disastrous. This can be changed only by resorting to other nonconventional sources of chemicals that are either naturally available in local areas or are liability of an operation site because it is a waste of by-product of other activities.

4.2.6 Gas Injection

Even thought CO_2 injection has been discussed in a previous section, this section is introduced in order to familiarize readers with the basic of gas injection projects. EOR gas flooding has been the most widely used recovery methods of light, condensate, and volatile oil reservoirs. Both miscible and immiscible gas injection schemes hold tremendous potential for future applications in the mainly untapped reservoirs of the world, especially in regions where the reserve/production ratios are high. However, a gas injection process can be severely flawed if the displacement front is not stabilized. It is essential to take advantage of the gravitational forces. Recently, horizontal wells have been proposed to enhance gravity segregation while maintaining stable displacement fronts.

Numerous field reports in Canada (Shell and Husky Oil) show that horizontal wells can be used to successfully conduct stable miscible displacement of both light and heavy oil. The key to success in miscible or immiscible gas injection appears to be the accurate prediction of the frontal stability, which is very sensitive to reservoir heterogeneity.

Another method of reducing viscous fingering during gas injection is the use of foam. The foam increases the viscosity of the displacing gas phase to the extent that an otherwise unstable front (with gas only) can become stable. Also, foam has a homogenizing effect when the displacement front encounters a heterogeneous spot in the reservoir. Even though no study has been reported on the topic of frontal stability in a heterogeneous medium, foam is likely to eliminate some of the problem associated with frontal instability due to heterogeneity.

Launched in 2000, the Weyburn–Midale CO_2 Project in Saskatchewan, Canada, is the world's largest full-scale, in-field study of CO_2 injection and storage in depleted oil fields. When completed, the 11-year International Energy Agency project (funded in part by DOE) will permanently store 40 million metric tons of CO_2 while increasing oil production by 18,000 bbl/day.

Although nitrogen (N_2) injection has been proposed to increase oil recoveries under miscible conditions favoring the vaporization of light fractions of light oils and condensates, today few N_2 floods are ongoing in sandstone reservoirs. Immiscible N_2 floods are reported in Hawkins Field (Texas) and Elk Hills (California) based on the Moritis EOR survey in 2008. No new N_2 floods in sandstone reservoirs have been documented in the literature during the last few years. HPAI schemes and their success tell us that N_2 injection is not an effective option and certainly not economically and environmentally sustainable.

Hydrocarbon gas injection projects in onshore sandstone reservoirs have been employed for many decades. Initially, they were used as a means of pressure maintenance in order to arrest the pressure decline that occurs in any naturally depleting reservoir (Figure 4.38). These projects made a relatively marginal contribution in terms of total oil recovered in Canada and the United States other than on the North Slope of Alaska, where large natural gas resources are available for use that do not have a transportation system to markets. Conventionally, the term "EOR" gas methods include mainly hydrocarbon gases such as water alternating gas (WAG) injection schemes, enriched gases or solvents and its combinations. Even though pressure maintenance is not considered to be an EOR technique, the process remains the same and therefore one must investigate the possibility of additional oil and gas recovery with pressure maintenance. Most of immiscible and miscible EOR hydrocarbon gas floods in the United States are on the North Slope of Alaska while in Canada a miscible gas flood is reported in Brassey Field. The situation of hydrocarbon gas injection

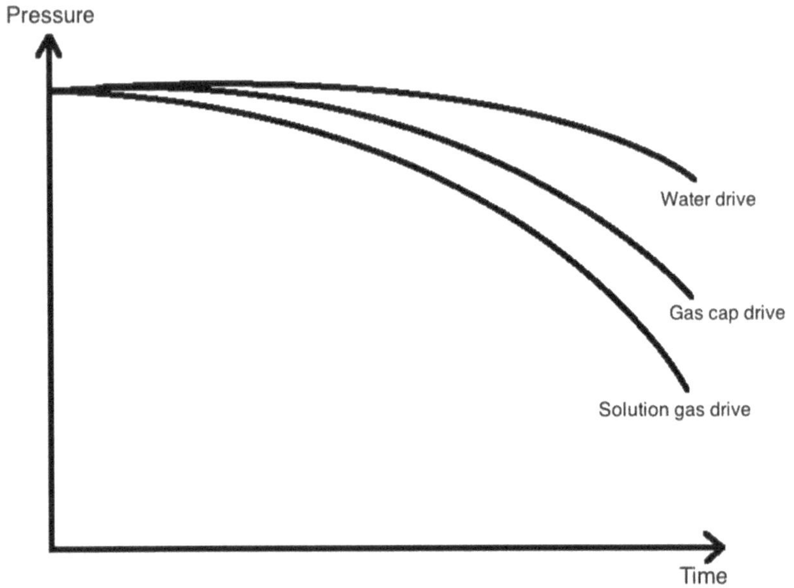

Figure 4.38 Pressure maintenance program involves artificially boosting pressure–time curve.

projects is different in offshore sandstone reservoirs. This aspect will be discussed in a later section. Conventionally, it is said that if there is no other way to monetize natural gas, then a more practical use of natural gas would be to use it in pressure maintenance projects or in WAG processes. However and if available, the substitution of hydrocarbon gases by non-hydrocarbon gases (N_2, CO_2, acid gas, air) oil recovery will make more natural gas available for domestic use or export while still maintaining reservoir pressure and increasing oil recoveries. This recommendation applies to both EOR and EGR.

On the other hand, CO_2 flooding has been the most widely used EOR recovery method for medium and light oil production in sandstone reservoirs during last decades, especially in the United States due to the availability of cheap and readily available CO_2 from natural sources. There has been an increasing trend in the number of CO_2 field projects in the United States during the last decade in both, sandstone and carbonate reservoirs.

The number of CO_2 floods is expected to continue to grow in the United States sandstone reservoirs. Some examples of planned CO_2-EOR projects in the United States include Cranfield Field, Heidelberg West (from anthropogenic sources) and Lazy Creek Field in Mississippi and

Sussex Field in Wyoming. Number of CO_2 floods in Wyoming sandstone reservoirs are also expected to increase based in a recent evaluation presented by Wo et al. (2009). This particular project depends on the CO_2 availability. In all these projects, availability of CO_2 is considered as the determining factor of EOR application.

Additionally, Holtz (2008) reported an overview of sandstone gulf coast and Louisiana CO_2-EOR projects to estimate EOR reserve growth potential in the area including sandstone reservoirs in the Gulf of Mexico. Table 4.5 summarizes major features of certain CO_2-EOR projects. Moritis EOR survey (2008) reports up to nine active immiscible CO_2 floods operating since mid-1970s. The experience of various countries represent different lessons that can be learned. For instance, Midale project in Canada uses trucked CO_2 from another Canadian field, whereas Weyburn projects imports CO_2 that is specifically generated for this project and pipelined (over 200 km) from the United States. Yet, Weyburn is the only project worldwide that has the classification of being an EOR and sequestration

Table 4.5 Selected Projects Involving CO_2 Injection

Country	Type of formation	Field name	CO_2 source	Incentive
Brazil	Sandstone	Buracica and Rio Pojuca	Anthropogenic, ammonia plant	EOR and storage
Canada	Sandstone	Pembina and Joffre	Anthropogenic	EOR
	Limestone	Weyburn	CO_2 from coal burning	EOR and storage
	Limestone	Midale	Transported truck	EOR
				PTAC EOR–EGR storage multipurpose
Croatia	Sandstone	Ivanic Field	Transported truck	EOR
Hungary	Sandstone	Budafa and Lovvaszi	Anthropogenic	EOR
	Sandstone	Szank	Sweetening plant	EOR and storage
Trinidad	Sandstone		Ammonia plant	EOR

project simultaneously. The experience of Trinidad is noteworthy. This project uses waste gas from a nearby Ammonia plant. As discussed earlier, waste gas from a chemical plant represents very high economic boon as well as environmental sustainability. Yet, this project is not considered to be a model for greenhouse gas sequestration. Canada's PTAC project represents the most comprehensive application of CO_2 projects. In terms of CO_2 EOR, Alberta projects 3.6 billion bbl additional oil recovery over the next two decades. At the same time, a significant amount of greenhouse gas would be sequestered. In addition, new industries that make use of CO_2 as a commodity would be developed.

Due to the lack of rigorous scientific investigation, the designs of EOR projects involving CO_2 or other greenhouse gases have been flawed from technical, environmental, and economical perspective. For instance, on the technical side, most flow rates selected, even with horizontal wells, are high enough to induce viscous fingering, with the only exception being the projects involving injection of gas from the top of an anticline or highly dipped formation. The problem is further accentuated for a heterogeneous formation or a formation previously flooded with water or with high water cut. Flow instability in these cases diminish or eliminate the possibility of maintaining a stable front. All laboratory experiments, however, are conducted under stable conditions, thereby, represent optimistic conditions that will not prevail in field. Any field or pilot design based on these experiments is likely to yield disappointing results (Islam et al., 2010, 2012).

Based on laboratory tests, it is often proposed that pure CO_2 be used for maintaining miscibility. Because of flawed definition of environmental integrity, it is suggested that the use of purified CO_2 and sequestration of it would be beneficial to the environment. A truly scientific criterion would indicate that processing CO_2 makes it lose its natural properties that make it a principal player of the photosynthesis process (Khan and Islam, 2007; Chhetri and Islam, 2008). As a consequence, "purified" CO_2 is toxic to the environment at the same time being costly. Ironically, the need for pure CO_2 arises from the premise that miscibility prevails in the reservoir—a premise that does not apply to most CO_2 applications. In presence of immiscible flow, there is no need for maintaining high concentration of CO_2 in the injected gas. Also, if miscibility is not achieved, there is little advantage to having pure CO_2, and even waste gas, produced gas, etc., would suffice. Pure CO_2 in high water cut zones would only result in a loss of valuable CO_2. On the other hand, the injection of greenhouse gas or produced gas would increase accessibility of the untapped oil.

The economics of any EOR project is unacceptable with expensive chemicals (e.g. pure CO_2, surfactant, polymer, alkali). While pure CO_2 is technically capable of recovering additional oil, it has been demonstrated through numerous field trials that conventional chemical floods do not yield acceptable results. This is because, these chemicals do not travel more than a few meters within the reservoir. This is the reason, chemical flooding has failed to recover any additional oil and the chemical techniques have been discontinued for several decades. Today, only China applies chemical flooding techniques and only a handful of operations have been tested in the North Sea. This is mainly because these operators produce their own chemicals.

In terms of emerging technologies, HPAI is the method with greatest potentials. This method combines the positive effects of CO_2 injection (through oxidation of in situ oil) as well as thermal methods. In case, CO_2 or other gases are not locally available, HPAI should be considered.

The following sequential screening is recommended.

1. Screen the type of fluid (gas or water) available.
2. Consider stability with both water and gas.
3. Consider miscible injection only with stable cases (in presence of natural dip for which CO_2 or other gases can be injected from a structurally higher position).
4. Consider reinjection of produced gas (including sour gas in original concentration), flue gas, and finally air, depending of availability.
5. Laboratory tests should be performed using the above fluids and not using idealized fluid.
6. Numerical simulation should be considered only for well placement, injection protocols, and similar strategic issues. One must note that reservoir simulators are incapable of modeling unstable flow.

For scaling an EOR process or to determine stability of the displacement front, the following steps are required:

1. Determine the end-point permeabilities.
2. Determine the capillary pressure curve.
3. Determine IFTs.
4. Estimate flood pattern dimensions.
5. Estimate frontal velocity from the injection well.
6. Calculate the gravity number.
7. Calculate the instability number.
8. If the instability number is less than π^2, follow the conventional approach (velocity matched with that of the field as per the scaling requirement).

9. If the instability number is greater than π^2, calculate the laboratory velocity such that the instability number in the laboratory matches with that of the field.

Khan and Islam (2007) give full details of these steps, including the definition of the instability number, capillary number, mobility ratio, etc.

It is important to note that, for gas injection, the instability number continues to affect the recovery, meaning the higher the flow rate the less recovery there will be. This is because the pseudo-stable regime is never reached with gas. Gravity plays an important role during gas injection because the value of the gravity number can stabilize a process.

During miscible displacement, the displacement front develops a transition zone that can vary in length significantly. If the crude oil in question is not light, the transition zone can be much wider. The problem with a wide transition zone is that the miscibility can be lost altogether. The lack of miscibility or the extension of the transition under any displacement situation would translate into an inadequate sweep of the reservoir, resulting in low oil recovery. The effect of the transition zone length has not been studied in the past. Inherently related to this problem is the storage or mitigation aspect of CO_2 displacement. Unless efforts are made to define the miscible/immiscible system, the performance prediction is bound to be inaccurate. Also, it is important to predict the sustenance of a miscible front. The lack of miscibility may in turn lead to the onset of viscous fingering.

4.2.7 EOR by Lithology

Reservoir lithology is one of the screening considerations for EOR methods, often limiting the applicability of specific EOR methods. Figure 4.39 shows that most EOR applications have been in sandstone formations. There are several reasons for this.

Typically, carbonate reservoirs show high recovery factors with primary and secondary modes. Similar to what happened with gas reservoirs, there are few incentives for applying EOR to carbonate reservoirs. Thermal recovery techniques are most applicable to heavy oil and tar sand reservoirs. Such reservoirs are exclusively sandstone or unconsolidated sand. The very few thermal projects applied in carbonate formations are not typical thermal operations nor are they with heavy oil. Nevertheless, thermal recovery projects are successful in carbonate reservoirs. The second most commonly used EOR technique for sandstone formations is the chemical technique. However, chemical floods are largely known as technical failures with nil

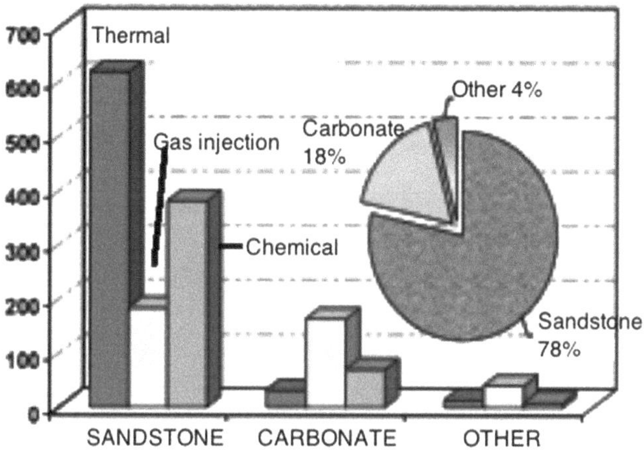

Figure 4.39 Distribution of EOR methods in various lithologies (1500 EOR projects). *From Alvarado and Manrique (2010).*

incremental recovery reported after the tax incentive for EOR was removed in 1980s. Chemical floods do work in the laboratory trials but have yielded no tangible result in the field. This demonstrates the incompetence of experimental design of scaled models. There have been fewer applications of chemical techniques in carbonate formations. However, most of them are tagged as "successful." This conclusion is based on misunderstanding the chemical process involved. Most "successful" recovery processes in chemical methods is the polymer flood technique. Polymer flood does improve oil production rates, but it is not because polymer flood contributes to the reservoir displacement process. It is mainly because polymer, while applying locally (not as drive process), reduces water cut drastically.

For sandstone reservoirs, gas injection is the least frequently used technique. However, incremental recovery is higher in gas injection than in chemical injection. Several decades were wasted in search of incremental recovery with chemical injection but with unfavorable results. In contrast, gas injection is the most commonly used in carbonate reservoirs. Among various gas injection techniques using natural gas, nitrogen and CO_2, CO_2 injection has been the most successful. In this chart pressure maintenance with gas is not included even though scientifically speaking they also constitute EOR.

4.2.7.1 EOR in Sandstone Formations

Most EOR techniques other than chemical flooding have been successful in sandstone formations. Most pilot and commercial projects have yielded favorable results. In several field applications, combination of more than one recovery technique have been implemented with success. Buracica and Carmopolis (Brazil), and Karazhanbas (Kazakhstan) are good field examples that have been subject to several EOR technologies at pilot scale in sandstone formations:

- Buracica is an onshore light oil (35 °API) reservoir with reported air injection (1978–1980), immiscible CO_2 injection (1991), and polymer flooding (1997) pilot projects. Immiscible CO_2 injection was expanded in the field using CO_2 captured from an ammonia plant. This particular project shows the usefulness of impure CO_2, which is far more economical than pure CO_2 cases.
- Carmopolis is an onshore heavy oil (22 °API) reservoir with reported ISC (1978–1989), polymer flooding (1969–1972 and 1997), steam injection (1978), and microbial EOR or microbial EOR (2002) pilot projects. The field has been developed mainly by waterflooding.
- Karazhanbas is an onshore heavy oil (19 °API) reservoir with documented polymer flooding, steam injection, ISC and ISC with foam injection as conformance strategy. Karazhanbas Field was developed by waterflooding, CHOPS (Cold Heavy Oil Production with Sands), and steam injection.

4.2.7.2 EOR in Carbonate Formations

It is well known that a considerable portion of the world's hydrocarbon endowment is in carbonate reservoirs. Some estimates put carbonate reservoirs to hold more than 60% of the world's proven oil reserves and 40% of the world's gas reserves. However, carbonate reservoirs are mostly fractured with low matrix porosity. Carbonate reservoirs usually exhibit low porosity and may be fractured. These two characteristics along with oil-to-mixed wet rock properties usually result in lowered hydrocarbon recovery rates. When EOR strategies are pursued, the injected fluids will likely flow through the fracture network and bypass the oil in the rock matrix. The high permeability in the fracture network and the low equivalent porous volume frequently result in early breakthrough of the injected fluids. This can be effectively resolved by proper characterization of a carbonate reservoir. The following features must be considered:

- predominant strike and dip of fractures,
- fracture density,
- fracture connectivity (extent of secondary cementation)
- horizontal and vertical variation of fractures.

A large number of EOR field projects in carbonate reservoirs have been referenced in the literature during the last decades. Traditionally gas injection has been the most commonly used EOR technique for carbonate reservoirs. This is not to say that this is the most suitable EOR technique. It turns out other methods (even chemical) works equally well. In terms of chemical methods, polymer travels more easily in carbonate formations, making it easier to control water cut. This is the reason many consider chemical methods to be effective in carbonate reservoirs. Similarly, thermal flooding has been used very rarely in carbonate reservoirs, mainly because the target reservoirs of thermal methods are heavy oil that is non-existent in carbonate reservoirs. However, when HPAI, a simpler version of ISC, was used in carbonate reservoirs, it gave positive results, as discussed earlier.

In contrast with sandstone reservoirs, there are few fields where different EOR technologies have been evaluated successfully at pilot scale demonstrating technical applicability of different EOR methods in carbonate formations. Yates Field (Texas) is a good example of a carbonate formation where different EOR processes have been tested successfully at different scales (from pilots to large-scale applications). Some of the EOR processes evaluated in Yates Field that have been documented in the literature include:

- Nitrogen (N_2) injection began in the mid-1980s as a reservoir pressure maintenance strategy.
- Steamflooding pilot was initiated in the late 1998 as a potential strategy to improve vertical gravity drainage process.
- Dilute surfactant well stimulation pilot test was reported in the early 1990s as a strategy to increase oil recovery by IFT reduction, gravity segregation of oil, and wettability alteration, among others mechanisms.
- In March 2004 Yates Field started replacing N_2 injection with CO_2 injection as a pressure maintenance strategy and enhanced gravity drainage.

Manrique et al. (2007) presented a comprehensive review of EOR field experiences in U.S. carbonate reservoirs. The same data were published by NETL of the DoE (2004). They summarized various EOR applications in carbonate reservoirs of the United States in the Table 4.6.

Not surprisingly, CO_2 injection has been the most successful of EOR operations in carbonate reservoirs. From the current US active CO_2 floods, 67% (48 projects) are in carbonate reservoirs, mostly located in the state of Texas. Table 4.7 shows some of the CO_2 floods employed in the United States. In the past, projects were custom designed based on selected fluids (e.g. purified CO_2, nitrogen, rich hydrocarbon gas). However, after 1990s,

Table 4.6 EOR in Various Carbonate Reservoirs in the United States

Location	Field	Pay zone/ reservoir	Formation	φ (%)	K (mD)	Depth (ft)	Gravity (°API)	Oil Viscosity (cp)	Temperature (°F)
Kansas	Hall-Gurney	LKCC	Limestone	25.0	85.0	2900.0	39.6	3.0	99.0
Michigan	Dover 36	Silurian-Niagaran	Limestone/ Dolomite	7.0	5.0	5500.0	41.0	0.8	108.0
Michigan	Dover 33	Silurian-Niagaran	Limestone/ Dolomite	7.1	10.0	5400.0	43.0	0.8	108.0
New Mexico	Maljamar	Grayburg/San Andres	Dolomite/ Sandstone	10.2	18.0	4000.0	36.0	1.0	90.0
New Mexico	East Vacuum	San Andres	Dolomite	11.7	11.0	4400.0	38.0	1.0	101.0
New Mexico	Vacuum	San Andres	Dolomite	12.0	22.0	4550.0	38.0	1.0	101.0
New Mexico	North Hobbs	San Andres	Dolomite	15.0	13.0	4200.0	35.0	0.9	102.0
North Dakota	Little Knife	Mission Canyon	Dolomite	18.0	22.0	9800.0	43.0	0.2	240.0
Texas	Anton Irish	Clearfork	Dolomite	7.0	5.0	5900.0	28.0	3.0	115.0
Texas	Bennet Ranch Unit	San Andres	Dolomite	10,0	7.0	5200.0	33.0	1.0	105.0
Texas	Cedar Lake	San Andres	Dolomite	14.0	5.0	4700.0	32.0	2.0	103.0
Texas	Adair San Andres Unit	San Andres	Dolomite	15.0	8.0	4852.0	35.0	1.0	98.0
Texas	Seminole San Andres Unit	San Andres	Dolomite	13.0	20.0	5100.0	34.0	1.2	101.0
Texas	Seminole Unit, ROZ Phase 1	San Andres	Dolomite	12.0	62.0	5500.0	35.0	1.0	104.0
Texas	Levelland	San Andres	Dolomite	12.0	3.8	4900.0	30.0	2.3	105.0
Texas	North Cowden	Grayburg/San Andres	Dolomite	12.0	5.0	4300.0	34.0	1.6	94.0
Texas	Wasson (ODC Unit)	San Andres	Limestone	9.0	5.0	5100.0	32.0	1.3	110.0

Texas	Slaughter (HT Boyd Lease)	San Andres	Dolomite	10.0	4.0	5000.0	31.0	–	108.0
Texas	Slaughter (Central Mallet)	San Andres	Limestone/ Dolomite	10.8	2.0	4900.0	31.0	1.4	105.0
Texas	Slaughter Estate Unit (SEU)	San Andres	Dolomite	10.5	4.3	5000.0	28.0	1.7	105.0
Texas	Slaughter Frazier	San Andres	Limestone/ Dolomite	10.0	4.0	4950.0	31.0	1.4	105.0
Texas	Wasson-Willard	San Andres	Dolomite	10.0	1.5	5100.0	32.0	2.0	105.0
Texas	University Waddell	Devonian	Dolomite	12.0	14.4	8500.0	43.0	0.5	140.0
Texas	McElroy	San Andres	Dolomite	11.6	1.5	3850.0	31.0	2.3	86.0
Texas	Goldsmith	San Andres	Dolomite	10.0	10.0	4200.0	32.0	1.2	94.0
Texas	Kelly Snyder (SACROC Unit)	Canyon Reef	Limestone	9.4	19.4	6700.0	41.0	0.4	130.0
Texas	South Welch	San Andres	Limestone	9.3	9.0	4890.0	34.0	2.2	96.0
Texas	Huntley	San Andres	Dolomite	16.0	5.0	3180.0	33.0	2.5	104.0
Texas	South Cowden	San Andres	Carbonate	13.0	3.0	4100.0	35.0	1.0	100.0
Texas	Wasson (Cornell Unit)	San Andres	Dolomite	8.6	2.0	4500.0	33.0	1.0	106.0
Texas	Wasson	San Andres	Dolomite	13.0	6.0	5100.0	33.0	1.0	110.0
Texas	GMK South	San Andres	Dolomite	10.0	3.0	5400.0	30.0	3.0	101.0
Texas	Slaughter	San Andres	Dolomite	10.0	3.0	5000.0	32.0	2.0	107.0
Texas	Slaughter (East Mallet)	San Andres	Dolomite	12.5	6.0	4900.0	32.0	1.0	110.0
Texas	Sharon Ridge	Canyon Reef	Limestone	10.0	150.0	6600.0	40.0	1.0	125.0
Texas	Means (San Andres)	San Andres	Dolomite	9.0	20.0	4300.0	29.0	6.0	97.0
Texas	Salt Creek	Canyon	Limestone	20.0	12.0	6300.0	39.0	1.0	125.0
Texas	Hanford	San Andres	Dolomite	10.5	4.0	5500.0	32.0	1.4	104.0
Texas	Hanford East	San Andres	Dolomite	10.0	4.0	5500.0	32.0	1.0	106.0
Texas	West Brahaney Unit	San Andres	Dolomite	10.0	2.0	5300.0	33.0	2.0	108.0
Texas	East Penwell (SA) unit	San Andres	Dolomite	10.0	4.0	4000.0	34.0	2.0	86.0

Continued

Table 4.6 EOR in Various Carbonate Reservoirs in the United States—cont'd

Location	Field	Pay zone/ reservoir	Formation	φ (%)	K (mD)	Depth (ft)	Gravity (°API)	Oil Viscosity (cp)	Temperature (°F)
Texas	Garza	San Andres	Carbonate	18.0	5.0	3000.0	36.0	3.0	80.0
Texas	Welch (North & south)	San Andres	Dolomite	11.0	4.0	4900.0	34.0	2.0	98.0
Texas	Crossett	Devonian	Limestone	22.0	5.0	5300.0	44.0	0.4	106.0
Texas	Wasson (Bennett Ranch)	San Andres	Carbonate	13.0	10.0	4900.0	32.0	1.4	107.0
Texas	Wasson (Denver Unit)	San Andres	Dolomite	12.0	5.0	5200.0	33.0	1.3	105.0
Texas	Wasson South	Clearfork	Carbonate	6.0	2.0	6700.0	33.0	1.2	105.0
Texas	Reinecke	Cisco Canyon Reef	Limestone/ Dolomite	10.4	170.0	6700.0	43.5	0.4	139.0
Texas	Slaughter Sundown (SSU)	San Andres	Dolomite	11.0	6.0	4950.0	33.0	1.0	105.0
Texas	Mabee	San Andres	Dolomite	9.0	4.0	4700.0	32.0	2.3	104.0
Texas	Wellman	Wolfcamp	Limestone	9.2	100.0	9800.0	43.5	0.5	151.0
Texas	Dollarhide (Clearfork Unit)	Clearfork	Dolomite	11.5	4.0	6500.0	40.0	–	113.0
Texas	Dollarhide (Devonian Unit)	Devonian	Dolomite	13.5	17.0	8000.0	39.5	0.4	125.0
Texas	Sable	San Andres	Dolomite	8.4	2.0	5200.0	32.0	1.0	107.0
Texas	Cogdell	Canyon Reef	Limestone	13.0	6.0	6800.0	40.0	0.7	130.0
Texas	T-Star (Slaughter Consolidated)	Abo	Dolomite	7.0	2.0	7850.0	28.0	1.9	134.0
Utah	Aneth	Ismay Desert Creek	Limestone	14.0	5.0	5600.0	41.0	1.0	125.0
Utah	Greater Aneth Area	Desert Creek	Limestone	12.0	18.3	5700.0	42.0	1.5	129.0
West Virginia	Hilly Upland	Greenbrier	Limestone/ Dolomite	14.0	3.0	1950.0	42.0	1.7	77.0

Table 4.7 In Situ Combustion Projects in Carbonate Reservoirs of the United States

Location	Field	Pay zone/ reservoir	Formation	φ (%)	K (mD)	Depth (ft)	Gravity (°API)	Oil viscosity (cp)	Temperature (°F)
North Dakota	Horse Creek	Red River	Dolomite	16.0	20.0	9500.0	32.0	1.4	198.0
North Dakota	Medicine Pole Hills	Red River B & C	Dolomite	18.9	15.0	9500.0	38.0	1.0	230.0
North Dakota	West Medicine Pole Unit	Red River B & C	Dolomite	17.0	10.0	9500.0	33.0	2.0	215.0
North Dakota	Cedar Hills North Unit	Red River	Dolomite	16.0	6.0	8300.0	30.0	2.9	200.0
South Dakota	Buffalo	Red River B	Dolomite	20.0	10.0	8450.0	31.0	2.0	215.0
South Dakota	West Buffalo	Red River B	Dolomite	20.0	10.0	8450.0	32.0	2.0	215.0
South Dakota	South Buffalo	Red River B	Dolomite	20.0	10.0	8450.0	31.0	2.0	215.0

Manrique et al., 2004, 2007.

most projects are selected based on available fluids. Even though not formally recognized, US EOR projects that are commercially successful employed this criterion. Canadian Weyburn project is the first one that was employed in recent time based on selected fluid. This was possible because the project was tagged as an experimental project for determining feasibility of CO_2 sequestration during EOR. By contrast, US projects are tied to the availability of natural sources of CO_2 and CO_2-transporting pipelines relatively close to the oilfields under this recovery method. One prime example is the Permian Basin. The Permian Basin is the largest consumer of CO_2, mostly through a vast network of pipelines (also CO_2 trucks). The majority of the CO_2 consumed in the West Texas and New Mexico Permian Basin are from commercial natural reservoirs in Colorado (the McElmo Dome and the Sheep Mountain fields), New Mexico (the Bravo Dome region) and Wyoming (La Barge Field). Similarly, successful Canadian projects have employed local CO_2 or at least trucked local CO_2. For instance, very successful Midale project trucks one-third of the total supply from Lyodminster in Canada at a cost of $70/ton.

After the oil price collapse in 1990s, it has become mandatory to minimize all costs of EOR fluids. That developed a corporate attitude of using local gas and possibly use waste gas. For instance, reports indicate that CO_2 floods in West Texas can be economically attractive at oil prices of $ 18/bbl assuming that CO_2 prices remains less than $ 1/Mscf. Even though the rising oil prices have made any CO_2 attractive, one must not give up on the scientific knowledge gained during the low-oil price era that showed that low-quality CO_2 with proper design can still be technically appealing, thereby, shifting the economics of the overall project.

As mentioned previously, CO_2 projects have become even more appealing because of the incentive related to greenhouse gas mitigation and sequestration. Even though the science of sequestration is flawed, it is reasonable to consider CO_2 sequestration as an added bonus to CO_2 projects. In addition, environmental concerns add to the argument that purified CO_2 is not warranted nor is it economically prudent to do so. Latest scientific investigation shows purification of CO_2 actually adds to the environmental footprint and defeats the purpose of the CO_2 sequestration. With increasing benefits of environmental greening, the economics of CO_2 looks the most attractive. Figure 4.40 shows an example of total costs and potential incomes of carbon capture storage projects with EOR (CO_2-EOR storage).

The economics (costs and revenues) involved in a carbon capture storage project can be broken down in many ways. However, it depends on

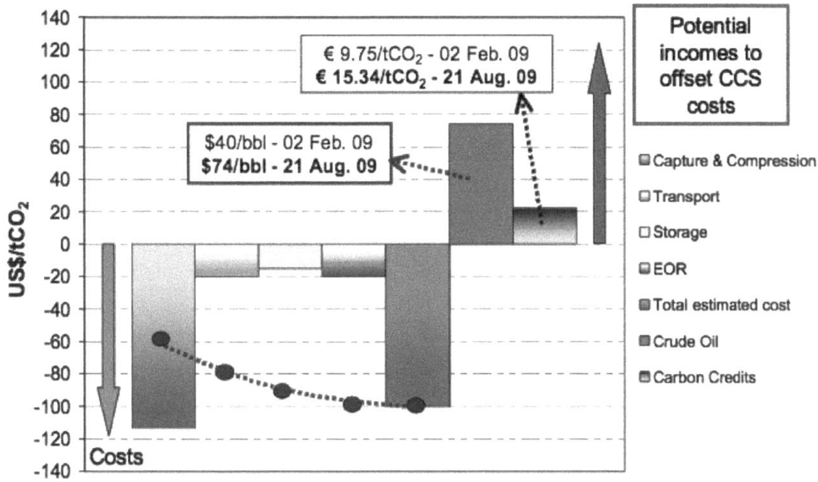

Figure 4.40 Potential benefits of coupling CO_2 EOR with storage. CCS, carbon capture storage.

the source of CO_2 (e.g., petrochemical plants vs coal-fired power plants) and where it will be injected (e.g., EOR vs saline aquifer). Assuming a scenario of CO_2 capture from a coal-fired power plant, we can divide the main economic variables into four categories: CO_2 capture and compression, CO_2 transportation, CO_2 storage (including wells and monitoring), and possible revenues (e.g., oil recovery and/or carbon credits), depending on the application (CO_2 EOR vs saline aquifers). Cost shown in Figure 4.40 represents average costs obtained from a comprehensive literature review completed in 2008. However, it is important to recall that costs such as transportation and compression of CO_2 will vary depending on the distance between the emission source (e.g., power plant) to the geologic sink (e.g., oil reservoir or saline aquifer) and its depth. In all, the economics become very attractive if waste gas is used.

Overall, CO_2 injection holds great promises from all aspects, namely economical, technical, and environmental. CO_2 works because it combines the benefit of chemical flooding while avoiding high chemical adsorption, costs, and toxicity associated with chemical floods with the benefit of thermal flood while avoiding high costs of thermal methods.

ISC is the oldest thermal recovery method. It has been used since 1920s with many successes and failures. It was not until 1990s that one recognized that ISC is least applicable to very heavy and tar sand reservoirs, the conventional targets of the technology. It is now well recognized that ISC is more successful in lighter and carbonate reservoirs. Recently, ISC has resurfaced as HPAI, which is exclusively applied to light oil reservoirs, particularly suitable for carbonate formations. HPAI has been very successful and holds great promises in the future. At present, there are seven active air injection projects in the United States, six of them in light crude oil (>30°API) carbonate reservoirs in North and South Dakota. Horse Creek, South and West Buffalo, and Medicine Pole Hill are good examples of combustion projects in light crude oil carbonate reservoirs (Table 4.7). The success and expansion of Buffalo's and Medicine Pole Hill in North and South Dakota demonstrates the feasibility of air injection in carbonate reservoirs to improve oil recovery and revitalize mature and waterflooded fields. They confirm previous observations in ISC that neither pure oxygen nor induced combustion is necessary for the success of ISC. These projects also give us an incentive to inject waste gas along with air whenever there is a need to dispose of waste gas. These projects also demonstrate the uselessness of high pressure nitrogen injection that gained popularity in the 1990s. Those projects were abandoned because of the poor economics but HPAI shows that the injection of nitrogen is not technologically attractive either.

Currently, air injection is considered an alternative for offshore and onshore mature fields with no access to CO_2 sources, specially mature fields in the Gulf of Mexico given the limitation of space available in platforms and also because CO_2 injection from onshore power generation plants and industrial sources would probably not be economic in the short term. An additional benefit of air injection projects is the generation of flue gases for pressure maintenance that also can be reinjected in the same or reservoirs close by. Production results of recent air injection projects in North and South Dakota (Williston Basin) may dictate the future of this recovery method in carbonate reservoirs in the United States.

In summary, ISC or HPAI are capable of recovering additional oil because of the following facts:

1. It combines the benefit of thermal with no investment involving steam generation;
2. It produces flue gas that can induce the effect of chemical injection, similar to CO_2 injection;
3. At high pressure, it induces the benefits of miscible flood;
4. Has the lowest investment cost involved in fluid purification and/or completion renovation.

Neither cyclic nor continuous steam injection has been widely used in carbonate reservoirs. The Garland Field in Wyoming and Yates Field in

Texas represent two of the few steam driven projects in carbonate formations documented in the United States. Some of the steam injection projects documented in carbonate formations outside Canada and the United States include:

- Steam drive pilot at Lacq Superieur Field, France.
- Steamflood Pilot at Ikiztepe Field, heavy oil fracture reservoir in Turkey.
- Cyclic steam pilot in Cao-32 Field, fracture limestone heavy crude oil in China.
- Steamflood pilot and full-field implementation in Qarn Alam Field, Oman.
- Cyclic steam injection pilot in Issaran heavy oil field, Egypt.
- Steamflood pilot at the giant Wafra Field, Kuwait—Saudi Arabia

Injection of nitrogen, especially under miscible conditions, had been in use since 1960s through 1990s. Because of the miscibility conditions, nitrogen injection saw applications only in deep light oil reservoirs. On the other hand, nitrogen has been used also under immiscible conditions, particularly for the purpose of pressure maintenance. During the last 40 years over 30 nitrogen injection projects have been developed in the United States, some of them in carbonate reservoirs in Alabama, Florida, and Texas (Table 4.8). At the present time there are only two active nitrogen injection projects in carbonate reservoirs in the United States, the WAG in Jay Little Escambia (N_2-WAG) and as a pressure maintenance project in Yates Field (Table 4.8). In the case of the N_2-WAG in Jay Little Escambia, this is a mature project started in 1982, while N_2 at Yates started in mid-1980s as a reservoir pressure maintenance strategy.

Recent scientific discoveries as well as the environmental impetus of CO_2 has removed the focus of EOR from nitrogen toward CO_2. For all practical purposes, N_2 injection is an obsolete technique. Far better efficiency and economics as well as environmental integrity is achieved by injecting CO_2 or using HPAI. For instance, Yates field of Texas has replaced nitrogen injection with immiscible CO_2 injection with expected better recovery than nitrogen.

Both miscible and immiscible hydrocarbon gas injection schemes have been used in carbonate reservoirs. Recent Oil and Gas Journal survey indicates that all eight active hydrocarbon miscible reported projects are in sandstone reservoirs, six of them in Alaska. Table 4.9 shows eight hydrocarbon injection projects developed in US carbonate reservoirs between early 1960s to mid-1980s. Historically, natural gas injection has not been

Table 4.8 Miscible and Immiscible Nitrogen Floods (Continuous or WAG) in the United States Carbonate Reservoirs

Location	Field	Pay zone/ reservoir	Formation	φ %	K (mD)	Depth (ft)	Gravity (°API)	Oil viscosity (cp)	Temperature (°F)
Texas	Block 31	Devonian	Limestone	12.0	5.0	8600.0	46.0	0.3	130.0
Alabama	Chunchula Fieldwide Unit	Smackover	Dolomite	12.4	10.0	18500.0	54.0	0.0	325.0
Florida	Blackjack Creek	Smackover	Carbonate	17.0	105.0	16,150.0	50.0	0.3	290.0
Texas	Andector	Ellenburger	Dolomite	3.8	2000.0	8835.0	44.0	0.6	132.0
Fla./ Alab.	Jay Little Escambia Creek	Smackover	Limestone	14.0	35.0	15,400.0	51.0	0.2	285.0
Texas	Yates	Grayburg/ San Andres	Dolomite	17.0	175.0	1400.0	30.0	6.0	82.0

Table 4.9 Hydrocarbon Injection Projects in Carbonate Reservoirs of the United States

Location	Field	Pay zone/ reservoir	Formation	φ (%)	K (mD)	Depth (ft)	Gravity (°API)	Oil viscosity (cp)	Temperature (°F)
Alabama	Chatom	Smackover Lime	Dolomite	22.0	12.0	15,900.0	54.0	—	293.0
North Dakota	Carlson	Madison	Limestone	11.0	0.1	8500.0	42.0	11.0	135
North Dakota	Red Wing Creek	Mission Canyon	Limestone	10.0	0.1	9000.0	40.0	—	241.0
Texas	Levelland	San Andres	Dolomite	10.2	2.0	4900.0	30.0	2.3	105.0
Texas	Slaughter	San Andres	Dolomite	10.5	4.3	5000.0	28.0	1.9	105.0
Texas	McElroy	San Andres	Dolomite	11.6	1.5	3856.0	31.0	2.3	86.0
Texas	Fairway	James	Limestone	12.6	11.0	9900.0	48.0	—	260.0
Texas	Wolfcamp University Block 9	Wolfcamp	Limestone	10.2	14.0	8400.0	38.0	0.3	140.0

economic unless there is no other way to monetize the gas and it has to be flared. Even then, it is far better to utilize the gas in a power plant to generate electricity and capture the flue gas to reinject. Because a power plant can vary in size and capacity, this option should be examined before reinjecting gas into the reservoir. This has become even more true considering the soaring price of gas ever since the gas crisis in Europe in 2008.

Steam injection has been exclusively used in heavy oil reservoirs. However, in 1990s, scientific investigation made it clear that steam has great potentials in light oil formations. However, due to low oil prices as well as relative success of simpler schemes (e.g., HPAI, CO_2 injection), only few steam injection schemes saw application in light oil reservoirs. For instance, for continuous steam injection in carbonate reservoirs, only two projects are currently active in Garland and Yates fields (Table 4.10). The Garland Field (Big Horn Basin, Wyoming) steam drive was developed in the Madison limestone formation, while the Yates (Grayburg/San Andres) steamflood project has been one of several EOR projects tested in this Texas giant field.

As it was mentioned above, polymer flooding has been the most used EOR chemical method in both sandstone and carbonate reservoirs. To date, more than 290 polymer field projects have been referenced or reported in the literature. The number of polymer floods in United States peaked in 1986 with 178 active projects. Ever since the repealing of tax rebates of EOR, the number of polymer as well as chemical projects has fallen sharply. Most of the polymer floods used water-soluble polyacrylamides and biopolymers (polysaccharides and cellulose polymers) to a lesser degree. Polymer concentration of as little as 50 ppm and as high as 3.7% was reported. Related incremental oil recovery was reported to be anything from 0% to 18%. Most successful operations involved mobility control and high water cut oil reservoirs. Table 4.11 shows some of the chemical floods that have been developed in US carbonate reservoirs during the period between 1960 and 1990. Following is a summary of various chemical flooding projects. They are described here along with the lessons learnt.

4.2.7.2.1 Eliasville Caddo Unit

The Eliasville field was discovered in 1920. The field produces from the Caddo limestone at 3250–3350 ft. having approximately 40 ft. of net pay. Eliasville Caddo Unit (ECU) has a paraffinic light crude oil (39 °API). The

Table 4.10 Examples of Steamfloods in carbonate reservoirs of the United States

Location	Field	Pay zone/ reservoir	Formation	φ (%)	K (mD)	Depth (ft)	Gravity (°API)	Oil viscosity (cp)	Temperature (°F)
Texas	Yates	Grayburg/San Andres	Dolomite	17.0	175.0	1400.0	30.0	6.0	82.0
Wyoming	Garland	Madison	Limestone/ Dolomite	15.5	10.0	4250.0	22.0	29.0	140.0

Table 4.11 Chemical Floods in Carbonate Reservoirs of the United States

Location	Field	Pay zone/ reservoir	Formation	φ (%)	K (mD)	Depth (ft)	Gravity (°API)	Oil viscosity (cp)	Temperature (°F)
Arkansas	Wesgum[a]	Smackover	Limestone	26.7	36.0		21.0	11.0	185.0
Illinois	Tonti	Renoist Auxvases McClusky	Limestone	47.3	358.0	2050.0	39.5	4.0	83.0
Kansas	Trapp	Lansing/Kansas City	Limestone	18.4	150.1	3215.0	38.0	1.4	97.0
Kansas	Bates Unit	Mississippi	Limestone	15.5	19.7	3700.0	42.0	0.6	117.0
Kansas	Harmony Hill	Lansing/Kansas City	Limestone	12.5		3130.0	38.6	3.7	105.0
Louisiana	Old Lisbon	Pettit	Carbonate	16.0	45.0	5300.0	34.9	2.5	178.0
Nebraska	Dry Creek	Lansing/Kansas City	Limestone	13.0		4100.0	31.0	9.0	120.0
New Mexico	Vacuum	San Andres	Dolomite	10.6	21.0	4700.0	37.0	1.5	100.0
New Mexico	Vacuum	Grayburg/San Andres	Dolomite	11.5	17.3	4500.0	37.0	1.2	101.0
New Mexico	Vacuum	San Andres	Dolomite	11.6	8.5	4720.0	38.0	1.5	105.0
North Dakota	Blue Buttes	Madison	Limestone	9.6	22.0	9400.0	42.0	0.3	240.0
Oklahoma	Fitts	Viola	Limestone	13.6	18.5	3900.0	39.0	3.2	119.0
Oklahoma	Fitts (E. Fittts Unit)	Cromwell 60, Hunton, Viola	Limestone/ Sandstone	17.5	6.6	3250.0	40.0	4.0	115.0
Oklahoma	Balko South	Kansas City	Limestone	21.0	535.0	6100.0	40.0	1.8	125.0
Oklahoma	Fitts	Cromwell, Viola, Hunton	Carbonate	17.5	750.5	3250.0	40.0	4.0	115.0
Oklahoma	Stanley	Burbank	Carbonate	18.0	300.0	3000.0	39.0		105.0
Oklahoma	Osage–Hominy	Miss. Chat	Limestone	30.0	27.0	2880.0	38.7	3.0	100.0
Texas	C-Bar	San Andres	Dolomite	10.0	6.0	3350.0	36.0	5.0	107.0

Texas	Dune	San Andres	Dolomite	14.0	28.0	3350.0	32.0	3.5	95.0
Texas	Goldsmith 5600	Clearfork	Dolomite	15.0	28.0	5600.0	32.0	3.5	100.0
Texas	McElroy	Grayburg	Dolomite	13.0	37.0	2800.0	32.0	2.7	88.0
Texas	Garza	San Andres	Limestone	19.8	4.1	2900.0	36.0	2.5	90.0
Texas	Westbrook	Clearfork	Dolomite	7.4	6.3	3000.0	26.0	9.1	90.0
Texas	Lucy N.	Pennsylvanian	Limestone	9.7	30.0	7640.0	40.0	0.4	140.0
Texas	Salt Creek	Canyon Reef	Limestone	12.0	13.2	6300.0	39.2	0.9	129.0
Texas	Stephens County Regular	Caddo (ECU)	Limestone	13.2	9.0	3200.0	39.0	2.7	113.0
Texas	Slaughter	San Andres	Dolomite	11.2	6.0	5000.0	31.0	1.5	110.0
Texas	S. Robertson	Glorieta/Clearfork	Dolomite	7.9	38.6	5800.0	34.0	1.0	107.0
Texas	Cogdell	Canyon Reef	Limestone	9.6	5.0	6800.0	41.7	0.6	128.0
Texas	Levelland	San Andres	Dolomite	10.0	0.6	4720.0	30.5	1.5	107.0
Texas	Cowden North	Grayburg/San Andres	Lime./Dolomite	10.1	3.6	4450.0	34.0	1.6	94.0
Texas	Mabee	San Andres	Dolomite	10.5	1.6	4700.0	32.0	2.4	106.0
Texas	Jordan	San Andres	Dolomite	10.5	6.0	3600.0	34.0	2.8	95.0
Texas	McElroy	Grayburg/San Andres	Dolomite	11.0	5.0	3000.0	32.0	2.6	95.0
Texas	Penwell	San Andres	Dolomite	11.0	2.2	3800.0	32.0	4.5	108.0
Texas	Herris	Glorieta	Dolomite	8.6	3.0	5818.0	30.8	3.1	115.0
Texas	Dollarhide (Clearfork)	Clearfork	Dolomite	11.6	8.5	6500.0	37.0	0.6	110.0
Texas	South Cowden	Grayburg	Dolomite	13.0	3.1	4500.0	34.0	3.5	103.0
Texas	Smyer	Clearfork	Carbonate	8.3	10.5	5900.0	27.0	5.0	112.0
Texas	North Riley	Clearfork	Carbonate	7.7	12.0	6300.0	32.0	2.6	104.0
Texas	Salt Creek	Canyon Reef	Limestone	12.0	20.0	6500.0	39.0	6.0	130.0

Continued

Table 4.11 Chemical Floods in Carbonate Reservoirs of the United States—cont'd

Location	Field	Pay zone/ reservoir	Formation	φ (%)	K (mD)	Depth (ft)	Gravity (°API)	Oil viscosity (cp)	Temperature (°F)
Texas	Headlee North	Headlee North	Limestone	4.1	0.3	12000.0	47.0	0.7	190.0
Texas	Foster	San Andres	Dolomite	12.0	5.8	4200.0	34.0	1.2	101.0
Texas	Wichita County Reg.[a]	Gunsight	Carbonate	22.0	53.0	1750.0	42.0	2.2	89.0
Texas	Stephens County Regular	Caddo Lime	Limestone	14.5	12.0	3200.0	40.0	2.3	106.0
Texas	Bob Slaughter Block[a]	San Andres	Dolomite	12.0	5.9	5000.0	31.4	1.3	109.0
Texas	Robertson	Clearfork	Carbonate	7.8	2.0	6450.0	34.0	1.1	110.0
Texas	Sand Hills	Tubb	Carbonate	12.0	27.0	4500.0	35.0	2.5	148.0
Texas	McCamey	Grayburg-San Andres	Carbonate	14.0	18.0	2100.0	26.0	28.0	80.0
Texas	Keystone	Colby	Dolomite	12.0	5.0	3300.0	37.0	6.0	87.0
Utah	Aneth Unit	Paradox	Limestone	10.6	18.3	5300.0	47.0	0.6	134.0
Wyoming	Byron	Embar/Tensleep	Limestone/ Sand.	13.9	41.3	5600.0	23.0	17.0	121.0
Wyoming	Grass Creek	Phosphoria	Carbonate	21.6	20.0	4300.0	24.0	15.0	105.0
Wyoming	Oregon Basin	North Embar	Limestone	20.2	68.0	3370.0	22.5	9.8	108.0
Wyoming	Oregon Basin	South Embar	Limestone	19.5	39.0	3600.0	20.9	15.7	110.0

[a]Micellar polymer flood.
After Manrique et al., 2004.

main reservoir properties are shown in Table 4.11. The waterflood started in 1966 with poor results. A large polymer flood (16 well patterns, 57 producers) was proposed and started in December 1980. The polymer used was a hydrolyzed polyacrylamide (HPAM) to viscosify a fresh injection water (1200 mg/l TDS, total dissolved solids) in a reservoir with a salty connate water (165,000 mg/l TDS). The polymer was injected over a period of 34 months (December 1980 to November 1983). The viscosity of the injected polymer solution was reduced from 40 cp at the beginning to 5 cp at the end of the injection. A total of 12.9% PV (pore volume) polymer slug (30 million lbs) was injected having a good production response. Oil production increased from 375 BOPD (barrels of oil per day) (October 1981) to an all unit high of 1622 BOPD (August 1984). The cumulative HPAM injected (54 lb/acre-ft) at ECU was greater than most US polymer injection projects. Finally, polymer retention by the limestone reservoir was 50 lb/acre-ft and an estimate of 0.46 bbl of incremental oil per pound of polymer injected were reported. This value is markedly less in limestone than in sandstone.

4.2.7.2.2 Byron Field

Embar-Tensleep oil was discovered in Byron (Big Horn Basin, Wyoming) in 1929. The Embar and Tensleep reservoirs are limestone and sandstones formations, respectively. The waterflood operation began in 1974. In December 1982, the polymer flood started as a strategy to improve waterflood sweep efficiency. The project area covered 1500 acres with 36 injectors and 47 producers. The polymer flood was concentrated in Tensleep, where most of the oil field reserves are. Successful operations, however, were mainly in carbonate formations. The Embar formation is a limestone/dolomite reservoir with an average pay zone of 22 ft with a crude oil of 23 °API. Major reservoir properties are listed in Table 4.11. The polymer flood at the Byron field considered a tapered sequence of three slugs of 10% PV each starting with partially hydrolyzed polyacrylamide (PHPA) solutions of 1000 ppm, 600 ppm and 330 ppm, followed by the drive water. The polymer flood ended on December 1, 1985 after the injection of 0.37 PV of polymer. The project has significantly improved oil recovery measured by total field production and water–oil ratio.

4.2.7.2.3 Vacuum Field

The Vacuum (Grayburg, San Andres) Field (New Mexico) was discovered in 1924. Production on the Phillips' Hale and Mable leases started in 1939.

Grayburg is a dolomitic formation with an average net pay of 148 ft for the 320 acres of both leases. Water injection was initiated in May 1983 and polymer injection started three months later (August 1983). The Hale-Mable leases are one of three polymer floods developed at the Vacuum Field. Particularly PHPA polymer solution started in late August 1983. Polymer solutions were prepared with fresh water (387 ppm TDS) produced from the Ogallala formation. Although the injection was to be performed increasing polymer concentration from 50 ppm to 200 ppm, the polymer slug was kept at 50 ppm due to an underestimated injectivity reduction (from 13,000 to 10,000 BWPD). The original plan considered the injection of 15% PV of a 200-ppm polymer solution (676,000 lbs of active polymer) over a period of two years. However, the project was developed considering a total injection rate of 10,000 BWPD (12 injectors) of a polymer solution of 50 ppm until the end of 1984 (16 months). During this period of time, production peaked and remained almost constant at 3500 BOPD (14 producers) and 1985 production started to decline and an increase of water production was reported. The polymer floods at the Hale and Mable leases were declared as successful projects in terms of increasing ultimate oil recovery. Finally, a polymer retention/absorption of 94.5 lbs/acre-ft was reported based on laboratory experiments. This value is much less than conventional estimate of field trials.

Micellar polymer flooding, also known as SP flooding, has been the second most used EOR chemical method in light and medium crude oil reservoirs in the United States up until the early 1990s. However, reported field projects are relatively low in comparison with polymer floods. Until 1990 at least 30 field micellar polymer floods have been referenced or reported in the literature. Although this recovery method was considered as a promising EOR process since the 1970s, the high concentrations and cost of surfactants and co-surfactants, combined with the low oil prices during mid-1980s limited its use.

In most of the field cases reviewed the type of surfactants used in micellar polymer floods were petroleum sulfonates and synthetic alkyl sulfonates, which usually requires the use of co-surfactants (non-ionic surfactants) or co-solvents, mostly alcohols. Additionally, to reduce potential surfactant-formation brine incompatibilities and potentially reduce chemical adsorption in some cases a preflush of fresh water was required. Water-soluble polyacrylamides have been the most common polymer used in these projects with a few cases using biopolymers. Although some projects reported significant oil recoveries (Loudon, Big Muddy, Henry West, and Bingham), oil recoveries were less than expected. This is because scaling of chemical processes are very difficult and laboratory models show only greatly exaggerated oil production and greatly marginalized surfactant retention. Such difficulty arises from improper time scaling of chemical flood experiments.

Few of numerous chemical injection projects have used micellar polymer flooding (Table 4.11). They are Wesgum Field (Arkansas), Wichita County Regular and Bob Slaughter Block in Texas. The Bob Slaughter Block Lease (BSBL) is a San Andres dolomite reservoir. This lease is under production since late 1930s with waterflooding operation starting in the 1960s. The BSBL reservoir is at a depth of 5000 ft and has a reservoir temperature of 109 °F. The reservoir thickness is about 100 ft and contains a crude oil of 31 °API. The first surfactant pilot test reported in this reservoir was in 1974 and, based on those results, two micellar polymer pilots were developed in the early 1980s. Micellar polymer formulations were based on petroleum sulfonates. Water injection at the first well-pair test (86 ft well spacing) started in April 1981. Surfactant injection commenced on August 26 and consisted in an emulsion formulation containing a mixture of petroleum sulfonates and an alkylaryl ether sulfate as a solubilizer. A total of 12,846 bbl of surfactant was injected in a period of 171 days (February 1982). The surfactant slug was followed by the biopolymer slug (1000 ppm) dissolved in fresh water from the Ogallala formation. The polymer injection finished on July 16 (5840 bbl) continuing with the injection of fresh water until November 8, 1983 when the injection was switched to field brine. The pilot reported high recovery efficiency (77%) with a low retention of surfactant and polymer. About 65% and 55% of surfactant and polymer were recovered, respectively. With regard to the second well-pair pilot test (101 ft well spacing), brine injection began on April 21, 1981. The injection of the non-emulsion surfactant system started at the end of July 1982. The surfactant formulation consisted of a mixture of petroleum sulfonates and an alkyl ether sulfate solubilizer. The surfactant injection ended in late September after injecting 5058 bbl over 61 days at an average of 83 bbl/day. The surfactant slug was immediately followed by the polymer injection (1,000 ppm) for 45 days at an average injection rate of 72 bbl/day. Fresh water injection continued after the end of the polymer slug until November 1983, switching to the injection of field brine. Although oil recovery efficiency (43%) was lower than the previous well-pair pilot test, results were considered promising. Surfactant and polymer retention were also low, 41% and 58% of the chemical additives were recovered, respectively.

ASP combines the key mechanisms from each of the EOR chemical methods. Generally, ASP formulations use moderate pH chemicals such as sodium bicarbonate ($NaHCO_3$) or sodium bicarbonate (Na_2CO_3) rather than sodium hydroxide (NaOH) or sodium silicates. Main functions of alkaline additives are to promote crude oil emulsification and increase ionic strength decreasing IFT and regulating phase behavior. The alkaline additives also help to reduce the adsorption of anionic chemical additives by increasing the negative charge density of mineral rocks and at the same time making the rock more water-wet. Thus, the use of alkaline agents contributes to reduce the surfactant concentrations making ASP formulations less costly than conventional micellar formulations. With regard to the surfactants; the most common products that have been used are petroleum sulfonates. The main function of the surfactants is to reduce IFT between the oil and the injected aqueous formulation. The injected surfactants may sometimes form mixed micelles (at the oil–water interface) with in situ natural surfactants, broadening the alkali concentration range for minimum IFT. On the other hand, the polymer (usually polyacrylamides) is used to reduce water mobility and sweep efficiency by increasing the solution's viscosity and decreasing effective solution permeability when it is adsorbed onto the formation.

ASP flooding is an oil recovery method that has traditionally been applied to sandstone reservoirs and until now no field tests in US carbonate reservoirs have been reported in the literature. Not surprisingly, ASP has been a failure in the field trials despite numerous laboratory tests that showed it to be effective. Once again, the problem lies within improper scaling of chemical flooding (Taylor et al., 1990; Islam and Farouq Ali, 1990).

Surfactant injection is the only chemical method used recently as a well stimulation and wettability modification of carbonate reservoirs. In fractured reservoirs, spontaneous water imbibitions can occur from the rock matrix into fractures. Subsequently, this mechanism leads to oil drainage from the matrix towards the fracture network, making surfactants attractive to improve oil recovery in oil-wet carbonate reservoirs by changing rock wettability (to mixed/water-wet) and promoting the imbibition process.

Although surfactant or micellar flooding field projects in carbonate reservoirs are not currently reported in the United States, surfactant injection has been tested in carbonate reservoirs as chemical stimulation methods (Huff & Puff) in the Cottonwood Creek and Yates fields. The Yates Field (Texas) was discovered in 1962. The Yates San Andres reservoir is a naturally fractured dolomite formation and several EOR methods have been evaluated in this prolific field with a cumulative production over 1.3 billion bbl of a 30 ° API crude oil. San Andres is a 400-ft thick formation with average matrix porosity and permeability of 15% and 100 mD, respectively (Table 4.12). Marathon Oil Co. started dilute surfactant well stimulation pilot tests in the early 1990s. Surfactant slugs were injected into the oil water transition zone considering single and multi-well injection strategies. Once the surfactant slug was injected the well was shut in (soak time) for a brief period of time. The well was returned to production increasing the recovery of oil mainly due to the reduction of IFT, gravity segregation of oil, and water between the fractures and the matrix, and wettability alteration, although to a lesser extent. The surfactant used in Yates pilots was a non-ionic ethoxy alcohol. The surfactant solutions injected were prepared with produced water in high concentrations (3100–3880 ppm), well above the critical micelle concentration (CMC). Field results were reported as economically encouraging. As an example, the average oil production rate for one of the pilot wells increased from 35 to 67 bbl/day with an incremental of 17,000 bbl.

Another example is the Cotton Creek Field. It is located in the Bighorn Basin of Wyoming. Cottonwood Creek is a dolomitic class II reservoir. Class II reservoirs have low matrix porosity and permeability. The matrix provides some storage capacity and the fractures provide the fluid flow pathways. Typically, these types of reservoirs produce less than 10% of the OOIP by primary recovery and exhibit low additional recovery factors during waterflooding. Cottonwood Creek produces from the dolomitic Phosphoria formation. Reservoir thickness varies from 20 to 100 ft and average porosity and permeability of 10% and 16 mD, respectively (Table 4.12). The reservoir produces a sour 27 °API crude oil.

Continental Resources Incorporated started, in August 1999, single well surfactant stimulation treatments at Cottonwood Creek. Well treatments consider the injection of 500 to 1500 bbl of a surfactant solution slug depending on the perforated interval. Typically the injection period takes 3 days and the shut-in period (soak time) about a week. Surfactant solutions were prepared using the non-ionic poly-oxyethylene alcohol at a concentration of 750 ppm, almost twice the CMC. Initial well treatments considered an acid cleanup with HCl (15%) to remove iron sulfide (FeS) from the wellbore to avoid or reduce surfactant adsorption. However, production results were below expectations. The initial results led to the elimination of the acid pretreatment and the increase of the surfactant

Table 4.12 Examples of Chemical Floods in Carbonate Reservoirs of the United States (during 1990–2000)

Location	Field	Pay zone/ reservoir	Formation	φ (%)	K (mD)	Depth (ft)	Gravity (°API)	Oil viscosity (cp)	Temperature (°F)
Texas	Yates	San Andres	Dolomite	15.0	100.0	1400.0	30.0	6.0	82.0
Wyoming	Cottonwood Creek	Phosphoria	Limestone	10.4	16.0	7900.0	30.0	2.8	150.0

concentration to 1500 ppm (to allow for potential losses by adsorption to FeS) in subsequent surfactant stimulations. The response has been mixed. Wherever successful, the success has been attributed to alteration of wettability.

Overall, chemical flooding holds promises in carbonate reservoirs. However, for a project to be technically, environmentally, and economically sustainable, synthetic chemicals must be avoided. If injection fluid selection is based on waste chemicals or readily available chemicals, rather than synthetic chemicals, the overall project becomes attractive. Unlike sandstone formations, carbonate formations are less vulnerable to high rate of adsorption. However, injectivity is low in carbonate formation and most injected fluid will travel through fractures. With careful selection of cheap injection fluids, one can design a chemical flood project to maximize invasion of unflooded zones. Because certain chemicals can alter the wettability, there is significant possibility of reducing rock wettability. However, such change cannot be invoked with economic benefits if the chemical in question is purchased. As discussed before, waste chemicals or naturally occurring chemicals hold the only promise of a successful chemical flooding scheme.

4.2.8 Offshore EOR

The situation in offshore fields represents a different and very challenging situation. Major offshore discoveries concentrate at water depths beyond 1200 m. The latter, of course, narrows down the number of possible EOR alternatives, mainly focusing on reservoir management optimization choices, combined with well architectures such as horizontal or highly deviated wells to yield maximum return from those fields. Only recently, EOR projects are being considered for offshore applications.

Since waterflooding is the main offshore activity in Brazil, water management becomes an important issue. By 2002, 12 fields were under water injection, mainly for pressure maintenance, while for other seven, water injection plans were underway. From the 888,000 BWPD injected by 2005, and 330,000 BWPD produced, it was expected that roughly 3,145,000 BWPD will be injected by 2006. Several problems are associated with water injection. Lost of injectivity is an impairing problem because of the subsea wells completions that tend to dominate production schemes in the Brazilian offshore. Intervention for stimulation purposes in injection wells becomes rather expensive, requiring floating rigs. Several alternatives to alleviate water management problems have arisen. Open-hole completion in water injectors has been successful, but remedial activities for water diversion are then difficult. Raw water injection (or lower quality water) above the fracture threshold has been put forth as a serious option, and Petrobras PRAVAP program has dedicated efforts in this direction (PRAVAP is Petrobras corporative technology program that covers all aspects of the EOR activity, including monitoring programs such as time-lapse seismic). Another issue, now for producer wells, is inorganic salt deposition, mainly Barium and Strontium sulfates.

Offshore heavy oil reservoirs (API gravity lower than 19 and/or oil viscosity greater than 10 cp at reservoir conditions) are a challenge for operators in Brazil. The current proposed alternative would be cold production through high productivity wells (long-reach extended horizontal wells), plus associated water injection. Well-planned well architectures to delay/minimize water production and increase sweep efficiency are being proposed to make exploitation feasible. However, no more than 20% recovery factor is expected at present. Some of these projects are described below.

Albacora is one of the offshore giant fields, contains an estimated STOIIP (stock tank oil initially in place) of 4.4 billion bbl (by 1989, time of the development plan) at water depths ranging from 230 m up to 1900 m. The field was expected to develop in three phases for a peak production of 288,000 BOPD from 188 completed wells. The idea being to prepare exploitation phases for successively deeper water, as technology development and learning curves required progress. This is the tendency in offshore operations in the Campos Basin, because water depths grow substantially reaching ultradeep waters in some of the new discoveries.

Seven oil reservoirs were detected: Namorado (typical Cretaceous turbiditic sandstone in Campos Basin), and Eocene, Oligocene 1, Oligocene 2, Oligocene 3, Oligo-Miocene and Miocene, Tertiary. Namorado is a representative turbidite in the Brazilian offshore, like Brent for the North Sea, present in many of the Campos Basin reservoirs. At the time (1989), the field represented 10% of Brazilian STOOIP and 15% of Campos Basin STOOIP.

Phase I comprised the production of six exploratory wells, at water depths ranging from between 252 and 419 m. At the time, oil and gas production reached 33,000 BOPD and 430 m^3/day, respectively. Phase II would add 95 wells, completed in Namorado, Eocene and Oligocene 1, 2, and 3 reservoirs, and a few in Miocene and Oligo-Miocene units, to gather information. Thirty nine injectors will be activated. This phase was divided in two steps. First, all possible alternatives and economical screening was used. The remaining cases were then optimized.

The challenge for developing offshore deep water oil fields led to a paradigm that differs from the 1970s and 1980s view, when closely spaced vertical wells was the way to go. This is something that characterizes most of the onshore developments and early cases in the offshore, such as Namorado Field, in relatively shallow waters. In the new scenario, with a need to reduce the number of wells, largely spaced or multilateral wells are required. Another important challenge was to describe some of the internal heterogeneities of turbiditic reservoirs. Although facies can be described from cores, scarce information is available for interwell areas. One feature of Campos Basin turbidities is the lack of outcrops that would facilitate the finding of analogues.

To illustrate the internal complexity of some offshore reservoirs that may have an impact on EOR activities, one can cite the example of the turbiditic reservoir Albacora, identified with the Namorado sandstone, a Cretaceous in age sediment. A relatively detailed stratigraphic description of Namorado sequences is reported. From the point of view of dynamics, calcite cementation (1–53 vol%) is the most important porosity and permeability control of the Namorado sandstone, which range $1.8\% < f < 32\%$ and $0.1 < K < 1624$ mD.

At this time, more than 50% of the Brazilian oil production came from the offshore fields at water depths over 1000 m. By 2001, Marlim Field was producing 85,000 m^3/day (535,000 BOPD), from 60 producers and 32 water injectors. This medium oil field presents excellent petrophysical properties and good vertical communication. The oil is sub-saturated, with viscosity values between 4 and 8 cp. This combination of properties makes water pressure maintenance an efficient process. However, paraffin deposition in flowlines represents a problem for Marlim Field. To sustain the roughly 540,000 BOPD, 640,000 BWPD were being injected. The recovery factor by 2001 reached 7.2%.

Pressure maintenance is carried out by injecting seawater in an alternate line drive pattern in the oil leg. Due to good reservoir properties and high vertical permeability, water injection tends to be stable, and hence efficient in the Marlim complex. Although water injection is the only EOR process used in the Marlim Complex, polymer injection and WAG have been studied alternatives, but not implemented.

Several EOR projects are underway in offshore North Sea. They will be discussed under field case studies. It is interesting to advance the idea that CO_2 flooding has been given serious consideration in offshore environments. An example of a detailed study is associated with the Forties Field.

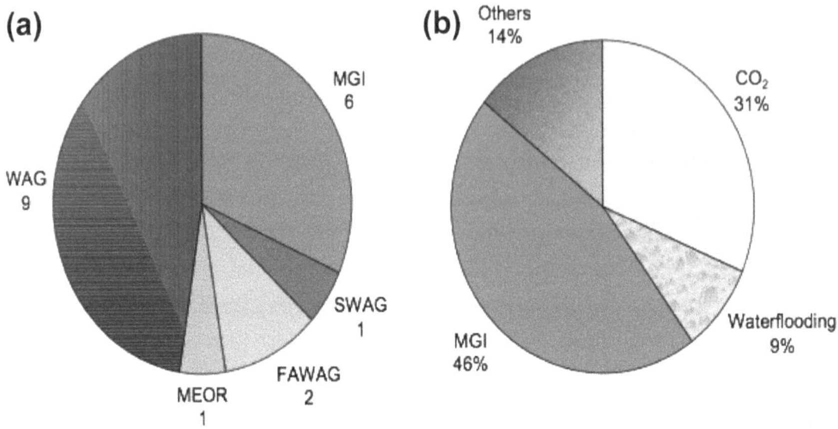

Figure 4.41 Examples of EOR in offshore fields: (a) North Sea EOR Projects *(From Awan et al., 2008)* and, (b) EOR Opportunities Offshore Malaysia *(From Samsudin et al., 2005)*.

The field is located 170 km northeast of Aberdeen. Significant incremental could, in principle, be obtained in a process like this. Internal complexities of turbiditic reservoirs must be resolved in order to mitigate uncertainties in this depositional environments.

Figure 4.41 shows the statistics of oil recovery projects in the North Sea as well as EOR opportunities in offshore Malaysia. In this figure, SWAG stands for simultaneous water alternating gas injection whereas FAWAG stands for foam assisted-WAG.

The main task so far surrounds the placement of the injection well. Figure 4.42 shows one such configuration.

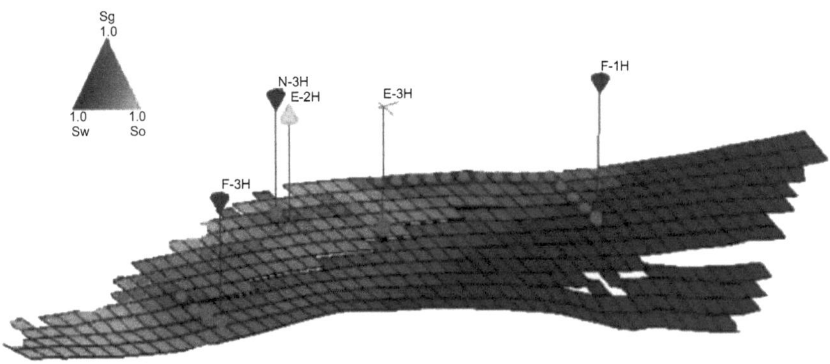

Figure 4.42 Placement of injection well strategically is the most important task in offshore developments.

After placement, proper strategies involve selection of SWAG or WAG. From Figures 4.43a and 4.43b, it can be seen that case C had better performance in terms of residual oil recovery for both SWAG and WAG. The main difference in implementing the two techniques is related to the pressure profile of the reservoir. When SWAG injection is considered, the pressure profile for the case C lies between 250 and 290 bar, instead WAG injection pressure profile decrease rapidly with time, achieving the

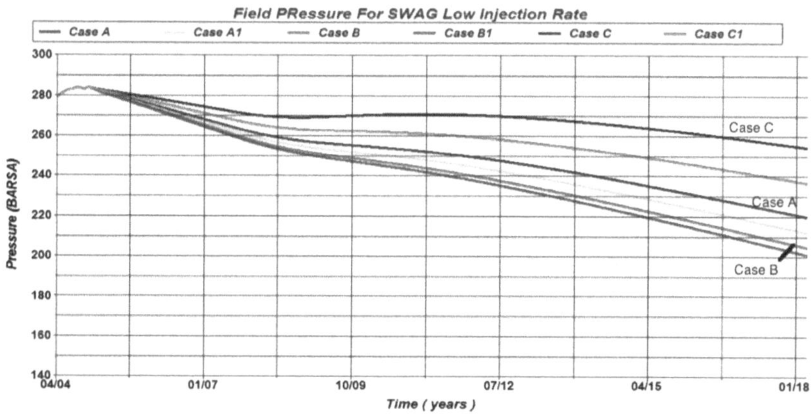

Figure 4.43a Field pressure of SWAG with low injection rate. *From Nangacovie, 2009.*

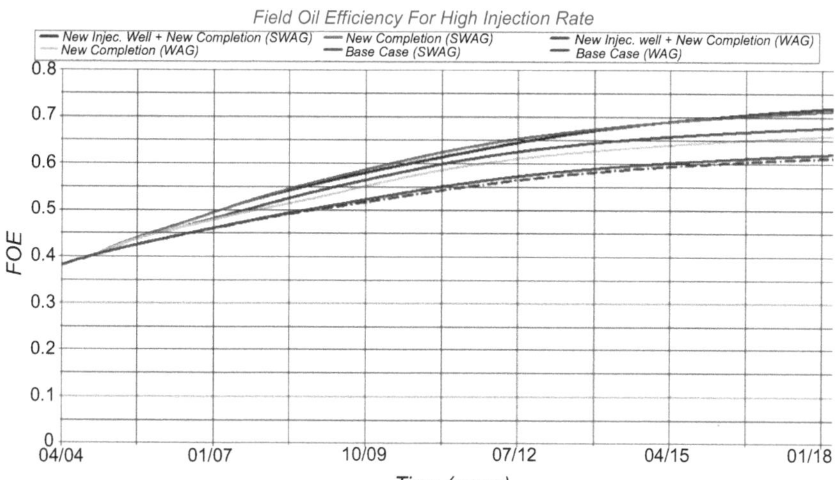

Figure 4.43b Comparison between WAG and SWAG in North Sea formation. *From Nangacovie, 2009.*

minimum pressure value of 220 bar approximately for case C and the minimum pressure values for cases different from case C are below 220 bar.

Figure 4.44 shows oil production in the U.S. Gulf of Mexico in shallow and deep waters since 1990 and the forecast until the year 2013. Gulf of Mexico oil production is mainly supported by water and/or gas injection.

Mexico is another example of offshore reservoir production supported by gas injection, bottom water drive, and/or water injection. Regarding gas injection, it is important to note that Cantarell/Akal represents the largest N_2 injection project in the world.

Similar to previous examples, most offshore environments are under continuous optimization strategies of gas and waterflooding to extend field production life and maximize oil recovery.

4.2.9 Microbial Enhanced Oil Recovery

Microbial enhanced oil recovery (MEOR) has received a great deal of attention over last few decades. However, the field application in the United States has not been very successful, and very little oil production has been attributed to MEOR. In this regard, Russia has implemented several MEOR projects with more success. Canada has reported the use of microbes for plugging water-producing zones (Jack et al., 1991). Despite extensive background laboratory work, the project was a failure because of

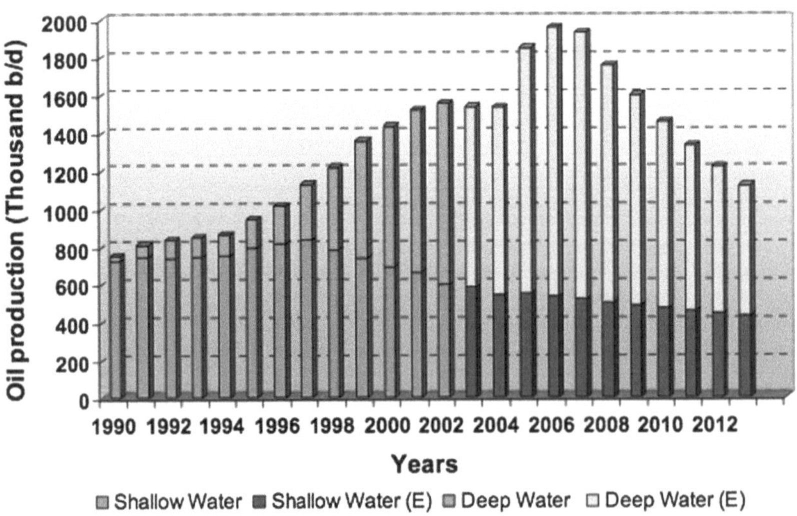

Figure 4.44 Historic and forecast oil production rates of U.S. Gulf of Mexico shallow and deep water projects. E = Estimated. *From MMS, 2005.*

scaling up problems. It was not later that it was discovered that linear model results could not be used directly to design a field project. It was found that the same strain of bacteria reduces permeability in a linear core significantly while making practically no difference for a three-dimensional system. Microbial EOR has enjoyed significant scientific research and it is time that some of the engineering problems associated with the scheme be addressed before field implementation.

One specific aspect of MEOR appears to be unexplored—the possibility of using thermophilic bacteria along with hot water injection. Hot water alone can reduce oil viscosity significantly (Islam et al., 1992). If thermophilic bacteria are added, additional improvement can be caused due to decomposition of inorganic carbonates, evolution of viscosity-reducing gases, and IFT reduction with surface active bio-agents. Al-Maghrabi et al. (1999) have conducted some preliminary studies with this scheme. This study also opens up prospects of using bacteria in combination with other EOR techniques, such as surfactant-water flooding and others. In the combined process, it is important to ensure that the selected surfactant is compatible with the bacteria to be used (Sundaram et al., 1994). It is possible that the strain of bacteria will eventually use the surfactant as a nutrient that can add synergy to the system.

4.3 CARBON SEQUESTRATION ENHANCED GAS RECOVERY

It was discussed earlier that EGR has great potentials, despite being ignored as a commercial project. It is also presented that Alberta's CO_2 backbone template includes EGR as part of overall CO_2 sequestration and oil and gas recovery. This section presents further discussion on EGR.

It has been recognized for decades that depleted natural gas reservoirs are promising targets for carbon dioxide sequestration. However, the same reservoirs are not devoid of methane. In addition, the possibility of using CO_2 sequestration to EGR should be investigated.

The carbon sequestration enhanced gas recovery (CSEGR) process consists of collecting CO_2, for example, by scrubbing CO_2 from flue gases at fossil-fueled power plants or collecting by-product CO_2 from refineries, pressurizing the CO_2 to supercritical conditions for transport in a pipeline, transporting the CO_2 to a depleted natural gas reservoir, injecting the CO_2 into the reservoir, and enhancing the production of CH_4 from the reservoir. After some period of enhanced CH_4 recovery, the production wells would be sealed and the reservoir would be filled with CO_2 up to initial reservoir pressure. The injected CO_2 would then be sequestered in the gas reservoir just as CH_4 was stored over geologic time prior to its production as an energy resource. A schematic illustrating the CSEGR process for a gas-fired power plant is shown in Figure 4.45.

Even though depleted natural gas reservoir is cited above, the technique equally applies to active gas reservoirs. The use of a power plant is a quick utilization natural gas. Depending on the size of the reservoir or the overall flow rate, the utilization scheme can be any other potential use, such as fertilizer, cement factory, etc. The idea is to maximize the value addition of the natural gas.

The flue gas typically is processed in order to capture high-quality CO_2. Once again, purification of CO_2 does not need to be carried out as most reservoirs are amenable to reinjection of low-quality CO_2 that will offer similar benefit as pure CO_2 but without having to spend more on CO_2 purification. As discussed earlier, purification with expensive and toxic solvents actually increases the footprint of the process, thereby, defeating the purpose of CSEGR.

During the injection process, following factors must be considered:

1. injectivity of CO_2 in a gas reservoir;
2. effects of CO_2 injection pressure on injectivity and flow;
3. cooling around the injection well due to phase change and Joule-Thomson effects;
4. flow of CO_2 within the reservoir;
5. mixing of CO_2 and CH_4 in the reservoir; and
6. repressurization and production of CH_4.

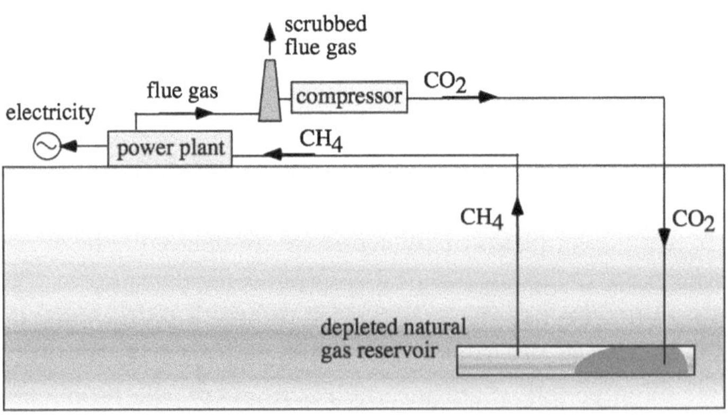

Figure 4.45 Various components of CSEGR.

Conventionally, CO_2 is injected under supercritical conditions. This has implication on the design of compressor. The supercritical condition is also subject to strong cooling at the injection port due to (1) flashing of supercritical liquid-like CO_2 to gas, and (2) Joule-Thomson cooling as the CO_2 gas expands in the low pressure reservoir. In addition, the injected CO_2 dries the formation, another potential heat consuming process. Given that the formation and residual gas and liquid are at somewhat elevated temperature ($T > 40\ ^\circ C$), heat will be available for the expanding gas. Eventually however, the temperature around the well may become quite low leading to the possibility of hydrate formation and associated decreases in injectivity. Pure carbon dioxide hydrate can form at approximately 0 °C at 20 bar pressure (Haneda et al., 2000).

Assuming there is sufficient permeability, the injected CO_2 should flow in the reservoir due to pressure gradient and gravitational effects. If there is liquid CO_2 immediately around the wellbore, it will flow strongly downward through the gas reservoir due to its large density. Such gravity segregation must be accounted for during considerations of the well placement in a structurally inclined formation. It is preferable that high pressure CO_2 be injected through a structurally lower well so that gravity stabilization of the front takes place. Once flashed to gas, CO_2 is also notably denser than CH_4 at all relevant pressures (see Figure 4.46) and will tend to flow downwards, displacing the native CH_4 gas and repressurizing the reservoir. Because CO_2 gas is more viscous than CH_4 (see Figure 4.47), the displacement will be stable.

The reservoir processes of CO_2 injection and enhanced CH_4 production are shown schematically in Figure 4.48. As observed in the figure, CO_2 injection can deflect the water table, giving rise to repressurization at a large distance from the injection well. CO_2 can be used effectively to minimize water coning. Selection of injection well should be made in such a way that water coning is minimized while gas mobilization and displacement are maximized. Heterogeneity in the formation may lead to preferential flow paths for the injected CO_2. This phenomenon may be favorable for injectivity and carbon sequestration in that it allows greater amounts of CO_2 to be injected. However, preferential flow may lead to early breakthrough and is therefore detrimental to EGR. Furthermore, the development of larger gas composition gradients and subsequent mixing by molecular diffusion is enhanced by preferential flow.

Heterogeneity in the gas reservoir plays a dual role during CO_2 injection. CO_2 gains access to high permeability zones, thereby increasing

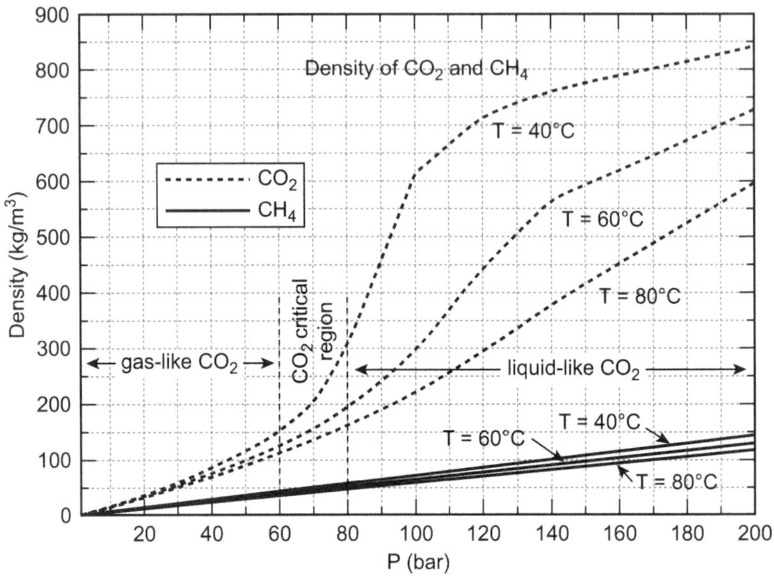

Figure 4.46 Density of CO_2 and CH_4 as a function of pressure for various temperatures.

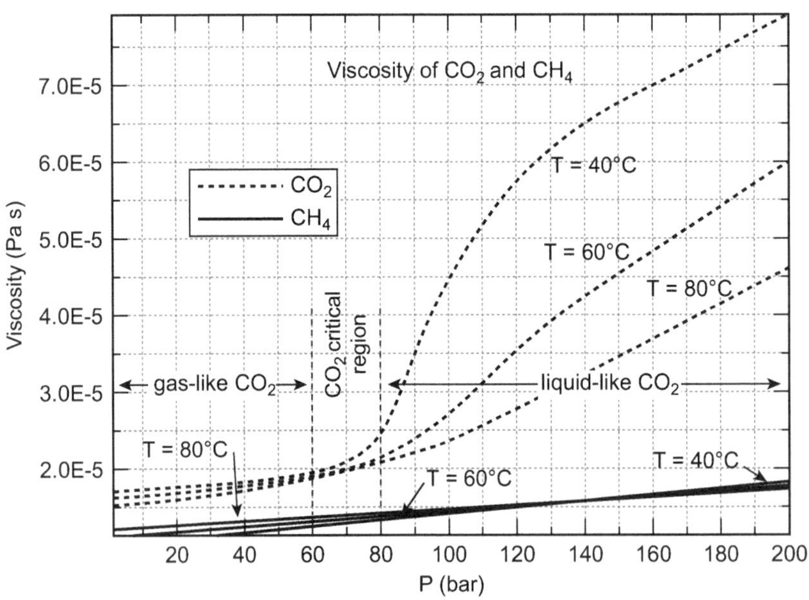

Figure 4.47 Viscosity of CO_2 and CH_4 as a function of pressure for various temperatures.

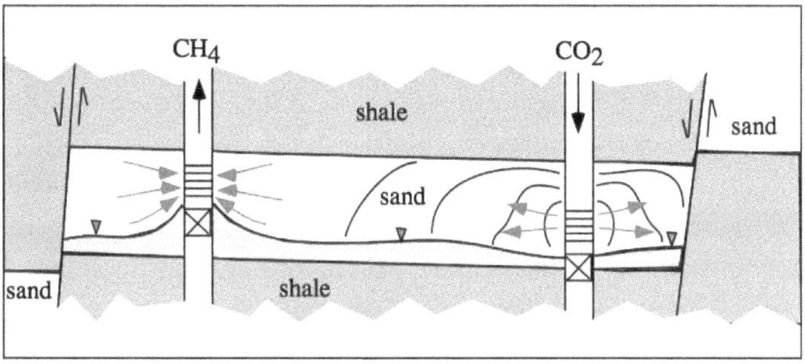

Figure 4.48 Crosssection of the CSEGR site.

injectivity. On the other hand, heterogeneity may lead to channeling or fingering, which would lead to early breakthrough. This case can be alleviated by reducing the injection pressure or by using horizontal well for gas injection. In case there is a dip, CO_2 should be injected through the lower section of the reservoir in order to take advantage of gravity.

Prior reservoir simulation studies show favorable results for CSEGR in terms of the feasibility of reservoir repressurization and enhanced production of CH_4. While simulations demonstrate that reservoir heterogeneity promotes early breakthrough, standard techniques for controlling preferential flow may be used in the field where required. Our simulations for a particular system, combined with the inherently favorable attributes of CO_2 in general, such as a comparatively high density and viscosity, suggest that CSEGR is feasible. Further progress in evaluating the feasibility of CSEGR requires field testing. We propose a phase I field test to investigate the reservoir processes involved in injecting high-pressure CO_2 into depleted natural gas reservoirs. Key processes will be the injection of CO_2 and associated cooling of the formation, flow, and transport of CO_2 gas through the reservoir, repressurization of the reservoir, and the enhanced production of CH_4. The Rio Vista Gas Field in California is a promising target for the field test, however, we will consider other locations if they have better potential to satisfy the objectives of the pilot study.

Heterogeneity in the formation may lead to preferential flow paths for the injected CO_2. This phenomenon may be favorable for injectivity and carbon sequestration in that it allows greater amounts of CO_2 to be injected. However, preferential flow may lead to early breakthrough and is therefore detrimental to EGR. Furthermore, the development of larger gas composition gradients and subsequent mixing by molecular diffusion is enhanced by preferential flow.

In case, flow instability occurs, adheres to high pressure injection should be avoided. Instead of trying to achieve miscibility at high pressure, it is recommended that immiscible injection at lower pressure conditions be planned. Numerical simulation results shows that such scenario will eventually produce much of the residual natural gas as long as well placement is designed with flow instability in mind. The use of chemicals in order to viscosity CO_2 or to create gels in situ should be avoided. Field trials of these schemes under the auspices of oil production have been dismal.

One aspect of CO_2 sequestration is rarely considered. If CO_2 is injected in natural porous media (limestone or sandstone), its properties are altered after it is stored. In case the injected CO_2 has contaminants, such as heavy metals or toxic additives (e.g. glycol, diethanolamine, triethanolamine), the porous medium is likely to adsorb the contaminants, purifying CO_2. The resulting CO_2 is more amenable to photosynthesis than CO_2 that is produced during combustion of refined crude oil.

4.4 ENHANCED GAS RECOVERY

There has been just one published field test of EGR. This test was conducted during 1986–1994 in the Hungarian field Budafa Szinfelleti, a weak water drive sandstone reservoir of 5–40 mD permeability (Papay, 1999). For that pilot study, EGR started when the natural gas recovery was 67% OGIP (original gas in place) and the injected gas was an impure CO_2 stream, consisting of 80% CO_2 and 20% CH_4 from an adjacent natural CO_2 pool.

The incremental gas recovery was 11.6% OGIP, or 35% of the gas in place at the initiation of CO_2 injection. CO_2 breakthrough occurred 1.5 years after the start of injection; the distance between injection and production wells was 500 m. This recovery is in sharp contrast to 70% recovery expected from core flood tests. The discrepancy can be explained through the existence of viscous instability and/or loss in sweep efficiency due to heterogeneity. Because the density of CO_2 is 3–9 times higher than that of CH_4, instability would occur if CO_2 is injected from the top of a formation.

Turta et al. (2007) reported laboratory test results of sweep efficiency of CO_2 and flue gas in an EGR scheme. These results are shown in Table 4.13. This table confirms previous postulation that flue gas can significantly improve the economics of the EGR process due to its

Table 4.13 Comparison between Flue Gas and CO_2 Efficiency

Test#	Injection Gas	Swi	Methane recovery at 1% N_2 contamination	Methane recovery at 10% N_2 contamination	Methane recovery at 20% N_2 contamination
		%	% OGIP	% OGIP	% OGIP
11	CO_2	18	61	71	—
12[a]	14% CO_2 in N_2	18	66	76	84
13	14% CO_2 in N_2	20	75	82	87
18[a]	CO_2	19	62	70	—

Swi = Intial water saturation.
[a]velocity reduced.

relative abundance and lower cost as compared to pure CO_2. In addition, the delay in the production of CO_2 when flue gas was used (as the displacing agent) means that the equipment is less vulnerable to corrosion. The problem will arise in the fact that the flue gas contains nitrogen concentration higher than the allowable limit in pipelining. This is not a technical issue but an issue of policy, which needs to be revised for unconventional gas applications.

4.4.1 Challenges and Opportunities of EGR

Everything about unconventional gas is challenging, mainly because conventional tools of characterization do not apply to unconventional gas. From origin to diagenesis, that is the overall history of most unconventional reservoirs is different from conventional ones. Only tight gas and tight oil formations are equivalent to low-permeability sandstone. As seen in the previous chapter, tight gas and oil are low-permeability version of sandstone. However, origin, depositional setting, stratigraphy, structure, geochemistry, geomechanics, seismic character, and petrophysical properties are different for other unconventional reservoirs. Some of the biggest challenges are:

- understanding the origin of the gas, i.e., how and where these rocks are charged with gas;
- determine a pattern of the highly-productive "sweet spots";
- determine what factors influence the large variation in both drainage area and permeability; and
- determine a pattern of fracture distribution and orientation.

Even though it is commonly assumed that shale gas is also the source rock, there are numerous scenarios that have shale gas interbedded with sandstone or even limestone. Determining the origin of this gas is not a trivial task. The age of this gas can be determined by correlating gas properties with time. Arising from the premise that natural processing increases the quality of petroleum fluids, a scenario depicted in Figure 4.49 can be developed. The actual numbers in this figure will arise from database of the region. In absence of such data, geological age of the rock can be determined and gas property data be evaluated in order to assess the origin of the gas.

Because the processes involving the formation of unconventional gas are different, the quality at the beginning of the gas generation is different. Even if it is assumed that the original gas is biogenic for each of these cases, gas hydrate methane is the purest form of methane from the beginning. It is because the original gas is purified with water and only a pure form of methane gas can form hydrate structures. This process is diametrically opposite to the formation of diamond, albeit in a different time scale. This discussion of diamond formation is relevant for both explaining purity of gas hydrate methane as well as the formation of CBM.

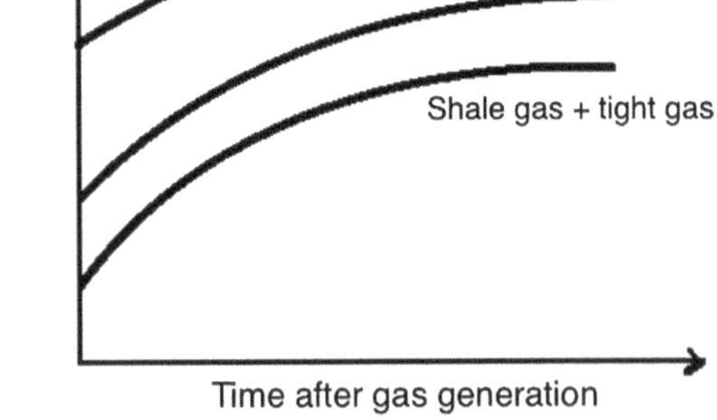

Figure 4.49 Variation of gas qualities as a function of time. CBM, coalbed methane.

Coal is considered to be the fossilized form of large volumes of biomass. One theory suggests that coal is formed in presence of water and with a low thermal gradient. Any gas that forms has the ability to escape or remain absorbed in the aqueous phase as well as adsorbed in the coal mass. Diamond, on the other hand, is formed when the cooling rate is much faster. Crystals are formed with a very high bonding energy without water molecules. As a result coal oxidizes at a relatively low temperature (similar to biomass oxidation temperature). On the other hand, diamond only oxidizes at a very high temperature, making it impractical to be an energy source for human consumption.

Volcanic rocks can be yet another form of unconventional oil/gas source. Vegetation entrained in ash flows may contain enough water to protect it from heat of emplacement. Subareal volcanism may create lakes and swamps with kerogen-rich sediments, and the volcanically warmed water in these basins can trigger nutrient growth, further enhancing the production of organic material. While there has been little consideration of exploring the potential of volcanic oil and gas reservoirs, it is well known that many conventional reservoirs have volcanic bedrocks that enhance oil and gas recovery from them. Once hydrocarbons are found in igneous reservoirs, assessing hydrocarbon volumes and productivity presents several challenges. Volcanic reservoirs are extremely hard. Seismic and logging fail to characterize these formations. The difficulty arises from the fact that volcanic rocks act as a shield for transmission of practically all signals that are used in logging. Log interpretation in igneous reservoirs often requires adapting techniques designed for other environments. For instance, resistivity logs that are based on conventional reservoirs will be less suitable for highly resistive matrix, low porosity, shale-free clean formations. Similar anomalies will occur with density log and neutron logs. Also, magnetic elements are more due to igneous origin. So nuclear magnetic resonance tool is ineffective in such formations. While sonic logs have better chance of being functional, volcanic rocks are so highly reflective, that they would not allow these signals to travel through.

Furthermore, because mineralogy varies greatly in these formations, methods that work in one volcanic province may fail in another. Usually, a combination of methods is required. In terms of porosity, direct measurement of volcanic rock porosity is most preferred.

The formation of volcanic rocks and entrapment of fluid within it can be explained by following the history of volcanic rocks. Figure 4.50 depicts formation of volcanic rocks and emplacement of igneous rocks. Plutonic

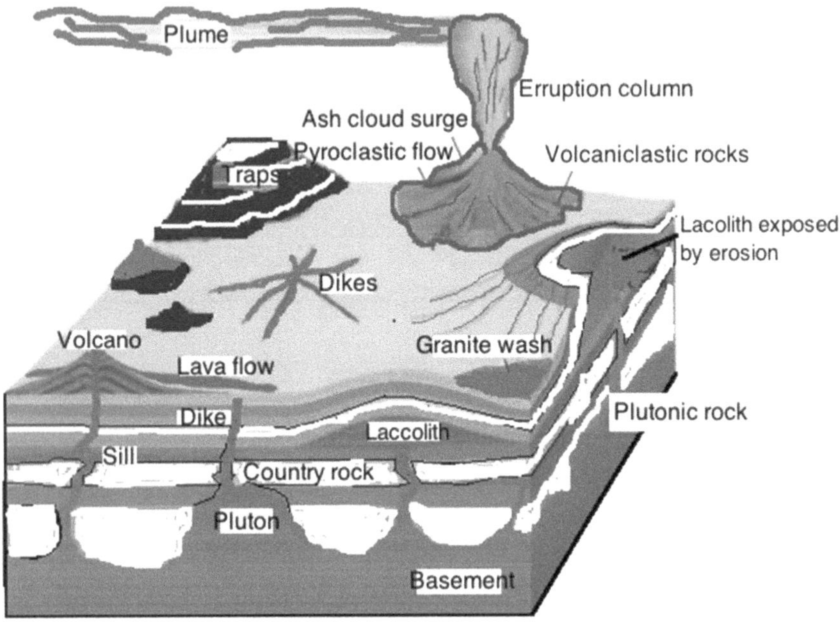

Figure 4.50 Formation of volcanic rocks. *Redrawn from Schlumberger, Oilfield Review, 2009.*

rocks, formed by cooling of magma within the Earth, display well-developed crystals with little porosity. Plutons and laccoliths—bulging igneous injections into sedimentary layers—are examples of plutonic rock. Volcanic rocks, formed when magma extrudes onto the surface and cools rapidly, show very fine crystalline or even glassy textures. While during the cooling process, cooling of volcanic fluids either underground or at the surface, and agglomeration of fragments ejected during explosive eruptions will develop different tendency of fracturing. This explains why volcanic rocks do not show well-defined patterns in terms of fractures. The onset of fractures itself is distinctly different in volcanic rocks from others, such as sedimentary rocks. When magma contains large amounts of water and gases, buildup of excessive pressure will cause sudden eruption, leading to the development of fissures and eventual fractures. During cooling of volcanic rocks, tremendous restoring force is developed internally, because rock is a poor thermal conductor and there is differential stress that develops along the temperature variation line. This restoring force or stress depends on the nature of cooling. Nature and tendency of internal force guides tendency or direction of fracturing.

During the cooling process the neighboring rocks are also affected. Volcanic eruption near sedimentary basins will intrude into sediments extending over several kilometers and during cooling will create fractures and subsequently faults. This process essentially creates the basis for cyclic nature of fracture, fissures and faults. Figure 4.51 shows the cyclic nature of the process. In the subsurface, volcanic events trigger earth quakes that can create faults within consolidated rocks. Any fault is accompanied with fissures and cracks that eventually can grow into fractures. These fractures themselves make a rock system vulnerable to faults that lead the way to volcanic fluid invasion and earth quake.

This is in conformance with what has been known as burial-thermal diagenesis. The diagenetic evolution of the middle member of the Bakken Formation is depicted in conjunction with the reconstructed burial-thermal history of the unit in the deep part of the Williston Basin (Figure 4.52). The burial-thermal model takes into account differences in deposition and erosion, as well as variations in thermal regime during the basin's history. Thermal data constrain the maximum amount of erosion during the late Tertiary to about 500 m, which agrees closely with previously reported erosional estimates (Webster, 1984; Sweeney et al., 1992).

Figure 4.53 highlights the permeability-porosity characteristics of volcanic and conventional rocks. Volcanic rocks have much lower porosity of the matrix and the permeability of this rock is much higher than permeability encountered in conventional reservoirs. In contrast, other unconventional gas reservoirs, such as tight gas and shale will have higher porosity but lower permeability. Of course, this is due to the fact that shale and tighter formations have high surface area but little effective porosity. It is, therefore, difficult to penetrate shale gas and tight sand altogether and similarly it is difficult to penetrate the matrix of a volcanic reservoir. In

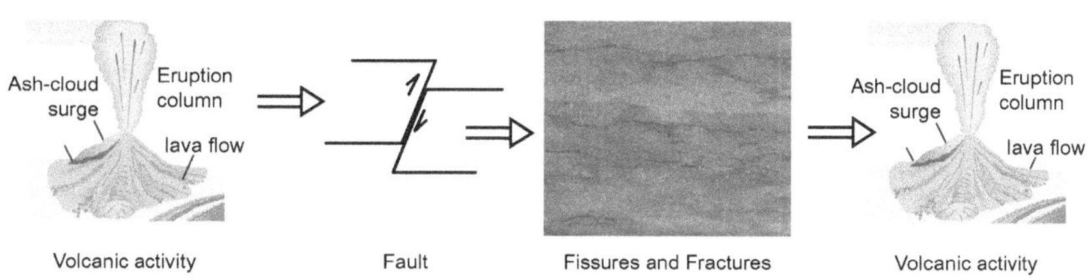

Figure 4.51 Volcanic activities and their impact on sedimentary rock create cycles of tectonic events.

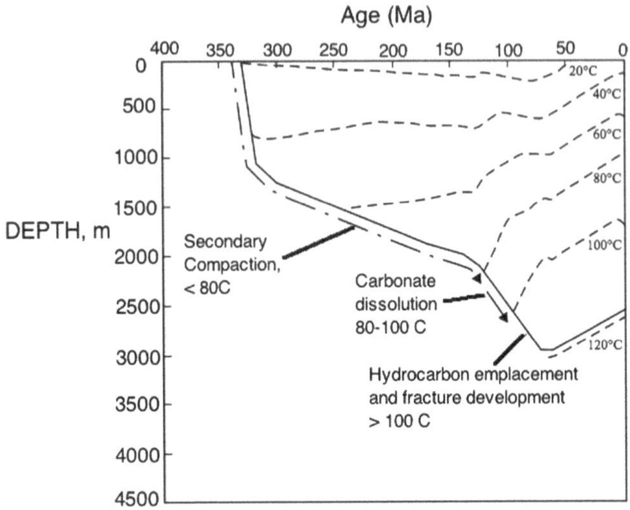

Figure 4.52 Reconstructed burial and thermal history curves of the Bakken Formation showing relative timing of major diagenetic events in the deep part of the Williston Basin. *Redrawn from Oilfield review, Schlumberger.*

terms of fluid production, shale gas or tight sand form a typical "permeability" as shown in Figure 4.53.

While Figure 4.54 highlights the difficulties associated with EGR, it also unlocks potential techniques that would work in such situations. It is known that fracturing as well as drilling horizontal wells recover significant amount of oil and gas from shale and tight sand. Both these processes increase gas saturation within fractures and the wellbore, thereby increasing the relative permeability to gas. The same effect can be invoked by using ISC (or equivalent to HPAI) or selected gas injection (e.g. with CO_2 or flue gas).

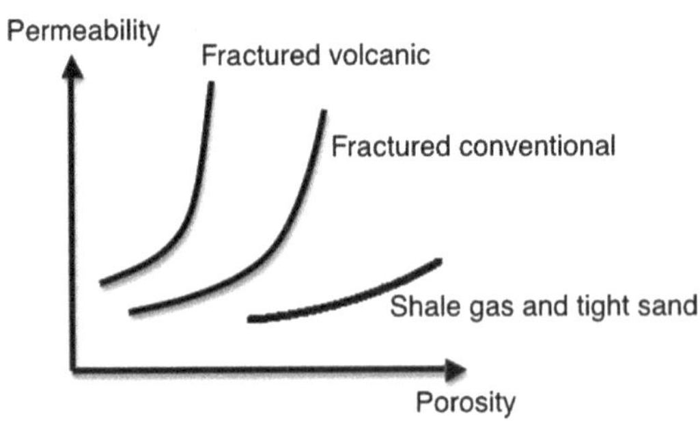

Figure 4.53 Fractured volcanic reservoir properties compared with fractured conventional reservoir.

For very tight but fractured formations, the possibility of using cyclic thermal fluid injection should be considered. Because diffusive heat and mass transfer is the most prominent mechanism prevalent in these formations, hot gas and/or steam injection can be beneficial. Heat injection can also trigger thermal cracking.

Figure 4.54 also shows how injection of gas can move the actual placement of a reservoir toward the left side, rendering the gas phase mobile. The presence of fissures and fractures can facilitate this process, as gas injectivity is moderate under such conditions.

For CBM, the most effective recovery technique is production of water that contains almost the entire amount of methane in absorbed state. The possibility of EGR for CBM is not well explored. However, it is well known that the production can be greatly enhanced with multilaterals, which have been widely successful in recent years. It is also well known that the addition of heat can increase the sweep efficiency of water tremendously. Steam injection in such cases is not economically feasible. However, ISC holds great promises. Coal burning is slow and conforms to the slow mode of steam generation within the reservoir. Combustion of coal will lead to the formation of CO_2 that itself will enhance the displacement of gas as well as penetration of fractures. Figure 4.55 shows how such scheme would work.

Similar scheme is also applicable to oil-bearing volcanic reservoirs. However, for such reservoirs, the fuel is the hydrocarbon and not the matrix itself as is the case for CBM.

For gas hydrate formations, typically thermal, depressurization, and chemical stimulation have been considered for enhancing gas recovery. Among these, chemical stimulation is the least desirable. Chemical

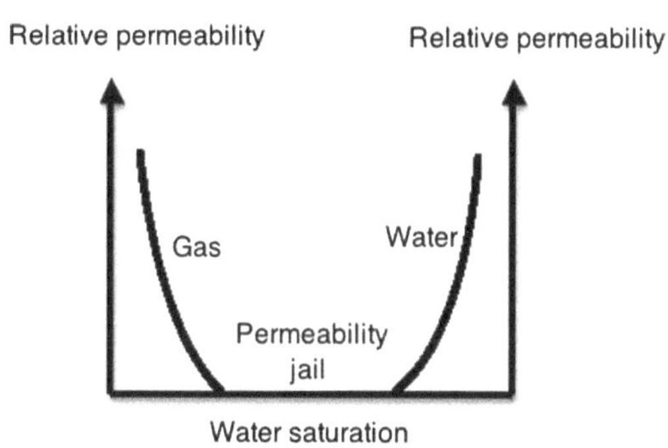

Figure 4.54 Existence of "permeability jail" in tight gas and shale formation.

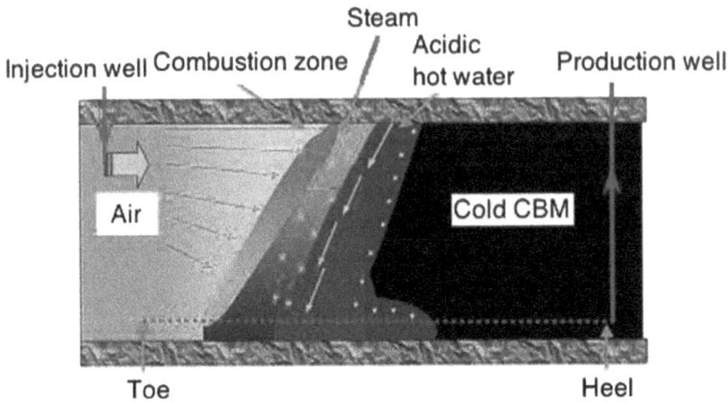

Figure 4.55 In situ combustion in coalbed methane (CBM).

stimulations are economically impractical and the efficiency of the process is poor. Depressurization is effective if in situ pressure is high. Otherwise, depressurization does not yield reasonable gas production. Figure 4.56 shows various types of gas traps within a gas hydrate reservoir. Two of the three types of traps are structural whereas the third one is stratigraphic. In case, there is a gas trap, it becomes easier to initiate depressurization by putting the trap gas in production. This production destabilizes the hydrate zone and stream of gas is released making the process commercially viable. In case the volume of the trapped gas is not sufficient to create instability within the hydrate body, an innovative technique would be to initiate ISC by injecting air into the trap. This would trigger immediate instability due to sudden thermal change (Figure 4.57). This initial instability is followed by generation of CO_2 that acts as yet another

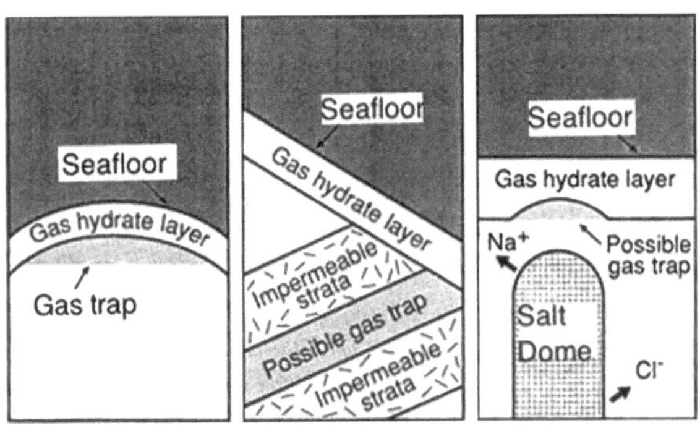

Figure 4.56 Different types of gas traps. *From Kvenvolden, 1993.*

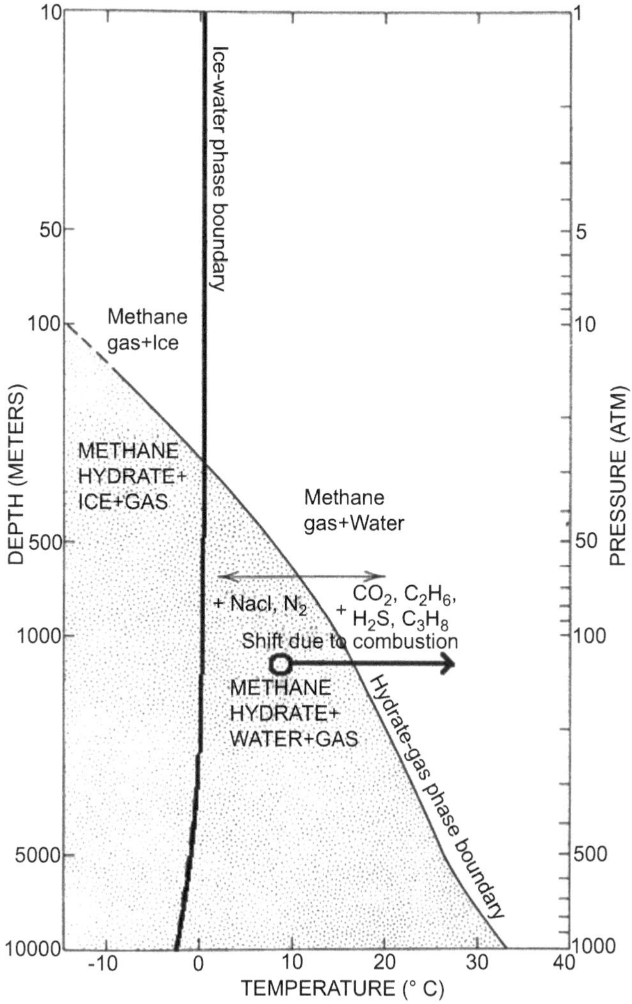

Figure 4.57 In situ combustion can trigger instability in gas hydrate. *From Kvenvolden, 1993.*

stimulus for further hydrate instability (Figure 4.58). At this point, production of methane can be resumed because all three modes of hydrate dissociation have been activated. Even though, the use of geothermal energy has been discussed in the literature as a means of stimulating hydrate production, air injection to trigger in situ heat generation is the most effective technique.

At present, three ways of transport of gas from hydrates are known: (1) by conventional pipeline, (2) by converting the gas hydrates to liquid

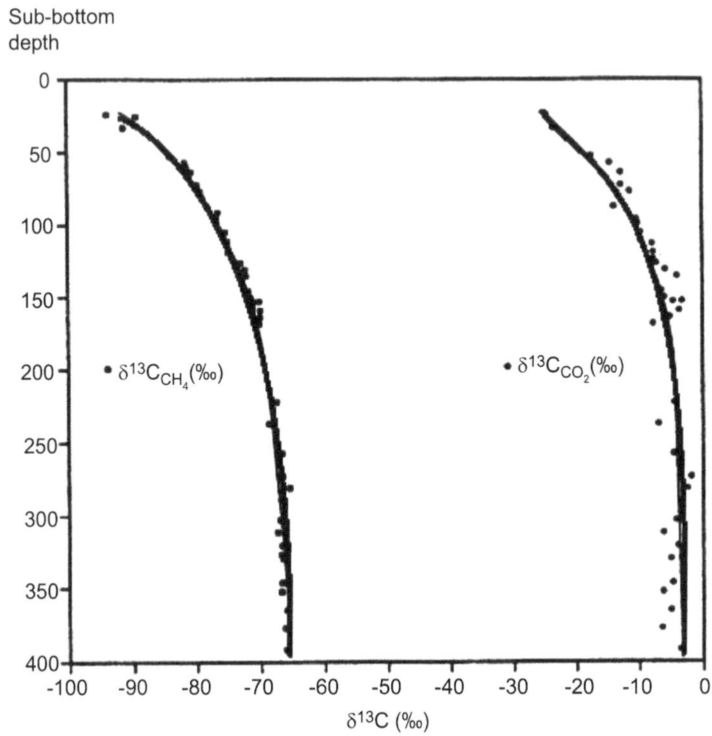

Figure 4.58 Typical distribution of carbon isotopic compositions with depth. *From Kvenvolden, 1993.*

middle distillates via the newly improved Fischer-Tropsch process and loading it onto a conventional tanker or barge, or (3) by reconverting the gas into solid hydrate and shipping it ashore in a close-to-conventional ship or barge. The latter option was proposed in 1995 by a research team at the Norwegian Institute of Technology, which determined that the use of natural gas hydrate for the transportation and storage of natural gas was a serious alternative to gas liquefaction since the upfront capital costs are 25% lower. Yet another positive factor is that it is far safer to create, handle, transport, store, and regasify natural gas hydrate than liquefied natural gas.

Because natural gas hydrates are often found in shallow reservoirs, mining them is also an option. More importantly, natural gas hydrates offer the best candidates for downhole electricity generation. Efficiency of such process would be much higher than conventional electricity generation with gas turbines.

4.4 SUMMARY AND CONCLUSIONS

Unconventional reserves are typically undervalued. The potential of these resources is far greater than what is set out to be the norm, with multilaterals and fracking. The true potential lies within the development of custom-designed schemes based on the nature of resources. The flow chart presented in Figure 4.59 offers as a guideline for determining these schemes. This flow chart does not contain natural gas hydrate but includes tight carbonate formations. Natural gas hydrate options are presented in the section above.

The first question to ask is if there is natural productivity. Of course, if the answer is yes, then the reservoir falls under the realm of conventional technology. However, if there is no natural productivity or if the productivity is too low to sustain economic developments, the next question to be asked is if there is fracture network in place. Even though, unconventional gas reservoirs are mostly fractured, whenever they do not have clear fracture networks, injectivity is a major issue. The case becomes akin to tar sand that has very low injectivity. For this set of reservoirs, SAGD equivalent schemes are recommended.

As an example, the case of fracture-free hydrate reservoirs can be cited. Figure 4.60 shows how natural gas hydrates can invade permeable and porous beds that are surrounded by fractured formations. It is difficult to penetrate these formations with an injected fluid due to poor injectivity. However, access can be gained through fractures in order to trigger dissociation of the natural gas hydrate. Once set in motion, such dissociation process snowballs and the reservoir can produce high volume of in situ gas.

In a fracture-free formation, the next most suitable scheme is called "thermal cracking equivalent." If for whatever reason the formation in question is not accessible, ISC should be considered in order to trigger production. Unlike conventional ISC that needs special completion in order to deal with high temperature, air injection is recommended. Air injection can trigger combustion that would lead to steam formation, CO_2, generation and creation of thermal cracks, all of which will increase gas recovery or render the gas recovery process economically attractive. Figure 4.61 shows this particular feature of unconventional gas production.

The lowest priority option of fracture-free unconventional gas reservoir is displacement-type scheme. Because of the low injectivity of these

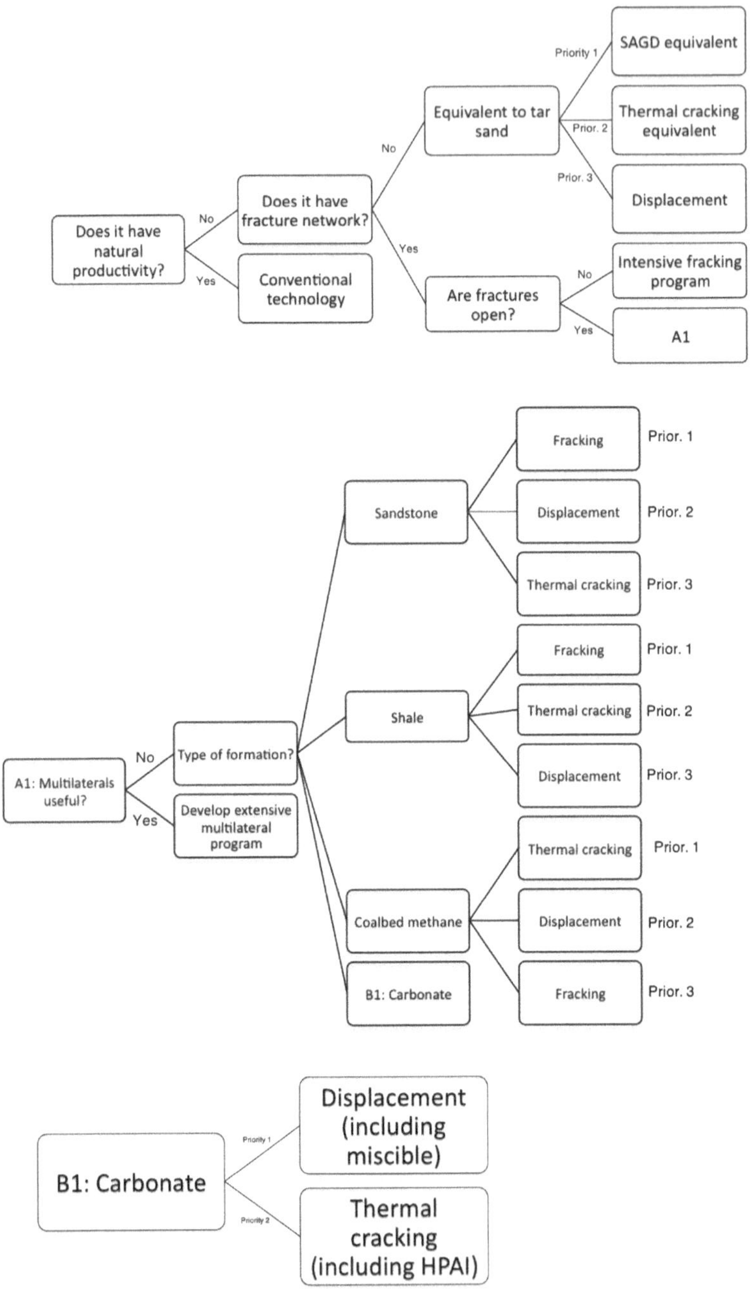

Figure 4.59 Flow chart for determining appropriate unconventional gas production schemes.

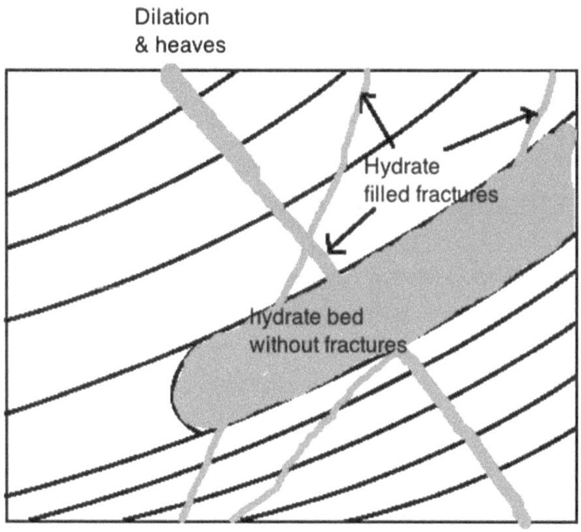

Figure 4.60 Example of SAGD-like recovery scheme for natural gas hydrate (yellow marks hydrate). For interpretation of the references to color in this figure legend, the reader is referred to the online version of this book.

reservoirs, a displacement-type recovery scheme can be achieved only with chemically active water or gas. For instance, alkaline water or carbon dioxide can both become effective displacement agents. In case of CO_2 injection, high solubility of CO_2 leads to the formation of carbonic acid that itself causes leaching effects. Because of the tightness of the formation, the injection rate has to be slow. However, diffusive forces will eventually

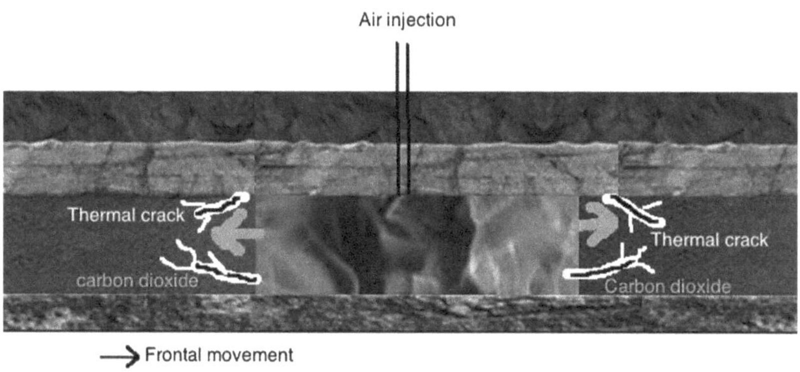

Figure 4.61 Air injection can trigger several enabling features that will render a recovery process economically attractive.

drive the injected fluid to affect a large area that "softened," thereby being able to produce gas, either from a different well or through huff and puff mode with the same well. This is depicted in Figure 4.62.

While all the above mentioned options are applicable to unconventional reservoirs without a fracture network, the presence of fracture networks calls for a new line of strategy. In this case, if the fractures are open, it is recommended that intensive fracking be used. Note that open fractures and low productivity is typical of shallow unconventional reservoirs. Fracturing does not work well in deep formations, for which fracturing is both prohibitively expensive and least effective. Hydraulic fractures would be formed in a direction orthogonal to the principal direction of natural fractures. However, prior to creating new fractures, the injected fluid would open existing fractures. In terms of volume, these newly opened fractures can account for very significant increase in gas production. Once hydraulic fractures are created and proppants placed, the overall productivity is sustained at a high value. The placement of hydraulic fractures is more challenging with a horizontal well (Figure 4.63). However, this challenge is easily overcome when the formation is shallow. For deeper formations, hydraulic fractures are more difficult to create and to sustain.

For reservoirs with open fractures, the question to be asked is if multilaterals are useful. Most of the cases, multilaterals translate into increased productivity. For a multilateral to be effective, following conditions have to be fulfilled:

1. horizontal well should intersect most of the fractures;
2. $k_h/k_v > 10$, in the event of lower vertical permeability, hydraulic fracturing is desirable in case they are feasible (e.g. when natural fractures are predominantly horizontal) and liner cementing can be avoided;

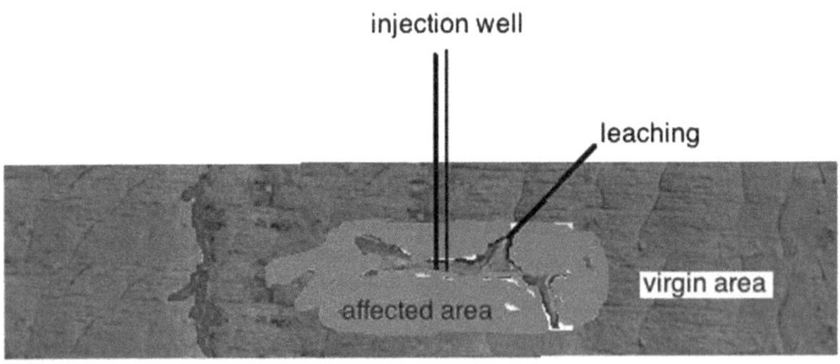

Figure 4.62 Chemically active water or gas can affect a large area in a tight reservoir.

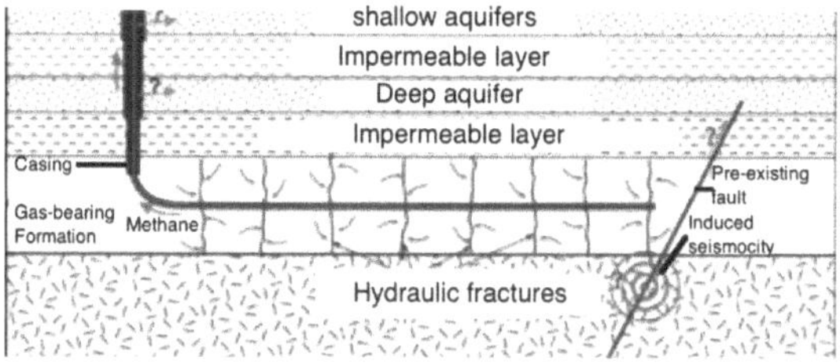

Figure 4.63 Hydraulic fracturing is most effective in shallow unconventional formations.

3. flow near wellbore is Darcian (as non-Darcy effects can reduce flow effectiveness);
4. pressure drop in the horizontal wellbore is not significant, in which case telescopic liner can be used to complete the horizontal well.

Figure 4.64 shows how the productivity improvement factor is generally impacted for various cases of reservoirs. Naturally fractured formations offer the best probability of success over vertical wells, even for the unconventional gas reservoirs as long as open fractures are present.

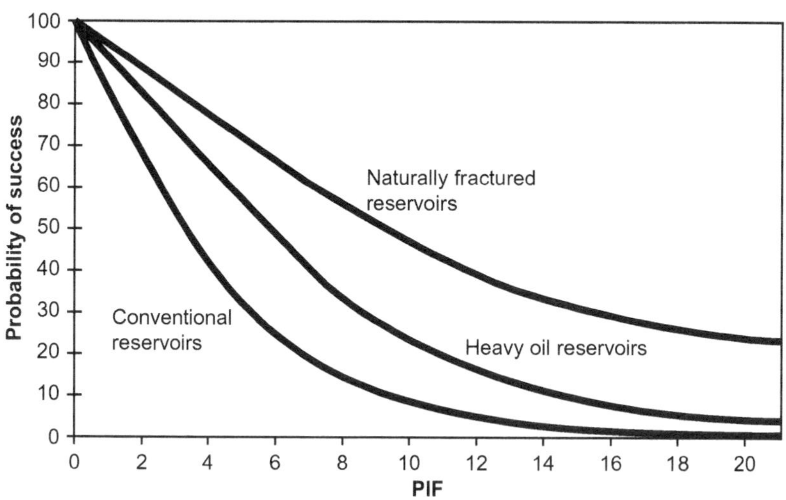

Figure 4.64 Probability of success with horizontal wells for various types of reservoirs.

Table 4.14 Various Types of Unconventional Reservoirs with Different Priorities

Type of formation	Priority 1	Priority 2	Priority 3
Sandstone	Fracking	Displacement	Thermal cracking
Shale	Fracking	Thermal cracking	Displacement
Coal bed methane	Thermal cracking	Displacement	Fracking
Carbonate	Displacement	High pressure air injection	

For cases for which multilaterals fail to yield satisfactory results, different types of schemes are recommended for different types of formations. Table 4.14 summarizes the final recommendation for this category.

CHAPTER 5

Reservoir Characterization of Unconventional Gas Formations

5.1 SUMMARY

Unconventional gas reservoirs have shifted the focus to special considerations of fundamentals of reservoir engineering. All reservoir characterization tools as well processing techniques are uniquely designed for conventional reservoirs. Using the tools and techniques to unconventional reservoirs has the risk of conclusions that are irrelevant and random at best. The main mechanism of fluid flow in unconventional reservoirs is through fractures, natural or induced, whereas existing techniques do not integrate this aspect to reservoir characterization. This chapter presents a new reservoir characterization technique that is suitable for unconventional reservoirs.

5.2 INTRODUCTION

All consolidated petroleum reservoirs contain natural fractures or fissures. Natural fractures result from the interaction of earth stresses; whereas artificial fractures result from drilling activities, increase in pore pressure in injection operations, reservoir cooling during water flooding, redistribution of earth stresses in the field as a result of injection and production practices, etc.

Maximizing economic recovery from naturally fractured reservoirs is a complex process. It requires a thorough understanding of matrix flow characteristics, fracture network connectivity, and fracture–matrix interaction. It involves knowing the geological history. Therefore, key to successful reservoir characterization is in connecting with geologists that can construct an overall picture of the reservoir history. Construction of this history is pivotal. This construction must be scientific, following objective systematic abstraction. This is shown in Figure 5.1. The abstraction process has to be bottom up. It involves collecting data in its raw form. These data have to be collected in proper time sequence and at each step, verified from

Unconventional Gas Reservoirs
ISBN 978-0-12-800390-9
http://dx.doi.org/10.1016/B978-0-12-800390-9.00005-0

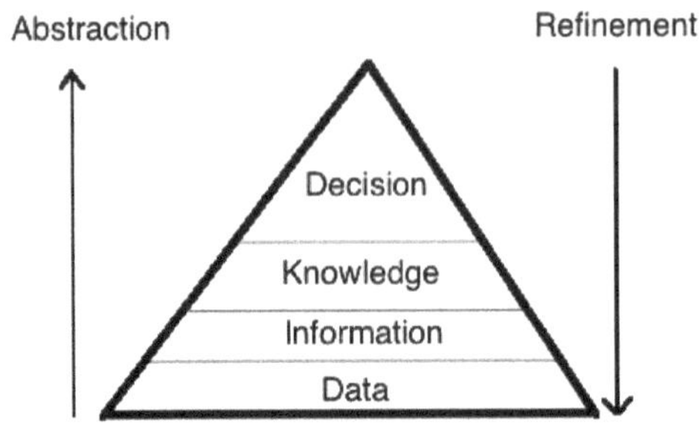

Figure 5.1 The knowledge model: The abstraction process must be bottom up.

multiple sources. Medieval scholar, Alkindus famously said, "Multi-source information is a treasure." Reservoir description should rely on information from many sources. For petroleum reservoir applications, abstraction starts with collection of geological data. In the first phase of abstraction, depiction of the subsurface strata is made. In order to create this picture, geologists must collect data from any available source, such as outcrop, regional subsurface maps. Based on this geological map, the decision to conduct geophysical survey is made. During the geophysical survey, decisions to implement a certain grid, type of survey, etc., have to be made. The process of geophysical survey offers an example of how multisource data must be integrated. During this process, geological data are used as a baseline, whereas geophysical data are later used to refine geological data. At the end, decision to drill is made only after several cycles of abstraction and refinement.

Reservoir description should rely on information from many sources including static data (well logs, cores, petrophysics, geology, and seismic) and ultimately on dynamic data (formation evaluation well tests, long-term pressure transient tests, tracer tests, and longer term reservoir performance).

This process begins even before a geological survey begins. For instance, economic policies dictate to what extent exploration should be performed. During exploration, geologists become involved and need to be made integral part of the policy making. In order to refine geological findings, geophysical measurements are made and geophysical analysts are brought into the equation. Only after repeated abstraction and refinement, one decides to drill the first well. At this point, drilling engineers, reservoir engineers, and petroleum geologists become involved. Data collected during drilling already form integral basis for further abstraction and refinement of data. After a drilling operation is complete, coring, logging, and drill stem test (DST) are performed in order to refine static data and abstract dynamic data.

The idea behind reservoir characterization is to know the past. That means, knowing the following:

- Origin of fluid
- Origin of the reservoir
- Origin of the fractures
- Process of erosion, transport, and deposition
- Process of faulting, fracturing
- Process of secondary activities (leaching, cementation, etc.)

The next task in reservoir characterization is to know the present. It involves collection of mud data, drill cutting, oil and gas "shows," pressure data, temperature data, and core data. These dynamic data are then assembled with static data in order to refine data on the following.

- Stratigraphy
- Lithology
- Fracture density
- Fracture orientation
- Nature of fractures
- Shale breaks
- "Sweet spots"

Initial static data collected during exploration process are the most valuable because it serves as the foundation. As soon as the first well is drilled, data must be collected even during the drilling process. Data on mud loss, rate of penetration (ROP), cuttings, azimuth, and others offer valuable insight for refining the subsurface picture.

The next set of data are generated during coring. These data include both cores and fluid sample. Even though seldom practiced, fluid data and core data should be treated like geological and geophysical data that are used for abstraction and refinement in a cyclical mode.

The core data offer the first set of direct evidence of fractures. These data should be compared with the fault network predicted in the geological map that was refined after the geological survey. Cores also give one an opportunity to determine fracture frequency, nature of fractures (e.g., plugged or open), and thus, lead toward developing the rose diagram. The development of the rose diagram is of utmost importance because that would dictate the direction of flow.

The DST data are the first set of dynamic data available. Any discrepancy between core and DST permeability would indicate the existence of fractures. As a well is put on production, it continues to generate dynamic data that are invaluable for the abstraction and refinement process, outlined above.

During the production cycle of a petroleum reservoir, it continues to provide one with valuable data on fracture characterization. Table 5.1 shows how fracture characteristics can be predicted with the data available at each stage of petroleum operations.

5.3 ORIGIN OF FRACTURES

Characterization of fracture networks is a pragmatic process that relies heavily on experience and empiricism and very little (to date) on systematic approaches. Reservoir description should rely on information from many sources including static data (well logs, cores, petrophysics, geology, and seismic) and ultimately on dynamic data (formation evaluation well tests, long-term pressure transient tests, tracer tests, and longer term reservoir performance).

Unconventional gas reservoirs are unique in many respects. Even though they are lumped into one category having a permeability value less than 0.1 mD, each type is unique and in need of independent characterization tool. It is so because none of the characterization tool is universally applicable based on only one property, i.e., permeability. For instance, sandstone has conventional oil and gas even though the formation itself is unconventional. The main difficulty in sandstone is that fracture network is not well developed due to lack of proper method that includes fracture properties. For instance, not a single core analysis technique can shed light on fracture properties of a rock. In fact, the presence of fractures or even fissures disqualifies core analysis plugs from being considered for further analysis. While several logging tools have emerged that can identify fractures, there is no systematic process to integrate that information into a reservoir characterization tool.

Shale gas plays differ from conventional gas plays in that the shale formations are both the source rocks and the reservoir rocks. There is no migration of gas as the very low permeability of the rock causes the rock to trap the gas and it forms its own seal. The gas can be held in natural fractures or pore space or can be absorbed onto the organic material. Apart from the permeability, total organic content (TOC) and thermal maturity are the key properties of gas potential shale. Generally, it can be stated that the higher the TOC, the better the potential for hydrocarbon generation. In addition

Table 5.1 Various Stages of Fracture Data Collection

Stage	Well locations	Fracture description	Abstraction/ refinement
Exploration	No well present	Based on faults	Geology/ geophysics
Delineation	Wells used to refine data	Based on faults + nearby wells	Static/dynamic data from nearby wells
Drilling	Wells used to correlate	Drilling data	Static/dynamic data
Logging and coring	Direct evidence	Images and visual inspection	Static data refinement
Primary recovery	Optimization can lead to infill drilling	Based on dynamic data	Static/dynamic data
Enhanced oil/gas recovery	Injected fluid can act as a tracer	Pathway selected by injected fluid	Static/dynamic data

to these characteristics, thickness, gas in place, mineralogy, brittleness, pore space, and the depth of the shale gas formation are other characteristics that need to be considered for a shale gas reservoir to become a successful shale gas play. The organic content in these shales, which is measured by their TOC ratings, influences the compressional and shear velocities as well as the density and anisotropy in these formations. Consequently, it should be possible to detect changes in TOC from the surface seismic response.

Coalbed methane (CBM) is a gas formed as a part of the geological process of coal generation and is contained in varying quantities within all coal. CBM is exceptionally pure compared to conventional natural gas, containing only very small proportions of "wet" compounds (e.g., heavier hydrocarbons such as ethane and butane) and other gases (e.g., hydrogen sulfide and carbon dioxide). Coalbed gas is over 90% methane and is suitable for introduction into a commercial pipeline with little or no treatment.

From the earliest days of coal mining, the flammable and explosive gas in coalbeds has been one of mining's paramount safety problems. Over the centuries, miners have developed several methods to extract the CBM from coal and mine workings. CBM well production began in 1971 and was originally intended as a safety measure in underground coalmines to reduce the explosion hazard posed by methane.

The primary (or natural) permeability of coal is very low, typically ranging from 0.1 to 30 mD. Because coal is a very weak (low-modulus) material and cannot take much stress without fracturing, it is almost always highly fractured and cleated. The resulting network of fractures commonly gives coalbeds a high-secondary permeability (despite coal's typically low-primary permeability). Groundwater, hydraulic fracturing fluids, and methane gas can more easily flow through the network of fractures. Because hydraulic fracturing generally enlarges preexisting fractures and rarely creates new fractures, this network of natural fractures is very important for the extraction of methane from the coal. This is in stark contrast to what happens in sandstone formations.

Yet, gas hydrates, another unconventional natural gas source, make up completely different set of properties. Gas hydrates can be found on the seabed, in ocean sediments, in deep lake sediments, as well as in the permafrost regions. The amount of methane potentially trapped in natural methane hydrate deposits may be significant, which makes them of major interest as a potential energy resource. This methane is also of high quality.

In this chapter, a reservoir characterization technique is proposed that is applicable to all categories with the exception of gas hydrate.

5.4 SEISMIC FRACTURE CHARACTERIZATION

As discussed in Chapter 4, any rock deformation is the result of tectonic events that are a unique function of time. With time, events such as magma movement, faulting, earthquake, and fracturing occur in a cyclical form. It is recognized that the state of stress changes with time, affecting rock deformation directly. The stress–strain relationship is different in different zones, depending on the formation, its contents, and temperatures. Two zones are identified broadly: the shallower zone (Zone 1) where any stress translates into active reaction and changes in strain, and the deeper zone (Zone 2) that is more resilient and the strain deformation is narrow. Figure 5.2 shows a schematic of this relationship.

In Zone 1, deformation causes brittle failure and rock strength is limited by frictional strength of preexisting faults or fractures, whereas in Zone 2, the prevalent temperature makes it more resilient and faults and fractures can endure greater stress. In this zone, the temperature helps make the flow more ductile with rock strength declining exponentially with increasing temperature. Few studies have investigated the deformation of rock as a function of stress and temperature. However, it is important to understand

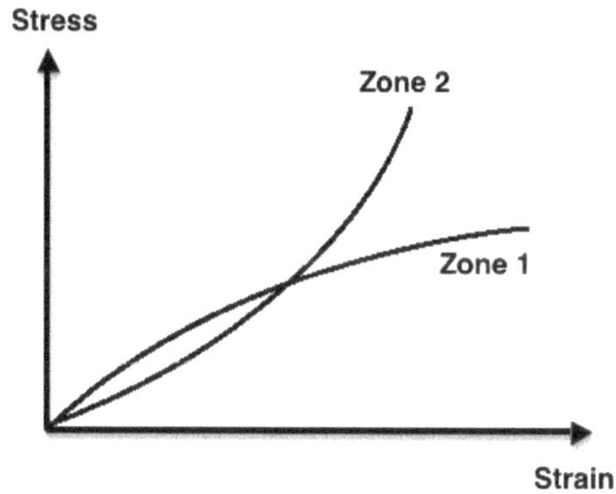

Figure 5.2 Schematic of the two zones on the earth's crustal region.

the nature of deformation as a function of these variables in order to properly characterize fractures in a reservoir.

Natural fractures develop at lower temperatures with minimal stress. This is how all outcrops of consolidated rocks exhibit natural fracture networks. The existence of fracture eventually leads to the development of a fault when lateral displacement is large enough to invoke such movement across more than one sedimentary bed. This onset of fracture immediately follows the development of fissures with orientation orthogonal to the direction of fault. Depending on the nature of stress, fissures develop into fractures. These fractures become the main vehicle for hydrocarbon transport. The same fractures can onset thermal convection of water. Such flow has tremendous implication on eventual hydrocarbon generation and transport.

Two phenomena add to the complexity of fracture flow: the occurrence of shale breaks and the secondary cementation of fractures. Shale breaks decrease overall vertical permeability whereas cementation reduces overall permeability of the reservoir. For the latter, fracture orientation becomes the most important feature of fluid flow. Shale breaks, on the other hand, serves as a barrier to vertical flow and has the capacity to become a storage site for the so-called shale gas and oil. While most of the shale gas and oil reservoirs are believed to contain source rock, the shale breaks as well as caprocks also contain significant amount of gas, albeit being trapped within very low-permeability shales.

Because deeper fractures are mainly oriented normal to the direction of minimum in situ compressive stress, the presence of seismic anisotropy can signal certain pattern of fractures. While this requires true understanding of the relationships between the response of seismic anisotropy and fracture properties, the existence of seismic anisotropy offers one with the baseline for fracture characterization.

Several theoretical studies of fracture-induced anisotropy have been reported in the literature (O'Connel, R., and Budiansky, B., 1976; Budiansky, B., and O'Connel, R., 1976; Hoenig, 1979; Hudson, 1980, 1981, 1986; Crampin et al., 1986; Hudson, et al., 1996). How the presence of the fracture sets affects the elastic modulus of fractured rocks has been discussed in the literature (Schoenberg and Sayers, 1995; Thomsen, 1995; Liu et al., 1996). Based on the simplifying assumption of linear and elastic behavior, it is the background elastic modulus and fracture parameters (fracture density, aspect ratio, and saturating fluid) that determine the behavior of seismic waves, which propagate through, and are reflected from, the reservoirs.

Fracture models are based on major physical features such as the azimuthal P- and S-wave velocity variations with fracture parameters. The azimuthal variations in P-wave velocity and reflectivity in homogeneously fractured reservoirs as a function of anisotropic parameters have been described by many researchers (Tsvankin, 1996; Ruger, 1998; Al-Dajani and Tsvankin, 1998). These anisotropic parameters are related to fracture parameters through the elastic stiffness tensors. In principle, azimuthal amplitude versus offset (AVO) responses can be used to detect fractures. Successful use of azimuthal variation of P-wave AVO signatures to determine the principal fracture orientation and density have been reported in the literature (Lynn et al., 1995, 1996; Perez and Gibson, 1996).

In principle, the distribution of fractures in reservoir zones can be treated as homogeneous. In practice, the number and variability of small-scale fractures are so large that recovering significant information from P-wave seismic data requires the distribution characteristics to be averaged over the reservoir zone, which leads to a statistical representation. This is equivalent to the notion of assigning pseudo-homogeneity. Heterogeneity due to spatial variations of fracture density could result in spatial variations of velocity anisotropy. It is important to identify the coherent features in reflected seismic data. These features are often used as major seismic characteristics in exploration geophysics. The small incoherent arrivals that occur between the major reflections also contain information about the media. They are currently treated as "noise," but contain valuable information that can alter the reservoir description if filtered correctly. Currently used techniques are not capable of analyzing these signals with scientific accuracy (Charrette, 1991). The 3D finite-difference modeling technique is effective in studying the azimuthal AVO response and scattering characteristics in heterogeneously fractured media. However, they allow no room for including "noises." Although grid memory requirements make computations very expensive and grid dispersion effects limit finite-difference models to small regions, this approach has been successfully used to model energy diffracted at highly irregular interfaces (Lavender and Hill, 1985; Dougherty and Stephen, 1987), to study fluid-filled bore-hole wave propagation problems in anisotropic formations (Cheng et al., 1995), and to study the scattering in isotropic media (Frankel and Clayton, 1986; Coates and Charrette, 1993; Zhu, 1997). Additionally, unlike boundary integral techniques, lateral velocity variations can be easily incorporated (Stephen, 1984, 1988).

Azimuthal AVO variations have been used in fracture detection and density estimation (Perez, 1997; Ramos and Davis, 1997). However, the sensitivity of reflected P-waves to the discontinuity of elastic properties at a reflected boundary and to the spatial resolution makes it difficult to interpret this attribute unambiguously. The motivation of this thesis is to explore the efficiency, benefits, and limitations of using P-waves to characterize fractured reservoirs, theoretically and practically. Shen (1998) studied the possibility of using P-waves to investigate properties of fractured reservoirs and the diagnostic ability of the P-wave seismic data in fracture detection. This study also considered rheological behavior of rocks at a crustal scale, based on observation and modeling of continental deformation, in particular, deformation of the Tibetan plateau. The Tibetan plateau is an ideal location that features continental topography, resulting from the N–S convergence between the Indian and Eurasian plates.

5.4.1 Effects of Fractures on Normal Moveout (NMO) Velocities and P-Wave Azimuthal AVO Response

Shen (1998) investigated the effects of fracture parameters on anisotropic parameter properties and P-wave NMO velocities, based on developed effective medium models and crack models. Anisotropic parameters of the pseudo-transversely isotropic medium model, S(v) and E(v), have different characteristics in gas- and water-saturated, fractured sandstones. When fractures are gas-saturated, $\delta^{(v)}$ and $\varepsilon^{(v)}$ vary with the fracture density alone. In water-saturated, fractured sandstones, both $\delta^{(v)}$ and $\varepsilon^{(v)}$ depend on fracture density and crack aspect ratio. $\delta^{(v)}$ is related to the V_p/V_s of background rocks and $\varepsilon^{(v)}$ is a function of the V_p of background rocks.

Studies show that the shear wave splitting parameter, $\gamma^{(v)}$, is most sensitive to crack density and insensitive to saturated fluid content and crack aspect ratio. Properties of P-wave NMO velocities in a horizontally layered medium are the function of $\delta^{(v)}$. The effects of fracture parameters on P-wave NMO velocities are comparable with the influences of $\delta^{(v)}$.

P-wave azimuthal AVO variations are not necessarily correlated with the magnitude of fracture density. Shen (1998) showed that the elastic properties of background rocks have an important effect on P-wave azimuthal AVO responses. Results from 3D finite-difference modeling show that azimuthal AVO variations at the top of gas-saturated, fractured reservoirs, which contain the same fracture density, are significant in the reservoir model with small Poisson's ratio contrast. Analytical solutions indicate that azimuthal AVO variations are detectable when fracture-induced reflection coefficients can generate a noticeable perturbation in the overall reflection coefficients. Varying fracture density and saturated fluid content can lead to variations in AVO gradients in off fracture strike directions. Shen's numerical results also show that AVO gradients may be significantly distorted in the presence of overburden anisotropy caused by vertical transverse isotropy media, which suggests that the inversion of fracture parameters based on an individual AVO curve would be biased without correcting this influence. He recommended that azimuthal AVO variations could be effective for detecting fractures; model analysis studies and combination of P-wave NMO velocities are more beneficial than using reflection amplitude data alone.

5.4.2 Effects of fracture Parameters on Properties of Anisotropic Parameters and P-Wave NMO velocities

In exploration geophysics, NMO describes the effect that the distance between a seismic source and a receiver (the offset) has on the arrival time of a reflection in the form of an increase of time with offset. The relationship between arrival time and offset is hyperbolic, typically described by a wave equation. This relationship is the principal criterion that a geophysicist uses to decide whether an event is a reflection or not. The NMO depends on complex combination of factors including the velocity above the reflector, offset, dip of the reflector, and the source receiver azimuth in relation to the dip of the reflector. Of concern is the role of fractures; to understand the effect of fracture parameters on NMO velocities, one needs to understand effects of fracture parameters on anisotropic parameter properties.

Table 5.2 Elastic Parameters Used by Shen (1998)

Sandstone no.	V_p (m/s)	V_s (m/s)	ρ (g/cm^3)	V_p/V_s	References
No. 1	3368	1829	2.50	1.84	Thomsen (1986)
No. 2	4405	2542	2.51	1.73	Thomsen (1986)
No. 3	4539	2706	2.48	1.68	Thomsen (1986)
No. 4	4476	2814	2.50	1.59	Thomsen (1986)
No. 5	4860	3210	2.32	1.51	Teng and Mavko (1996)

Table 5.3 Elastic Parameters and Fracture Parameters of Model 1 and Model 2

Model	V_p (m/s)	V_s (m/s)	ρ (g/cm^3)	Fracture density(%) and aspect ratio	Poisson's ratio	Type of rocks
Model 1	4358	3048	2.81		0.021	Mesaverde shale
	3368	1829	2.50	10 0.01	0.291	Taylor sandstone
Model 2	4561	2988	2.67		0.124	Shale
	4860	3210	2.32	10 0.01	0.113	Sandstone

Shen (1998) studied various elastic parameters of five sandstones, whose characteristics are summarized in Tables 5.2 and 5.3. He studied for two aspect ratios, i.e., 0.01 and 0.05. Figure 5.3 shows $\delta^{(v)}$ as a function of fracture density for a gas-saturated sandstone with fracture aspect ratio of 0.01. Results are shown for an aspect ratio of 0.05 (Figure 5.4). These results show that for gas-saturated sandstones, $\delta^{(v)}$ is insensitive to aspect ratio. For the water-saturated case, however, absolute values of $\delta^{(v)}$ increase with fracture density and crack aspect ratio. This latter case (Figures 5.5 and 5.6) also shows a range of values for different samples, as compared to gas-filled case that show little dependence on sample types. It is also noted that $\delta^{(v)}$ is dependent on V_p/V_s of isotropic, unfractured sandstones. The smaller the V_p/V_s, the larger the absolute value of $\delta^{(v)}$ obtained.

$\varepsilon^{(v)}$ shows similar characteristics to $\delta^{(v)}$. The difference is that $\varepsilon^{(v)}$ is the function of the V_p of the background medium. In gas-saturated, fractured sandstones, $\varepsilon^{(v)}$ is sensitive to fracture density alone. In water-saturated, fractured sandstone the absolute value of $\varepsilon^{(v)}$ increases with both fracture density and aspect ratio. The smaller the Vp, the smaller the absolute value

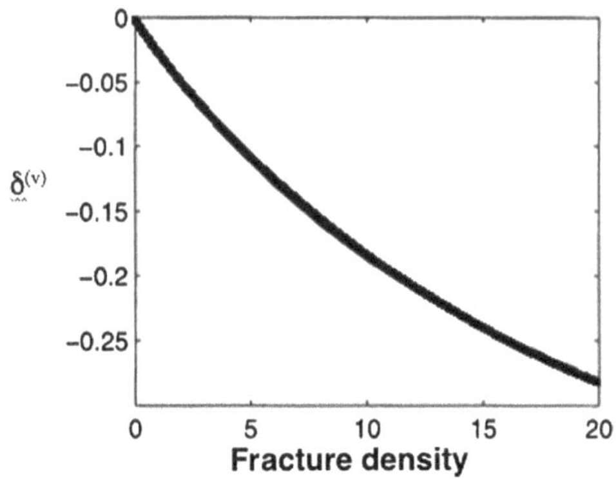

Figure 5.3 Variation in anisotropic parameter as a function of fracture density for gas-saturated sandstone and fracture aspect ratio of 0.01. *Redrawn from Shen, 1998.*

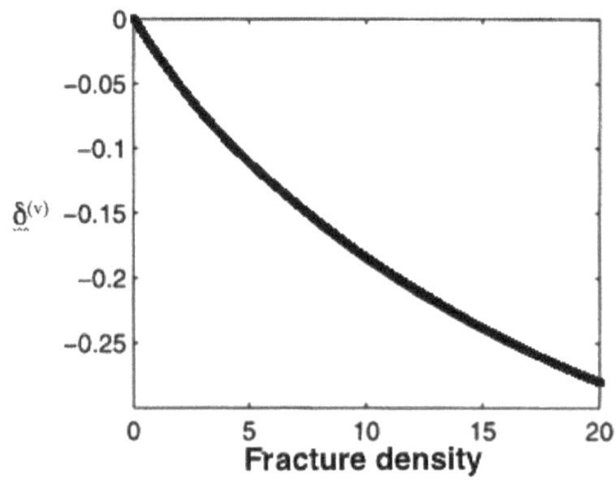

Figure 5.4 Variation in anisotropic parameter as a function of fracture density for gas-saturated sandstone and fracture aspect ratio of 0.05. *Redrawn from Shen, 1998.*

$\varepsilon^{(v)}$ obtained. $\varepsilon^{(v)}$ as a function of fracture density and crack aspect ratio in gas- and water-saturated, fractured sandstones is shown in Figures 5.7 through 5.9.

Parameter $\gamma^{(v)}$, is different from both $\delta^{(v)}$ and $\varepsilon^{(v)}$ in that it measures the degree of shear wave splitting at vertical incidence. Figures 5.10 through 5.13 show that $\gamma^{(v)}$ has little dependence on fluid bulk modulus and crack aspect ratio and is the parameter most directly related to fracture density. Therefore, for parallel, penny-shaped cracks, the shear wave splitting parameter, $\gamma^{(v)}$, can provide direct information about fracture density with least ambiguity.

These findings show that the variations of parameters $\delta^{(v)}$ and $\varepsilon^{(v)}$ are sensitive to fluid content. For gas-saturated fractures, $\delta^{(v)}$ and $\varepsilon^{(v)}$ vary with

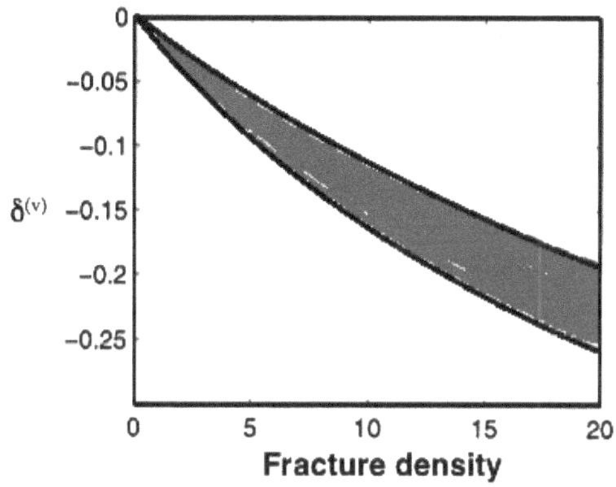

Figure 5.5 Range of variation in anisotropic parameter as a function of fracture density for water-saturated sandstone and fracture aspect ratio of 0.01. *Redrawn from Shen, 1998.*

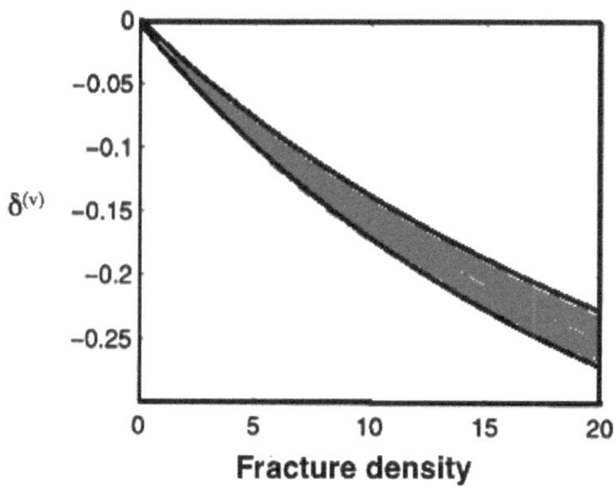

Figure 5.6 Range of variation in anisotropic parameter as a function of fracture density for water-saturated sandstone and fracture aspect ratio of 0.05. *Redrawn from Shen, 1998.*

the fracture density alone, making them an effective indicator of fractures. On the other hand, for water-saturated fractures, the magnitudes of $\delta^{(v)}$ and $\varepsilon^{(v)}$ depend on fracture density, crack aspect ratio, and elastic properties of background rocks. Whereas the shear wave splitting parameter, $\gamma^{(v)}$, is insensitive to both fluid content and crack aspect ratio. It is the parameter most related to crack density. P-wave NMO velocity is controlled by the vertical P-wave velocity, angle between crack normal and survey line, and the parameter $\delta^{(v)}$.

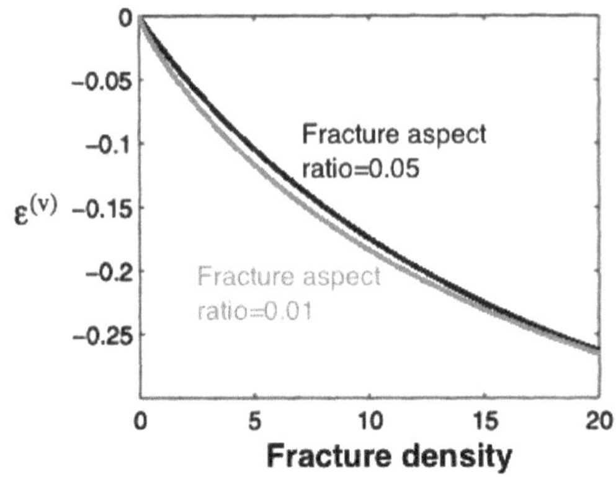

Figure 5.7 Variation in anisotropic parameter as a function of fracture density for gas-saturated sandstone and fracture aspect ratio of 0.01 and 0.05. *Redrawn from Shen, 1998.*

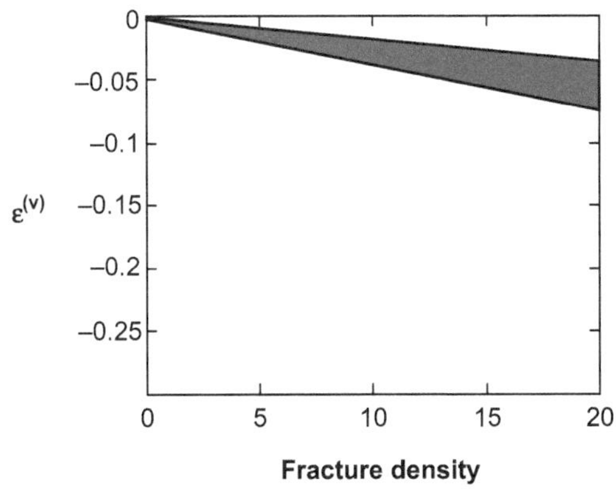

Figure 5.8 Range of variation in anisotropic parameter as a function of fracture density for water-saturated sandstone and fracture aspect ratio of 0.01. *Redrawn from Shen, 1998.*

5.5 RESERVOIR CHARACTERIZATION DURING DRILLING

It is generally understood that vast unconventional gas resources are accessible only with more advanced techniques. Cost-effective drilling techniques and well completion strategies constitute the most successful technological development in the petroleum industry. At present, 90% of the new wells drilled are horizontal. These wells are likely to intersect natural fractures that are predominantly vertical in unconventional reservoirs. Drilling and completion of horizontal wells and multilaterals have

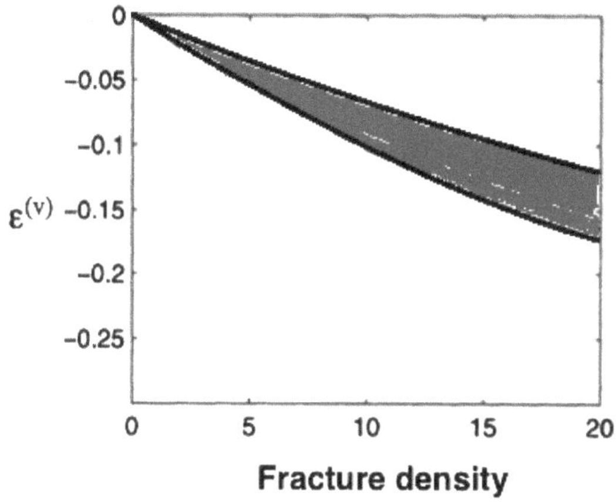

Figure 5.9 Range of variation in anisotropic parameter as a function of fracture density for water-saturated sandstone and fracture aspect ratio of 0.05. *Redrawn from Shen, 1998.*

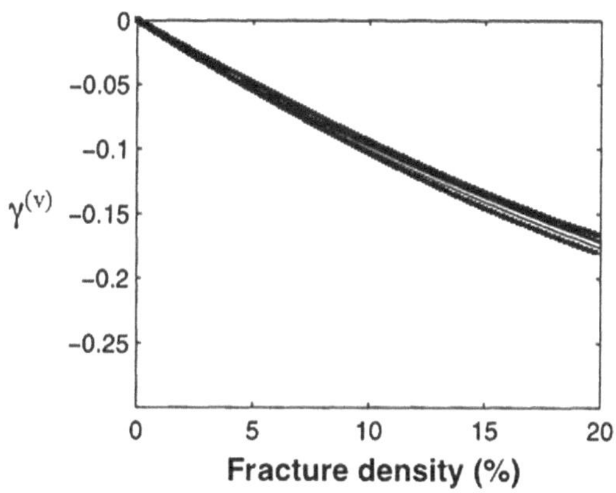

Figure 5.10 Range of variation in shear wave splitting parameter as a function of fracture density for gas-saturated sandstone and fracture aspect ratio of 0.01. *Redrawn from Shen, 1998.*

been the cornerstone of successful unconventional gas recovery schemes. Drilling in this type of formations is often optimized with the so-called managed pressure drilling (MPD). MPD uses tools at the surface such as a choke to control the drilling fluid flow rate and bottomhole pressure. The various drilling operations that MPD comprised of provide economical solutions to many drilling problems such as managing gas kicks, lost circulation and other well control issues, improving ROP, minimizing formation damage, and enabling dynamic reservoir characterization from

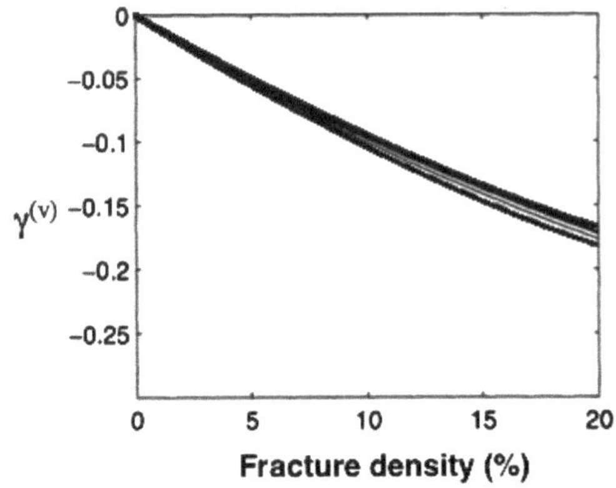

Figure 5.11 Range of variation in shear wave splitting parameter as a function of fracture density for gas-saturated sandstone and fracture aspect ratio of 0.05. *Redrawn from Shen, 1998.*

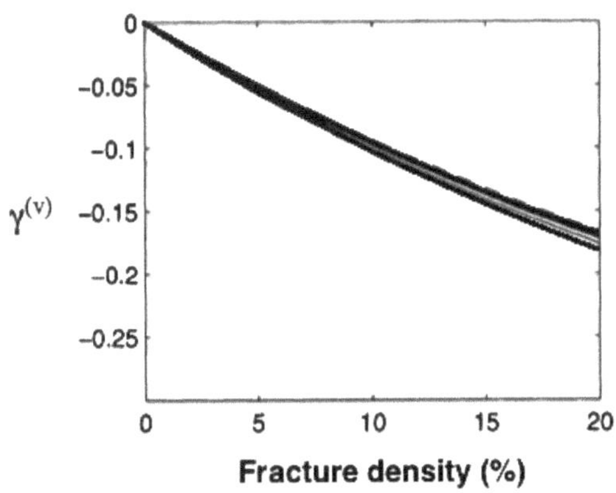

Figure 5.12 Range of variation in shear wave splitting parameter as a function of fracture density for water-saturated sandstone and fracture aspect ratio of 0.01. *Redrawn from Shen, 1998.*

real-time mud log data (Ramalho et al., 2009). Underbalanced drilling (UBD) is a form of MPD that is particularly useful when drilling horizontal wells in tight gas formations. It also generates data that can turn it into a dynamic reservoir characterization tool.

As shown in Table 5.1, the first set of direct data is produced during drilling. As soon as drilling is commenced, the drilling log becomes available. The drilling log contains information about the progress of the well such as measured depth (MD), true vertical depth, inclination, weight on bit, ROP, and gamma ray. It also provides information about the drilling

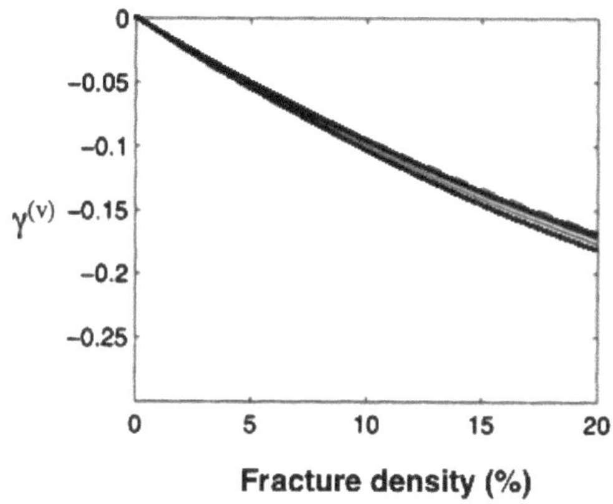

Figure 5.13 Range of variation in shear wave splitting parameter as a function of fracture density for water-saturated sandstone and fracture aspect ratio of 0.05. *Redrawn from Shen, 1998.*

mud circulation system such as mud pit volume, pump pressure, and mud flow rate. Each of these data is valuable for description of lithology as well as fracture system of a formation.

Any drilling process also accompanies the mud log. This log is generally created by the on-site geologist as the well is drilled. Mud logs contain valuable information regarding formation geology and hydrocarbon in place. As the drill bit penetrates the formation, the rock is crushed, and these cuttings are flushed from the well and carried to the surface by the circulating drilling mud. A geologist routinely examines the cuttings and describes the lithology of the formation being penetrated. This information is recorded in the mud log on a depth basis as the well is drilled in order to create a geologic profile of the entire well. At the same time, total gas measurements are also made and recorded. Total gas measurements indicate the relative concentration of hydrocarbons (methane, ethane, propane, etc.) present in the circulating drilling mud at any given time.

During conventional drilling operations, the drilling mud density is maintained above the reservoir pore pressure for wellbore stability issues. While, during this overbalanced drilling useful data are generated that can indicate the presence of fractures and sweet spots (e.g., mud loss, fluid loss, high ROP), information regarding fluid in place is limited. In such system, produced fluids can only occur when an unexpected overpressured zone is encountered or when a transient mud pressure reduction occurs as the drill string is raised (swabbing). Ever since the advent of UBD that uses mud pressure lower than pore pressure, the possibility of extracting in situ fluid in order to characterize the reservoir has been increased drastically. During UBD, hydrocarbon production will occur consistently during drilling, whenever a sweet spot is penetrated. Such "sweet spots" can be the result of natural fractures or otherwise high-permeability zones. Produced fluid as a result of the underbalanced pressure condition is a major focus of this investigation. Recycled fluids may occur if the gas contained within the circulating mud is not entirely released at the surface. In this instance, the remaining gas will be recirculated through the system and will be detected again on the next pass. Finally, contamination will always occur due to various unavoidable causes. Reasons for contamination of the total gas readings include petroleum products intentionally added to the drilling mud, chemical reactions, and degradation of organic mud additives, and even emissions from construction equipment on-site. It is very important to understand all of these processes so that one can effectively interpret the mud log analysis.

The drilling fluid circulation system is essentially a closed-loop system in which the mud is pumped down the well through the center of the drill string, flows through the drill bit nozzle, and is then forced back up to the surface through the annular section. This process ensures that the drill bit stays cool, creates a hydrostatic pressure that is exerted against the wellbore wall for well stability issues and flushes the cuttings to the surface. At the surface, the mud is transported through a shaker table to remove the cuttings and then released into a mud pit to complete the cycle. To obtain the total gas measurements, a gas trap is installed at the mud pit that is able to capture a sample of gas from the mud. An impeller agitates the mud releasing gas into the air. A mixture of this gas–air sample is then sent to the mud logging unit for analysis. The total gas reading is a measure of the relative concentration of all hydrocarbons combined. This concentration is recorded along depth of the drill bit. A correction may be necessary to account for the mud travel time.

While "gas shows" are routinely used to delineate productive zones as well as plan completion strategies, they can also serve as a tool for formation characterization. This is particularly important for unconventional gas reservoirs.

Most unconventional reservoirs are known to contain a high density of natural fractures and production is in fact dominated by fracture flow. Typically, it is presumed that vertical natural fractures exist in situ. Therefore, to produce economically from these types of reservoirs it is most efficient to drill horizontal wellbores. The lateral sections of these wellbores are often drilled underbalanced. During UBD operations, gas is expected to flow into the wellbore consistently throughout the drilling process. The combination of low-permeability matrix, high-permeability natural fractures, and the underbalanced pressure condition leads to the result of highly complicated mud logs. However, a thorough analysis of the mud log data can reveal critical information about the natural fracture system near the wellbore.

The fundamental premise of this analysis is that for unconventional reservoirs, the bulk of the fluid flow takes place through fractures. Such premise is justified based on Darcy's law:

$$\vec{v} = -\frac{k}{\mu}\nabla P \tag{5.1}$$

Here $\vec{v}$ is the velocity, k is the permeability, μ is the viscosity, and P is the pressure. Permeability has the dimension of L^2, which means it is exponentially higher in any fracture than the matrix. For a system with very low permeability, fracture flow accounts for 99% of the flow, whereas in terms of volume fractures account for 1% of the volume of the void (or total porosity). This is significant, because in classic petroleum engineering, governing equations are always applied without distinction between storage site (where porosity resides) and flow domain (where permeability is conducive to flow). In unconventional reservoirs, fluid flow equations apply to the fracture network whereas storage volume applies to the matrix that has little permeability. In fact, permeability values are so low in the matrix, typical Darcy's law does not apply to this domain. It is recommended that Forchheimer equation can be used to describe gas flow. This equation is given by:

$$-\nabla P = \frac{\mu}{k}v + \rho\beta v^2 \tag{5.2}$$

In the above equation, β is an additional proportionality constant that depends on rock properties. For a system with predominant fracture flow, β would depend on the fracture density and aspect ratio.

In case, fracture network is insignificant and the reservoir matrix permeability is very low, flow in such system is best described with Brinkman equation, described as:

$$-\frac{\partial P}{\partial x} = u\frac{\mu}{k} - \mu\frac{\partial^2 u}{\partial x^2}. \tag{5.3}$$

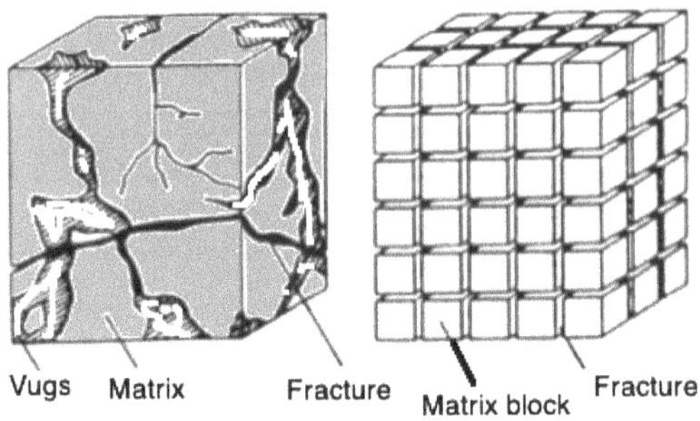

Figure 5.14 Depiction of Warren and Root model.

To date, the most commonly used model is that proposed by Warren and Root (1965). This so-called dual porosity model (Figure 5.14) assumes that two types of porosity are present in the formation, one arising from vugs and fracture system, whereas the other from matrix. For unconventional reservoirs, the matrix permeability is negligible compared to fracture permeability (hence depicted with shades). Warren and Root invoked similar assumptions even for a matrix with relatively high permeability. The approach operates on the concept that fractures have large permeability but low porosity as a fraction of the total pore volume. The matrix rock has the opposite properties: low permeability but relatively high porosity. This approach describes the observation that fluid flow will only occur through the fracture system on a global scale. Locally, fluid may flow between matrix and fractures through interporosity flow, driven by the pressure gradient between matrix and fractures.

Fracture flow is described by Snow's equation (1963), given below:

$$\frac{Q}{\Delta P} = Cw^3 \tag{5.4}$$

Here w is the fracture aperture and C is a proportionality constant that depends on the flow regime that prevails in the formation. Snow's equation emerges from a simple synthesis of parallel plate flow (Poiseuille law) that assumes permeability to be $b^2/12$, where b is the fracture width.

Fracture geometries are often idealized to simplify modeling efforts. In most cases the width is assumed to be constant, and the fracture is usually considered either a perfect rectangle or a perfect circle. In reality, fracture geometries are very complex (Figure 5.14), and many different factors could affect the behavior of fluid flow. In Figure 5.14 that was originally published by Warren and Root (1965), vugs are shown prominently. It is no surprise that they introduced the concept of dual porosity. Indeed, porosity in vugs and in matrix is comparable. For unconventional reservoirs, however, the vugs are nonexistent and most fractures have very little storage capacity, making their porosity negligible to that of the matrix. In determining sweet spots within an unconventional reservoir, the consideration of very high fracture to matrix permeability, k_f/k_m is of importance. For application in dynamic reservoir characterization using real-time mud log data, the term "sweet spot" is used when the drill bit intersects a transverse natural fracture. Such process is equivalent to numerous passes of history match in the context of reservoir simulation.

5.5.1 Overbalanced Drilling

Overbalanced drilling approaches for fracture characterization mostly consist of methods that take advantage of mud-loss data and the rheological properties of the circulating drilling fluid. Drilling mud is usually a non-Newtonian fluid that exhibits shear-thinning behavior. Shear thinning implies that the fluid viscosity decreases with an increase in the shear rate. During drilling, the drilling mud is constantly circulated through a closed-loop system. If the circulation is stopped, the drilling mud will develop into a thick gel. The mud will remain in this state until a pressure exceeding the mud's yield stress is applied, at which point it will return to its "fluid" state. This nonideal fluid behavior has allowed engineers to develop methods to characterize fracture permeability using mud-loss data. In a dynamic setting, mud rheology can be calibrated against mud circulation loss, which is directly related to fracture in an unconventional tight gas reservoir.

During overbalanced drilling, when a sweet spot is encountered, the mud pressure is greater than the fluid pressure contained in the fracture resulting in a flood of drilling mud into the fracture. When the drill bit intersects the fracture, drilling mud will flow into the fracture. Because the matrix permeability is low, leak off into the matrix is minimal and the mud loss is entirely due to fluid flow through fractures.

This mud flow is reflected in the rheology of the returning mud, thereby, creating a correlation between surface-observed rheology and fracture density as well as fracture geometry (Figure 5.15).

Lietard et al. (1999) provide type curves describing mud-loss volume versus time that can be used to determine the hydraulic width of fractures

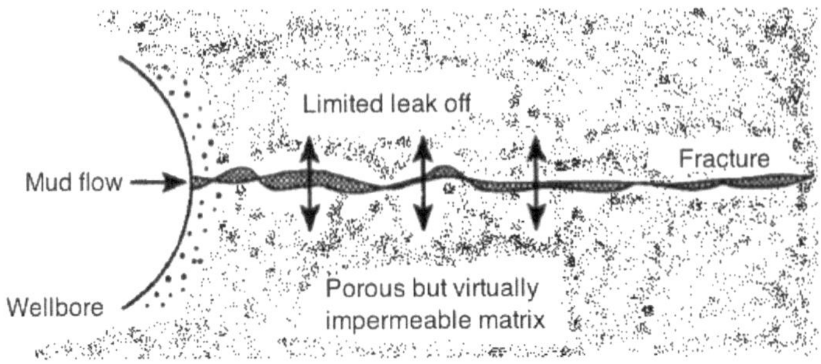

Figure 5.15 Schematic of mud flow in a tight formation with fractures. *After Dyke et al., 1995.*

through a curve matching approach. The type curves are based on an analysis of the local pressure drop in the fracture (Lietard et al., 1999):

$$\frac{dP}{dr} = \frac{12\mu_p v_m}{w^2} + \frac{3\tau_y}{w} \tag{5.5}$$

Where, v_m is the local velocity of the mud in the fracture, μ_p is the plastic viscosity of the mud, w is the fracture aperture, and τ_y is the yield stress of the mud. This equation was improved by Huang et al. (2010) that suggested the following equation:

$$\left(\frac{\Delta P_{OB}}{\tau_y}\right)^2 w^3 + 6R_w\left(\frac{\Delta P_{OB}}{\tau_y}\right) w^2 - \frac{9}{\pi}(V_m)_{\max} = 0 \tag{5.6}$$

In the above equation, R_w is the well radius and V_m is the maximum mud-loss volume. This equation is easier to use than the previously used type curve. However, it is recommended that such correlation be developed for each reservoir.

5.5.2 Underbalanced Drilling

During UBD operations, a low density drilling fluid is used in order to maintain a wellbore pressure profile that is lower than the pore pressure of the formation at all locations along the borehole. One major advantage of UBD over conventional drilling is that formation damage is reduced because a filter cake is not allowed to form near the wellbore. Wells completed with UBD have been shown to perform three to four times better than their conventional counterparts in the same formation. Among others, lost circulation is minimized with UBD. Overall, the ROP is increased significantly with UBD. Figure 5.16 shows how switching from

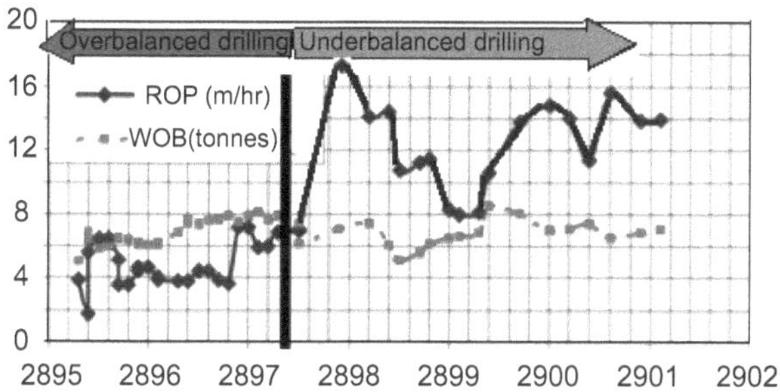

Figure 5.16 Data from a well drilled overbalanced until a certain depth and then switching to underbalanced operations. Immediately as UBD begins, the ROP greatly increases. *Redrawn from Woodrow et al., 2008.*

overbalanced drilling to UBD can drastically increase ROP. This is especially true for tight gas formations or any formation with harder than normal. This phenomenon is not well understood, but it is thought that the increased ROP can be attributed to the lower confining pressure on the formation rock under UBD conditions and to the fact that cuttings are more easily flushed from the bottom of the wellbore reducing the resistance on the drill bit.

The most useful aspect of UBD is in the insight gained during UBD. The deliberate underbalanced pressure difference between the drilling fluid and the formation pressure causes an inflow of formation fluid into the wellbore along the entire drilled section. This can act as a tracer for dynamic reservoir characterization.

The most important piece of data is the rate of fluid flow from the formation into the wellbore. Very rarely are flow rates actually measured at bottomhole. In almost all cases, the formation fluid-flow rate is estimated from surface measurements of inflow and outflow of the drilling fluid. The difference between the mud injected into the wellbore and the outflow of mud from the annular section is often estimated as the formation fluid-flow rate. Methods to account for the expansion of gas due to changes in temperature and pressure must be taken into account to obtain accurate data (Aremu and Osisanya, 2008). Other data that models tend to utilize include bottomhole pressure, ROP, formation porosity, wellbore diameter, and wellbore length. Models generally provide profiles of formation permeability and pore pressure versus depth. Norbeck (2012) presented field data, showing correlation between UBD data and fractures. He extracted field data originally reported by Myal and Frohne (1992). The report investigates the effectiveness of directional drilling in a tight gas

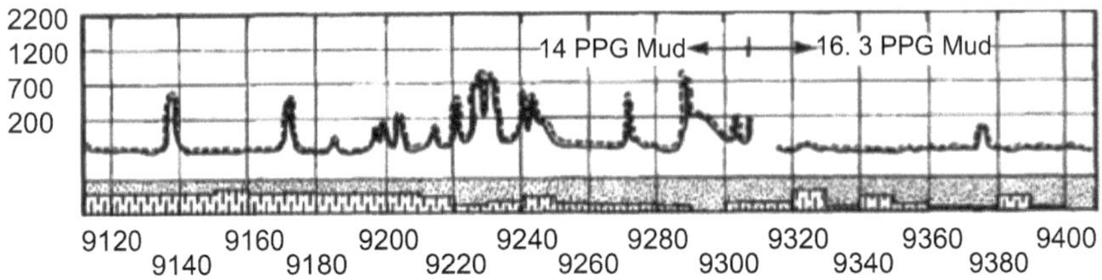

Figure 5.17 Mud log data from a portion of a well, drilled underbalanced in the Piceance Basin. The large gas peaks were attributed to natural fractures that were intersected by the wellbore. *Redrawn from Myal and Frohne, 1992.*

formation located in the Piceance Basin of western Colorado. The formation is known to be highly naturally fractured, and consequently the decision was made to drill a large section of the well underbalanced in order to reduce formation damage and lost circulation. As can be seen from Figure 5.17, at least 10 major gas shows were detected during drilling. These gas shows were attributed to the presence of natural fractures intersected by the wellbore. An increase in mud density not only led to suppression of the gas shows, but also prevented extraction of gas show data and their correlation with fracture distribution. This type of correlation can lead to depiction of the formation fracture network.

Norbeck (2012) proposed two criteria that can be combined to develop a correlation between mud log data and reservoir properties. They are:

Criterion 1: The first criterion is the use of total gas concentration measurements from mud logs. Using a gas chromatograph, the mud logging unit is able to determine the concentration of gas present in the drilling fluid at any given time. As natural fractures are intersected by the drilling bit, gas flows into the wellbore and the amount of gas is proportional to the average fracture permeability. These concentration values show up as spikes on the total gas concentration of mud log.

Criterion 2: The second criterion is based on observations of the mud pit volume. Observations of the mud pit volume at the surface also show potential as a fracture identification criterion. It is widely accepted that decreases in mud pit volume (mud losses) correspond to encounters with natural fractures while drilling overbalanced. It is logical to assume that the reverse is also true during UBD. It means that as a natural fracture is encountered with the drill bit, the formation fluid influx will cause a displacement of drilling fluid in the mud pit. This response is observable at the surface.

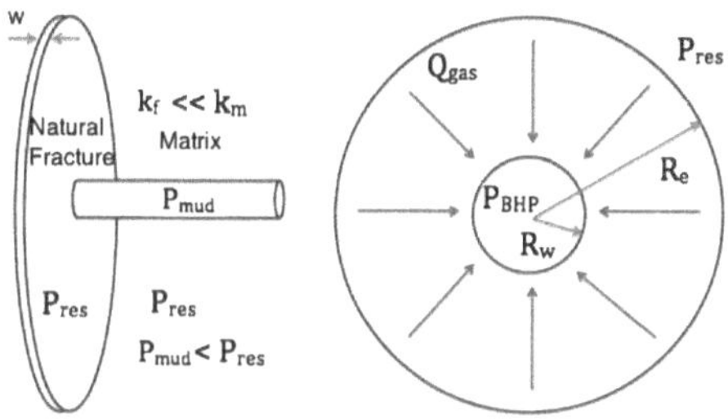

Figure 5.18 Schematic of the model used by Norbeck (2012).

Both criteria are related to open fractures that contribute to flow directly. Furthermore, fractures are uniquely correlated if the formation is tight with negligible permeability. In order to estimate natural fracture permeability, several assumptions have to be made:

1. All natural fractures that have been intersected by the wellbore are transverse to the wellbore and have circular geometry with finite extent (as depicted in Figure 5.18).
2. Natural fractures have constant aperture. At least, it must be assumed that an equivalent aperture is a reasonable and practical approximation. Tortuosity or other "eccentricity" factors can be introduced; however, simple geometry is a reasonable assumption.
3. Gas contained within natural fractures is composed 100% of methane. Methane density and viscosity remain constant while flowing through fractures. This approximation avoids the analysis of compositional effect on gas chromatography.
4. Fluid flow through fractures follows the cubic law relationship. This is typical of all existing fracture flow models. The cubic law relationship assumes steady state, laminar flow between two parallel plates. The cubic law can be derived from a force balance between the forces due to the pressure gradient and the shear resistance on the boundaries, as opposed to the diffusivity equation, which is derived using the principles of conservation of mass. As such, no compressibility term is present in the cubic law relationship. However, the high compressibility of gas will most likely have significant effects on the flow rate through the fracture. Nonetheless, it is assumed that at the high pressure conditions present in the reservoir, compressibility effects will be negligible over the relatively low magnitude pressure drop between reservoir and bottomhole pressures.
5. Matrix permeability is much lower than fracture permeability. For most unconventional reservoirs, this is a reasonable approximation.

6. No charging of the fractures occurs during the time spans considered. Considering low permeability of the matrix, this is a reasonable assumption.
7. The gas influx volume is equal to the mud pit volume increase. For most shallow reservoirs, this is a reasonable approximation. Because no direct measurements of gas flow rate at bottomhole are recorded on today's drilling rigs, it is assumed that the observed mud pit volume increase is equal to the volume of gas that entered the wellbore from the fracture. However, compressibility effects could be significant and add to the uncertainty of this analysis. Also, it is well known that methane is highly soluble in oil-based drilling mud, and it is has been reported that observations of mud pit volume increase as a response to a gas kick will be reduced because of solubility effects.

The following estimates can be obtained from the drilling and mud log data:

- Gas flow rate
- Pressure drop (underbalance)
- Methane viscosity
- Wellbore radius

If an assumption about the radial extent of the fracture can be made, then fracture aperture can be determined as follows:

$$w = \left(\frac{Q}{C\Delta P}\right)^{1/3} \tag{5.7}$$

Then, fracture permeability can be estimated by the following equation that is derived from Poiseuille law, as applied in parallel plate flow.

$$k = w^2/12 \tag{5.8}$$

Norbeck (2012) demonstrated through study of six wells that this technique is both practical and accurate. Data from six wells that were drilled underbalanced were collected. Conductive natural fracture zones are determined for each well, thereby estimating fracture permeabilities. The wells were from two tight gas shale formations, one located in the United States and the other in Canada. The lateral sections of these horizontal wells range from 3000 to 6000 ft. The lateral sections of all wells were drilled using oil-based mud.

In order to verify accuracy of the analysis, a special technique had to be applied because no borehole image logs were available. The main validation

Table 5.4 Length of Lateral Sections, Average True Vertical Depth (TVD) of Lateral Sections, and Average Reservoir Pore Pressures for Corresponding TVD for Wells A-1 and A-2

Well name	Length of lateral (ft)	Average TVD of lateral (ft)	Average reservoir pressure (psi)
A-1	4280	12,381.8	11,550
A-2	3131	12,389.8	11,575

From Corbeck, 2010.

technique used in that study took advantage of a commonly used horizontal drilling technique in which wells are drilled in parallel. Six wells selected for the study constituted three sets of parallel wells drilled to similar elevations. The spacing of these wells was between 500 and 800 ft. The parallel well sets should penetrate similar natural fracture systems. The results of the natural fracture identification analysis for each parallel well set are compared to determine if any patterns exist that may be indicative of the orientation of natural fracture planes.

In the first field case study, first two wells (Well A-1 and Well A-2) were chosen from the same field. The horizontal spacing between these two wells is roughly 800 ft. Well A-1 runs in S–N orientation and Well A-2 runs in the opposite direction. Well A-1 was drilled toe down and Well A-2 was drilled toe up. The geometric properties and the average reservoir pressure for each well, obtained from diagnostic fracture injection testing, are listed in Table 5.4. The targeted pay zone is roughly 175-ft thick.

The analysis of Well A-1 indicates that 10 conductive natural fractures were intersected during the drilling process (see Table 5.5). Two of these natural fractures are within very close proximity to each other and are considered a single conductive natural fracture zone. In total, nine natural fracture zones are present along the lateral of this well. As can be seen in Figure 5.19, the stretch of lateral between 14,000 and 15,500 ft MD contains no conductive natural fracture zones. This is a primary example of the insight that can be gained from this type of analysis. This zone can be selected for creating sweet spots through hydraulic fracturing. The fracture apertures range from 13 to 53 μm. The cross-plot indicates a general positive correlation between mud pit volume peak and gas peak for these conductive natural fracture zones (see Figure 5.20).

Similarly, a total of nine conductive natural fracture zones were identified for Well A-2. The relevant results are listed in Table 5.6. The zones are relatively evenly spaced along the lateral (see Figure 5.21). The fracture

Table 5.5 Results of Fracture Identification in Well A-1

Well A-1

Candidate fracture location (ft)	Fracture aperture (μm)	Fracture permeability (mD)	Mud pit volume peak (bbl)	Gas peak (units)	Underbalance pressure(psi)
12,533.0	30	7.500E+04	4.000	988	1093
12,826.0	35	1.021E+05	3.000	171	1081
12,835.0	53	2.341E+05	3.200	156	1081
12,950.0	44	1.613E+05	1.600	286	1080
13,235.0	42	1.470E+05	1.700	182	1143
13,529.0	37	1.141E+05	1.700	69	1138
13,884.0	36	1.080E+05	1.600	74	1135
13,960.0	42	1.470E+05	2.300	150	1133
15,613.0	35	1.021E+05	1.900	103	1178
16,156.0	13	1.408E+04	1.600	644	1234

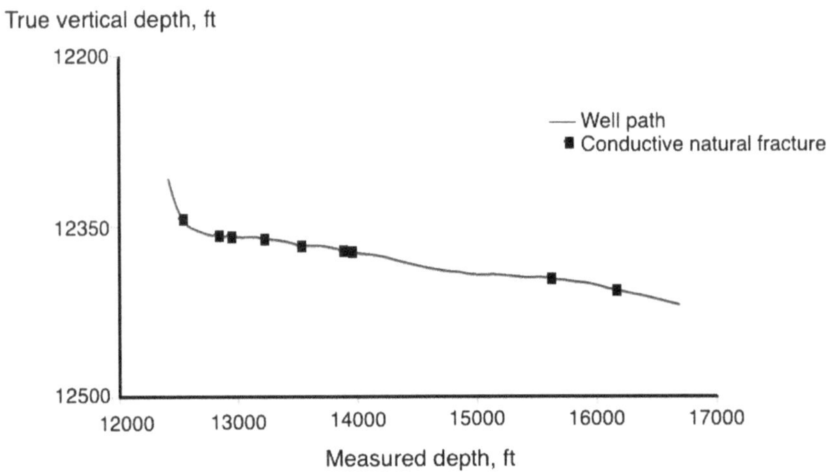

Figure 5.19 Locations of conductive natural fractures along the lateral of Well A-1. *Redrawn from Corbeck, 2010.*

apertures are generally smaller than for Well A-1, ranging from 15 to 38 μm. These estimates could be largely due to the higher level of underbalance maintained while drilling Well A-2. Additionally, the observed rise in mud pit volume due to the presence of these fractures is relatively low. The cross-plot does not show a strong trend for the relationship between mud pit volume peak and gas peak for these fractures, however, it is a positive correlation (see Figure 5.22).

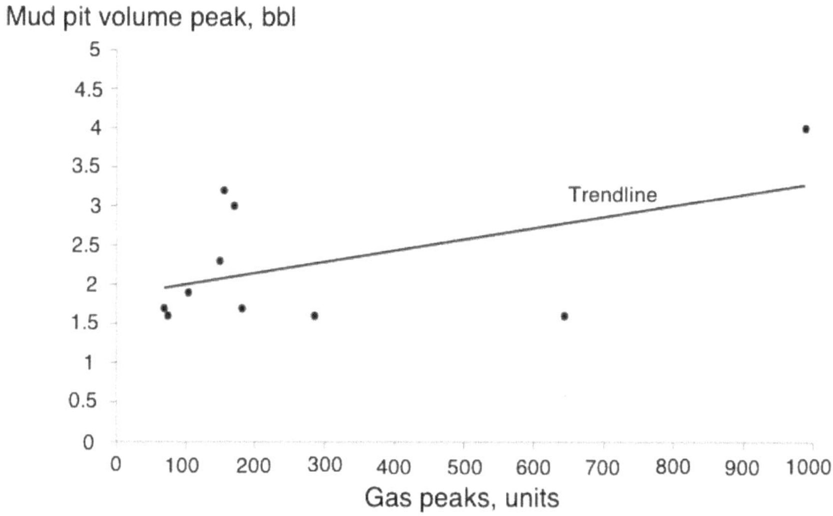

Figure 5.20 Cross-plot of mud pit volume peak versus gas peak corresponding to each conductive natural fracture location identified for Well A-1. *Redrawn from Corbeck, 2010.*

Table 5.6 Results of Fracture Identification in Well A-2
Well A-2

Candidate fracture location (ft)	Fracture aperture (μm)	Fracture Permeability (mD)	Mud pit volume peak (bbl)	Gas Peak (units)	Underbalance pressure (psi)
13,467.0	27	6.075E+04	1.177	88	1262
13,867.0	29	7.008E+04	0.811	53	1334
13,943.0	32	8.533E+04	1.000	75	1335
14,192.0	38	1.203E+05	1.176	67	1339
14,598.0	16	2.133E+04	0.774	682	1339
14,788.0	15	1.875E+04	0.790	328	1336
14,990.0	32	8.533E+04	1.170	924	1334
14,991.0	35	1.021E+05	1.460	715	1334
15,296.0	25	5.208E+04	0.768	70	1338
15,750.0	25	5.208E+04	0.930	88	1421

Similar results were obtained for other fields as well. On average, the computational tool identifies between nine and ten conductive natural fracture zones for each well. For each conductive natural fracture zone, the fracture aperture and fracture permeability was estimated. The estimated fracture apertures all lie within the expected range of values

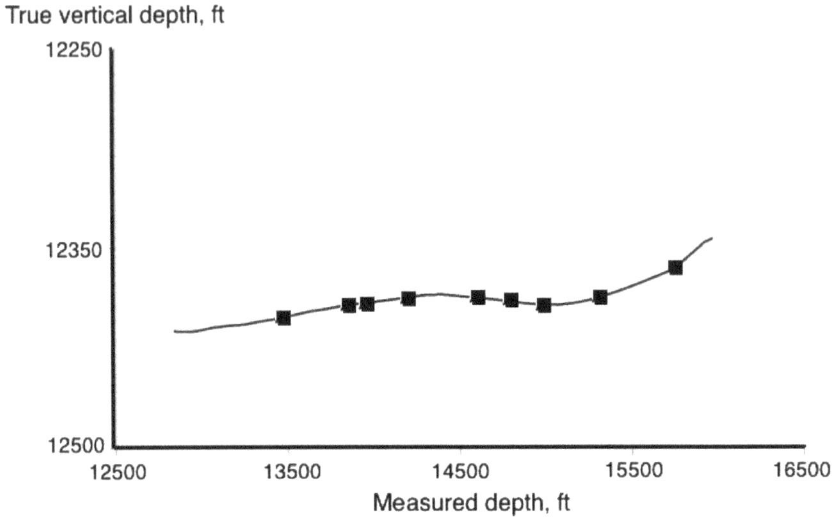

Figure 5.21 Locations of conductive natural fractures along the lateral of Well A-2. *Redrawn from Corbeck, 2010.*

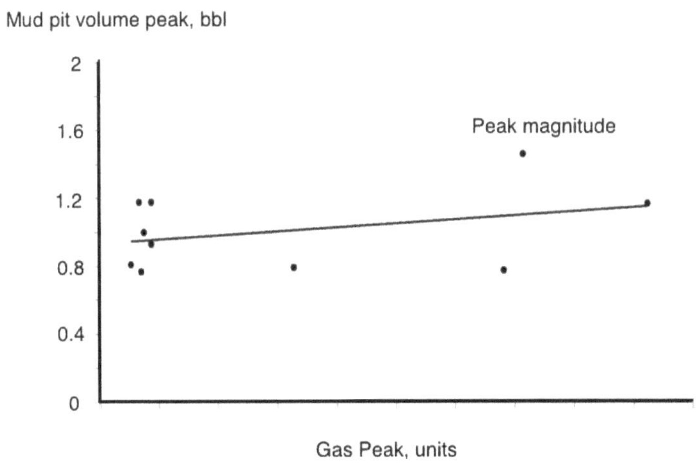

Figure 5.22 Cross-plot of mud pit volume peak versus gas peak corresponding to each conductive natural fracture location identified for Well A-2. *Redrawn from Corbeck, 2010.*

(i.e., 10–1000 μm). Overall, although the estimates of fracture aperture may not be entirely accurate, they should be considered as a lower bound for the true fracture apertures.

In absence of other means of validation, patterns in the locations of conductive natural fracture zones between wells were used. Existence of

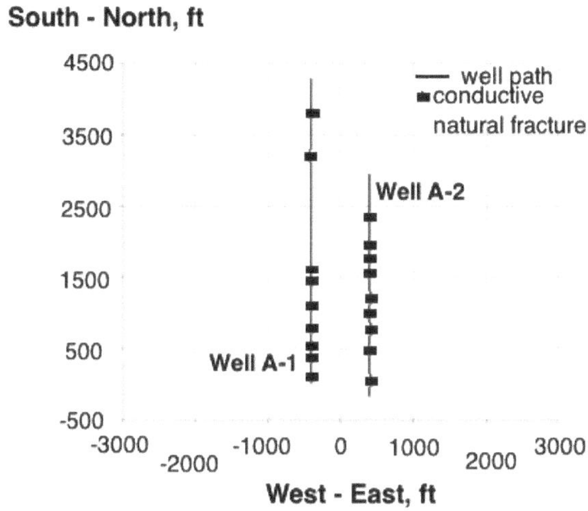

Figure 5.23 Plan view of Field A. Wells A-1 and A-2 are parallel wells drilled in the S–N direction. The lateral spacing between these wells is roughly 800 ft. *Redrawn from Corbeck, 2010.*

such patterns would confirm tectonic continuities that are essential for history matching of the diagenesis involved to validate the results, because each pair of wells should penetrate similar geologic conditions. Once validated the existence of fractures (or sweet spots), orientation of fractures could be used to refine reservoir characterization. For the case in question, two out of the three parallel well pairs exhibit strong patterns. From visual inspection of the results obtained for Field A, two dominant patterns can be observed, as can be seen in Figure 5.23. A total of seven pairs of natural fractures are aligned at an orientation of roughly N65°E (see Figure 5.24). Only one identified natural fracture from Well A-2 does not have a corresponding feature in Well A-1. For each of the seven natural fracture planes identified, the estimated apertures of the corresponding pair of natural fractures compare well, with the exception of pairs four and five (see Table 5.7).

5.6 RESERVOIR CHARACTERIZATION WITH IMAGE LOG AND CORE ANALYSIS

As presented in Table 5.1, image logging and core analysis present an important stage of reservoir characterization. Techniques for directly assessing near-wellbore fracture density, fracture aperture, and fracture orientation are presently available to the industry by means of image log

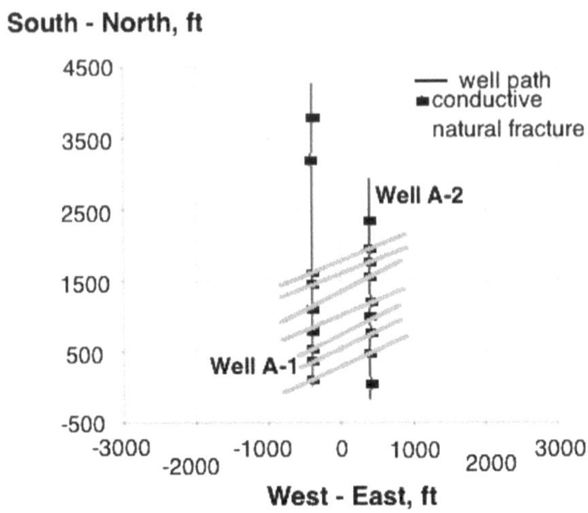

Figure 5.24 Natural Fracture System Orientation #1 for Field A. A dominant pattern exists that seems to indicate the presence of a natural fracture system oriented at N65°E.

Table 5.7 Comparison of Estimated Fracture Aperture between Pairs of Conductive Natural Fractures for Wells A-1 and A-2

Pair no.	(μm)	(μm)
1	42	29
2	36	32
3	37	38
4	42	16
5	44	15
6	44	34
7	30	25
Average	39	27

Pairs numbered from north to south.

testing. Examples of image log techniques include borehole video camera, acoustic formation image technology (AFIT), and resistivity image logs. These three methods are based on different fundamental principles and each has its own set of advantages and disadvantages. A common drawback is that the image resolution quality is generally too poor to be able to identify conductive features that are believed to be on the order of 100-μm wide. Circumferential Borehole Imaging Log (CBIL), when utilized for potential fractured layers already tagged by other techniques (as acoustic waveforms), has been proved as very effective and detailed. Each of the following logs

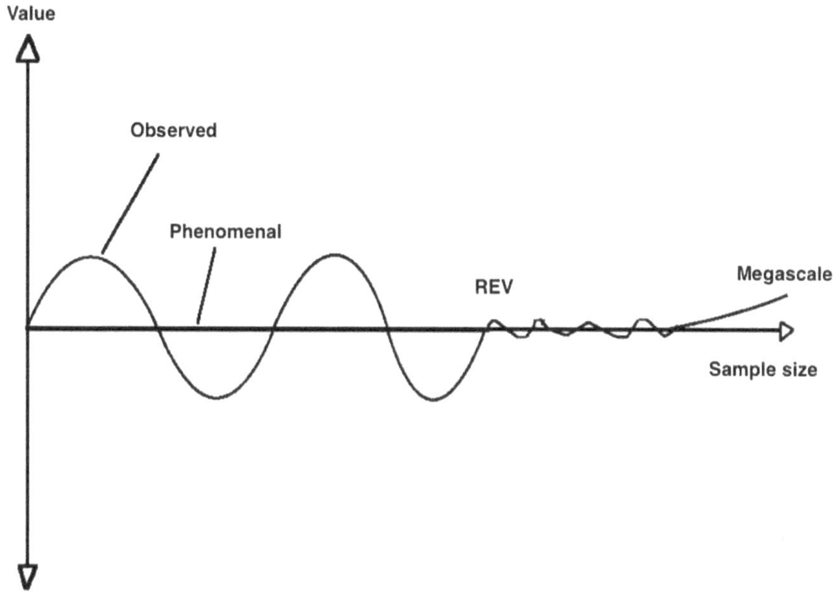

Figure 5.25 Representative elemental volume in fractured reservoirs is greater than core size. *Redrawn from Islam et al., 2014.*

also gives information that can lead to refinement of reservoir characterization.

- Spontaneous Potential
- Gamma Ray Log
- Density Log
- Neutron Log
- Dual-Induction Log
- Sonic Log

Following new array of logs has recently been introduced.

- Array Induction Log
- Array Sonic Log
- Electromagnetic Propagation Log
- Nuclear Magnetic Resonance

One should highlight here that the representative elemental volume (REV) for fractured reservoirs is greater than the core size as well as the depth of resolution of most imaging tools. As can be seen from Figure 5.25, below REV, fluctuations occur. Any correlation that is apparent must, therefore, be corroborated/refined with previously available data, starting from data acquired during geological survey.

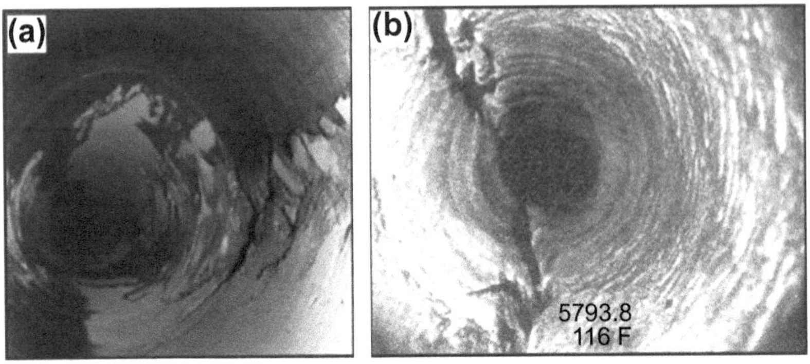

Picture 5.1 (a) Example of borehole breakout taken by a downhole camera. (b) Example of a borehole fracture observed on a downhole camera. *Figure 4(b) from Asquith and Krygowski, 2004.*

Visual observation with downhole camera is the most effective tool for gathering information on natural fractures. Conventionally, borehole video cameras have been used in oil and gas wells to investigate wellbore integrity, but they have also been utilized for the purposes of natural fracture characterization with limited success. One report by Overbey et al. (1988) presents a horizontal well drilled with air in which borehole video was used to identify fractures. It is important to have a clean borehole so as to facilitate borehole imaging. In this particular study by Overbey et al. (1988), more than 200 features were identified as natural fractures over the 2217 ft of wellbore that was surveyed with the video camera. The report presents an approach to interpret the fracture orientation based on the geometry of the observed feature. The report makes no attempt to quantify the aperture of the features identified or to distinguish between conductive and nonconductive features. The report concludes that borehole video cameras can be implemented as a natural fracture identification technique in air-drilled horizontal wells in low-pressured reservoirs. Optical image logging tools, such as the Optical Televiewer (from Schlumberger) and Downhole Video tool (from Downhole Video), are wireline tools that utilize cameras to directly image the wellbore wall. Picture 5.1 shows such an image.

5.6.1 Geophysical Logs

In general, following geophysical logs are routinely available for reservoir characterization.

- Gamma Ray (GR) Spectralog – this one is based on collecting gamma-ray signals from natural rocks in the reservoir. This log can be performed in open as well as cased holes and allows a detailed stratigraphic reconstruction for the entire depth of the well, even in case of cuttings absence due to total loss of circulation.

- Densilog and Acoustilog—contribute to the stratigraphic structural reconstruction of the well and are essential for the bulk density and seismic wave velocity determination in order to give calibration elements for the interpretation of surface gravimetric and seismic surveys. Furthermore these logs are fundamental to computing the formational elastic parameters and their variations in case of presence of fractures.
- Multiarm Caliper—is very useful not only for the imaging of the hole geometry, but also for structural reconstruction by means of breakout analyses.
- Borehole imaging log—allows the 360° mapping of the walls of the hole by analyzing the formational variation of both velocity and resistivity. This is the only specific tool for the direct fracture analyses in terms of nature and geometric parameters.

Usually, during the field recording phase it is possible to make a preliminary individuation of levels that can be potentially fractured. These are very often associated to:

- sharp decrease of bulk density and P-wave velocity (V_p);
- strong attenuation of the wave form;
- intense and very thin cavings in the walls of the hole;
- peaks of GR in case of mineralized fractures.

Borehole imaging tools provide an image of the borehole wall that is typically based on physical property contrasts. There are currently a wide variety of imaging tools available, though these predominately fall into two categories: resistivity and acoustic imaging tools.

Resistivity imaging tools provide an image of the wellbore wall based on resistivity contrasts (Ekstrom et al., 1987). Resistivity imaging tools consist of four- or six-caliper arms with each arm ending with one or two pads containing a number of resistivity buttons. Resistivity imaging tools provide one with the same information on borehole diameter and geometry as the older dipmeter tools, however, the resistivity buttons also allow high-resolution resistivity images of the borehole wall to be developed. There are a wide variety of wireline resistivity imaging tools available, some of the more common tools are the Formation MicroScanner (FMS; from Schlumberger), Formation MicroImager (FMI; from Schlumberger), Oil-Based MicroImager (OBMI; from Schlumberger), Simultaneous Acoustic and Resistivity tool (STAR; from Baker Atlas), Electrical MicroScanner

(EMS; from Halliburton) and Electrical Micro Imager (EMI; from Halliburton). Furthermore, recent years have seen the development of a range of logging while drilling (LWD) or measurement while drilling (MWD) resistivity image logging tools, such as the Resistivity At Bit (RAB; from Schlumberger) and STARtrak (from Baker Inteq). For more details on resistivity image logging tools see Ekstrom et al (1987) or Asquith and Krygowski (2004).

Acoustic tools, on the other hand, emit high frequency sonar waves. The acoustic imaging tool then records the amplitude of the return echo as well as the total travel time of the sonic pulse. The acoustic wave travel time and reflected amplitude are measured at numerous azimuths inside the wellbore for any given depth. These data are then processed into images of the borehole wall reflectance (based on return echo amplitude) and borehole radius (based on pulse travel time). There are a wide variety of acoustic imaging tools available, some of the more common tools are the Borehole Televiewer (BHTV; from Schlumberger), Ultrasonic Borehole Imager (UBI; from Schlumberger), Circumferential Borehole Imaging Log (CBIL; from Baker Atlas), Simultaneous Acoustic and Resistivity tool (STAR; from Baker Atlas), Circumferential Acoustic Scanning Tool-Visualization (CAST-V; from Halliburton) and the LWD/MWD Acoustic Caliper tool (ACAL; from Halliburton).

The use of AFIT has been implemented to characterize the permeability of feed zones in oil and gas and geothermal wells. A description of the AFIT tool is given by McLean and McNamara (2011):

> *As the AFIT tool is lowered and raised in the well an acoustic transducer emits a sonic pulse. This pulse is reflected from a rotating, concave mirror in the tool head, focusing the pulse and sending it out into the borehole. The sonic pulse travels through the borehole fluid until it encounters the borehole wall. There the sonic pulse is attenuated and some of the energy of the pulse is reflected back towards the tool. This is reflected off the mirror back to the receiver and the travel time and amplitude of the returning sonic pulse is recorded. Through the use of the rotating mirror ($\leq$ 5 rev/sec) 360° coverage of the inside of the borehole wall can be obtained.*

In practice, the interpretation of AFIT data is quite sophisticated. Planar natural fractures appear as sinusoids in the imaged data set, as shown in Figure 5.26.

Data processing software allows for characterization of geologic features including strike and dip, fracture aperture, and fracture density. The signal amplitude can be used to distinguish between open and closed fractures.

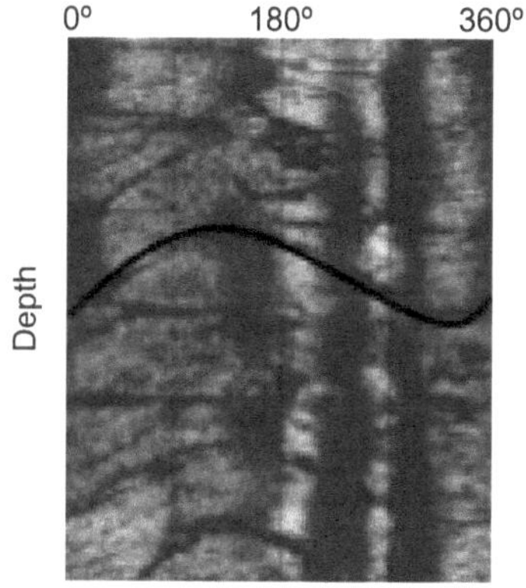

Figure 5.26 Example of an AFIT image log. The horizontal axis is azimuth around the wellbore. The sinusoids are interpreted as planar geologic features. *From Mclean and McNamara 2011.*

Low amplitude signals are seen as dark features on the acoustic image and often interpreted as open. High amplitude signals are seen as light features on the acoustic image and are thought to be attributed to mineral fill. These high amplitude features are usually considered as closed fractures that do not contribute to flow. While McLean and McNamara (2010) report a good level of correlation between measured feedzone fluid velocity and fracture aperture determined from AFIT, the fracture apertures reported range from several centimeters to greater than 50 cm. This implies that the resolution of AFIT can at best distinguish fractures of roughly 1–2 cm. This is nowhere near the level of resolution quality necessary for fracture characterization in highly fractured tight gas reservoirs.

Resistivity image logs, also called FMI logs, have been documented as an improved technique to characterize geologic features along the wellbore. These techniques make use of a tool that places an electrode at constant electrical potential against the borehole wall and measuring the current. Picture 5.2 shows the device. The tool is a small-diameter imaging tool that can be deployed with or without a wireline. The high sampling density of these tools (e.g., 120 samples per foot) provides extremely high resolution in the image quality. Microresistivity imaging tools have the ability to visualize features down to 2 mm in width. Careful interpretation of resistivity image logs can provide helpful information about the geologic conditions near wellbore, including dip analysis, structural boundary interpretation, fracture characterization, fracture description, and fracture distribution.

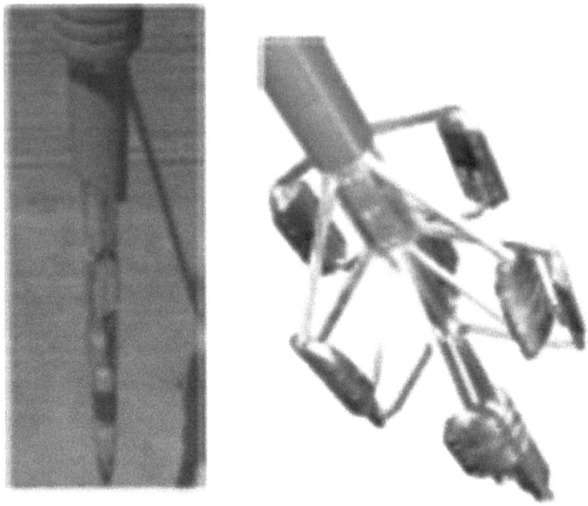

Picture 5.2 Photograph of a microresistivity imaging device (left) passing through drill pipe and (right) in the open position. *From Kalathingal and Kuchinski 2010.*

Data obtained from all three of these image log techniques can be analyzed to gain useful insight about the natural fracture system and state of stress system near wellbore. Barton and Zoback (2002) present an approach for discriminating natural fractures from drilling-induced fractures from different types of image logs, and using this knowledge to determine the state of stress in situ. It has been well documented that drilling-induced tensile fractures will form in the azimuth of the maximum horizontal principal stress. If the three in situ principal stresses can be determined and the formation fluid pressure is known, then a Coulomb failure analysis can be applied to the natural fractures identified in the image logs. Shear and effective normal stresses acting on each fracture plane can be determined from knowledge of the orientation of the fracture plane with respect to the orientations of the principal stresses. The Mohr–Coulomb failure envelope for fractures is determined from laboratory measurements on prefractured rock. The failure line is constructed assuming no cohesion and using the friction angle of the prefractured rock. Poles to fracture planes are then displayed on the Mohr diagram. Critically stressed fracture planes lie above the Mohr–Coulomb failure envelope (Figure 5.27). Barton and Zoback (2002) report a strong correlation between critically stressed fracture planes and hydraulic conductivity of the fractures. These findings indicate that only a small percentage of the total number of fractures are likely to contribute to flow.

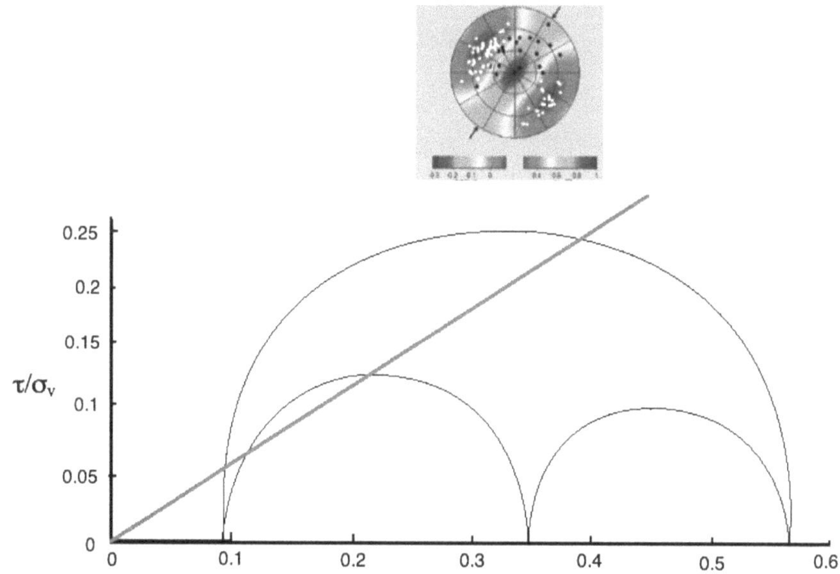

Figure 5.27 These figures illustrate the concept that critically stressed natural fractures predominantly contribute to fluid flow through reservoirs. *Redrawn from Barton and Zoback 2002.*

5.6.1.1 Circumferential Borehole Imaging Log

Recently, the CBIL, based on the digital acoustic imaging technology (McDouglas and Howard, 1989), has gained popularity among geophysical tools for characterizing fractured formations. All the processing steps are mainly aimed at pointing out all those variations of the rock physic characteristics that can be related to the presence of fracture systems.

The first processing phase involves the Densilog and Acoustilog in order to compute the acoustic impedance, the reflection coefficient, and the synthetic seismogram. The last one is particularly useful for a comparison with surface and well seismic profiles data, because seismic reflections have been proved to be very often a signature of fractured horizons.

The wave form analysis, recorded by means of advanced digital acoustic tool, allows to map the image of the instantaneous amplitude. This shows the wave form energy distribution and content evidencing very clearly that wave form attenuation is due to fractures. Furthermore, the S-wave velocity (V_s) and the V_p/V_s ratio are also computed from the wave form analyses. These parameters are combined with the density values and many elastic properties can be computed (see Figure 5.28). Among these elastic

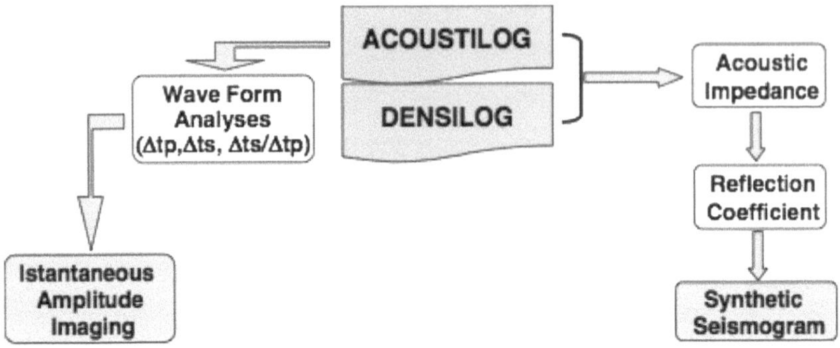

Figure 5.28 Processing flow chart of density and acoustic well logging data. *From Batini et al., 2002.*

parameters, the fracture toughness modulus is particularly sensitive to the presence of fractured levels.

The second processing phase (Figure 5.29) is aimed at the fracture characterization of both the nature and the structural pattern using data from Multiarms Caliper and CBIL (orientation-corrected in case of deviated wells). Rough structural information comes from the breakout analysis of the Multiarms-oriented Caliper that allows the definition of the minimum horizontal stress direction (σ3), which is orthogonal to the fracture planes considering a vertical direction of the maximum stress (σ1).

CBIL data allow detailed structural reconstruction. In the CBIL tool an acoustic transducer, continuously spinning 360° of the walls of the hole, emits an acoustic pulse directed into the formation and records both the amplitude and the travel time of the returning wave. The acoustic amplitude is mainly a function of the acoustic impedance of the formation, so that fractures and their nature (open, mineralized, foliation, etc.) can be clearly evidenced. Even though the depth of penetration is not impressive with CBIL, detection of fractures even at the hole surface is useful and practical for reservoir characterization.

Advanced CBIL processing techniques provide enhanced 360° acoustic amplitude images of the reflected wave. On these images, it is possible to distinguish different types of fractures as a function of both the acoustic impedance variation degree and of their shape and size. These "structural events" can be then picked and all the geometric parameters (i.e., strike, inclination, and dip direction) computed. Correlations can also be made with other data (e.g., downhole camera, gamma ray, geological data) in order to refine lithological data.

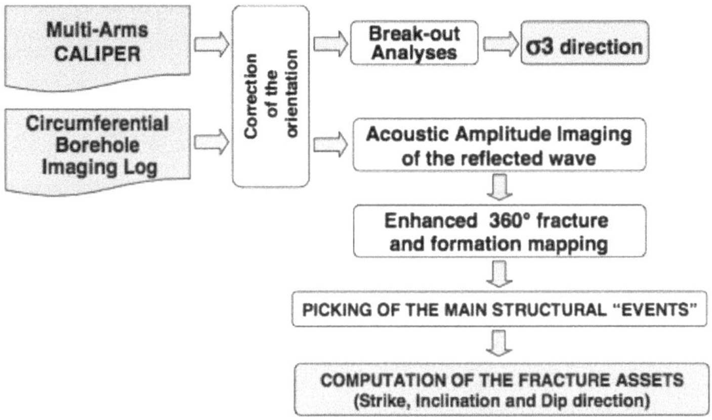

Figure 5.29 Processing flow chart for fracture analyses from well logging.

The most effective way to refine CBIL information is to use it in combination with well test data, which include thermal gradient and injectivity. The utilization of temperature and pressure log can be a useful tool for the identification of each productive zone in the well and the direct measurement of the injectivity. As stated earlier in this chapter, UBD creates a dynamic database. In absence of UBD, similar information can be extracted during an injection test. These test results are affected by the existence of different fractures inside the well, thereby generating data on fracture aspect ratio, density, and others. The temperature profile during an injection test will exhibit a change of slope of the thermal gradient where there is a change in the flow rate, i.e., where there is an adsorbing zone: the thermal gradient is proportional to the fluid that passes in the formation. For a gas well, the thermal change is more intense due to augmented Joule Thomson effect.

The permeability distribution of the reservoir must provide a hydraulic connection throughout all the system; a pressure change in a part of the reservoir (due to exploitation or injection) is propagated in all the system. The propagation velocity of the pressure wave depends on the so-called "hydraulic diffusivity." The well testing is the way for measuring the most important reservoir parameters, as well as the characteristics of the fluid motion. During the drawdown/injection tests, the pressure gauge is placed close to the productive zone and the pressure change is recorded while the well is operated at constant production/injection rate. From the shape of the curve it is possible to identify the reservoir's unique characteristics: the trasmissivity (the permeability–reservoir height product), the skin factor (the well–reservoir coupling factor), the deviation from the ideal radial flow (storage effects, closed or constant pressure boundaries, linear motion of the fluid along preferential paths).

During an interference test, the pressure change of a given well is recorded, while a drawdown/injection test of another one is performed. This is a very important way for measuring the average characteristics of the reservoir in the volume between the two wells, or for establishing a higher limit of the permeability in the case of negative response.

Batini et al. (2002) presented a comprehensive field case study that utilizes CBIL along with well tests and other geophysical logging. They reconstructed the stratigraphy in order to determine main rock physical properties for each geological formation of interest. Table 5.8 gives an example of geological characterization performed by means of the GR Spectralog, which gives a value of total GR and of its spectral components: Potassium (K), Thorium (TH), and Uranium (U).

Table 5.9 shows some of the physical characteristics of the rock. This is necessary for further analysis of data. They also conducted core analysis on two core samplings within the interval in question. These data can be utilized to refine static data gathered independent of core sampling. The bulk density of 2.6 g/cm^3 can be compared with the previous indirect measurement from geophysical logs of 2.77 g/cm^3 for the mica schists reservoir rock. Table 5.10 lists these data.

The case study involves a geothermal well that was drilled during May 8, 2000 through September 12, 2000. The formation is cased and the open hole started at a depth of 2202 m. The first important fractured zone has been highlighted at 2600 m; after acidification and hydraulic stimulation an injection test measured a low injectivity: 1.6 $m^3/h/bar$. Subsequently, a T&P = temperature and pressure log has been recorded during another stimulation (with 80 kg/s for 2 5 h), followed by another medium-duration injection test (with 8 kg/s). Three adsorbing zones have been identified, but, due to the low overall injectivity, it was decided to deepen the well, until the final depth of 4002 m was reached. The following tests were performed:

- Build up immediately after drilling;
- A 17-day production test (the well production could be estimated as 4 kg/s at 1.6 MPa well head pressure);
- Two T&P logs during the production test, with an indication of six productive fractured zones (Fig. 5.30);
- An interference test, showing a linear motion connecting the two wells;
- Final build up after production test.

Table 5.8 Geological Characterization from GR Spectralog

Lithology	Depth interval (m)	GR (GAPI)	K (%)	TH (ppm)	U (ppm)
Neogene Sediments	0–280	44.0 ± 3.8	1.1 ± 0.1	3.9 ± 0.6	2.0 ± 0.5
Flysch	280–550	63.0 ± 5.3	1.9 ± 0.1	2.8 ± 0.77	2.8 ± 0.7
Tectonic Wedges	550–1900	48.0 ± 7.6	1.92 ± 0.1	6.7 ± 1.2	2.5 ± 0.7
Phyllites	1900–2220	93.5 ± 12.5	2.38 ± 0.6	11.7 ± 1.9	3.2 ± 1.1
mica schists	2220–3800	109.8 ± 35.5	2.40 ± 0.9	12.7 ± 4.6	4.4 ± 1.6
Gneiss	3800–4000	N/A	N/A	N/A	N/A

GAPI = gamma ray unit of API (American Petroleum Institute)

Table 5.9 Physical Characteristics of the Reservoir Rock

Parameter	Value
V_p	4.87 ± 0.32 (km/s)
V_s	2.81 ± 0.20 (km/s)
V_p/V_s	1.7 ± 0.1
Density	2.77 ± 0.07 (g/cm^3)
Acoustic Impedance	12.7 ± 1.6 (km/s g/cm^3)
Young's modulus	53.25 ± 10.2 (GPa)
Poisson's Coefficient	0.2 ± 0.06
Fracture Toughness	0.006 (GPa^{-1})

Batini et al., 2002.

Table 5.10 Core Analysis Results

Core sample	Depth interval	Grain density (g/cm^3)	Bulk density (g/cm^3)	Porosity (%)	Heat capacity (J/g°C)
mica schists	3085–3088	3.0	2.6	1.3	0.67
Gneiss	3830–3833	2.9	2.6	1.6	0.67

Batini et al., 2002.

Figure 5.30 shows the temperature profile. Unfortunately, the drawdown analysis does not give a clear indication of the reservoir characteristics, due to the superposition effects of each production zone. The final build-up shows a slight tendency toward a radial motion, with a stabilized flow rate of 2.2 kg/s at 1.6 MPa. Assuming 1000 m of reservoir height, the formation permeability can be estimated as 0.7 mD and a negative skin factor of −4.2. Table 5.11 shows results of the well test.

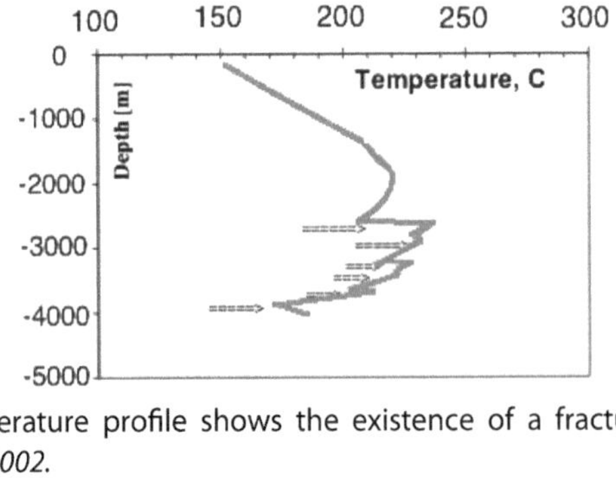

Figure 5.30 Temperature profile shows the existence of a fractured zone. *Redrawn from Batini et al., 2002.*

Table 5.11 Well Test Results

Depth (m)	First T&P flow rate (kg/s)	Second T&P flow rate (kg/s)
2640	2.08	1.77
2910	1.11	0.14
3240	N/A	0.33
3400	N/A	0.14
3660	N/A	0.33
3880	2.31	1.44
TOTAL	5.50	4.15

The CBIL results gave a means of most definitive confirmation of fractures, as well as quantitative information. The geophysical log processing confirmed that the six depth intervals preliminarily identified for the CBIL investigation were particularly affected by signatures related to the presence of fractures (Figure 5.31).

The CBIL analysis allowed the identification of different kinds of fractures and their geometrical parameters (Figure 5.32). These last were processed and mapped for each interval as "pole density of all the fracture planes," using the Wulf's lower hemisphere stereographical projection. It should be noted that visual inspection of cores is an integral part of the analysis (last column of Figure 5.32).

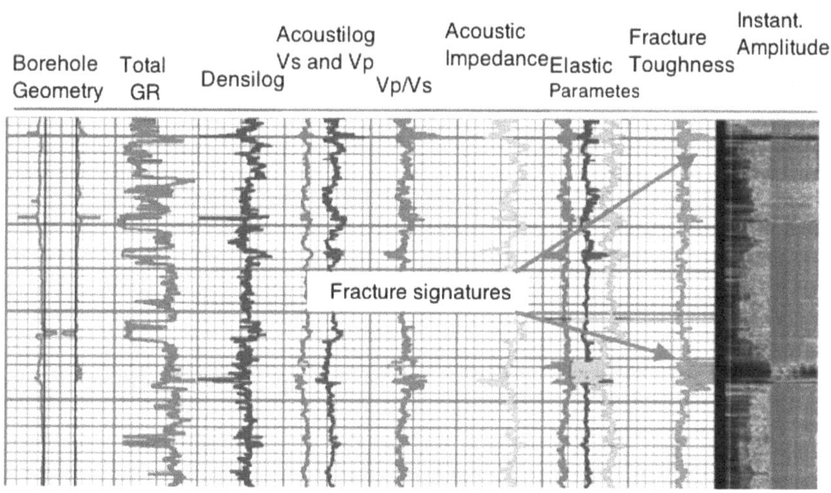

Figure 5.31 Fracture signatures from geophysical logs. *From Batini et al., 2002.*

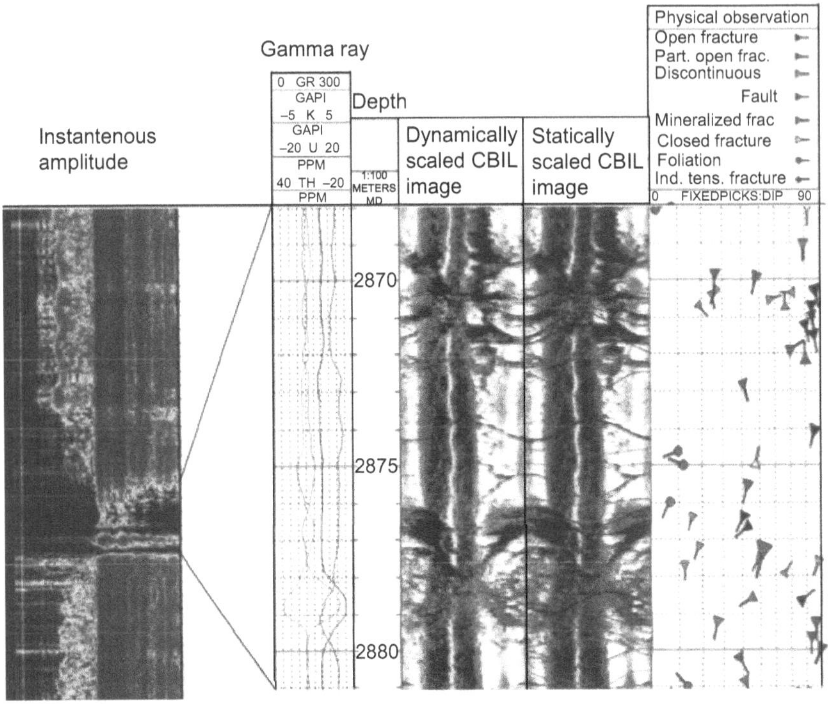

Figure 5.32 Fracture analysis from CBIL. *From Batini et al., 2002.*

For each interval, the pole density distribution for fractures and faults (foliations excluded) is shown in Figure 5.33 together with the most representative cycle-graphical traces. These are characterized by a prevalent E–W azimuth direction, the dip direction is almost variable, but the inclination shows a tight variation between 65° and 80°.

A comparison with core fracture analysis is possible only for cores extracted from the same metamorphic formation in the vertical well Sesta 6 bis. They are not oriented, so that the only reliable value is an average slope of about 70° measured on few samples of continuous joints.

A comparison between the fractures detected by geophysical logs and well testings is given in Table 5.12 together with a tentative correlation between fracture asset and productivity.

There is quite a correspondence with fractures detected by well testing in four out of six intervals characterized by geophysical fracture signatures. Excluding the deepest productive zone at 3880 m, not investigated by CBIL, the levels with higher productivity (1.77 and 0.33 kg/s) are

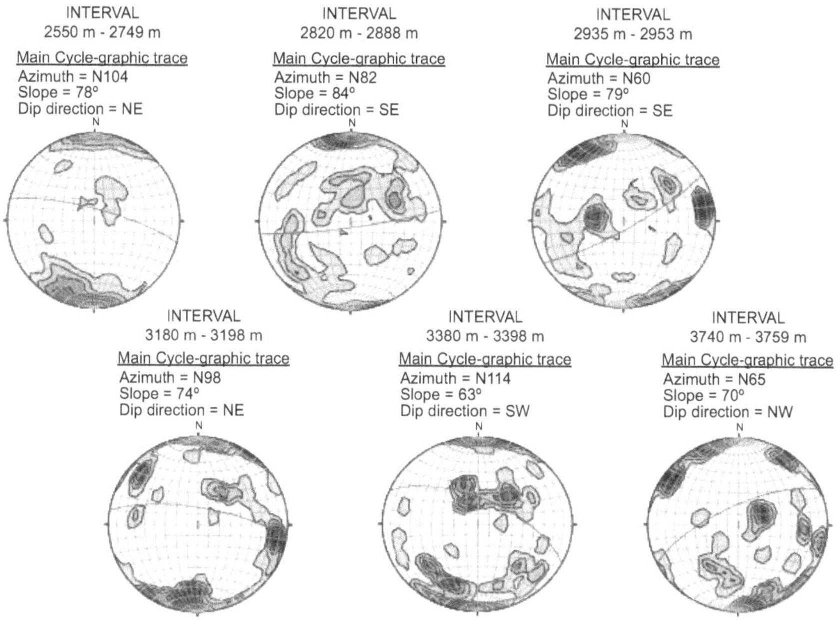

Figure 5.33 Fracture asset mapped as pole density. *From Batini et al., 2002.*

associated with subvertical fractures (inclination of 70–87°) with an E–W strike direction and northward dip direction.

5.6.1.2 Petrophysical Data Analysis Using Nuclear Magnetic Resonance

Recent years have seen a surge in the use of NMR-based logging tools. It is commonly perceived that NMR alone or in combination with conventional logs as well as special core analysis (SCAL) data can lead to better determination of petrophysical properties of heterogeneous tight gas sand reservoirs (Hamada, 2009). Hamada reported the determination of following parameters of a tight gas formation:

1. Detailed NMR porosity in combination with density porosity, Φ_{DMR};
2. NMR permeability, k_{BGMR}, which is based on the dynamic concept of gas movement and bulk gas volume in the invaded zone; and
3. Capillary pressure derived from relaxation time T2 distribution, with further possibility of its use in measuring formation saturations, particularly in the transition zone.

Hamada (2009) presented an interesting case study that is used in this section as a template. The case study involves a gas condensate field that produces from a Lower-Mesozoic reservoir. The reservoir is classified as a

Table 5.12 Comparison between Geophysical Logs and Well Testing

Fractured levels from CBIL				Fracture from well testing	
Depth (m)	**Strike direction**	**Slop and dip direction**	**Number of Samples**	**Depth (m)**	**Production flow rate (kg/s)**
2550–2750	E–W	87° N	242	2640	1.77
2820–2890	E–W	84° SE	72	Not detected	
	NNW–SSE	46° E	22		
	N–S	50° W	22		
2915–2975	N–S	27° E	36	2910	0.14
3180–3210	E–W	70° N	18	3240	0.33
3380–3410	WSW–ENE	24° SSE	30	3400	0.14
				3660	0.33
3730–3780	Not definable		Few	Not detected	
Bottom Log					
				3880	1.44

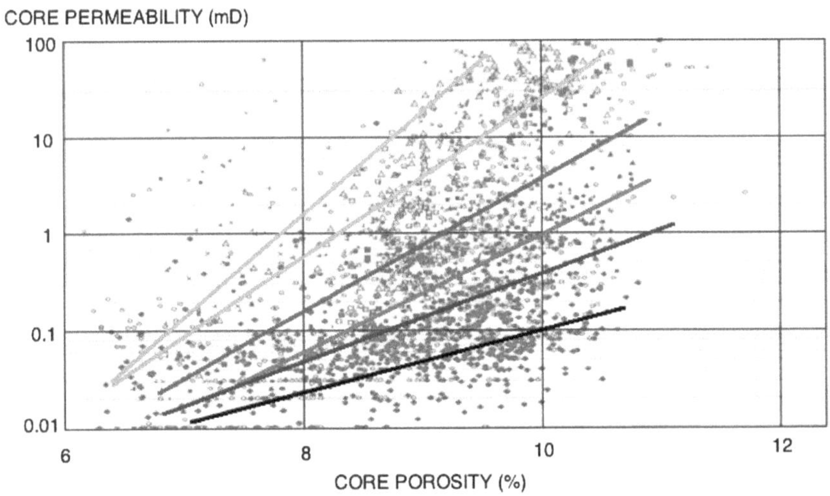

Figure 5.34 Core permeability versus core porosity for a heterogeneous formation. *From Hamada, 2009.*

tight heterogeneous gas shaly sands reservoir. Complex heterogeneity occurs both laterally and vertically due to diagenesis involving kaolinite and illite. As shown in Figure 5.34, the permeability ranges from 0.01 to 100 mD with a narrow band of porosity ranging from 8% to 10%. The petrophysical analysis indicates narrow 8–12% porosity range while wide permeability ranges from 0.01 to 100 mD. This cross-plot is not useful in its original form as there is no discernable trend. Because fractures are not accounted for in the core analysis, any extension of core data to field scale would be severely skewed. Figure 5.34 shows that the cloud points were subdivided into six subunits, ranging from high productivity (green) to very low productivity (black). The integration of NMR analysis as well as SCAL can lead to the establishment of facies-independent porosity and permeability models, thereby avoiding the use of lithology-independent T2 cutoff. Incidentally, lithology-independent cutoff is a standard for conventional reservoir characterization.

Hamada (2009) presented three-step procedure for integrating NMR data with SCAL and conventional core data. They are:

1. The application of density magnetic resonance (DMR) porosity technique for porosity calculation;
2. Bulk gas magnetic resonance permeability (k_{BGMR}), new technique for permeability calculation beyond the limits of oil-based mud filtrate;
3. Quantify the effect of oil-based mud filtrate on NMR data and then calibration for approximated capillary pressure from NMR.

Freedman et al. (1998) proposed a combination of density porosity and NMR porosity (Φ_{DMR}) to determine gas-corrected porosity formation and flushed-zone water saturation (S_{xo}). Density/NMR cross-plot is superior to density/neutron cross-plot for detecting and evaluating gas shaly sands. This superiority is due to the effect of thermal neutron absorbers in shaly sands on neutron porosities, which cause neutron porosity readings too high. As a result neutron/density logs can miss gas zones in shaly sands. On the other hand, NMR porosities are not affected by shale or rock mineralogy, and therefore density/NMR (DMR) technique is more reliable to indicate and evaluate gas shaly sands. Freedman et al. (1998) expressed true porosity as:

$$\phi = \left(\frac{\alpha}{\beta + \alpha} * \phi_D + \frac{\beta}{\beta + \alpha} * \phi_{NMR}\right) \tag{5.9}$$

where Φ_D is the apparent density porosity, Φ_{NMR} is porosity determined by NMR, and α and β are given as:

$$\alpha = \left(1 - (HI)_g P_g\right)$$

$$\beta = \frac{\rho_L - \rho_g}{\rho_m - \rho_L}$$

with $(HI)_g$ indicating hydrogen index for gas.

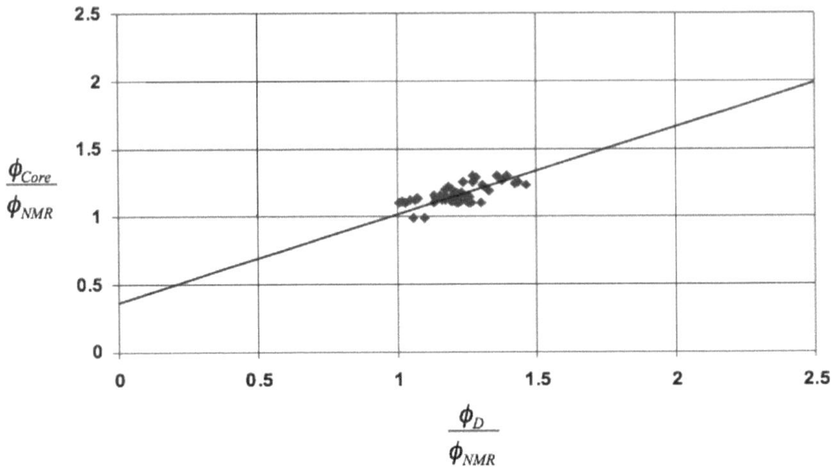

Figure 5.35 Developing filter out of NMR data.

Equation 5.9 can be rearranged as follows:

$$\frac{\phi_{Core}}{\phi_{NMR}} = A * \frac{\phi_D}{\phi_{NMR}} + B \tag{5.10}$$

where A and B can be extracted by cross-plotting core porosity with NMR porosity, thereby, creating a filter that can be used throughout the reservoir. Such cross-plot is shown in Figure 5.35.

Note that at $S_{gxo} = 0$, the pores are completely filled with liquid (mud filtrate and irreducible water), so the NMR porosity reading and density porosity should be correct and both should equal to core porosity. As a result, the trend line should intersect at control point, where□ $\Phi_{CORE}/\Phi_{NMR} = \Phi_D/\Phi_{NMR} = 1$. Fluid density for apparent □ Φ_D estimation is best fitting at 0.9 g/cc (in this particular case), which is a combination between formation water density and mud filtrate density.

For the above case, A and B are best fitted with the values 0.65 and 0.35, respectively, resulting in the following filter:

$$\phi_{DMR} = 0.65\ \ \phi_D + 0.35\ \ \phi_{NMR} \tag{5.11}$$

The results of Φ_{DMR} transform applications in the three wells A, B, and C showed very good match between Φ_{DMR} and core porosities as shown in Figures 5.36, 5.37, and 5.38. As a result, it is considered as an independent facies porosity model. These corrected porosities can be used in order to estimate permeability in gas bearing formations.

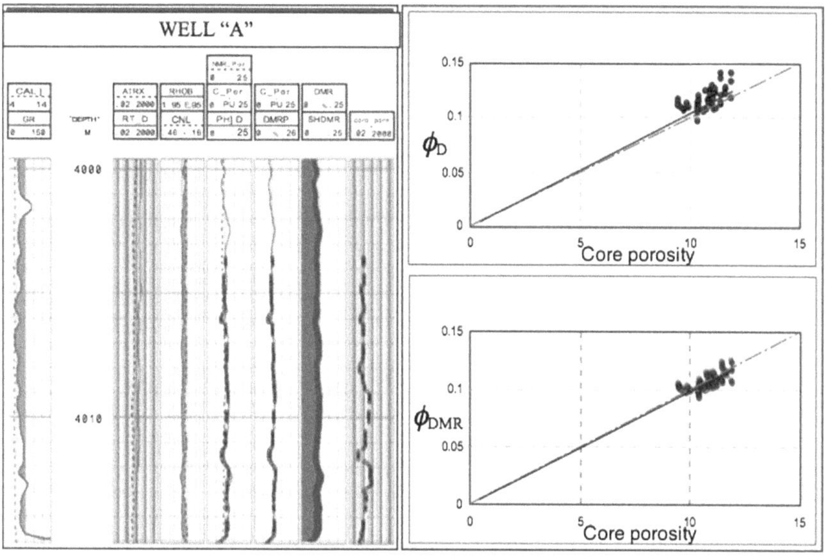

Figure 5.36 Filter for "Well A." *From Hamada, 2009.*

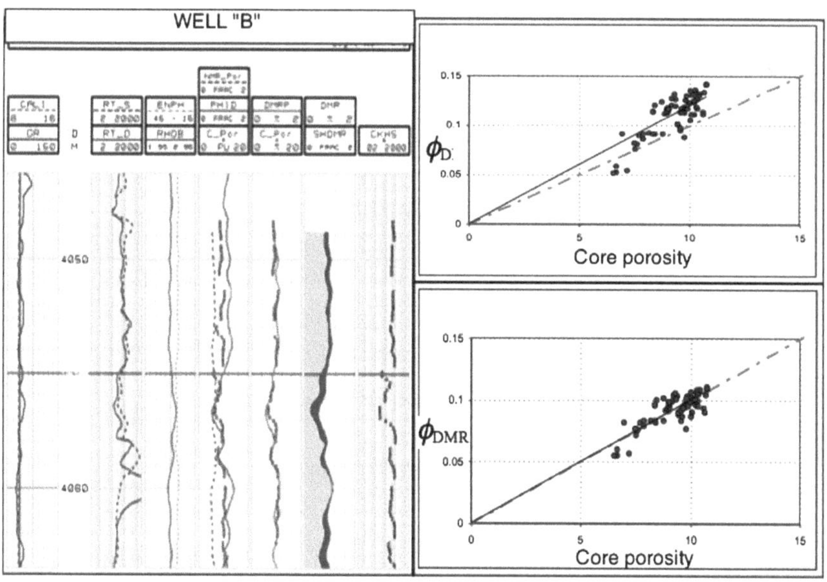

Figure 5.37 Filter for "Well B." *From Hamada, 2009.*

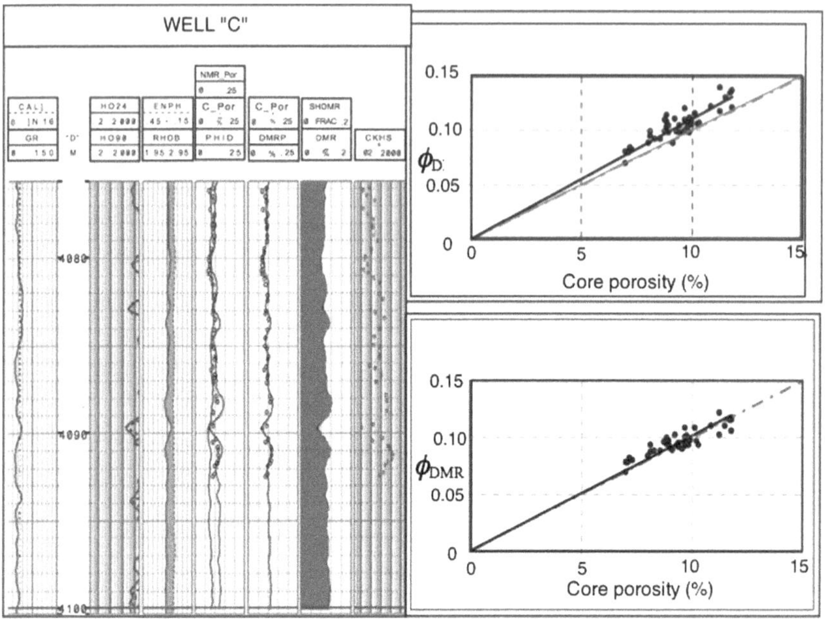

Figure 5.38 Filter for "Well C." *From Hamada, 2009.*

Figures 5.36, 5.37, and 5.38 present well logs, showing Φ_D and □ Φ_{DMR}. Gamma ray and Caliper curves are shown in the first track (GR&CALI); the second track shows depth in meters; the third track shows resistivity; the fourth track shows neutron-density logs; the fifth track shows comparison between core, density, and NMR porosities; the sixth track shows comparison between Φ_{DMR} and core porosity; the seventh track shows saturations of gas (green shadow) and water (blue shadow); and the last track shows core permeability in milliDarcys.

The next step involves correlation with permeability. Bulk gas magnetic resonance permeability (k_{BGMR}) is a new technique for permeability estimation in gas reservoirs. It is a dynamic concept of gas movement behind mud cake as a result of permeability formation, gas mobility, capillarity, and gravity forces. Because gravity forces are constant, capillarity depends mainly on permeability, and mobility depends on permeability and fluid viscosity, which is constant for gas; the gas reentry volume directly function in permeability.

The technique involves:

1. Calculation of gas volume in the flushed zone by using Differential spectrum (ΔT_w) Multi acquisition using different waiting times (T_w);

2. Solve diffusivity equation to solve for the volume of gas. Friedman et al. (1998) gave the following expression:

$$V_{g,xo} = \frac{DPHI - \frac{T_{NMR}}{(HI)_f}}{\left[1 - \frac{(HI)_g * P_g}{(HI)_f}\right] + \lambda}$$

$V_{g,xo}$ = gas volume in the flushed zone
$DPHI$ = formation porosity from density using filtrate fluid density
T_{NMR} = total NMR porosity
$(HI)_f$ = fluid hydrogen index
$(HI)_g$ = gas hydrogen index
P_g = gas polarization function = 1−exp (−W/$T_{1,g}$), where W is the wait time and $T_{1,g}$ is the longitudinal relaxation time for gas.

$$\lambda = \frac{\rho_f - \rho_g}{\rho_m - \rho_f}$$

3. Estimate invasion gas saturation, S_{gxo} with

$$S_{gxo} = \frac{(\phi_D - DMR) * (\rho_m - \rho_L)}{DMR * (\rho_L - \rho_g)}$$

4. Calculate gas volume approximately by ignoring the gas response in the NMR measurements especially in short T_w, and then the gas saturation in the invaded zone as:
Bulk gas volume (BG) = $\Phi_{DMR} - \Phi_{NMR}$

Figure 5.39 shows core permeability versus core porosity. It reflects how the permeability varies between facies to other within same porosity range. The same method is applied for the three wells A, B, and C, and then BG is plotted versus formation permeability, as shown in Figure 5.40. The correlation is normalized by dividing the gas volume by the total porosity of DMRP to be equal to S_{gxo}, as shown in Figure 5.40.

$$S_{gxo} = \frac{DMRP - \phi_{NMR}}{DMRP}$$

The correlation between S_{gxo} and permeability shown in Figure 5.41 has resulted in following permeability transform.

$$K_{BGMR} = 0.18 \ * \ 10^{(6.4 * S_{gxo})}$$

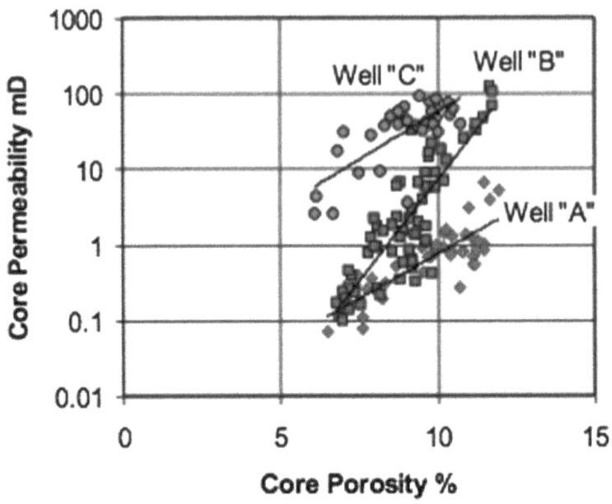

Figure 5.39 Correlation between core permeability and core porosity. *From Hamada, 2009.*

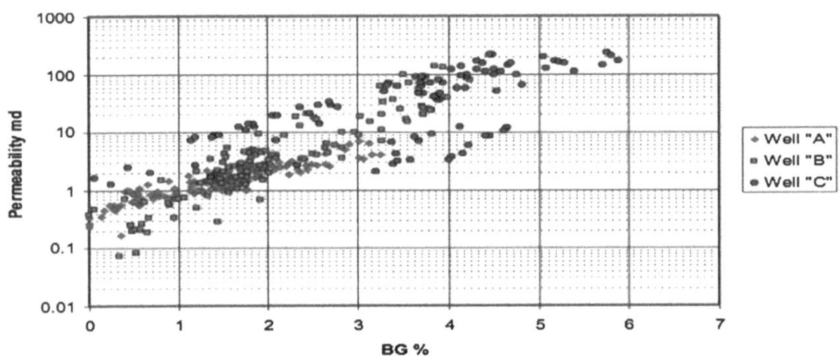

Figure 5.40 Correlation between permeability and BG. *From Hamada, 2009.*

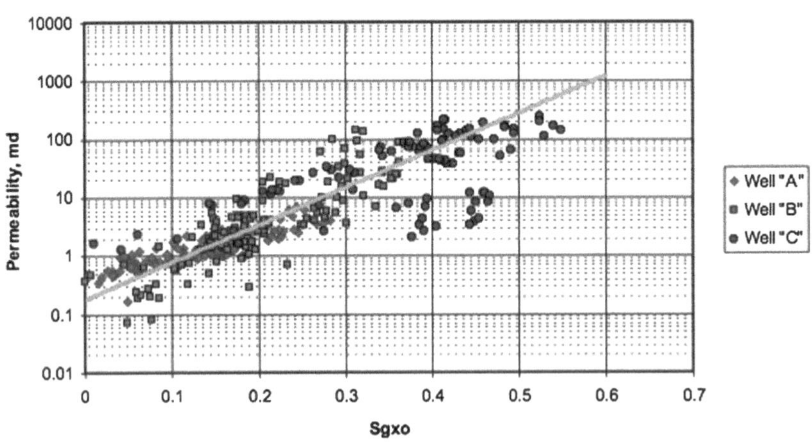

Figure 5.41 Correlation of permeability versus S_{gxo}. *From Hamada, 2009.*

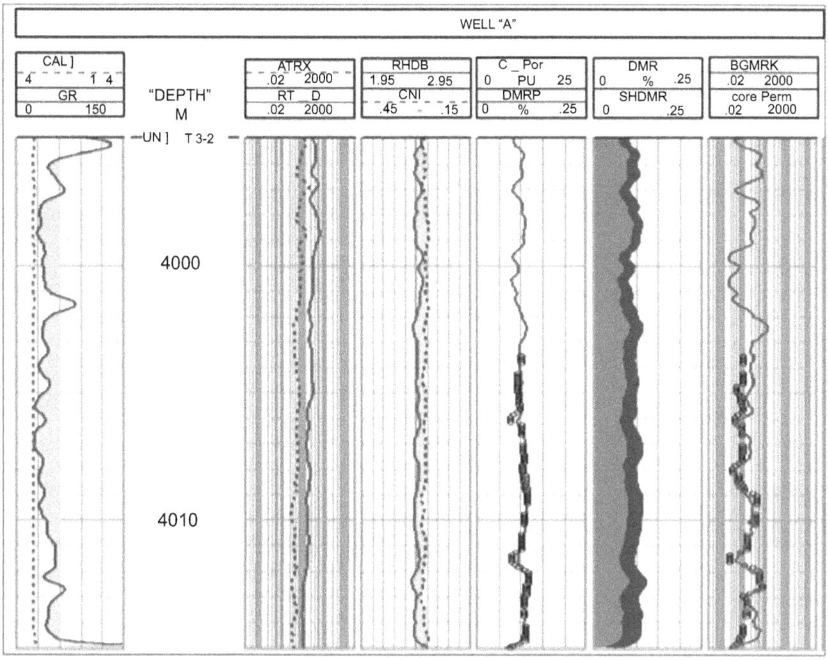

Figure 5.42 Permeability distribution (track 6) for "Well A." *From Hamada, 2009.*

Permeability derived using equation 5.11 in the three wells A, B, and C is shown in Figures 5.42 through 5.45. All the three wells A, B, and C show a good match between k_{BGMR} permeability with core permeability.

Similarly, correlation was achieved for capillary pressure by Hamada (2009). Figure 5.44 shows such correlation for "Well B," reported above.

5.6.2 Core Analysis

The most direct method for characterizing the natural fractures present in the reservoir is to perform core sample analyses. Fracture density and fracture spacing can be determined through visual analysis of the core (e.g., Gale et al., 2007; Kubik and Lowry, 1993). Plugs are typically taken for laboratory experiments to measure permeability and other flow properties. These experiments can only measure an "effective" permeability of the core sample. Generally, it is not possible to determine the natural fracture contribution to flow unless the fracture geometry is well known. The key issue, however, is that experience has shown that using laboratory measurements to represent reservoir-scale properties can be vastly misleading. Nonetheless, engineers are faced with the challenge of interpreting the few

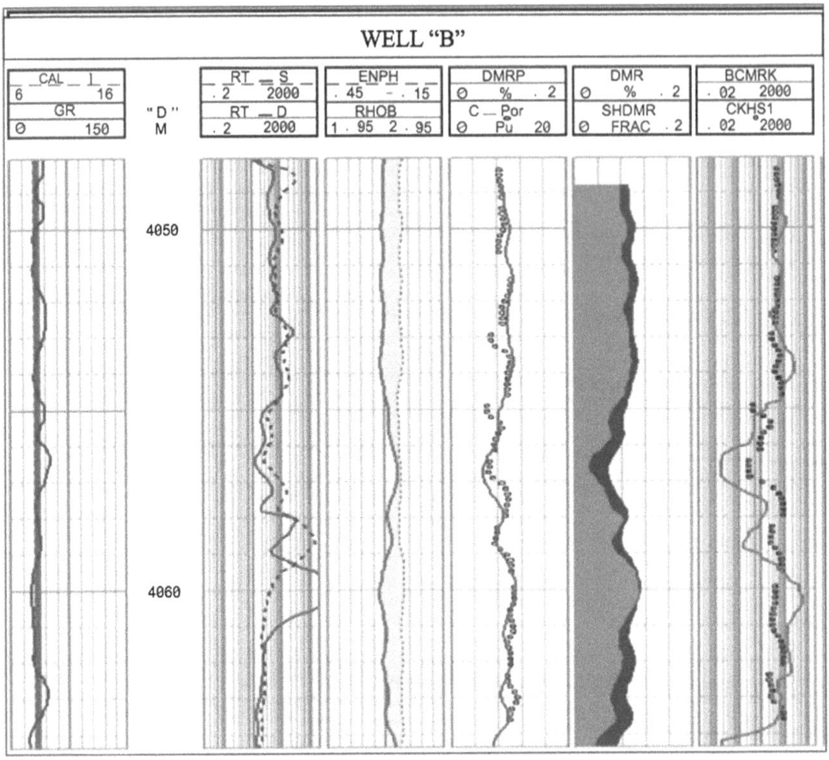

Figure 5.43 Permeability distribution (track 6) for "Well B." *From Hamada, 2009.*

available direct measurements of the reservoir rock and translating them to field-scale properties. The idea is to use core data to refine data already collected. In addition, several features, including fracture density, mineralization, fracture opening, shale breaks, etc., can be quantified only with visual inspection.

Of relevance is also the fact that total porosity can only be determined through core analysis. For unconventional gas perspective, this is of great consequence because as much as 50% recoverable gas can go unaccounted without definitive knowledge of porosity. Conventional assessment of porosity through GR analysis can be vastly misleading in unconventional reservoirs.

Practically all aspects of core analysis are different in unconventional reservoirs from conventional reservoirs. They are highlighted as follows:

1. Low-permeability (matrix and fractures alike) structure itself;
2. Low effective porosity but high total porosity;

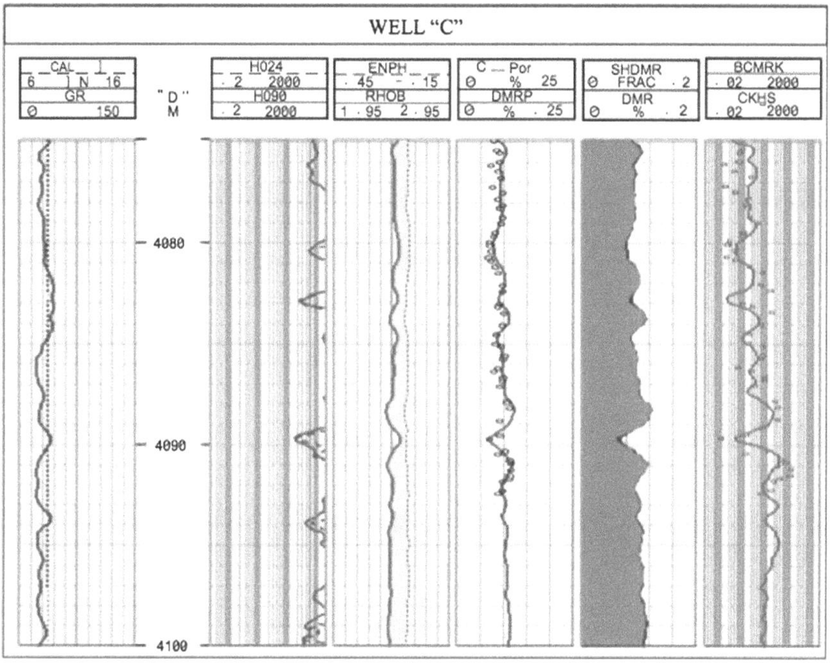

Figure 5.44 Permeability distribution (track 6) for "Well C." *From Hamada, 2009.*

3. Response to overburden stress;
4. Impact of the low-permeability structure on effective permeability relationships under conditions of multiphase saturation;
5. Capillary pressure data as well;
6. The role of fractures and fracture–matrix interactions.

The most prominent of the above is the one related to relative permeability graph. In a conventional reservoir, it is clear that there is relative permeability in excess of 2% to one or both fluid phases across a wide range of water saturation. In traditional reservoirs, critical and irreducible water saturation occur at similar water saturation values. Under these conditions, the absence of common water production usually implies that a reservoir system is at, or near, irreducible water saturation, low-permeability reservoirs, however, one can find that over a wide range of water saturation, there is less than 2% relative permeability to either fluid phase. In these reservoirs, the lack of water production cannot be used to infer irreducible water saturation. In some very low-permeability reservoir, there is virtually no mobile water phase even at very high water saturations. The term "permeability jail" coined by Shanley et al., 2004 describes the

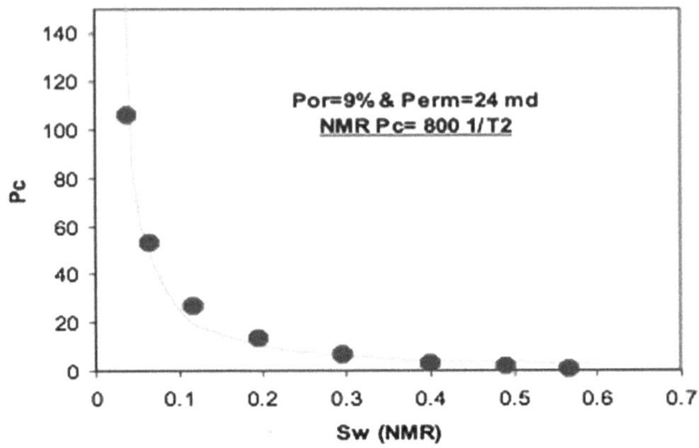

Figure 5.45 Correlation between core Pc (core capillary pressure dots) and NMR Pc (core capillary pressure line).

saturation region across which there is negligible effective permeability to either water or gas.

Low-permeability reservoirs are usually characterized by high to very high capillary pressures at relatively moderate wetting-phase saturations (Figure 5.46). In many cases, wetting-phase saturations of 50% (close to critical gas saturation, S_{gc}) are associated with capillary pressures in excess of 1000 psia, suggesting that a large number of pore throats are less than 0.1 μm in diameter and are of the micro- to nanoscale. In many low-permeability sandstone reservoirs, wetting-phase saturation continues to decrease with increasing capillary pressure.

The relationship between relative permeability, capillary pressure, and position within a trap in an unconventional reservoir are shown in Figure 5.47 (Shanley et al., 2004). It shows a reservoir body that thins and pinches out in a structurally updip direction. In this case, significant water production is restricted to very low structural positions near the free water. In many cases, the effective permeability to water is so low that there is little to no fluid flow at or below the free water level. Above the free water level, a wide region of little to no fluid flow exists. Farther updip, water-free gas production is found.

In terms of porosity, overburden pressure plays a significant role in unconventional reservoirs. This is unlike carbonate reservoirs or conventional sandstone reservoirs. Figure 5.48 shows little change in porosity with net increase in stress.

For unconventional reservoirs, such as shale formation, tight sand, etc., porosity is affected by overburden stress. Mechanical compaction is thus an

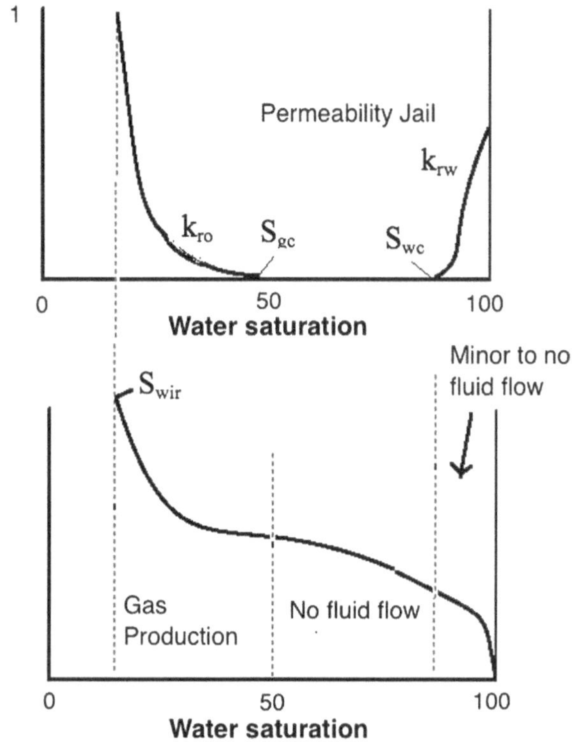

Figure 5.46 Typical relative permeability and capillary pressure curve for an unconventional gas reservoir.

inevitable consequence of burial and basin evolution. Often, this effect is modeled as a function of depth that is most affected by the overburden. However, depth is only a position coordinate that specifies present-day location. Consequently, depth is a poor measure of the processes that have acted upon a sedimentary section through time, which is the primary function that governs all features of a formation. In that sense, everything, including diagenetic process, remains dynamic. Conventional models of shale compaction relate porosity to effective stress using the empirical relationship between void ratio and effective stress established in soil mechanics (Burland, 1990; Yang and Aplin, 2004). Okiongbo (2011) explored the effect of petroleum generation by evaluating the variation in porosity and effective stress in the Kimmeridge clay formation (KCF) above and within the oil window.

Figure 5.49 shows porosity and effective stress relationship for above and below oil window. A common interpretation of the effective

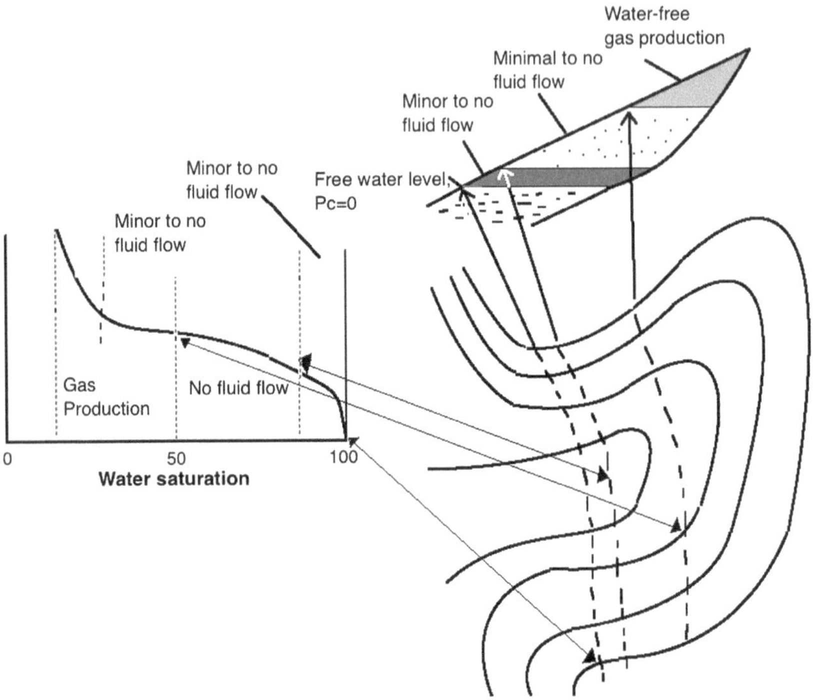

Figure 5.47 Representation of the relationships between relative permeability, capillary pressure, and position within a trap in an unconventional reservoir.

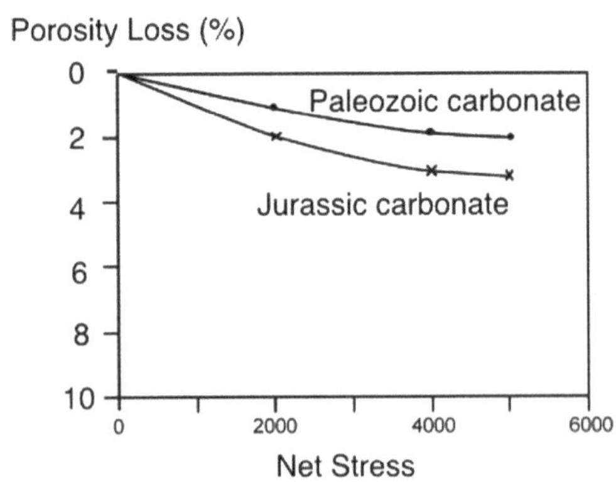

Figure 5.48 Porosity is only slightly affected by net stress for carbonate formations. *Data from Lucia (2007) and Hariri et al. (1995).*

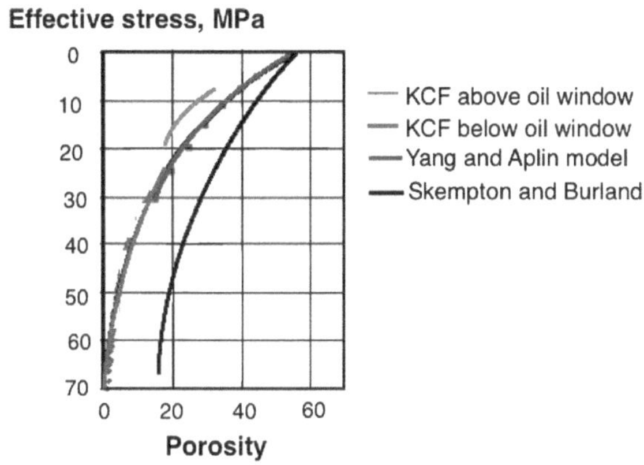

Figure 5.49 Porosity variations with effective stress. *After Okiongbo, 2011.*

stress–porosity relationship above and within the oil window is that the loss of porosity is faster in the pregeneration zone than within the oil window. In addition, cementation and deeper burial create a relatively stiffer matrix for the KCF sediments within the oil window making it more difficult to compact at high effective stresses. In presence of organic matter, porosity may be impacted by biogenic products. The presence of fractures and fissures alters the dynamics of porosity and pathways for escape of organic gaseous products. The exact nature of the origin of porosity is not known, but there are factors that are not well understood, particularly in the context of unconventional reservoirs. Figure 5.49 also shows a correlation developed by Yang and Aplin (2004) as well as Skempton (1970) and Burland (1990). These models express porosity as a unique function of clay and its natural, fine-grained clastic sediments. Yang and Aplin's (2004) model is based on well data from the North Sea as well as data derived from the studies of Skempton (1970) and Burland (1990). Although the clay content is assumed to be the same (~65%) in all the data sets, a remarkable difference exists in terms of their organic carbon content. Skempton (1970) and Burland (1990) data sets are low TOC samples (TOC < 1 wt%), the North Sea data set used by Yang and Aplin, (2004) have TOC ranging between 1% and 5wt%, while Okiongbo (2011) data set that are portrayed with red and orange lines have TOC ranging between 5% and 10wt%. In addition, Yang and Aplin's (2004) data set exclude chemically compacted sediments. Above the oil window, the Figure 5.49 indicates similarity between the trend of Yang and Aplin (2004) and that from this study on extrapolation to low effective stresses (<5 MPa). Significant variation only occurs at effective stresses less than five.

Within the oil window, the Figure 5.49 shows a close agreement between Yang and Aplin's data and Okiongbo data. Overall, the following factors play a role:

- TOC
- Mechanical compaction during diagenesis
- Diagenesis
- Thermal effects

While the first three components have been investigated with some details, thermal effects have not received much attention. Ehrenberg et al. (2009) related geological age of rocks with their diagenesis in order to assess the impact of temperature on porosity. Figure 5.50 shows the effect of geological age on porosity. The impacts of age and lithology are captured in this figure. For petroleum reservoirs, however, the presence of hydrocarbon adds complications. The release of gas increases pore stress, and any escape and migration by hydrocarbon fluids add complications to the nature of porosity. This dependence is different in carbonate formations from silicate

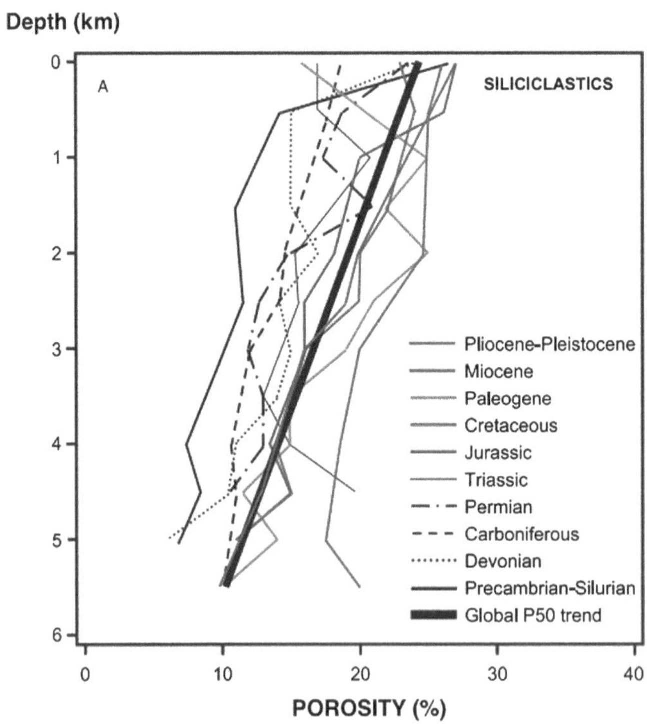

Figure 5.50 Effect of geological age on porosity. *From Ehrenberg et al., 2009.*

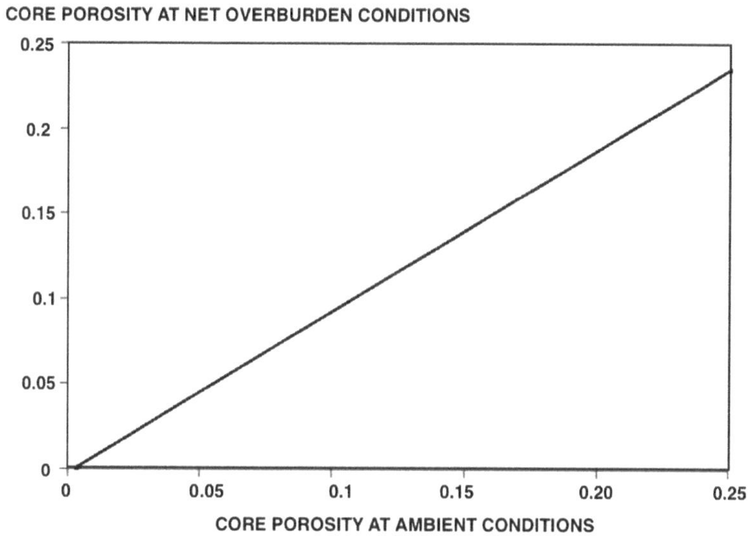

Figure 5.51 Porosity variation under net overburden conditions. *Modified from Petrowiki.spe.org.*

formations (including clay and shales). While depth adds to the overburden stress adding to the compaction of pores, it also adds temperature that affects hydrocarbon-bearing pores in a manner different from water-bearing pores. Adds to this is the chemical compaction and cementation that are directly affected by thermal conditions as well as overburden stress.

Figure 5.51 shows overburden stress alone would decrease the porosity. One must caution, however, unconventional reservoirs cannot be modeled uniquely with overburden stress. The degrees of "stress sensitivity" will be a function of the lithology and the pore throat size distributions. Rocks with "slot pores" will be more stress sensitive than rocks with more round pore throats. Slot pores are created by quartz overgrowths during diagenesis. Whenever fractures (natural or induced) are present, the presence of fractures can alter the flow behavior very differently from conventional reservoirs. This has tremendous impact on permeability. Figure 5.52 shows gas permeability at net overburden pressure as a function of gas permeability at ambient pressure (both Klinkenberg corrected). For high-permeability (10–100 mD) core plugs, the permeability under the original overburden pressure is slightly less than the value of unstressed permeability for that same core plug. However, as the permeability of the core plugs decreases, the effect of net overburden pressure on the core plug increases substantially. For the core plugs that had values of unstressed permeability of

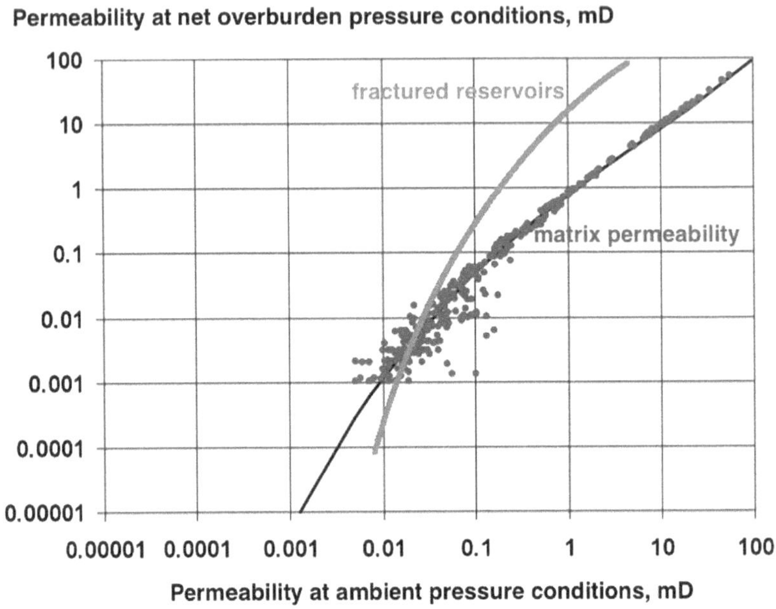

Figure 5.52 Effect of overburden stress on matrix and fracture permeability

around 0.01 mD, the values of permeability under net overburden stress were about an order of magnitude lower, or 0.001 mD. The lower permeability rocks are the most stress sensitive because the lower permeability core samples have smaller pore-throat diameters than the higher permeability rocks. As overburden stress increases, the diameter of the pore throat decreases. Because the permeability of a rock is roughly proportional to the square of the diameter of the pore throat, the permeability reduction in low-permeability rocks is much more dramatic than in high-permeability rocks. This behavior would explain the trapping mechanism presented earlier in this section. It also explains why core permeabilities have little relevance when it comes to fractured formations. Figure 5.52 also shows that a fractured formation would show much steeper dependence on overburden stress. This behavior indicates that flow rates would be increased under low stress conditions.

5.7 MAJOR FORCES OF OIL AND GAS RESERVOIRS

There are three main forces prevalent in any petroleum reservoir. They are capillary forces, gravity forces, and mobility forces. It is often useful to study these forces in terms of dimensionless numbers. Three main dimensionless numbers have been identified for many decades. They are capillary number, mobility ratio, and gravity numbers. The interplay of these three numbers is the essence of oil and gas recovery. In 1980s, another number, which is a modified form of "mobility ratio" has been introduced by Peters and Flock (1981) and Bentsen (1978).

Capillary number, N_c is defined as:

$$N_c = \frac{v\mu}{\sigma}$$

The following formula is also valid:

$$N_c = \frac{k\left(\frac{\Delta\rho}{l}\right)}{\sigma}$$

Where:

$\mu \rightarrow$ Displacing fluid viscosity

$v \rightarrow$ Darcy's velocity

$\sigma \rightarrow$ Interfacial tension (IFT) between the displacing and the displacing fluids

$k \rightarrow$ Effective permeability to the displaced fluids

$\frac{\Delta\rho}{l} \rightarrow$ Pressure gradient

The denominator would contain the term cos θ, where θ is the contact angle, which is a function of rock wettability to certain fluids. The capillary number shows what can be done in a reservoir, the idea being to maximize its value so that more irreducible oil can be recovered. Figure 5.53 shows

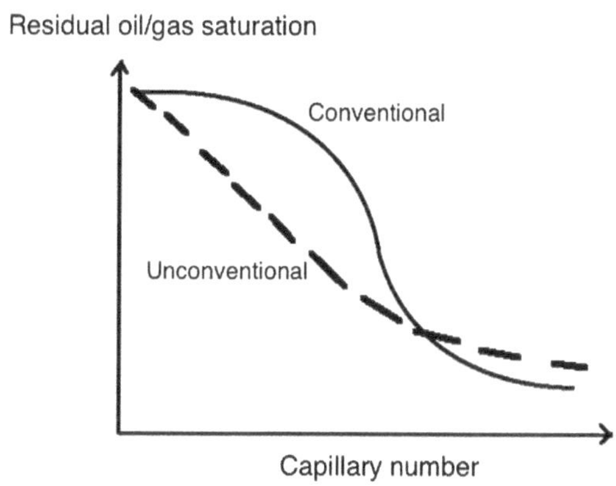

Figure 5.53 General trend of N_c versus residual saturation.

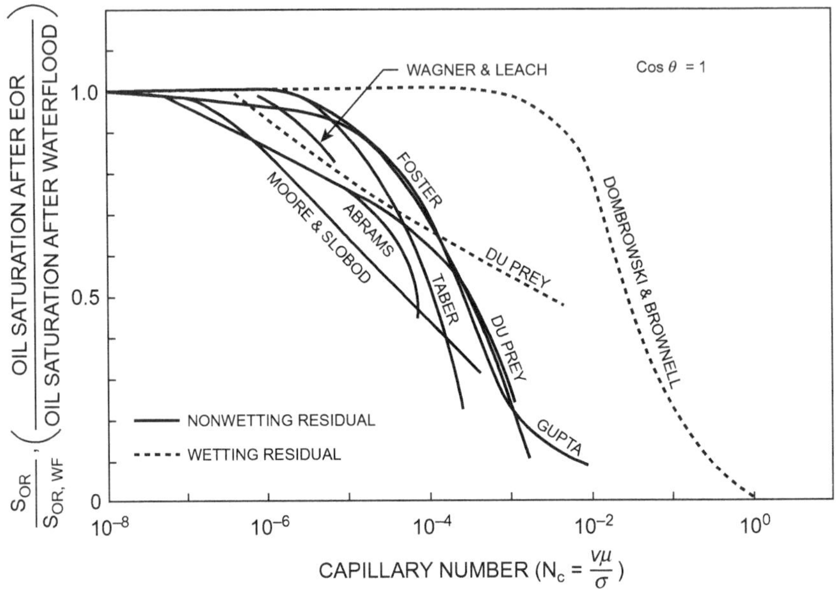

Figure 5.54 Several correlations between capillary number and residual oil saturation.

how capillary number is linked to residual hydrocarbon saturation. Note that x-axis is in log scale. This trend means that N_c has to be increased several orders of magnitude before any impact on residual oil saturation is invoked. However, for unconventional reservoirs, residual saturation bears a different meaning. These reservoirs can be exploited continuously by increasing N_c. Figure 5.54 shows some of the correlations reported in the literature. This graph is also valid for gas reservoirs and is of significance in unconventional reservoirs, where trapping of gas is the main mechanism of gas saturation.

The above equation shows that in theory the displacing phase viscosity alone will increase N_c. If the increase is manifold, this increase will be sufficient to decrease the residual oil saturation. For instance, by using polymer the displacing phase viscosity can be increased some 1000 fold. This would explain why in laboratory scale, polymer injection has yielded high recovery efficiency. However, large increase in aqueous phase leads to lowering of injectivity. This can be severe for reservoirs that need enhanced oil recovery (EOR), hence creating a dilemma for operators.

The next component that can be manipulated is the Darcy velocity. However, the extent of increase of this velocity is limited due to the fact

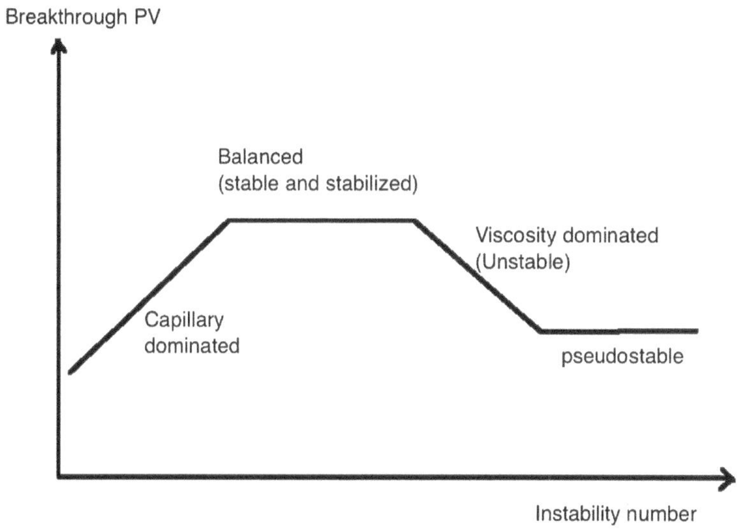

Figure 5.55 General trend of breakthrough recovery and instability number.

that flow instability in porous media can be triggered with high displacement rates, causing early breakthrough. In fact, Bentsen's work from 1980s shows that the following relationship exists between instability number (which includes Darcy velocity) and breakthrough recovery. Figure 5.54 shows such correlation. Note that at very low velocity, capillary forces dominate and very little oil is recovered because of the travel of injected water through water films in the porous media. This velocity is unrealistic in field applications and has been demonstrated only in laboratory models. Figure 5.55 shows how a balance between viscous and capillary forces leads to the formation of stable and stabilized front. This segment has the highest breakthrough oil recovery. This segment corresponds to the classic Buckley–Leverett profile. Practically all laboratory models are operated at this regime. For instance, unsteady state relative permeability measurement is performed using flow rates that would correspond to this flow regime. This fact is heavily consequential as none of these relative permeability graphs applies to a flow regime other than stable and stabilized. In case the flow regime is unstable, viscous forces dominate the process and early breakthrough takes place. As the instability number is increased, the breakthrough recovery declines. For very high instability number values, however, the decline is arrested and breakthrough recovery becomes insensitive to instability number.

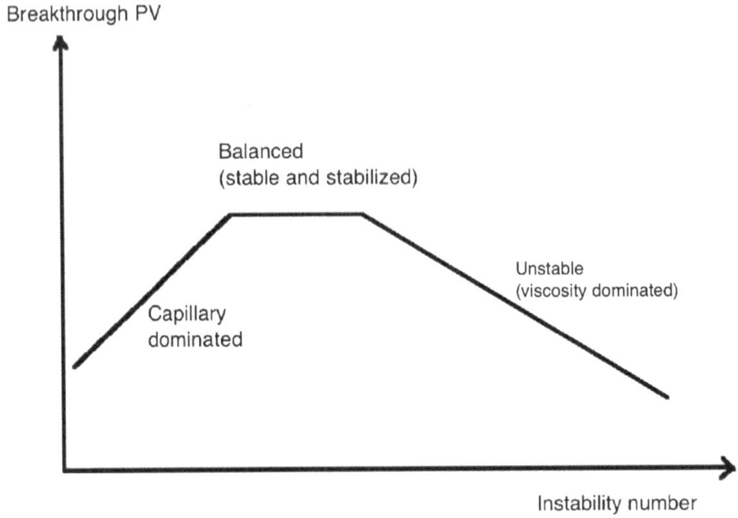

Figure 5.56 Instability number versus breakthrough recovery for immiscible gas injection.

For gas injection, laboratory models and theoretical interpretation thereof show that a different trend emerges. This is shown in Figure 5.56. Note that the stable region is very small and the unstable region is extended to very large instability numbers. The pseudo-stable region is missing within practical limitations of instability numbers.

In general, it is understood that increasing velocity of displacement is not an option for reducing residual oil saturation or increasing recovery through EOR. However, the support for this conclusion came much later than the original conclusion. Until now, the science of instability is little understood. The science of scaling up laboratory results of such cases is in its infancy.

The next factor considered is IFT. Reduction of IFT is at the core of all chemical flooding techniques. For perfect miscibility, the IFT is reduced to zero, signaling total recovery. However, such is never the case in the reservoir. Even when the operating conditions (e.g., injection pressure greater than minimum miscibility pressure) are met, natural media does not offer conditions conducive to instant or perfect miscibility. There is always a transition zone for which the concentration of each component goes through a gradient.

It is conventionally known that if miscibility conditions prevail, the recovery is very high, even if perfect miscibility is not achieved. The most important condition, however, is the stability of the displacement front. Stability was historically connected to mobility ratio, given by the following equation.

$$M = \lambda_{ing}/\lambda_{ed},$$

Where:

λ_{ing} → Mobility of the displacing fluid, $\left(\frac{k_{ing}}{\mu_{ing}}\right)$

λ_{ed} → Mobility of the displaced fluid, $\left(\frac{k_{ed}}{\mu_{ed}}\right)$

Several authors studied the relation between capillary number and residual oil saturation that is presented in the Figure 5.54. Thus, the significant increase in oil/gas recovery was due lowering the IFT between the fluids and pressure gradient promoting a reduction in the capillary number.

Since mobility ratio is less than one ($M < 1$), the displacement is stable, a fairly sharp "shock front" separates the mobile oil and water phases, and the permeability to water stabilizes fairly quickly. In other case, where the mobility ratio is slightly greater than one ($M > 1$) is considered unfavorable, because it indicates that the displacing fluid flows more readily than the displaced fluid (oil), and it can cause channeling of the displacing fluid, and as a result, bypassing of some of the residual oil. Under such conditions, and in the absence of viscous instabilities, more displacing fluid is needed to obtain a given residual oil saturation. However, if the value is close enough or equal to unity, the displacement is nearly pistonlike, and is denoted by a favorable mobility ratio. Mobility ratio influences the microscopic (pore level) and macroscopic (areal and vertical sweep) displacement efficiencies.

For waterflood, it is M that is the unique function comes from Buckley–Leverett.

The mathematical relationship between microscopic and macroscopic recoveries efficiencies are represented using the oil recovery factor (R_f), and it can be given by the following formula:

$$R_f = E_v \times E_h \times E_m$$

Where:

$E_m = E_I(S_{oI} - S_{or})$

The macroscopic sweep efficiency is defined by the horizontal and the vertical sweep efficiencies. The horizontal sweep efficiency is related to the mobility ratio and the vertical sweep efficiency depends on viscous to gravity forces ratio. The vertical sweep efficiency is related to difference in

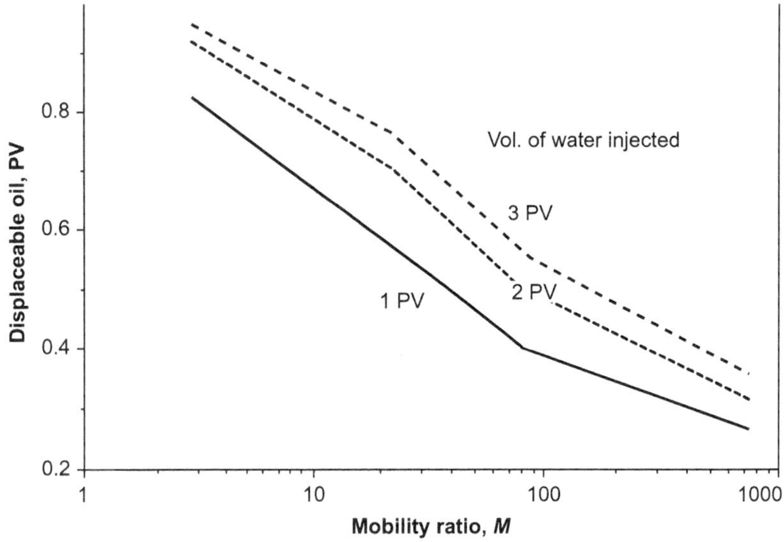

Figure 5.57 Correlation of mobility ratio with oil recovery for waterflood.

density of the injected fluid and in situ fluid. From early days on, there have been efforts to relate recovery efficiency to mobility ratio. However, until 1980s, it only considered mobility ratio as a unique function of recovery efficiency. While this is a simplistic representation, this was the only correlation available for decades. Until now, many operators use this correlation in order to design both waterflood and chemical flood projects. Figure 5.57 shows one such correlation. This equation is a deduction from Buckley–Leverett equation that shows the following relationships to hold. Here f_w and f_o are fractional flow of water and oil, respectively.

$$f_o = \frac{1}{1+M}, \quad f_w = \frac{1}{1+\frac{1}{M}}$$

The idea is to manipulate the injection fluid viscosity or mobility in order to maximize oil recovery potential. This correlation does not include the effect of viscous fingering. Such fingering is a possibility when the mobility ratio is greater than one, unless there is a gravity advantage to the injected fluid. For instance, if a gas is injected from structurally higher locations, it would stabilize the displacement front. This phenomenon was long recognized but not included in a dimensionless number until the early works of Flock and Bentsen. Even after these pioneering works, the petroleum industry continued to use older version that included capillary number, N_c, concept or its variation to delineate the onset of fingering. For instance, see the following equation outlined by Lake (1989)

$$N_L = \left(\frac{\phi}{K}\right)^{1/2} \frac{\mu_w v L}{k_{rwe}\sigma \cos \theta}$$

The above criterion was applied by several researchers and all determined that this definition of instability is grossly insufficient. A far better set of results were obtained using Peters and Flock (1980) criteria. Peters and Flock (1980) worked with velocity potential to come up with a stability criterion for a cylindrical system. It was for immiscible fluid and for the first time they introduced a variable that contains the dimension of the reservoir (or model thereof). It meant, the same system that would show no fingering in lab scale will show fingering in field scale. While in chemical engineering the dimension is always incorporated in determining stability of flow in an open duct, for petroleum reservoir applications it was new. For petroleum reservoirs, the characteristic dimension has always been considered to be pore diameter and not the reservoir dimensions.

Peters and Flock performed a stability analysis of the equations of two-phase flow to determine under what conditions viscous fingers tend to grow and propagate through a porous medium during immiscible flow. Bartley and Ruth (2002) conducted a series of experiments and demonstrated that Peters and Flock's number correlate broadly with their experimental observation of water breakthrough and none supported the criterion offered by Lake (Figure 5.58).

No such correlation existed in terms of capillary number or the modified capillary number as proposed by Lake. See Figure 5.59.

Note that the work of Peters and Flock is based on the earlier theoretical work of Chuoke et al. (1959). Values of the Peters and Flock number, Ns, that are less than 13.56 indicate stable displacements, and values of Ns greater than 13.56 indicate unstable displacements, meaning viscous fingering occurs. The Peters and Flock stability number depends on the mobility ratio, M, a wettability number, N_w, and a calculated value for a characteristic velocity, V_c. In the context of Peters and Flock's work, "stable" means that extensive viscous fingers do not form during a waterflood, or they become damped out; "unstable" means that viscous fingers continue to grow and propagate through the porous medium. They do not include gravity effects. It turns out as a significant omission and would form the basis for modification of the theory and the approach by Bentsen (1978).

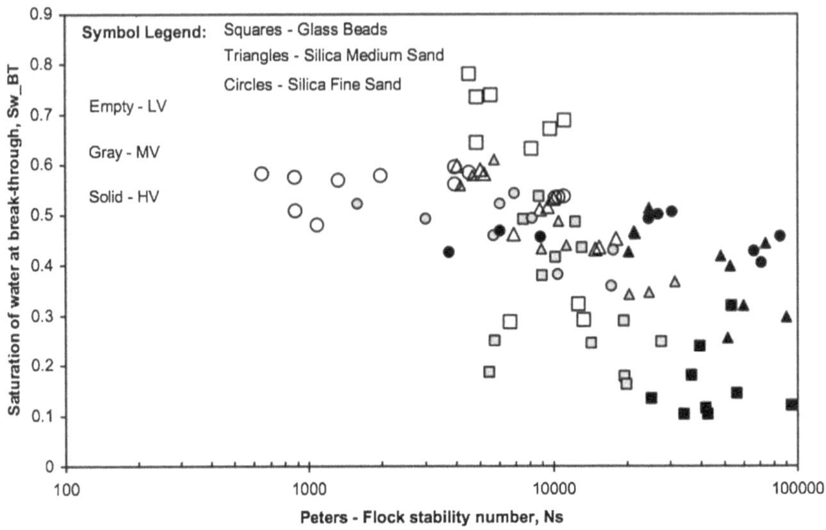

Figure 5.58 Correlation between breakthrough recovery and Peters–Flock stability number.

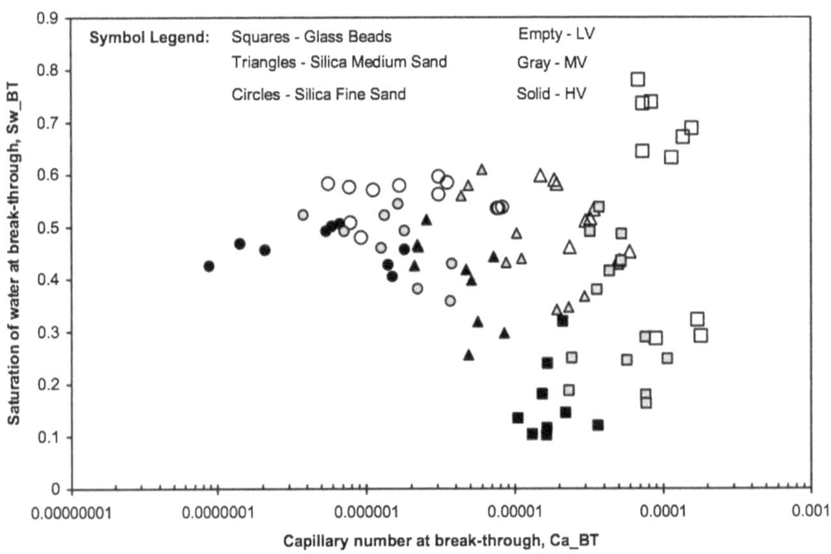

Figure 5.59 There is no correlation between capillary number and water breakthrough.

In this process, fractures play an intermediary role. Older theories did not include the role of macroscopic dimensions and considered pore geometry only. Later theories considered macroscopic dimensions but did not include the role of fractures. Saghir and Islam (1999) showed through a series of theoretical work that fractures are important because they trigger instability. Because conventional theories do not distinguish between fracture dimension and pore geometry and interplay thereof, the role of fracture is ignored. This aspect will be discussed in latter sections.

Bentsen (1978) developed a different instability criterion based on force potential as applied on a rectangular system. He introduced the concept of pseudo-capillary pressure. Both these criteria have gravity numbers in them. Following is the expression developed by Bentsen (1978).

$$I_{sr} = \frac{\mu_w v(M - 1 - N_g)}{k_{wr}\sigma_e} \times \frac{M^{5/3} + 1}{(M + 1)(M^{1/3} + 1)^2} \frac{4L_x^2 L_y^2}{L_x^2 + L_y^2}$$

Where N_g is the gravity number defined as:

$$N_g = \frac{\Delta \rho g k_{or} \cos \alpha}{\mu_o v}$$

Note that for a vertical injection N_g assumes the largest value possible. In case N_g is larger than the expression, (M−1) the displacement is unconditionally stable. This one shows the value of a gravity-stabilized displacement process that occurs when gas is injected from the top or water is injected from the bottom. This is the case for both miscible and immiscible displacement processes.

In the above expression, σ_e is the pseudo-IFT, which is:

$$\sigma_e \frac{C_1 \sigma \cos \theta}{\phi} = \frac{2 - v}{1 - v} \bar{d} A_c.$$

A_c is the area under the capillary pressure curve and v is a parameter dictated by the curvature of the capillary pressure.

Many enhanced oil recovery schemes involve the displacement of oil by a miscible fluid. If a scheme is not stable, miscibility cannot occur and all design criteria fail. Stability criteria for miscible processes were developed by Chuoke et al. (1959) but were not modified until Peters and Flock (1980) extended that theory to immiscible displacement processes. Whether a displacement is stable or unstable has a profound effect on how efficiently a solvent displaces oil within a reservoir. That is, if viscous fingers are present, the displacement efficiency and, hence, the economic return of the recovery scheme is seriously impaired because of macroscopic bypassing of the oil. As a consequence, it is of interest to be able to predict the boundary that separates stable displacements from those that are unstable.

Coskuner and Bentsen reported a series of stability numbers that deal with miscible displacement. They used linear perturbation in order to obtain the scaling group. The new scaling group differed from those obtained in previous studies because it had taken into account a variable unperturbed concentration profile, both transverse dimensions of the porous medium, and both the longitudinal and the transverse dispersion coefficient.

It has been shown that stability criteria derived in the literature are special cases of the general condition given here. The stability criterion is verified by comparing it with miscible displacement experiments carried out in a Hele-Shaw cell. Moreover, a comparison of the theory with some porous medium experiments from the literature also supports the validity of the theory. The stability criterion is given below.

$$\frac{U\dfrac{d\mu}{dC} - kg\dfrac{d\rho}{dC}\sin\gamma}{\overline{\mu}D}\,\frac{\partial\overline{C}}{\partial x}\,\frac{L^2}{\Omega}\cdot\left[\left(\frac{1}{\Omega}+\frac{D'}{D}\right)\left(\frac{1}{\Omega}+1\right)\right]^{-1} > \pi^2,$$

$$\text{with } \Omega = \frac{L^2(B^2+H^2)}{B^2H^2}$$

$$\left(\text{for a two}-\text{dimensional system in which } H=0, \Omega=\frac{L^2}{B^2}\right) \text{ and}$$

Where $U =$ displacement velocity
$\mu =$ viscosity of mixture
$\rho =$ density of mixture
$C =$ injectant concentration
$k =$ permeability
$g =$ gravitational acceleration
$\gamma =$ dip angle
$\Phi =$ porosity

Picture 5.3 shows general shapes of viscous fingering in a porous medium. Note that all fingers have similar shape at the onset. As a finger starts to propagate, a dominant one moves faster than the rest of the fingers, leading to bypassing of oils and early breakthrough. This feature is not included in conventional analysis.

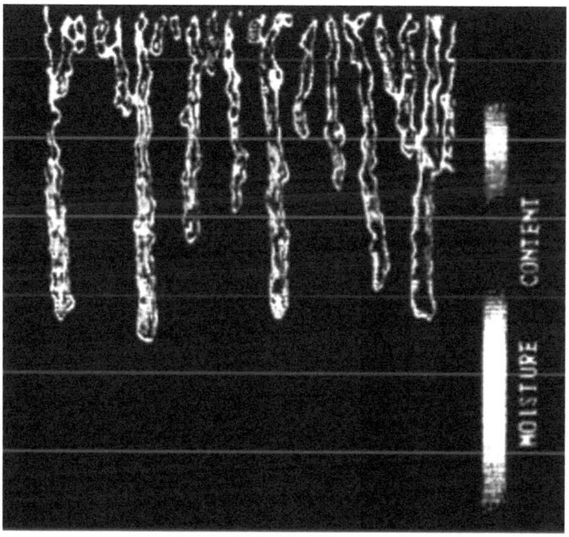

Picture 5.3 Viscous fingering in a miscible displacement process.

When it comes to miscible flood, a mixed scenario of immiscible and miscible displacement processes emerges. It is because every reservoir contains water and an immiscible process is inherently present. This fact is best utilized in designing water alternated with gas (WAG) processes that minimize the use of expensive chemicals. One such example is CO_2 miscible injection. Figure 5.60 shows the predominant displacement fronts in a typical CO_2 WAG injection process.

Note how the displacement front between oil and miscible gas is likely to be unstable because of the unfavorable mobility ratio, whereas the

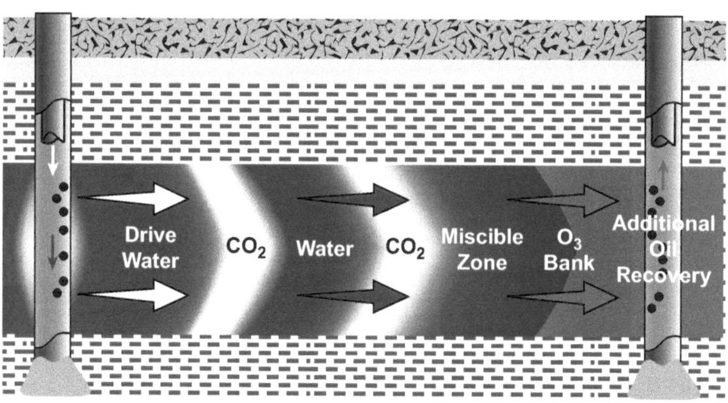

Figure 5.60 Typical CO_2 WAG process.

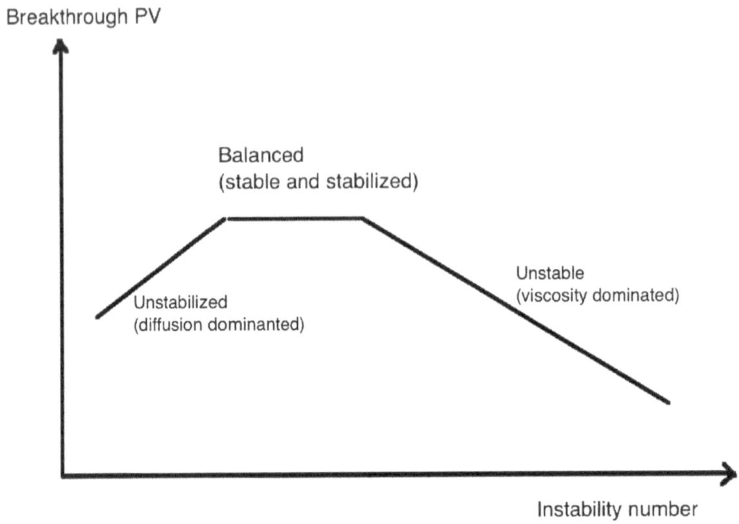

Figure 5.61 Breakthrough recovery versus instability number for miscible flood.

displacement front between water and miscible gas (CO_2 in this case) is likely to be stable. This is one of the greatest advantages of WAG that is often overlooked. In addition, WAG makes it possible to access different segments of trapped oil, as the wettability natures of gas and water are different. The combination of two different types of flooding has other advantages as well.

Figure 5.61 shows overall trends of breakthrough recovery with instability number for miscible flood processes. This is a theoretical graph. However, field results as well as scaled laboratory results indicate that this is prevalent. The number for which stability breaks down is important and is generally elusive. All theories are simplified and in need of adjustments due to heterogeneity of the formation. While instability numbers have such components as reservoir dimension and permeability, no one yet added a coefficient that would include heterogeneity in the equation. Yet, most candidates for miscible injection are heterogeneous.

It is likely that the graph moves toward the left for fractured formations and toward the right if those fractures are plugged. This can be corrected if the permeability of the reservoir is filtered and is representative of the reservoir. This aspect will be discussed in a later section.

The Capillary number expression shows lowering the IFT and/or increasing the contact angle will increase the capillary number. This is the basis of chemical as well as gas injection (along with various types of gas/water injection). Through the injection of a chemically active fluid, the IFT of the displacement front is decreased, leading to the recovery of residual oil. On the other hand, if rock wettability is changed, similar impact of lesser intensity can occur.

The property of a fluid is directly related to the viscosity of crude oil within the reservoir. These properties are determined by standardized laboratory procedures. Unfortunately the test results do not represent a general characteristic of the reservoir, because the samples are taken from different sites within the reservoir. The prediction of the reservoir fluid properties becomes even more complex when the prevailing conditions within the reservoir change as a result of undergoing processes, leading to unexpected reaction during injection and production operations.

Variations in rock–fluid interaction with changing conditions in a reservoir result in wettability variations, which in turn affect flow parameters such as capillary pressure and relative permeability, affecting both dynamic recovery as well as ultimate recovery from such reservoirs.

Figure 5.62 shows how end-point permeability values affect residual oil saturation. This figure shows how the use of different gas as well as WAG can affect the ultimate recovery.

For the dynamic part of the displacement process, the most profound impact is through alteration of permeability graphs. Figure 5.62 shows some examples of how lowering of IFT alters effect permeabilities of both water

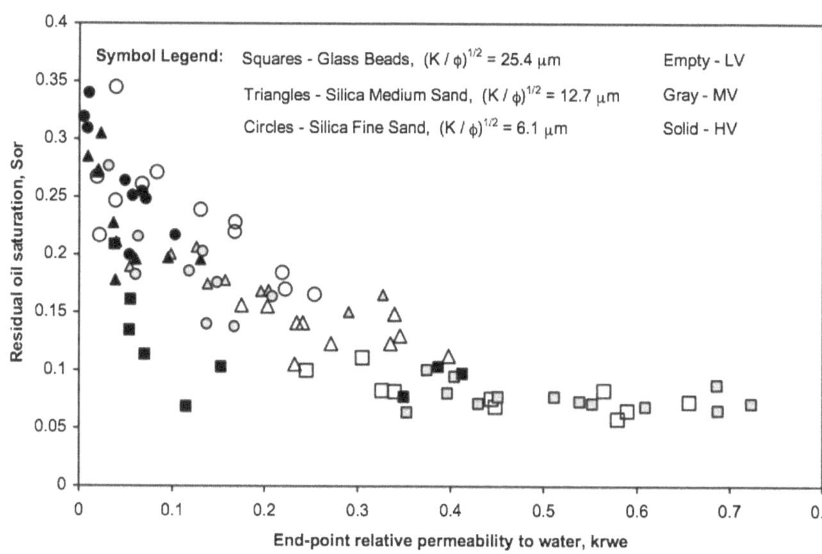

Figure 5.62 End-point relative permeability correlates with residual oil saturation.

and oil. The permeability values corresponding to lowest IFT form straight lines. While this is true in a core flood test, it rarely occurs in the field and discrepancy between laboratory data and field results emerges. In reservoir simulation studies, it has been demonstrated that the oil recovery with straight line permeability curves do not show an oil recovery curve markedly better than others. This happens even after one overcomes the stability problem that occurs when such straight lines are assigned as relative permeability values of an immiscible system (e.g., oil and gas).

The most frequently encountered saturation end points are as follows:
- Residual oil saturation
- Irreducible water saturation
- Trapped-oil and trapped-gas saturations
- Critical gas and condensate saturations

Residual oil, irreducible water, and trapped-gas and trapped-oil saturations all refer to the remaining saturation of those phases after extensive displacement by other phases. Critical saturation, whether gas or condensate, refers to the minimum saturation at which a phase becomes mobile.

The end-point saturation of a phase for a specific displacement process depends on the following:
- The structure of the porous material
- The wettabilities with respect to the various phases
- The previous saturation history of the phases
- The extent of the displacement process (the number of pore volumes injected)

The end-point saturation also can depend on IFTs when they are very low and on the rate of displacement when it is very high. For unconventional gas reservoirs, lowering of IFT, any chemical treatment, any thermal alteration, or onset of fractures, leads to the alteration of relative permeability graphs. Figures 5.63 and 5.64 shows how such movement can translate into gas production.

Results reported by Chatzis et al. (1983) give general insight on the combined effects of wettability and porous structure on residual saturations. In tests with an unconsolidated sand of nonuniform grain size, the wetting phase (oil) was displaced by a nonwetting phase (air) from an initial saturation of 100% to a residual value. In general, it is known that
- Residual saturation of a wetting phase is less than the residual saturation of a nonwetting phase.
- Residual saturation of a nonwetting phase is much more sensitive to heterogeneities in the porous structure.

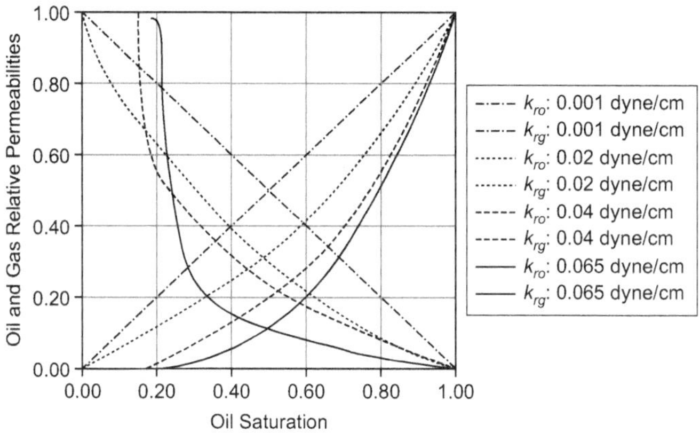

Figure 5.63 Relative permeability curves are altered by lowering of interfacial tension.

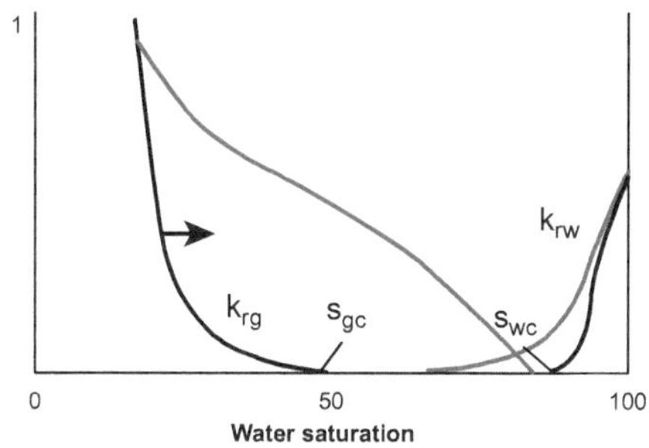

Figure 5.64 Permeability jail can be removed with thermal or chemical alteration in an unconventional reservoir.

General conclusions on the effects of wettability are useful, but the diverse array of wetting alternatives suggests caution, especially in oil/water reservoir systems. This wide range of wetting possibilities is an obstacle to interpreting or predicting the effect of wettability on end-point saturations. Indeed, conflicting results for different porous media are likely to occur. For example, Jadhunandan and Morrow (1995) report that residual oil saturation displays a minimum value for mixed-wet media as wettability shifts from water-wet to oil-wet—contrary to the results of Bethel and Calhoun (1953) that reported a maximum for media of uniform wettability.

For gas reservoirs, such analysis is important because of the critical gas saturation. The critical gas saturation is the saturation at which gas first becomes mobile during a gas flood in a porous material that is initially saturated with oil and/or water. If, for example, the critical gas saturation is

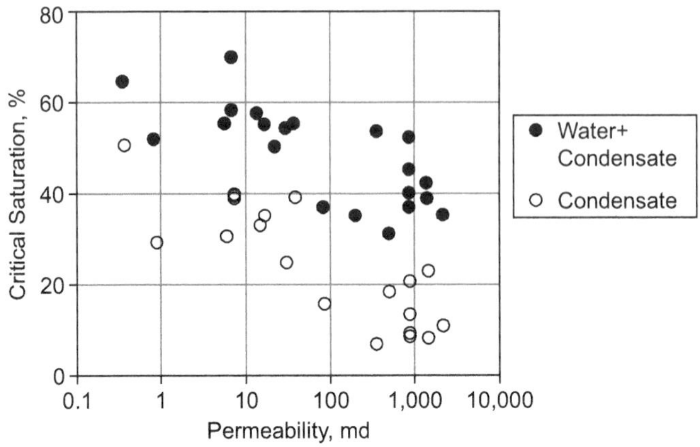

Figure 5.65 Critical gas saturation for various permeability values of a gas condensate reservoir.

5%, then gas does not flow until its saturation exceeds 5%. Values of S_{gc} range from 0% to 20%. For gas condensate reservoirs, this is of utmost importance. Interest on the mobility of condensates in retrograde gas reservoirs developed in the 1990s, as it was observed that condensates could hamper gas production severely in some reservoirs, particularly those with low permeability. The trend of increasing critical condensate saturations with decreasing permeability, as summarized by Barnum et al., (1995), is reproduced in Figure 5.65.

This information is important for designing of WAG processes. The WAG ratio can also be defined as the ratio of the volume of water injected within the reservoir compared to the volume of injected gas. It plays an important role in obtaining the optimum value of the recovery factor corresponding to an optimal value of the WAG ratio. This optimal WAG ratio is reservoir-dependent because the performance of any WAG scheme depends strongly on the distribution of permeability as well as factors that determine the impact of gravity segregation (fluid densities, viscosities, and reservoir flow rates). Studies made by John and Reid (2000) showed that the WAG ratio strongly depends on reservoir's wettability and availability of the gas to be injected. When the WAG ratio is high, it may cause oil trapping by water blocking or at best may not allow sufficient solvent–oil contact, causing the production performance behave like a waterflood. On the other hand, if the WAG ratio is very small, the gas may channel and the production performance would tend to behave as a gas flood, the pressure declines rapidly, which would lead to early gas breakthrough and high declination on production rate. To find the optimal WAG ratio, it is necessary to perform sensitivity analysis, proposing different relations of WAG ratio to study the effect on oil recovery.

Even though enhanced gas recovery has not been applied in the field at a commercial level, the fundamental mechanisms show that an enhanced gas recovery scheme is more useful and more likely to be successful than pressure maintenance or waterflood. It is because injected gas is less prone to creating channels or fingers in a gas reservoir than an oil reservoir. In addition, miscibility can be achieved in a gas reservoir under easily achievable conditions. Finally, the use of waste gas or gas that would be otherwise flared makes the economics of the project quite appealing.

5.8 RESERVOIR HETEROGENEITY

The degree of interconnection between the pores of an oil reservoir, are rarely evenly distributed due to nonuniformity of pore size, which gives rise to disordered and complex reservoir fluid flow behavior. Geologically speaking, this is known as phenomenon of heterogeneous permeability that can manifest different individual layers, forming different homogeneous layers within the oil reservoir with different permeability values.

Most consolidated formations are highly fractured under reservoir conditions. While the presence of fractures makes them ideal for oil and gas recovery because of very high permeability, the fractures are sealed with secondary cementing activities. Minor porosity of secondary origin occurs locally in sandstones and resulted from the removal of authigenic mineral cements and, to a lesser extent, detrital framework grains. In carbonate-cemented samples, evidence of dissolution includes corrosive contacts between successive carbonate phases and relict cement in pores. Carbonate dissolution features also are observed along the margins of some fractures. Collectively, the dissolution features in sandstones indicate that carbonate cements were previously more widespread before they were partially to extensively dissolved.

Most consolidated formations have fractures in both macroscopic and microscopic scales. Vast majority of these fractures are open without secondary mineralization (Pictures 5.4 and 5.5). The aperture widths commonly exceed 30 μm. One such example is shown in Picture 5.6.

An important characteristic of these fractures is that they typically form a dense network that is highly visible on wetted, slabbed rock surfaces, if the host sandstones and siltstones have high residual oil saturations (Picture 5.7).

Picture 5.4 Outcrops often show how fractures must be prevalent in consolidated reservoir formations.

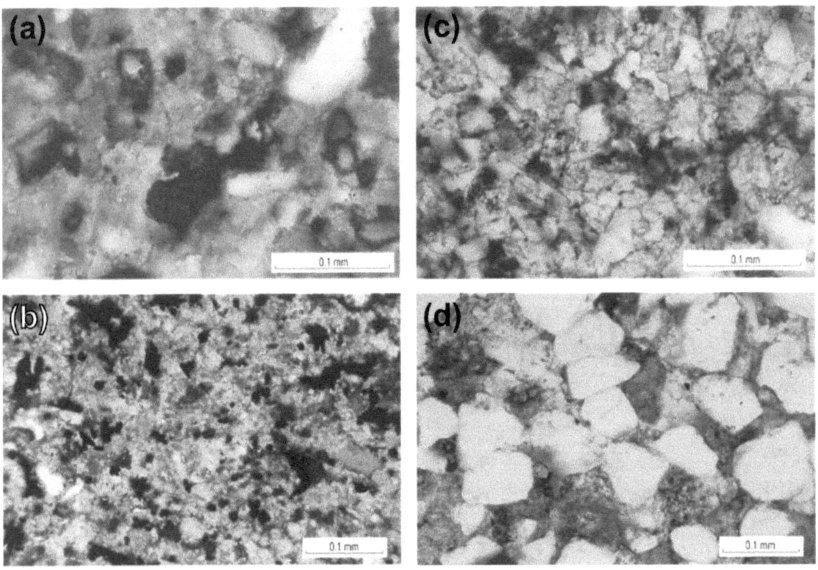

Picture 5.5 Thin section photomicrographs of sandstones illustrating, (a) occurrence and distribution of K-feldspar grains (stained yellow); (b) disseminated pyrite (dark grains); (c) bitumen (opaque material) filling pores and permeating matrix; and (d) secondary intergranular porosity with relict carbonate cement.

Such fractures are generally absent in rocks that have little or no residual oil. Often, factures are V-shaped and occluded with pyrite and fine- to coarse-crystalline calcite cement. These fractures, which tend to be small, resemble fluid-escape structures. Such cementation with quartz and cancite

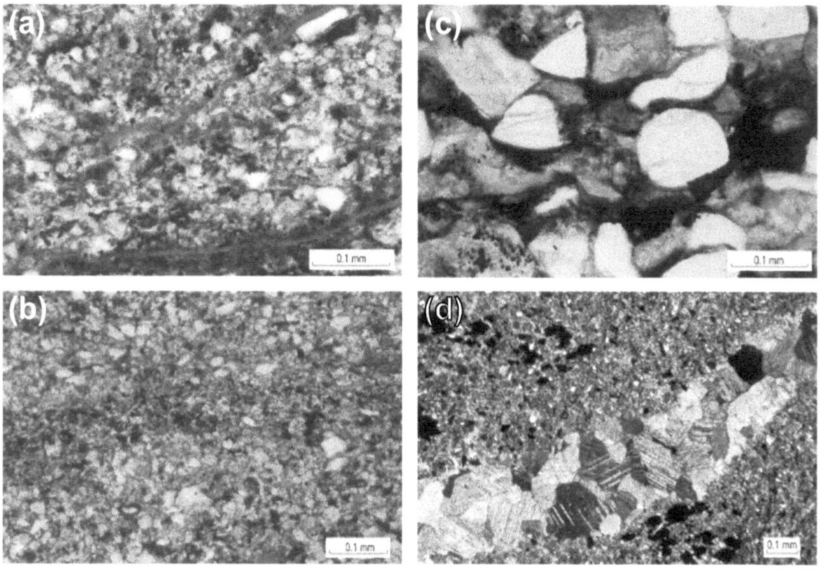

Picture 5.6 Thin section photomicrographs of sandstones depicting, (a) open (non-cemented), discontinuous fractures parallel to bedding. Such fractures are abundant and form a pervasive network in sandstones adjacent to mature shales; (b) secondary porosity associated with horizontal fracture swarms; (c) microscopic fractures cross-cutting framework quartz grains. Note bitumen filling secondary intergranular pores; and (d) calcite cemented vertical fracture.

(Picture 5.6) can lead to lowering of permeability below the matrix permeability. These types of formations exhibit productivity (or injectivity) lower than one would expect from core and log analysis.

The presence of secondary cementing activities make the porosity–permeability correlation skewed away from matrix permeability. It is, however, the matrix permeability that is routinely reported in core tests. For a fractured formation, matrix permeability has little to do with permeability of the reservoir. The permeability having a dimension of L^2, a correlation between permeability and porosity turns out to be linear, as long as the porous medium is homogeneous. This linearity no longer holds if the formation is fractured, as shown in Figure 5.66. If the fractures are open, the medium will have markedly higher permeability than the porosity correlation would indicate. This is consequent because log analysis is usually performed in order to determine point porosity of a core. This is later transformed into permeability values using one of numerous correlations available in the literature. Once the transformation is performed, little

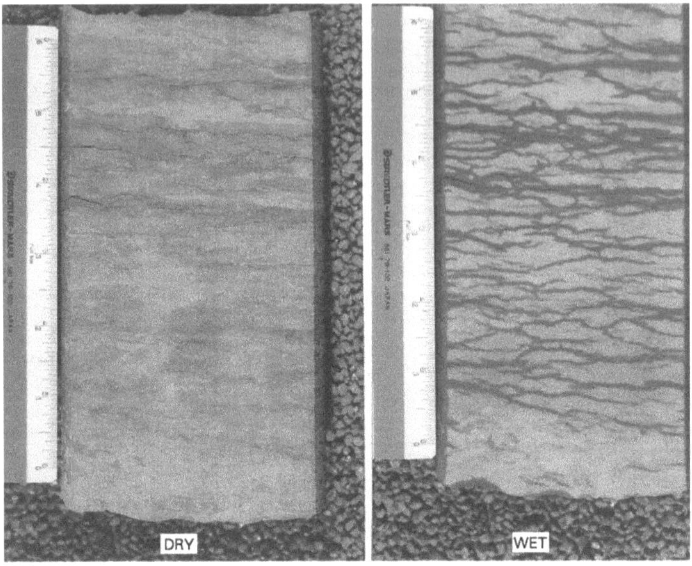

Picture 5.7 Slabbed sandstone displaying reticulated fracture network on wet surface. Note that the permeable nature and distribution of fractures are not apparent when surface is dry.

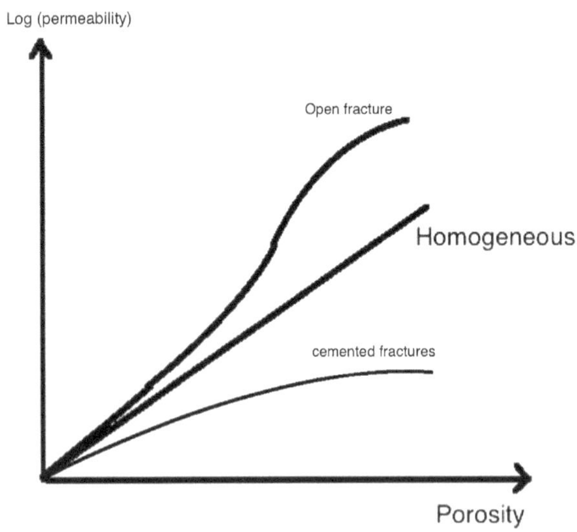

Figure 5.66 Permeability versus porosity correlation depends largely on the nature of heterogeneity.

attention is paid to the origin of these permeability values. Similarly, core tests as well as specialized core tests are performed on the homogeneous section of the core, leading to the determination of petrophysical properties that have little relevance to the reservoir. Figure 5.66 shows a typical correlation between porosity and log (permeability) of a homogeneous formation as compared to the ones that have fractures. An open fracture increases the reservoir permeability drastically, particularly for median porosity range. On the other hand, this trend is reversed if the fractures are closed due to secondary cementing phenomena.

In terms of anisotropy, it can be invoked due to the presence of fractures. Particular orientation of fractures would skew the directional permeability and the magnitude would depend on the aperture size, secondary cementation, and aperture length. Vertical permeability is affected by the extent of vertical fractures and whether they are open or cemented.

Figure 5.67 shows how porosity is correlated with permeability for various types of formations. Figure 5.68 adds correction factors as a function of fracture frequency. This graph applies to open fractures.

Figure 5.69 shows the trend of vertical permeability (k_v) over horizontal permeability (k_h) with fracture frequency. While fracture frequency is an important factor in determining the anisotropy, the nature of fractures is crucial for determining magnitude of the permeability ratio. When hundreds of wells and square kilometers of area are considered in order to estimate flow from reservoirs containing billions of barrels, the accuracy of k_v/k_h analysis can mean a difference of millions of dollars every day throughout the production history.

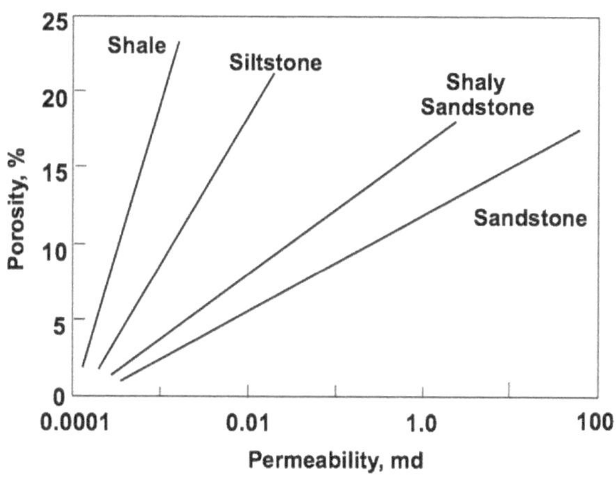

Figure 5.67 Correlation of porosity versus permeability for various types of formation.

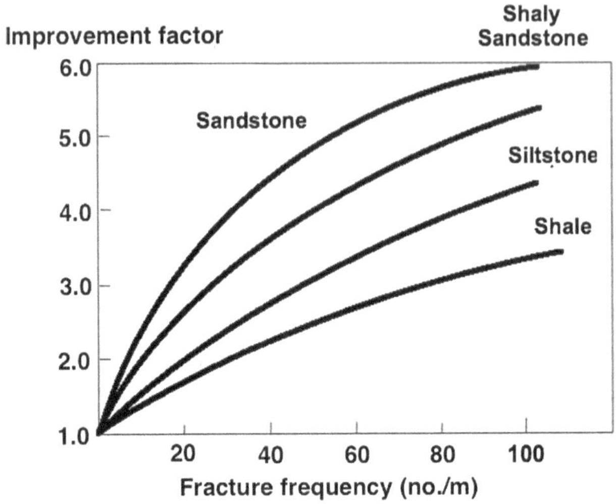

Figure 5.68 Improvement factor due to open fractures.

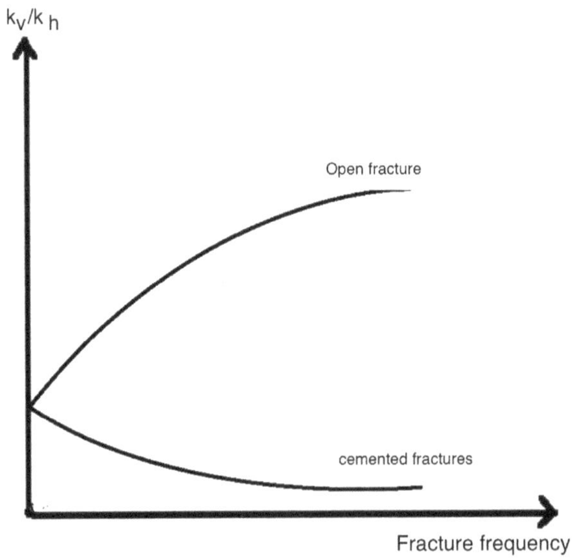

Figure 5.69 The effect of fractures on k_v/k_h.

The effects of stratification and heterogeneity can be distinct in different reservoirs, affecting various parameters such as capillary pressure, relative permeability, and mobility ratios. The presence of anisotropy and heterogeneity in a reservoir affects the displacement of the native fluids by the injected fluid. Channeling of the solvent through high-permeability regions reduces the storage and displacement efficiency of the displacing solvent. In addition, it can offset viscous fingering through perturbation in case the mobility ratio is not favorable or gravity stabilization is not a dominant mechanism. In case, WAG is being considered, heterogeneity and anisotropy become the most dominant factors that affect oil recovery. It is so because they control the injection and sweep patterns, as well as vertical and areal sweep, viscous fingering, gravity stabilization, and dispersive forces. In addition, horizontal wells play a role different from vertical wells and all calculations pertaining to vertical wells become irrelevant.

In terms of EOR, heterogeneity has tremendous impact. To begin with, the location of trapped oil and the critical capillary number for mobilization of trapped oil would be different for different pore distribution, which is controlled by fracture characteristics of a reservoir. Figure 5.70 shows how different residual saturations would emerge for different pore distribution.

In order to include the influence of fracture and fracture distribution, one must characterize fractures properly. It involves determining frequency and orientation of prominent fractures as identified through examination of cores, micrologs, etc. Commercial software can be used to analyze typical fractures observed in FMS (electrical Formation MicroScanner), as shown in Picture 5.8.

In order to quantify the role of fracture, a rose diagram should be plotted for open fractures. If there is trend for closed or cemented fracture,

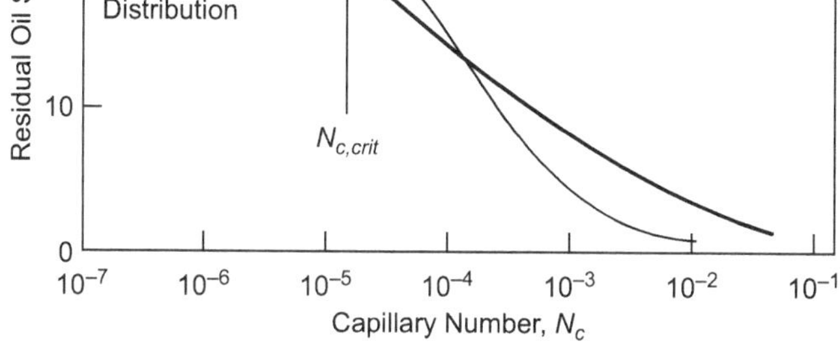

Figure 5.70 Pore size can be affected by fracture distribution and thereby impact residual oil mobilization.

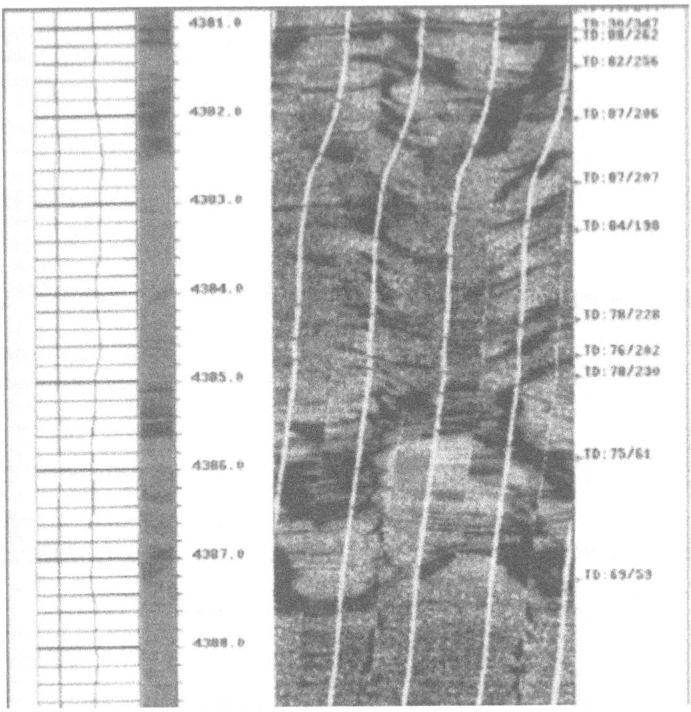

Picture 5.8 Commercial softwares can help identify fractures in FMS logs.

it should be included in the analysis. It is ideal to develop correlation that is specific to a field.

An example of the rose diagram is shown in Figure 5.71. It is a plot with each grouping of data being a petal of the rose. One starts the plot from a center point and draw a line outward (compass direction using a protractor) a distance that matches the number of recorded joints. Rose diagram includes both frequency and orientation with the assumption that all fractures have the same dimension. The rose diagram gives one the dominant orientation of fractures. This is of utmost importance in designing injection production strategies. It is useful to catalog core data, along with microlog information for each well. This information, then, can create isofrequency maps for the reservoir. It is useful for reservoir simulation.

The idea is to transit from macropore structure to a scalable model of the reservoir. Such model is essential in order to conduct useful reservoir modeling studies. This was done for the Weyburn CO_2 project—the project that is noted as the largest CO_2 sequestration project in history. Figure 5.72 shows an example from the Weyburn project.

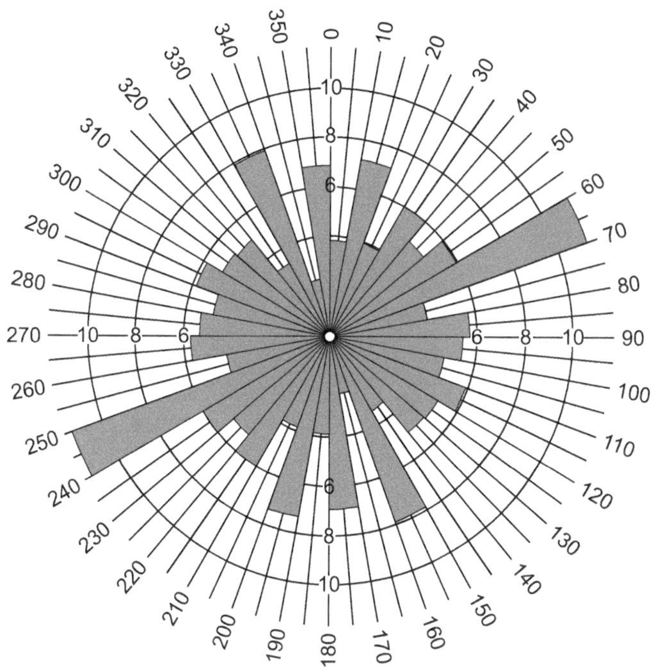

Figure 5.71 Rose diagram helps quantify the role of fractures.

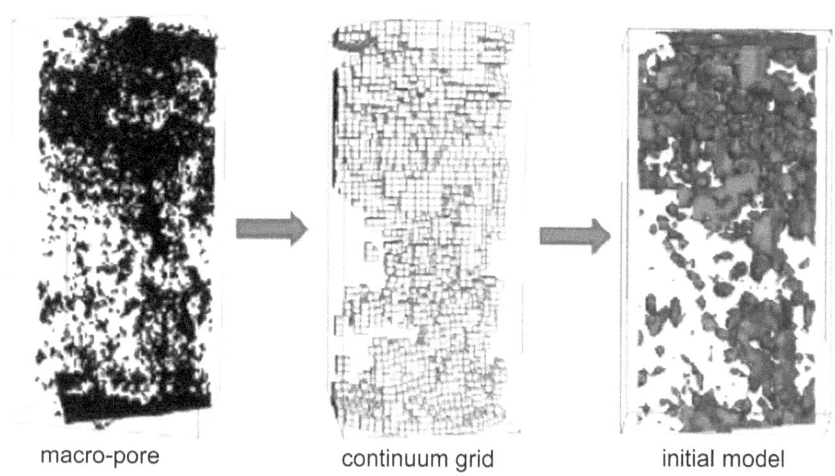

Figure 5.72 Transiting from macropore scale to an initial reservoir model, as experienced in Weyburn project of Canada.

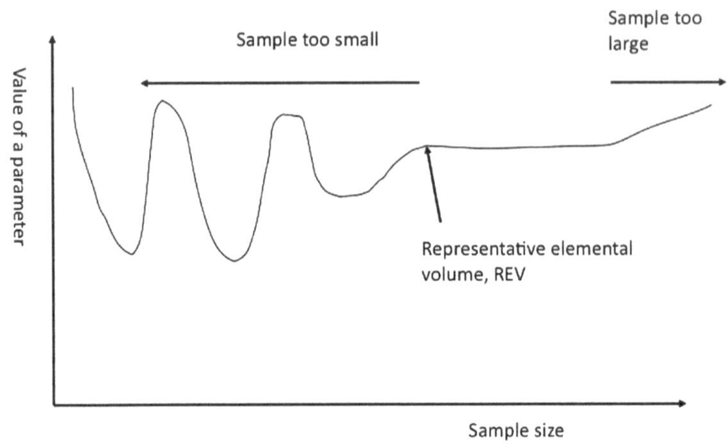

Figure 5.73 REV for a reservoir is much larger than the core samples collected.

5.8.1 Filtering Permeability Data

One of the most difficult problems in reservoir engineering is the fact that it is practically impossible to extract a sample that represents the reservoir. It is because the REV for petroleum reservoirs is much larger than the core size commonly encountered in a reservoir. Figure 5.73 highlights this difficulty. If the core size is below REV, all experiments conducted in it would have little relevance to the reservoir. This problem is further compounded in presence of fractures. The presence of fracture increases the size of REV and places a cored sample squarely in the oscillating region of Figure 5.73. In addition, coring in fractured formations is performed by avoiding the fractures as much as possible. This is because it is practically impossible to conduct fluid flow tests in a fractured core. For EOR design, this is of particular concern as the nature of fluid flow in a fractured formation is entirely different from that in a homogeneous system. Furthermore, flow that follows a certain regime (e.g., stable and stabilized) is likely to be different in presence of fractures. Because currently used instability number formulations do not use the presence of fractures (open or closed), it becomes even more difficult to predict the onset of fingering during an EOR process. However, fingering can lead to catastrophic failure of an EOR scheme.

In order to avoid this problem, filtering should be performed. Any such filter has to be custom designed for a specific field, if not specific well. For that, following data should be collected.

- Logs
- Core samples

Picture 5.9 The idea is to transit from microscopic to reservoir scale, following the correct scaling laws.

- Maps
- Production data
- Well completion history
- Waterflood-injection history
- Any other data that may be available

At first, the full 3D schematic of the reservoir should be constructed. As depicted in Picture 5.9, the idea is to transit from microscopic scale to reservoir scale, which involves developing the scaling laws.

This schematic is not quantitatively accurate but represents overall trend. In order to decide on the schematic as well as the type of flow that is expected in the reservoir, geophysicists, geologists, and engineers must work together.

Based on that, one of the following averages should be used to construct the permeability (k) × net pay (h) correlation.

For parallel flow, arithmetic average should be used:

$$k_{arith} = \frac{\sum_{i=1}^{n} k_i h_i}{\sum_{i=1}^{n} h_i}$$

For series flow, harmonic average should be used:

$$k_{harm} = \frac{\sum_{i=1}^{n} h_i}{\sum_{i=1}^{n} h_i / k_i}$$

For flow without a particular pattern (random flow), geometric average should be used:

$$k_{geom} = \sqrt[n]{k_1 k_2 k_3 \ldots k_n}$$

At this point, well test results should be gathered and history of permeability variation with time should be collected. Quite often, well test permeability becomes a function of time. Such behavior gives out useful clues as to the nature of fluid flow in the reservoir. The knowledge of the operators, combined with the construction of geological history, and fluid

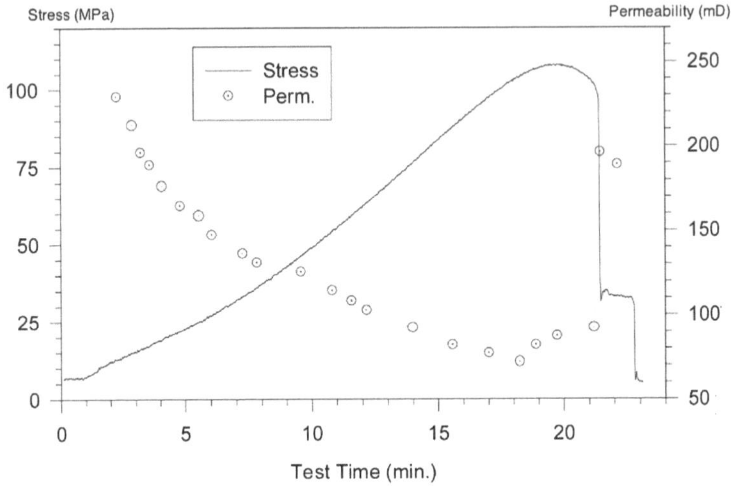

Figure 5.74 Laboratory test results under an overburden pressure of 50 MPa.

flow behavior all contribute to determining the true nature of reservoir. For instance, consistent decline in reservoir permeability, as evidenced from well test data can occur because of fracture closure, asphaltene deposition, or formation of gas pockets in the neighborhood of the well. It is important to conduct laboratory tests in order to determine natural reaction to cores to stress. Figure 5.74 shows an example of such a test. In this laboratory simulation with artificially consolidated cores, the original stress corresponds to initial conditions of a reservoir, which is subject to little stress because of overall balance of forces. This is expected because over geological time, any reservoir reaches a state that can be termed steady state in conventional terminology. This stress is increased over time as more and more fluid is extracted from the reservoir. During this time period, original permeability can be reduced to less than 50% of the initial permeability. Many field history support this observation. For instance, in several of the giant oil field of Hassi Messaoud, Algeria, similar decline in well test permeability has been observed. The decline is steeper for more heterogeneous zones, the ones that have profuse fracturing.

As can be seen in Figure 5.74, the core is crushed if the axial stress is very high (exceeding 100 MPa), resulting in sudden increase in permeability. Even then, however, original permeability is not restored. In theory, such occurrence of partial restoration of permeability can occur in the field. Several fields report such behavior. Almost all of these have fractures with intense network of secondary cementation.

In order to characterize a fractured formation, one has to know the following:

- The origin of formation
- The overall direction of flow
- Fracture distribution and frequency

One of the most useful tools of designing local filters is the compilation and plotting of *hk* from well tests and cores. In case of relatively homogeneous formation, the correlation between the two will follow the trend of a 45° straight line. Figure 5.74 demonstrates this point. On the other hand, if there are open fractures with significant aperture and length in the reservoir, the data points will fall over the straight line (meaning HKR is greater than unity, where HKR = ratio of $hk_{\text{welltest}}/hk_{\text{core}}$), the highest points representing maximum departure from the median line and highest frequency of open fractures. On the other hand, when points are located under the median straight line (meaning HKR less than unity), it signals the existence of secondary cementation that caused fractures to be plugged (Figure 5.75). For this case, the fraction of closed fractures over open fractures for various locations must be determined before a useful filter can be constructed.

This information is crucial to developing the filter. The filter uses the following data:

- frequency of open fractures
- frequency of closed fractures

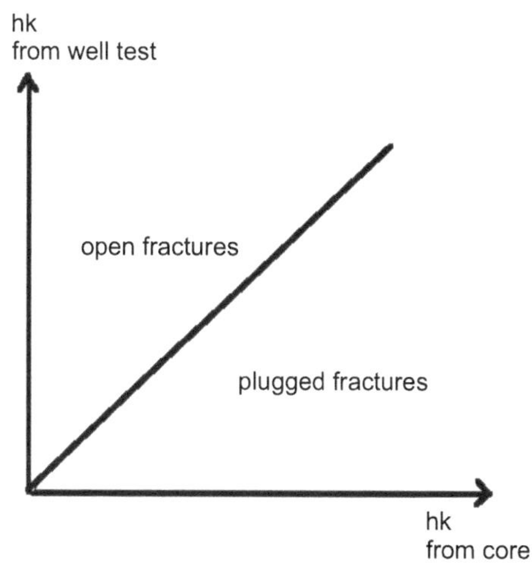

Figure 5.75 Determination of the nature of fractures from *hk* data.

- overall orientation of fractures
- overall orientation of sedimentation

5.8.2 Total Volume Estimate

Total volume estimate comes from the following sequential estimates.

1. Calculation based on geologic and seismic data;
2. Confirmation of hydrocarbon through well test (initial acidization and fracturing may be necessary);
3. The concept of net thickness does not apply;
4. Saturation cannot be estimated with logs in most cases;
5. Saturations must be confirmed with special core analysis;
6. End points are the only relevant points in a relative permeability curve;
7. Capillary pressure data is the basis for fine tuning total thickness, which is the determining factor for initial gas in place.

5.8.3 Estimates of Fracture Properties

TOC changes in shale formations influence V_p, V_s, density, and anisotropy and thus should be detected on the seismic response. To detect it, different workflows have been discussed by Chopra et al. (2012). Rickman et al. (2008) showed that brittleness of a rock formation can be estimated from the computed Poisson's ratio and Young's modulus well log curves. This suggests a workflow for estimating brittleness from 3D seismic data, by way of simultaneous prestack inversion that yields I_p, I_s, V_p/V_s, Poisson's ratio, and in some cases meaningful estimates of density. Zones with high Young's modulus and low Poisson's ratio are those that would be brittle as well as have better reservoir quality (higher TOC, higher porosity). Such a workflow works well for good quality data and is shown in flow chart below.

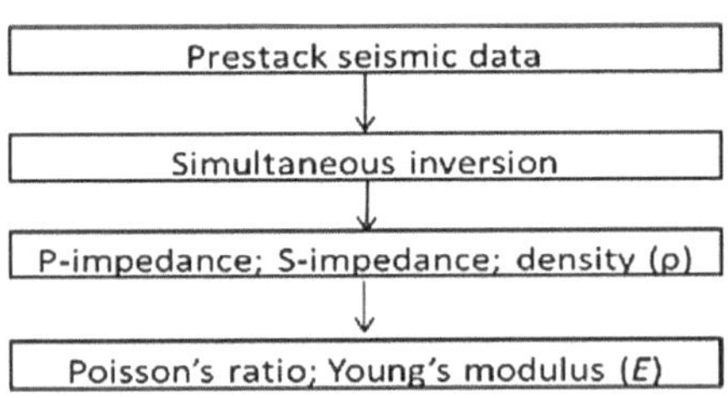

Natural fracture distribution map should be an indicator of how induced fracture will behave. Therefore, the data collected on natural fractures should be processed for designing an induced fracture.

Stress data collected during drilling as well as cuttings and well site logs should be all combined to develop enhanced understanding of reservoir fracture properties.

5.9 SPECIAL CONSIDERATIONS FOR SHALE

Some of the considerations of reservoir characterization must be altered for shale due to unique features of shale gas. Picture 5.10 shows how induced fracture would propagate in shale. Scenarios 1, 2, and 3 represent theoretically likely directions of fracture propagation resulting from increasingly higher levels of fracture-inducing force. As force increases in relatively unfractured rock, fractures propagate perpendicular to the direction of maximum principal stress. When they encounter a rock layer boundary (which is naturally weaker), the energy forcing the fracture dissipates laterally, making it harder for a fracture to continue across boundaries. In more fractured rock (scenario 3a), even with the same pressures as seen in 3, preexisting fractures can cause energy forcing a fracture to dissipate in other directions. Because of the many boundaries between rock layers and preexisting fractures in scenario 3a, the fractures propagated are wider and shorter than they would normally be if the same rock was relatively unfractured.

Passey et al. (1990) proposed a technique for measuring TOC in shale gas formations. This technique is based on the porosity–resistivity

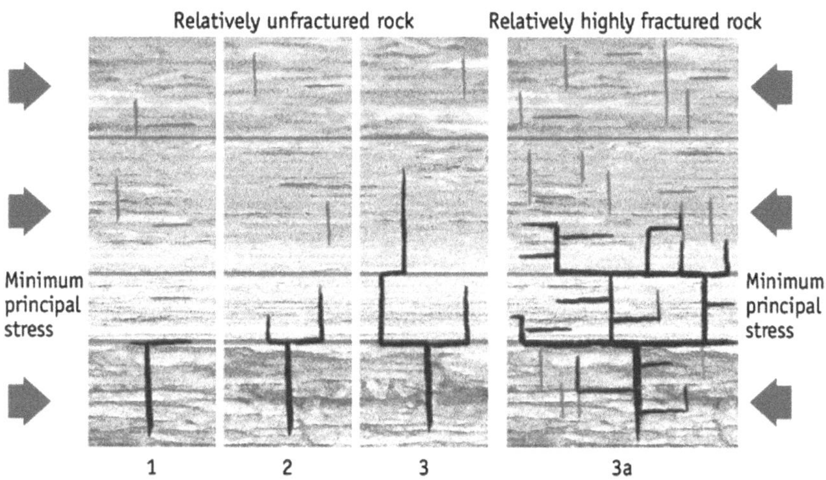

Picture 5.10 Different scenarios in fractured shale formation.

overlay to locate hydrocarbon bearing shale pockets. Usually, the sonic log is used as the porosity indicator. In this technique, the transit time curve and the resistivity curves are scaled in such a way that the sonic curve lies on top of the resistivity curve over a large depth range, except for organic-rich intervals where they would show crossover between themselves.

An integrated workflow in which well data as well as seismic data are used to characterize the hydrocarbon bearing shale can be developed as shown in the following workflow.

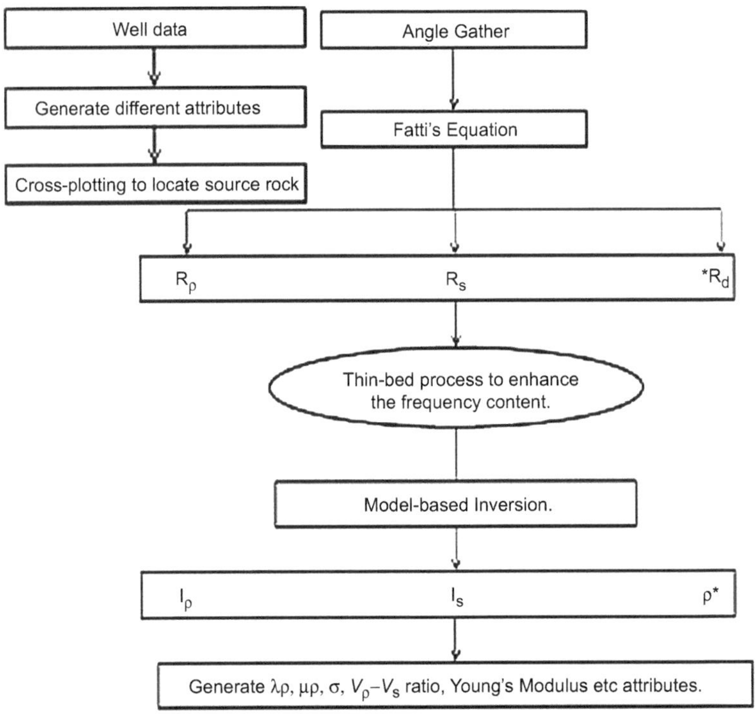

This compilation begins with the generation of different attributes from the well log curves. Then, using the cross-plots of these attributes one can identify the hydrocarbon bearing shale zones. Once this analysis is done at the well locations, seismic data analysis is picked up for computing appropriate attributes. Seismically, prestack data is essentially the starting point. After generating angle gathers from the conditioned offset gathers, Fatti's equation (Fatti et al., 1994) can be used to compute P-reflectivity, S-reflectivity, and density, which depends on the quality of input data as well as the presence of long offsets. Due to the band-limited nature of acquired seismic data, any attribute extracted from it will also be band-limited, and so will have a limited resolution. While shale formations may be thick, some high TOC shale units may be thin. So, it is desirable to enhance the resolution of the seismic data. An appropriate way of doing it is the thin-bed reflectivity inversion (Chopra et al., 2006; Puryear and Castagna, 2008). Following this process, the wavelet effect is removed from the data and the output of the inversion process can be viewed as spectrally broadened seismic data, retrieved in the form of broadband reflectivity data that can be filtered back to any bandwidth. This usually represents useful information for interpretation purposes. Thin-bed reflectivity serves to provide the reflection character that can be studied, by convolving the reflectivity with a wavelet of a known frequency band-pass. This not only provides an opportunity to study reflection character associated with features of interest, but also serves to confirm its close match with the original data. Further, the output of thin-bed inversion is considered as input for the model based inversion to compute P-impedance, S-impedance, and density. Once impedances are obtained, one can compute other relevant attributes, such as the $\lambda\rho$, $\mu\rho$, and V_p/V_s. These are used to measure the pore space properties and get information about the rock skeleton. Young's modulus can be treated as brittleness indicators and Poisson's ratio as TOC indicator.

5.10 SPECIAL CONSIDERATIONS FOR COALBED METHANE

Methane within coalbeds is not structurally "trapped" by overlying geologic strata, as in the geologic environments typical of conventional gas deposits. Only about 5–9% of the CBM is present as "free" gas within the joints and cleats of coalbeds. Most of the CBM is contained within the coal itself (adsorbed to the sides of the small pores in the coal). Picture 5.11 shows a typical CBM reservoir, along with fracturing operations.

Before CBM production begins, groundwater and injected fracturing fluids are first pumped out (or "produced" in industry terminology) from the network of fractures in and around the coal zone. The fluids are pumped until the pressure declines to the point that methane begins to desorb from the coal. CBM production initially requires pumping and removing significant amounts of water to sufficiently reduce the hydrostatic pressure in the subsurface so that methane can desorb from the coal before methane extraction can begin. CBM is produced at close to atmospheric pressure. The proportion of water to methane pumped is initially high and

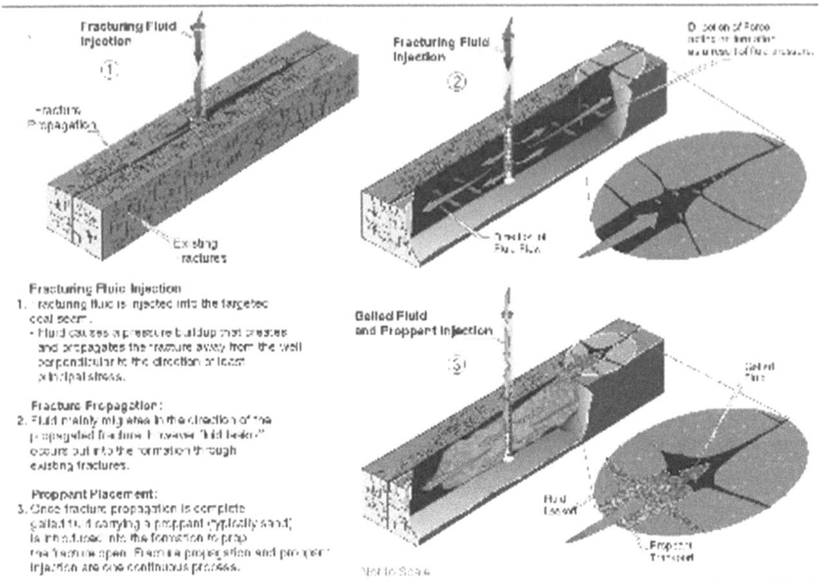

Picture 5.11 Schematic of a CBM with fracturing activities.

declines with increasing CBM production (Figure 5.76). In contrast, in the production of conventional petroleum-based gas, the production of gas is initially high, and as gas production continues over time and the gas resources are progressively depleted, gas production decreases and the amount of water pumped increases.

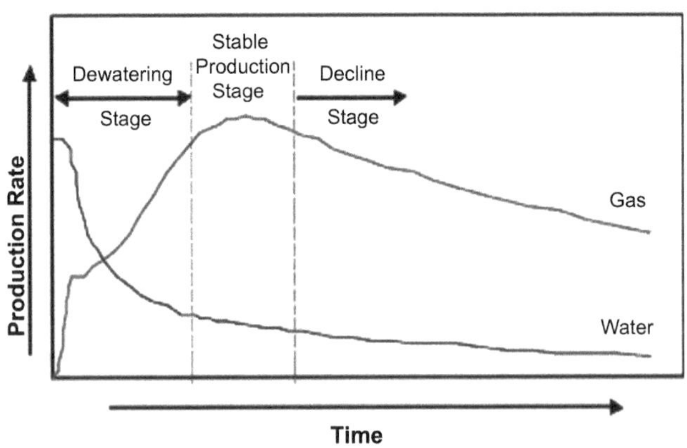

Figure 5.76 Production of water in CBM declines with time.

Almost every coalbed targeted for methane production must be hydraulically fractured to connect the production wellbore to the coalbed fracture network unless displacement or in situ combustion techniques are introduced. Many CBM wells are refractured at some time after the initial treatment in an effort to reconnect the wellbore to the production zones to overcome plugging or other well problems. Also, in response to site-specific coal geology and the economics of CBM production, where coal seams are thin and vertically separated by up to hundreds of feet of intervening rock, operators might design fracture treatments to enhance the vertical dimension and perform several fracture treatments within a single well to produce methane in an economically viable fashion.

CHAPTER 6

Scientific Characterization of Unconventional Gas Reservoirs

6.1 SUMMARY

Previous chapter presented reservoir characterization based on conventional science and engineering analysis. It is well known that New Science has numerous contradictions and unfounded premises that make it difficult to include new synthesis. This chapter introduces a discussion of fundamental science of mass and energy that enables readers to grasp the concept of natural energy and mass. This forms a background for characterizing unconventional gas reserves and rank them according to their sustainability and environmental integrity. When conventional oil and gas reservoirs are reconsidered based on their geology, they can be linked to added resources of unconventional oil and gas. It creates opportunities for new developments with little additional cost of reservoir development.

6.2 INTRODUCTION

Scientific characterization involves ranking in terms of energy contents. Energy is known to be the cause of actions that are ubiquitous. Scientifically, every action and movement has a driver. Because every object is in motion, that driver is ubiquitous. New Science has identified sun as the ultimate energy source for the earth. While this conclusion is true, the premise that defines energy in New Science is false (Islam et al., 2014). In this section, some of the scientific aspects of energy will be discussed.

Conventional notion of energy and the conservation of energy emerges from discrete description of mass and energy. It assumes that mass exists independent of energy. In addition, the "ability to work" is considered to be energy. The term "work" refers to displacement of an object. Therefore, if an object is moved around and brought back to the original place, no work has been performed on it. By definition, the pathway or the time function being mooted from the description of the process, one has lost track of actual work performed.

Unconventional Gas Reservoirs
ISBN 978-0-12-800390-9
http://dx.doi.org/10.1016/B978-0-12-800390-9.00006-2

In addition, any "work" is also related to "heat." This notion dates back to Lord Kelvin's notion of universe that in his view was constantly degrading to the point of being "heat dead" eventually. This tactic removes any dissimilarity between sunlight and solar heat from electric light and electrical heating, for instance. It also conflates energy from food with energy, say gasoline.

Core of this cognition has been in measuring blocks. For instance, Btu (British thermal unit) is defined as the amount of heat energy required to increase the temperature of one pound of water by 1 °F, at sea level. This definition assumes and imposes strictly linear property of water. It also conceals the chemical property of water. The hyperbolic extension does not stop here. This "Btu" is then transformed into energy from food in a strictly organic setting.

Conventionally, electricity does the same as the sunlight and New Science provides no basis for distinguishing electric energy from solar energy. This is one of the numerous disconnections between organic and mechanical systems. Interestingly, electricity and electromagnetism are based on the same atomic principle as the one used for describing mass and conservation of mass. Along with the assumption of spherical rigid balls, it is also assumed that each atom as well as subatomic particles are identical. After the discovery of some 69 subatomic particles, it is now commonly known that none of these particles are symmetrical, uniform, spherical, or rigid. However, the assumption of uniformity and identical form still holds, even when it comes to the "fundamental particle," most recently asserted as Higgs boson. While this notion won Nobel prize in 2013, scientists still do not have an answer to the questions, "If all Higgs Boson are identical and if there is nothing smaller than Higgs Boson, how are these particles moving? Does it mean then, there is certain space that is empty and devoid of anything?" This leads to the placement of Higgs boson as a static object. Static matters cannot impart energy transfer, thereby creating disconnection between mass and energy.

Higgs bosons are also considered to be "uniformly distributed" as well as "highly unstable." They are contradictory properties by themselves.

More dogmatic assertion involves the notion that photons from a radioactive substance "feel" the electromagnetic force as well as the weak force, but neutrinos only "feel" the weak force. This assertion makes neutrinos less reactive while more mobile within a material system. In order to remedy such an obvious logical gaffe more assertions are made that are equally aphenomenal. As such, it is stated that when a photon is emitted, it is attracted by the electromagnetic force that is generated by the atoms around it. While photons are attracted, neutrinos are considered to be deflected by the same atomic body. The aphenomenal assumption there is that the nucleus and electron are all ubiquitous to the extent that photons would "hit" them, whereas neutrinos would not, even though both types of particles are "mass less."

Other anomalies and contradictions also exist in terms of energy description. For instance, light is considered to be a collection of photons with finite speed (speed of light being the maximum possible speed by virtue of the assumption of zero mass of photons). This assertion disconnects light from its source, thereby removing the possibility of light pollution or the ability to distinguish between sunlight and artificial light.

It is also inferred that Higgs boson can travel through opaque objects at a speed close to the speed of light (some actually postulated it to be faster than light) whereas light can only travel through "transparent" bodies. This assertion does not appear as an anomaly in conventional analysis because of preexisting assumption that light and mass are discrete from each other.

The atomic model is used to describe mass and chemical reaction. This centuries old model used to assume that atoms are the elemental particles and are solid, spherical, and rigid. At later stage, such properties were invoked to neutron, proton, and electrons. It was hypothesized that certain atoms have loosely attached electrons. An atom that loses electrons has more protons than electrons and is positively charged. An atom that gains electrons has more negative particles and is negatively charged. A "charged" atom is called an "ion." Depending on the number missing electrons, an ion would be more prone to "bonding" with another element. This line of reasoning helped explain chemical reactions. However, the only way such reaction could be linked to energy is through "heat of reaction." Typically, this analysis satisfied the need of engineers, whose principal focus was heat. However, this disconnected "light" in general and artificial light in particular from being connected to chemical change. This remains a source of inconsistency in New Science.

In terms, energy generation through radiation, the concept of "unstable isotope" was introduced. The word "isotope" is defined as an atom that has unusual number of neutron. It is called stable isotope when the nucleon is not prone to breaking down. There are only a few stable isotopes recognized today. When an isotope is prone to breaking down spontaneously, it is called "unstable isotope." It is hypothesized that when unstable isotopes break down into new isotopes, they usually emit alpha, beta, or gamma radiation. The term "radioactivity" is synonymous with the emission of this radiation. This notion has been in existence since the early work of French physicist Henri Becquerel, who observed potassium-uranyl sulfate crystals on a film and concluded that the sun emits X-rays. Becquerel also found that all compounds of uranium and pure uranium behaved the same way. They all emitted what seemed to be X-rays, yet they did not need to be excited first with light or an electron beam. The uranium and its compounds could ionize gases, which permitted the gases to conduct an electric current.

The early work of Becquerel was further advanced by physicists Marie Sklodowska-Curie of Poland and Pierre Curie of France who conducted a series of experiments to determine which other elements and compounds emitted this mysterious radiation. They found that the element thorium behaved much like uranium. But the radiation from pitchblende, a uranium ore, was far greater than it should have been, considering the known percentage of its uranium content. They therefore suspected that the pitchblende contained some other previously undiscovered element. Beginning with a large sample of pitchblende, they employed a series of chemical separation techniques, always discarding the separated fraction, which did not emit the disproportionately high radiation. Eventually, they isolated a new radioactive element, which they called polonium in honor of Marie's home country. This was the beginning of "purification for nuclear energy."

Four years later, starting with 100 kg of pitchblende, and using similar techniques, they were able to isolate 0.1 g of an even more intensely radioactive substance, which they called radium. After Pierre's accidental traffic death in 1906, Marie was appointed in his place as a professor of physics at the Sorbonne in Paris. She was awarded the Nobel Prize in 1911 for her discovery of polonium and radium. She died in 1934 of leukemia, which was probably caused by overexposure to the radiation involved in her research. However, this connection was not made and until now, the failure to change the premise that separated mass from energy has made it impossible for scientists to find the root of energy pollution as well as cancer.

In the mean time, Ernest Rutherford, 1st Baron Rutherford of Nelson, a New Zealand-born British physicist became prominent for his work on radiation, eventually earning him the title "father of nuclear physics." His research focus was to measure the "penetrating power" of uranium's mysterious radiation. He discovered that the radiation was made up of three different types of "rays" with very different powers of penetration. The intensity of what he called alpha (α) rays, could be reduced to one-half as much by a very thin (0.005 mm) piece of aluminum foil. A similarly thin piece would cut the intensity by half again as much, to a total intensity of one-fourth; and a third piece would cut the total to one-eighth, etc. Beta (β) ray intensity could be reduced to one-half as much by a 0.5 mm aluminum sheet; and again each additional 0.5 mm sheet would cut the prevailing amount by one-half. In general, the thickness of a specific material required to reduce a certain type of radiation by one-half is called a half-thickness. The half-thickness for gamma (γ) the third type of uranium radiation was found to be 80 mm of aluminum.

Rutherford sealed a thin-walled vial of alpha-emitting radon gas inside a second glass tube. All the air was pumped out of the second outer tube before sealing. Rutherford attempted to ionize any possible remaining gas in the outer tube, and at first he was unsuccessful. However, as time passed, gas accumulated in the second outer tube. This was the beginning of light emission through excitation of ions. Today, this technique is promoted as the most effective lighting of buildings. They are dubbed as "energy savers" and many countries are considering making them mandatory.

This "discovery" of Rutherford became useful in explaining artificial electricity generation. It was postulated that electrons can be made to move from one atom to another, as long as they were kept in a state of instability. When those electrons move between the atoms, a current of electricity is created. The electrons move from one atom to another in a "flow." One electron is attached and another electron is lost.

Subsequent research in "creating" energy involves various ways to move electrons off of atoms. In another word, creating instability or imbalance became the only means to generate energy. The principal task becomes that of creating a system that generates large numbers of positive atoms and free negative electrons. Since positive atoms have affinity toward negative electrons so they can be balanced, they have a strong attraction for the electrons. The electrons also has an affinity toward the positive atom so there is an overall balance. This principle is the basis for electricity generation.

Even the briefest examination of the narrowness in the focus of the approach taken by New Science to the phenomenon of radiation—whether artificially induced or naturally occurring—uncovers interesting phenomena. Consider carbon.

Carbon is one of the most persistent elements to be found in the atmosphere and soils of the earth. Over time—especially geological periods of time—ordinary C–C bonds and especially C=C double bonds seem to have proven particularly resistant to the effects of radiation. This is "not" the case for most other elements connected chemically to such carbon bonds. Most elements apart from the noble gases seem vulnerable to radiation effects at the molecular level.

This particular feature of carbon bonding seems rife with many actual and potential consequences. Nevertheless, partly because there seems to be no consistent explanation afforded by the conventional treatment by New Science applied to sorting out this question, many geological transformations in the earth remain incompletely or incorrectly accounted for.

Before one can examine the science of energy that includes light and heat, one must review the existing theories and analyze their shortcomings. This will follow with proper characterization of energy with fundamentally sound premises.

6.3 SCIENCE OF NATURE VERSUS NATURAL SCIENCE

Table 6.1 shows the characteristic features of nature. These are true features and are not based on perception. Each is true insofar as no example of the opposite has been sustained. It is important to note that the following table describes everything in existence as part of universal order and applies to everything internal, including time, and HTM. However, the source of Human thought material (HTM), i.e., intention is not part of these features.

A lengthy discourse on the topic reveals that intention is external and the ownership belongs to every mature, awake, and sane adult (see for instance the chapter: Role of Intention in Zatzman and Islam (2007), Islam et al. (2013) and Islam et al. (2014)).

Table 6.2 lists the fundamental features of the external entity. The existence of an external entity is necessary condition to eliminating the notion of void that had been inherited from atomism philosophy and was carried forward first by Thomas Aquinas and then by subsequent scientists, without exception.

This external entity was first recognized as God (ancient Greek philosophers all the way to Avicenna and Averroes of Islamic golden era), then conflated as plenum and aether (Islam et al., 2013, 2014). While the existence of such entities has been denied and sometime "proven" to be nonexistent, the traits of these external entities have been included in all forms of "fundamental" particles, ranging from photon to Higgs boson. In addition, such features have also been invoked in galactic models in the

Table 6.1 Typical Features of Natural Processes that Are Part of Universal Order

Feature No.	Feature
1	Complex
2	Chaotic
3	Unpredictable
4	Unique (every component is different), i.e., forms may appear similar or even "self-similar," but their contents alter with passage of time
5	Productive
6	Nonsymmetric, i.e., forms may appear similar or even "self-similar," but their contents alter with passage of time
7	Nonuniform, i.e., forms may appear similar or even "self-similar," but their contents alter with passage of time
8	Heterogeneous, diverse, i.e., forms may appear similar or even "self-similar," but their contents alter with passage of time
9	Internal
10	Anisotropic
11	Bottom-up
12	Multifunctional
13	Dynamic
14	Irreversible
15	Open system
16	True
17	Self-healing
18	Nonlinear
19	Multidimensional
20	Zero degree of freedom[a]
21	Nontrainable
22	Continuous function of space, without boundary
23	Intangible
24	Open
25	Flexible
26	Continuous function of time
27	Balanced

[a]With the exception of humans that have freedom of intention.
From Islam et al., 2014.

form of various entities, ranging from "dark matter," "black hole" to "absolute void." Newton introduced this as "external" force and defined it as the originator of differential motion. The original Averroes concept, as supported by the Qur'an was that such originator of motion is the creator, whose traits are all different from the traits of creation.

Table 6.2 Features of External Entity

Feature No.	Feature
1	Absolutely external (to everything else)
2	All encompassing
3	No beginning
4	No end
5	Constant (independent of everything else)
6	Uniform
7	Alive
8	Infinity
9	Absolutely true
10	Continuous
11	All pervasive in space
12	All pervasive in time
13	Infinite degree of freedom
14	Irreversible
15	Unique
16	Open system
17	Dissimilar to everything else

Table 6.3 shows many currently used "laws" and theories, all of which emerged from the New Science after renaissance. Note how the first premises of practically all of these theories violate fundamental features of nature. Only conservation of mass that in fact has root in ancient times does not have an aphenomenal first premise. New Science has given us only theories and "laws" that have spurious first premise, as can be evident by using Averroes criterion for phenomenality (Zatzman and Islam, 2007a).

If all theories of New Science are based on premises that violate fundamental traits of nature, such laws and theories should weaken considerably or worse, implode, if applied as universal laws and theories. They can be applied only to certain fixed conditions that pertain to "idealized" situations not extant in nature. For example, it can be said that the laws of motion developed by Newton cannot explain the chaotic motion of nature due to its assumptions that contradict the reality of nature.

The experimental validity of Newton's laws of motion is limited to describing instantaneous macroscopic and tangible phenomena. However, microscopic and intangible phenomena are ignored. Classical dynamics as represented by Newton's laws of motion, emphasize fixed and unique initial conditions, stability, and equilibrium of a body in motion (Islam et al., 2010). With the "laws" and theories of Table 6.3, it is not possible to make distinction between the natural products and their corresponding

Table 6.3 Natural Features Are Violated in the First Premise of Various "Laws" and Theories of the Science of Tangibles

Law or theory	First premise	Features violated (see Table 6.1)
Conservation of mass	Nothing can be created or destroyed	None, but applications used artificial boundaries
Quantum theories	Anything can be created from nothing; everything has multiple history	4, 6, 22, 23, 26
Conservation of energy	No energy can be created or destroyed in isolation with mass	22, 23, 26
Big bang theory	14 billion years ago, there was a super hot entity of infinite mass and zero volume that has been expanding after the big bang, super hot	1,3,6,9,14,24,26
Big chill theory	14 billion years ago, there was a super chill entity of infinite mass and zero volume that cracked and exploded into infinite pieces	1,3,6,9,14,24,26
Saul Perimutter and Brian Schmidt (2011 Nobel Prize)	Universe is expanding with acceleration	1,3,6,9,14,24,26
Higgs boson (2013 Nobel Prize)	Uniform, discrete, symmetric, fundamental particles of zero mass, empty space in between	4, 7, 22, 23
Atomic theory	Uniform, symmetric, discrete, fundamental particles of finite mass	4, 7, 22, 23
Einstein's light theory	Photons of zero mass and constant speed	4, 7, 13, 22, 23, 26
Genetic theories	Genes—fundamental building blocks of living organism in isolation	4, 6, 13, 22, 23, 26
Defective genes	Inherent defects of genes in isolation	4, 6, 13, 17, 22, 23
Probability theories	Steady state, repetitive, and repeatable	4, 5, 6, 13, 16, 22, 23, 26
Relativity	Time function of perception, perception function of person	4, 7, 16, 22, 23, 26, 27
Gravitational	Force function of mass, steady state, time function of gravity	4, 7, 16, 22, 23, 26, 27

Continued

Table 6.3 Natural Features Are Violated in the First Premise of Various "Laws" and Theories of the Science of Tangibles—cont'd

Law or theory	First premise	Features violated (see Table 6.1)
Cosmic theories	Empty space between celestial bodies that are expanding or contracting	4, 7, 16, 22, 23, 26, 27
Lavoisier's deduction	Perfect seal	15
Phlogiston theory	Phlogiston exists	16, 22, 23, 26
$E = mc^2$	Mass of an object is constant	13, 22, 23, 26
	Speed of light is constant	13, 22, 23, 26
	Nothing else contributes to E	14, 19, 22, 20, 23, 24
Planck's law	If the medium is of homogeneous and isotropic constitution, then the radiation is homogeneous, isotropic, unpolarized, and incoherent	5, 8, 10, 17, 22, 23, 26
Aether theory	Zero mass, zero energy	16, 22, 23, 26
Quantum cosmic theory	Infinite mass, zero energy	16, 22, 23, 26
Charles law	Fixed mass (closed system), ideal gas, constant pressure	24, 3, 7
Boyle's law	A fixed mass (closed system) of ideal gas at fixed temperature	24, 3, 7
Kelvin's law	Kelvin temperature scale is derived from Carnot cycle and based on the properties of ideal gas	3, 8, 14, 15
Thermodynamics 1st law	Energy conservation (the first law of the thermodynamics is no more valid when a relationship of mass and energy exists)	22, 23, 26

Thermodynamics 2nd law	Based on Carnot cycle, which is operable under the assumptions of ideal gas (imaginary volume), reversible process, adiabatic process (closed system)	3, 8, 14, 15
Thermodynamics 0th law	Thermal equilibrium	10, 15
Poiseuille's law	Incompressible uniform viscous liquid (Newtonian fluid) in a rigid, noncapillary straight pipe	7, 22, 23, 25, 26
Bernouilli's law	No energy loss to the sounding, no transition between mass and energy	15, 22, 23, 26
Newton's 1st law	A body can be at rest and can have a constant velocity	13, 22, 23, 26
Newton's 2nd law	Mass of an object is constant; force is proportional to acceleration; external force exists	7, 13, 14, 16, 18, 22, 23, 26, 27
Newton's 3rd law	The action and reaction are equal	3, 22, 23, 26
Newton's viscosity law	Uniform flow, constant viscosity	7, 13, 22, 23, 26
Maxwell's equation	Uniform, spherical, rigid balls form energy	4, 7, 22, 23, 26
Newton's calculus	Limit $\Delta t \rightarrow 0$	22, 23
Fractal theory	Single pattern that repeats itself exists	1–4, 6, 8, 10

Table 6.3a Analysis of "breakthrough" technologies (revised from Khan and Islam, 2012)

Product	Promise (knowledge at t= 'right now'	Current knowledge (closer to reality)
Microwave oven	Instant cooking (bursting with nutrition)	97% of the nutrients destroyed; produces dioxin from baby bottles
Fluorescent light (white light)	Simulates the sunlight and can eliminate 'cabin fever'	Used for torturing people, causes severe depression
Prozac (the wonder drug)	80% effective in reducing depression	Increases suicidal behavior
Anti-oxidants	Reduces aging symptoms	Gives lung cancer
Vioxx	Best drug for arthritis pain, no side effect	Increases the chance of heart attack
Coke	Refreshing, revitalizing	Dehydrates; used as a pesticide in India
Transfat	Should replace saturated fats, incl. high-fiber diets	Primary source of obesity and asthma
Simulated wood, plastic gloss	Improve the appearance of wood	Contains formaldehyde that causes Alzheimer
Cell phone	Empowers, keep connected	Gives brain cancer, decreases sperm count among men.
Chemical hair colors	Keeps young, gives appeal	Gives skin cancer
Chemical fertilizer	Increases crop yield, makes soil fertile	Harmful crop; soil damaged
Chocolate and 'refined' sweets	Increases human body volume, increasing appeal	Increases obesity epidemic and related diseases
Pesticides, MTBE	Improves performance	Damages the ecosystem
Desalination	Purifies water	Necessary minerals removed
Wood paint/varnish	Improves durability	Numerous toxic chemicals released
Leather technology	Won't wrinkle, more durable	Toxic chemicals
Freon, aerosol, etc.	Replaced ammonia that was 'corrosive'	Global harms immeasurable and should be discarded

artificial substitutes (see below). Consequently, the same theories that formed the basis of engineering the artificial products cannot be called upon to make the reversal.

Wood → plastic
Glass → PVC
Cotton → polyester
Natural fiber → synthetic fiber
Clay → cement
Molasses → Sugar
Sugar → Sugar-free sweeteners
Fermented flower extract → perfume
Water filter (Hubble bubble) → cigarette filter
Graphite, clay → chalk
Chalk → marker
Vegetable paint → plastic paint
Natural marble → artificial marble
Clay tile → ceramic tile
Ceramic tile → vinyl and plastic
Wool → polyester
Silk → synthetic
Bone → hard plastic
Organic fertilizer → chemical fertilizer
Adaptation → bioengineering

The above transition is the hallmark of modern technological developments that have been characterized by Nobel laureate chemist, Robert Curl as "technological disaster." Table 6.3 shows how every promise made in technology development has been reversed in the most spectacular way.

6.4 DELINEARIZED HISTORY OF CHARACTERIZATION OF MATTER AND ENERGY

Figure 6.1 shows Aristotle's model for four fundamental elements of matter. This is the oldest form of phase diagram that can be found in Europe. This figure is in effect a steady-state model. The elimination of the time function made the diagram appear perfectly symmetrical, which is the essence of Atomism. Democritus is indeed most often cited as the source of the atomic theory of matter, but there's a strong argument/likelihood that what he had in mind was a highly idealized notion, not anything based on

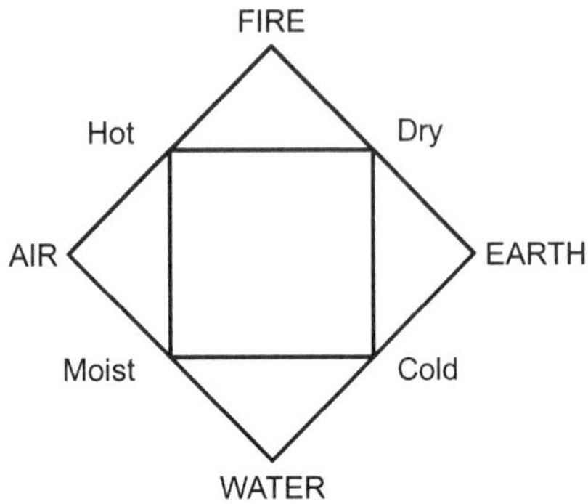

Figure 6.1 Aristotle's four-element phase diagram (steady state).

actual material structure. For the Greeks, symmetry was believed to be good in itself and was largely achieved by geometric rearrangement of (usually) two-dimensional space. There is an ambiguity as to whether Greek atomists thought of atoms as anything other than an infinite spatial subdivision of matter. Heraclitus' major achievement—which also marginalized him among the other thinkers of his time, unfortunately—was his incorporation of a notion of the effects of time as a duration of some kind, as some other kind of space in which everything played itself out.

On the matter of role of time sequence, and universal order, Heraclitus had a profound view that was considered to be a paradox and was rejected (Graham, 2006). Heraclitus wrote:

> *"This world-order [kosmos], the same of all, no god nor man did create, but it ever was and is and will be: everliving fire, kindling in measures and being quenched in measures."*

In this passage, he uses, for the first time in any extant Greek text, the word *kosmos* "order" to mean something perceived as "world." He identifies the world with fire, but goes on to specify portions of fire that are kindling and being quenched. Although ancient sources, including Aristotle as well as the Stoics, attributed to Heraclitus a world that was periodically destroyed by fire and then reborn, the present statement seems to contradict that view, as Hegel also noticed. If the world always was and is and will be, then it does not perish and come back into existence, though portions of it (measures of fire) are constantly being transformed. This contradiction and paradox are erased if "world-order" is replaced with "universal order" and the creation of time[1] is preceded to be before creation of everything else

[1] This time is a creation and is transient (changing).

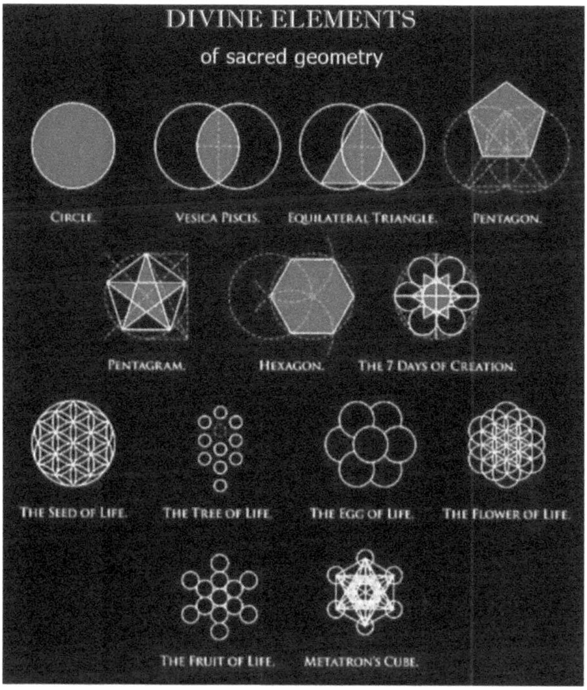

Figure 6.2 Divinity in Europe is synonymous with uniformity, symmetry, and homogeneity, none of which exists in nature.

(known as "matter" at later time). A consistent and nonparadoxical meaning emerges if the following sequence is used. Creator (absolute time) created the absolute plan (variable time) before creating everything as a function of time.

Figure 6.2 is a depiction of what is sacred in European narration of philosophy. Even though this depiction is attributed to Roman Catholic Church, it is clearly related to features seen in "absolute being" in ancient Greek and later transformed into all aspects of scientific and mathematical cognition. Aristotle said, "The mathematical sciences particularly exhibit order, symmetry, and limitation; and these are the greatest forms of the beautiful." This can very well be the beginning of reversing science from long-term to short-term through mathematization. Linearization is the ultimate version of such mathematization.

This fascination with homogeneity, symmetry, and other 'godly' attributes and assigning them to creation is uniquely European (Livio, 2005). The orient is known to have a different description of nature. At no time, ancient India, Babylonia, or ancient China conflated nature of God with

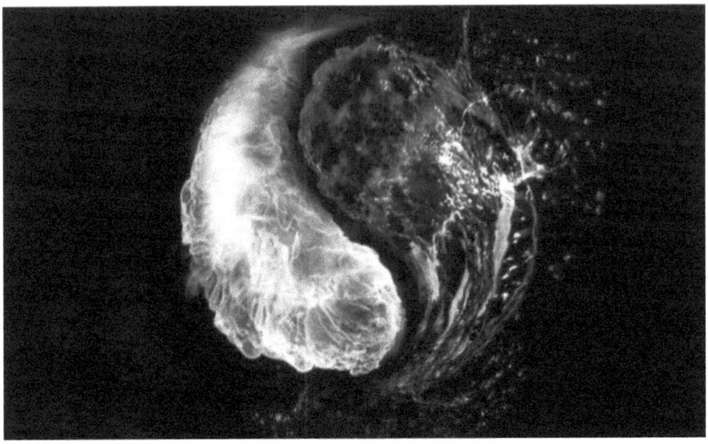

Figure 6.3 Water and fire are depicted through *taegeuk* (*yin–yang*).

nature of humans, even for *Avatars*. The most important "transformation" between creator and creation is through the notion of *Avatar* in India. An *avatar* is bound to change and is subject to all features of a mortal or natural objects.

Figure 6.3 shows depiction of essential elements of matter. This is typical of Yin–Yang concept that has been known in Korean peninsula for the entire recorded period of history. This figure shows the coexistence fire and water, the two main components of universe through asymmetrical pattern while keeping the complimentary nature of those components intact. The broader symmetry that is seen is more thematic than tangible. The central message of this picture is stated through the word of famous mythology professor, Joseph Campbell (1904–1974), who said, "The goal of life is to make your heartbeat match the beat of the universe, to match your nature with Nature." This is a theme that is in the core of Islamic faith that describes religion as *Deen* (natural trait) and defines Muslim (root word *salama* that means surrender and peace) to be the one in tune with nature and natural order. Note in Figure 6.3 how both fire and water parts contain global symmetry but have no local symmetry. Similarly, the flag of South Korea exhibits antisymmetry, or symmetry with a property reversal (Figure 6.4). The white background is a traditional Korean color. It traditionally represents peace and purity. Note that this circle is not tangible and it signifies the never-ending (sustainable) nature of nature. It is anti-symmetrical because of the red/blue interchange, both in color and deeper meaning. The deeper meanings represent opposing functions of the

Figure 6.4 Korean national flag contains ancient symbol of creation and creator.

universe. The most comprehensive meaning is the existence of tangible and intangible. These seemingly opposite or contrary forces are interconnected and interdependent in the natural universe that allows the existence of both in perfect harmony.

Such harmony is in sharp contrast with European notion that says the universe is constantly degrading or there is a continuous struggle between good and evil. The European notions emerge from a first premise that is similar to the dogmatic notion of "fall" and "original sin." Nature being a union of opposites in harmony comes from the first premise that nature is perfect and balanced (Khan and Islam, 2012). Many natural dualities (such as light and dark, woman and man, day and night, high and low, hot and cold, fire and water, life and death, and so on) are thought of as physical manifestations of the yin–yang concept.

This concept also reveals the fact that the duality is apparent and it is a matter of perception of the observer whereas it is absolute for the object. For instance, a blind person does not see the difference between night and day but that perception does not change anything about the night and day. This fact demonstrates the absurdity of quantum theory that makes reality a function of perception. Reality cannot be a function of perception, unless there is no such thing as reality. According to Aristotle, one of the mistakes of Zeno in his paradoxes of time and motion is that he did not distinguish between actual and potential infinities. Scientifically, it is the continuity of the time function that eluded Zeno. Aristotle "remedied" it by differentiating between actual (present) and potential (future). Then he asserted, "Everything is potential infinity and nothing is actual infinity." This in essence reversed the notion of reality concept of Plato. Such absurdity does not occur in oriental cognition that considers the time function properly. Furthermore, all objects (matter and energy) and events (time) in the world are expressed by the movement of "yin" and "yang." For instance, the moon is yin while the sun is yang; the earth is yin and the heaven (sky) is yang; a woman is yin and a man is yang; the night is yin and the day is yang; the winter is yin and the summer is yang; etc. Yin and yang are opposite, yet work in perfect harmony. This aspect is typical of Qur'anic philosophy as well as everything is reported to be created in pair (as in husband and wife).

The duality in yin–yang is further highlighted in the four markings, called trigrams or Pa-kua symbols, which are opposites of one another diagonally (Figure 6.4). Broken lines are unbroken and vice versa. The trigram together represents the principle of movement and harmony. Each trigram (hangul: kwae) represents one of the four classical elements, namely, heaven, earth, fire, and water. The word "hangul" represents heaven, while *kwae* means movement and harmony. Each "Kwae" consists of three bars of divination signs that can be either broken or unbroken bars. A broken bar stands for yin while an unbroken bar stands for yang. Numerous combinations are possible, but four basic elements correspond to heaven, water, earth, and fire.

If one replaces each unbroken line in the trigrams by 0 and each broken line by 1, one can see that the symbols represent numbers in the binary (base two) number system. The symbols and their meanings appear below. The binary numbers are read from bottom to top (Figure 6.5).

The top left corner of the flag is three unbroken lines and represent heaven (0). This is significant because both "heaven" and 0 signify the origin. Note that "heaven" is not the one that "righteous" people end up (in theological sense), this is the heaven that originated the entire creation. In English language, the use of the word "heaven" is full of ambiguity that comes from Roman Catholic Church's interpretation of origin of universe

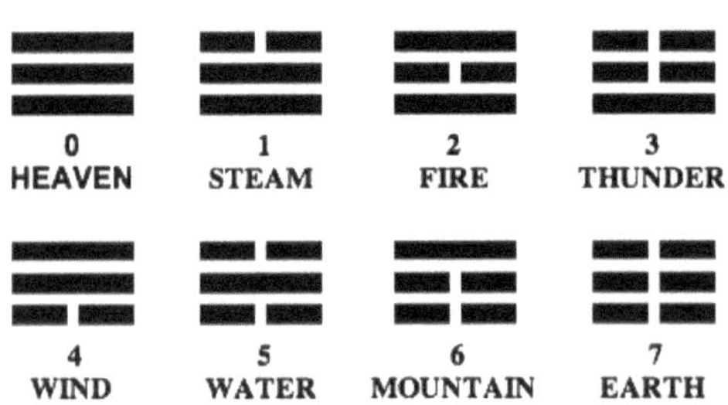

Figure 6.5 Combination of various fundamental elements make up the rest of the creation.

and men as well as the myths that purport the notion of "gods and angels" residing in heaven. However, in Arabic language, such ambiguity does not occur. The creator does not reside in *Jannah* (literally meaning "garden" and often translated as "heaven") as he is not constrained by space or time. He is also the originator of everything. For instance, Quran (2:117) states: "Originator of the skies and the earth. When He decrees a matter, He only says to it, "Be," and it is." The Arabic word for skies, *Samawah*, is often incorrectly translated as "heavens," highlighting confusion of the English language.

Equally important is the use of "zero" to denominate the source of everything. This "zero" can have two meanings, i.e., nothingness and origin. The nothingness would coincide with ancient Greek word, *Χαοσ*. However, this zero (*sunya* in Sanskrit and *cipher* in Arabic) never implies nothingness as in void. It rather implies the originator, who originated everything from nothing. This is consistent with Islam as well as notable oriental religions. For instance, in Sanskrit, *Maha Sunya* (Great Zero) refers to the outer space, which is anything but void as in nothingness and most often refers to the creator. Interestingly, the Arabic word "cipher," while recognized as the origin of the word "zero," does not represent void, it rather refers to "emptiness." In ancient Greek culture, however, void refers to the original status of the universe, which was thought to be filled with nothingness, in terms of tangibles, including time. Similarly, the equivalent word in Arabic is *As-sama'a*, which stands for anything above the earth, including seven layers of skies, is not "heaven," which is Jannah in Arabic and literally means "garden of paradise." Qur'an often refers to the creator as the one "in the sky" (e.g., Quran 67:16 says: Do you feel secure that He who is in the sky (*sama'a*) would not cause the earth to swallow you and suddenly it would sway?).

Opposite to "heaven" is the earth (designated by number Seven), placed at the lower right corner of the flag. This placement as well as the number Seven are both noteworthy and contain deeper meaning. While earth is known to be "just" a planet in European science, it holds a much deeper meaning the Qur'an that defines humans as the Viceroy (*khalifa*) of the creator (e.g., Chapter 2:30 of the Qur'an specifies man's role as the viceroy), charged with law and order on Earth. The Arabic word for "earth" is *Ardha*, which means "habitat for humans (the viceroy of the creator)." This outlook is clearly different from the Eurocentric notions, ranging from the vastly discredited "original sin" to widely accepted "evolution" theories (McHenry 2009; Hall, 2008) that detach human conscience from its functioning in a society. Overall, this has been the cognitive malfunction in European thought process. They confused God's traits with traits of creation (starting with Thomas Aquinas). This immediately cut off the relationship between creator and creation. This involved confusion in understanding what is natural—a confusion that continues until today. Then they confused creations (other than humans) with humans due to lack of logical premise defining the purpose of humanity. For them, humans are just another animal and the earth is just another planet. This latter confusion cut off conscience (and ownership of intention) from humanity. The ground became fertile for onset of various forms of aphenomenal cognition, some of which called itself atheism, naturalism, etc.

Water (number Five) is placed at the right top corner. The existence of water as a fundamental element is important. In every culture water is synonymous with life and liveliness. Qur'an places the existence of water before anything else. Opposite to water is fire (number Two) at the lower left corner. The role of fire is opposite to water, yet it is essential to life. Without fire, there is no carbon dioxide, the essence of plant, and therefore, life. Fire represents transition from cold to hot, from life to death, from tangible (water or liquid) to intangible (vapor or gas). This phase change is typical of creation. In fact, the very fact that everything is moving (a function of time) makes it essential to go through this phase of tangible and intangible. Overall, this continues in an eternal circle. Picture 6.1 shows how it is natural to have such dual characteristic in any object.

The next aspect of yin–yang is the existence of such transition in everything at all scales. Figure 6.6 shows how a 24-h clock of yin–yang allows continuous transition as a cycle of life.

Table 6.4 shows the tangible and intangible nature of yin–yang. Yin–yang shows contrast as well as interdependence. For instance, no matter is produced without energy and no energy is produced without matter. Water is needed for plant, which is then needed for fire. This logic also shows nothing is real unless it is part of the positive negative cycle. For instance, fire without water is not real. That would explain why diamond cannot be set on fire even though it is made out of carbon. Similarly, the presence of mass would indicate the presence of energy. This would make the existence of zero energy and infinite mass an absurd concept, even though new cosmic physicists routinely tout that notion (Krauss, 2012).

Figure 6.6 also shows how yin and yang encircle each other alternating as a continuous function of time. As time progresses, yin becomes yang and

Picture 6.1 Natural occurrence of yin–yang structure.

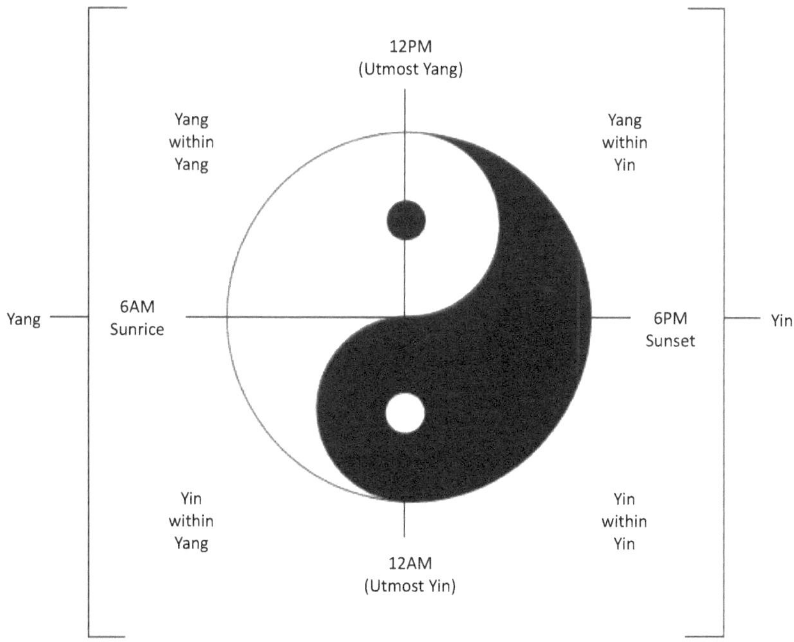

Figure 6.6 Evolution of yin and yang with time. *From Website 1.*

vice versa. This progression confirms the existence of characteristic time function for every object at every scale. Within the limitations of natural traits that is finite, the following picture depicts the role of two counteracting entities being held in harmony by the "mother." This

Table 6.4 The Tangible and Intangible Nature of Yin and Yang

Yin	Yang
Produces form	Produces energy
Grows	Generates
Substantial	Nonsubstantial
Matter	Energy
Contraction	Expansion
Descending	Ascending
Below	Above
Water	Fire

From Website 1.

mother is significant as in ancient Hindu culture, the supreme God is symbolized by "mother." In Islam, creator's first two traits literally mean "womb that is infinitely continuous in space" and womb that is infinitely continuous in time.

In large scale, the solar system offers an excellent manifestation of such characteristic time. The moon is moving around its own axis, then around the earth, while keeping pace with the earth that is orbiting around the sun and keeping pace with the sun that is moving both around its own axis as well as around an unknown object. Figure 6.7 shows the configuration of solar movement along with its planets and moons. It is reasonable to assume that such dependence of size on orbital speed reverses for invisible elements. Orbital speed of various known objects are plotted in the above figure as a function of size, along with a reverse relationship for invisible particles. If similar model is followed for invisible structure, smaller than dust speck, the Figure 6.7 emerges. In this figure, dust speck (*Dharra* in Arabic) is identified as the smallest object that is in line with the avalanche theory, recently advanced by Khan et al. (2008) and Khan and Islam (2012). From there, arises natural characteristic speed in the following scale.

Orbital speeds of various known objects are plotted in Figure 6.8 as a function of size. In this figure, dust speck represents reversal of speed versus size trend. For so-called subatomic particles, speed increases as the size decreases. Note that the actual speed in absolute sense is infinity. It is because each element has a speed in every dimension. This dimensionality is not restricted to Cartesian coordinate. As the number of dimension goes up, so does the absolute speed, approaching infinity for absolute dimension. The characteristic speed also increases as the size of the entity goes down.

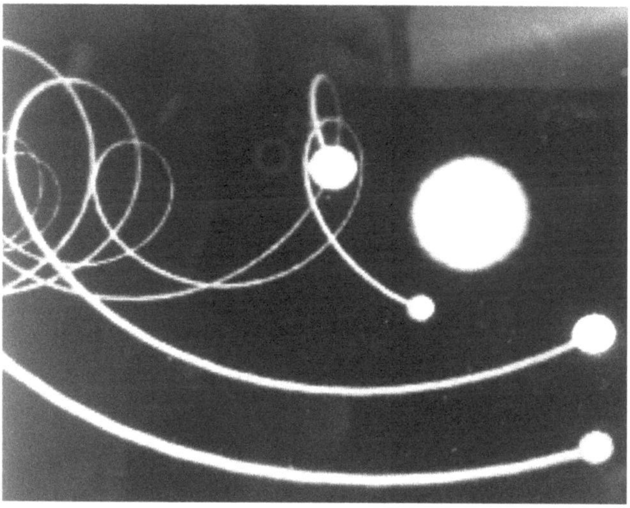

Figure 6.7 The sun, earth, and moon all are moving at a characteristic speed in infinite directions.

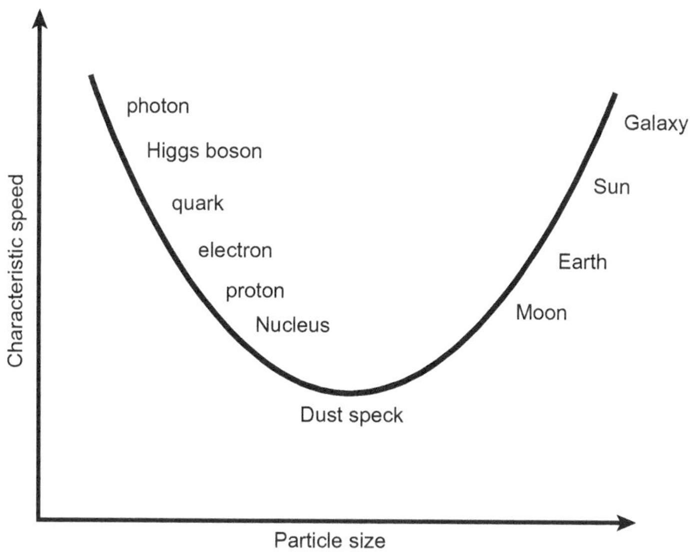

Figure 6.8 Orbital speed versus size (not to scale).

For infinitely small entity, the speed would approach infinity. This analysis shows how both small and large scales are in harmony with infinitude, associated with "void." In pre-Thomas Aquinas period, such "void" was synonymous with the creator within whom all the creation was believed to be embedded. Table 6.5 shows some of the characteristic speeds (and thus, frequencies) of various particles.

Such characteristic time exists for every object. The characteristic features change during phase transfer or when a life begins (from nonorganic to organic) or ceases for an individual (from organic to nonorganic). In this process, life and death are triggers or bifurcation points as associated time functions change drastically. It should be noted that such transition is subjective and the death of one entity only means death for that object. Here, it is the time function, $f(t)$ that defines the pathway of any entity, within the universal order. This time is not arbitrary and is tightly controlled by the external entity, the absolute time, as presented later in this paper. Following is a list of some of the characteristic time as related to humans:

Earth: day and night, year, pace with the sun

Humans: blink of an eye, sunrise, midday, sunset, sleep and wake, week, month, menstrual cycle; 40 days, 40 years

Society: 40 years, centuries, millennia

Geology: millennia

Cosmos: billion years

The heart rate is typical of natural frequency of humans. Even though, heart rate is frequently talked about in the context of both physical and psychological conditions, brain waves are also characteristics of human activities (Figure 6.9). Change in brain waves is evident during sleep, alertness, meditation, etc. Little is available on how such frequencies can affect overall human conditions, whereas most focus has been on how to alter natural frequencies. What makes it complicated is scientists have little knowledge of how they naturally vary with time as a person ages. Clearly, humans are in control of their brain waves, thereby consolidating the theory that humans are integral part of the universal order and their degree of freedom lies only within their intention.

Also, heart beats themselves change over time. As a sign of every characteristic frequency itself being a function of time, the graph in Figure 6.10 is produced. In this, the data on puberty and older people are extrapolated from published papers (e.g., Larson et al., 2013). Naturally, children are more dynamic and their body parts are renewed faster. This would necessitate faster replenishment of energy. The idea is to nurture natural frequencies, rather than fighting them. New Science does the opposite and every "treatment" is aimed at altering natural frequency, thereby countering natural forces.

Table 6.5 Characteristic Frequency of "Natural" Objects

Object	Nature of speed	Average speed sun	Comment
Sun	Orbital	240 km/s	Around unknown object, that is 2.55×10^{20} m away; estimated orbital time 200 million
	Drift	19 km/s	Due to variation in galactic rotation
	Spinning		Unclear
Earth	Escape	240 km/s	To match with the orbital speed of the sun
	Orbital	30 km/s	Around the sun
	Spinning	0.44 km/s	At equator
Moon	Broad escape	240 km/s	To keep up with the sun
	Escape	30 km/s	To keep up with the earth
	Orbital	1 km/s	
	Spinning	12 km/s	To keep the same face exposed to one side
Atom, radius 10^{-9} m		Unknown	Rigid ball assumption
Electron, 10^{-15} m		2200 km/s	Under nonexcited conditions (Bohr model)
Proton, 3. 10^{-15} m		Unknown	Rigid ball assumption
Quark		Unknown	Nonmeasurable dimension
Photon		300,000 km/s	Rigid ball assumption
Higgs boson		300,000 km/s	Rigid ball assumption

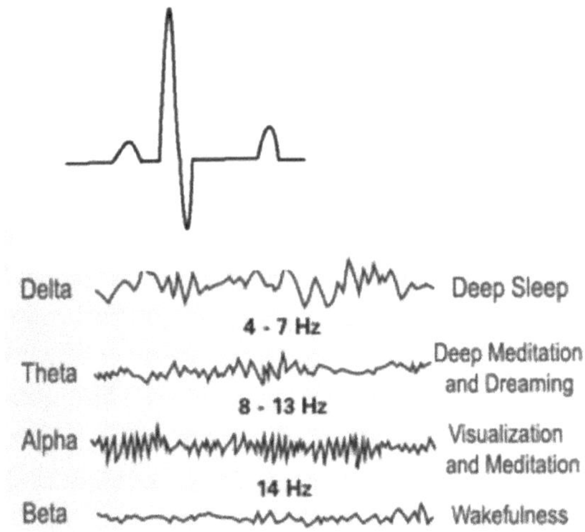

Figure 6.9 The heart beat (picture above) represents natural frequency of a human, whereas brain waves represent how a human is in harmony with the rest of the universe.

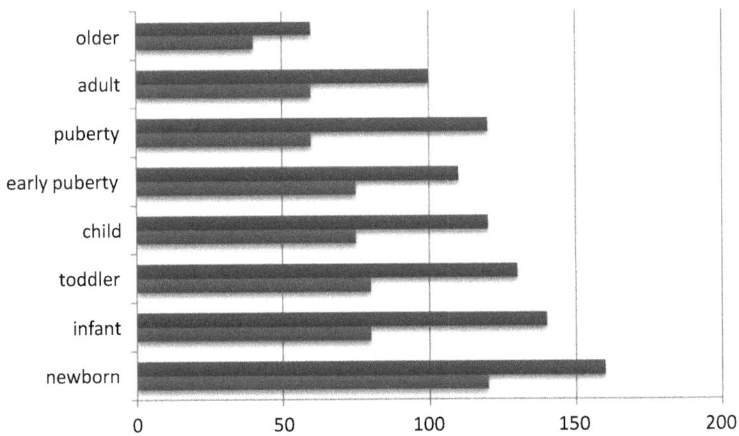

Figure 6.10 Maximum and minimum heart rate for different age groups.

There is what is characteristic but there is also fundamental. Hard to believe/accept, for example, that other natural frequencies in one's body are unrelated to heartbeat frequency. Thus, it would be difficult to believe that an individual's brain wave frequency, for example, could be entirely accounted for from investigation of phenomena occurring within the cerebral cortex alone by itself. Our theory is that we have no control over these frequencies. What we have control over is our intention. Thankfully that does not affect the universal order but doubly thankfully does affect an individual's long-term future. This connection of human intention with long-term future as well as disconnection of the universal order from human intervention is a notion that has been absent in every scientific cognition of post-Roman Catholic Church Europe.

The exchange between two dual and opposite objects continues and it is reasonable to assume that there is no distinction between energy particles and mass particles. The circle in the broader portion of the yin–yang represents the same shape that would house another yin–yang, which itself will house another set. Such trend continues until we reach a state that can be characterized as interface between absolute light (or absolute time), which is the external element. This is consistent with Qur'anic narration of divine traits and matches with pre-Newtonian narration. Figure 6.11 shows the depiction of such infinitude.

Figure 6.11 Tangible/intangible duality continues infinitely for megascale to nanoscale, from infinitely large to infinitely small.

Figure 6.11 shows how "pure light" that surrounds both tangible and intangible, surrounds everything at all times. Such is the connection between time, the creation and the absolute time, and "radiative light" and pure light. Both absolute time and pure light represent infinity.

Furthermore, the notion of male contributing to the life of female and female in turn giving birth to male becomes an integral part of life cycle of humans. With this cyclic balance, there is no superiority of any particular entity over other as long as they belong to the group of creation in general. In addition, every object is in perfect harmony with nature, except humans that are nonideal. This "nonideal" feature has been a crucial matter of contention in European history, ever since the concept of original sin was introduced. If that premise is removed then the premise that everyone is born perfect is consistent with the premise that nature is perfect. European dogma science defined Jesus Christ as the perfect man (a role model), but that is not consistent with his dual status as "son of god." In addition, very little is known about this "role model." In fact, modern scientists doubt he ever existed in the role of a messiah. European modern science does not define perfect human, abandoning the debate as "religious mumbo jumbo." In addition, it does not define good behavior or any purpose for humans, other than maximizing pleasure and minimizing pain. This is in contrast to Christian dogma but equally problematic as it gives rise to chaotic rules, akin to roller coaster ride (Islam et al., 2013).

In summary, both ancient oriental and Greek philosophers support the notion of separation of creator (external entity) from creation (internal entity), each of whom have completely different traits. The creation itself is further divided into tangible and intangible sets with opposite traits, each having continuously decreasing size and similar asymmetry (or duality). Distinction is made between pure light and radiative (or reflective) light. It is only pure light (PL) that is continuous in space and time. Everything else is embedded in PL. This "light" is not a collection of photons or any particle. It is pure energy and does not have any mass associated to it. The conventional denomination of light as a collection of photons each of zero mass does not account for the fact that the sun is losing thousands of tons of mass every second. With the proposed characterization, there is no contradiction in transition from mass to energy. This distinction between PL and radiative or reflective light is necessary but no sufficient condition to truly scientific description of mass and energy. Sequential functionality must be described as a sufficient condition. This requires proper characterization of mass and energy. This depiction is supported by Greek philosophers as well as Augustine of Roman Catholic Church.

6.5 CHARACTERIZATION OF EVERYTHING IN EXISTENCE

Following is a characterization of everything in existence.

1. Pure light (*Noor* in Arabic). *PL*, is absolutely intangible, nothing like rest of the items. For instance, this light does not change in space or time, not bound by time or anything else. This light has no source or boundary. This light is unique, it does not reflect nor does is it anything like other lights that we know (as per Qur'an 112:1–3). It does not have mass nor is it constrained by space. This light is all pervasive. Other "light" known as *Dha'a* in Arabic is "infected" or a different type of light. Note that this infected or deflected light is conflated with pure light in the sense that infected light is assigned zero mass, along with a finite speed, thereby, creating internal inconsistency. In addition, recent finding that neutrino has higher speed than light (Choi, 2011; Brumfiel, 2011) and subsequent refutation based on veracity of relativity theory (Brumfiel, 2012) highlights the need for having a consistent theory. Note that neutrinos are considered to be smaller than photon, whereas the mass of photon is already assumed to be zero. Further confusion arises from the definition of photon and neutrinos. Photons are considered to be particles of light. Neutrinos are particles of matter. This definition disconnects energy from mass. Qur'anic definition material eliminates this disconnection. Another misrepresentation occurs when photons are assumed to be traveling only through transparent material whereas neutrinos through opaque material. This definition makes neutrinos omnipresent (similar to aether) making it less mobile than light. This inherently makes a biased distinction between light and mass emitting from a radioactive substance. The current theory eliminates this mischaracterization.
2. Absolute time (*Dahr* in Arabic). As discussed before, in Arabic language the two words "Asr" and "Dahr" are very comprehensive. The former relates to creation whereas the latter relates to creator. This absolute time has no beginning or end. There is no English word to describe this absolute time and, therefore, we selected the word "absolute time". This Time is absolutely intangible, just like pure light.
3. Time (*asr* in Arabic), a function of pure light. This word is the one used by physicists of modern age. Conventional definition of time is in terms

of tangible movement in space. For instance, the unit of time, second, is chosen by dividing the entire year (measurement of the earth's movement around the sun) by the number of days, then hours, then minutes, and finally by the number of seconds. Similarly, atomic clock measures time by counting frequency of cesium radiation. With this definition of time (or unit of time), contradictions arise when Einstein's relativity theory is invoked. Such constraint or contradiction does not exist in Arabic language and time is independent of matter or space. The word *Asr* connotes the sweep of time or its swift flow. This time is intangible, but not in absolute sense as it can change, be initiated, ended, folded, all of which is supported by both relativity and quantum physics. However, the conventional approach is to make time a function of matter and thought material. For instance, gravitational forces are thought to be influencing time. Einstein invoked perception as another controller, whereas quantum physicists (e.g., Richard Feynman) assigned that to probability. Both these assumptions create contradictions in terms of multiplicity of the time function. This is equivalent to insulting the history, which is unique as per the fundamental logic of Aristotle (law of excluded middle). By contrast, Hadith Qudsi assigns time as a function of pure light. Pure light being unique and external to everything else (including time), $t = f(\text{PL})$ functionality does not create multiple solutions. This assumption leads to a single function of time from beginning to end of creation. Such uniqueness is in conformance with Qur'anic word, *Qadr*, that means unique pathway for every component of the universe. This word is often incorrectly translated as destiny. The existence of *Qadr* is recognized to be prepared much before creation of everything from nothing. Time also is something that has the following features:

a. it has a beginning and an end;
b. it can be folded, changed, altered;
c. it is a continuous function of pure light
d. it controls matter continuously

This time is a "tool" used to put material in motion. Time is intangible, but it becomes apparent to us through tangible expressions in material, events, HTM.

As stated earlier in this paper, characteristic time is necessary to drive everything in its cycles that connect to the rest of the universe.

4. Material. This is the original mass created from nothing (source of mass and energy). No distinction is made between "mass particles' and 'energy particles'. Therefore, contradictions such as the ones discussed earlier arise. We eliminate those contradictions by defining mass as the source of energy. However, we avoid the use of the word "mass," which has a specific meaning in modern science.

This material does not get created from nothing all the time. We conserve the concept of "conservation of mass" as well as "conservation of energy" by default. In this analysis, conservation of energy is inherent to conservation of mass as at no time a mass becomes zero. For instance, the current model that assumes both neutrino and photon have zero mass obfuscates the possibility that mass is the source of energy. In effect, such assertion violates the principle of conservation of mass. For instance, the sun loses approximately four million tons of mass every second because of which we receive sunlight. Yet, this mass cannot be accounted for if photon mass is set to zero. The denomination of material to describe both mass and energy preserves conservation of mass and energy simultaneously. The underlying logic is simple:

a. there is no energy without mass, energy being the outcome of mass;
b. energy generation from mass is continuous and is due to the continuous state of motion of every entity.

The material is made into a function of time. Therefore.

Material, $M = g(t)$. This functionality is reversed to the commonly introduced notion in Relativity, i.e., time is a function of perception and perception itself a function of the environment that contains material. By making material a function of time, the control is given to time, which remains intricately linked to every entity, from the largest galaxy to tiniest subatomic particle. This intricacy is shown in the previously presented yin–yang symbol.

HTM falls under this category. This analysis corrects the misconception that humans have freedom of thought. This false assumption leads to dogmatic contradiction and opens up controversies involving thought control (Ong, 2008).

5. Intention. This is the original entity created from nothing as a distinctive feature of humans (Islam et al., 2014). Intention is the one that defines conscience, which is the enabler of conscientious thoughts. Even though, thoughts are all part of universal order (material, as described in item four above), conscientious thoughts are in harmony with the universal order as they are rooted in the phenomenal intention (Khan and Islam, 2012).

6.6 ORGANIC AND MECHANICAL FREQUENCIES

Considered in its most general aspect, the universe comprising all phenomena can be comprehended as comprising two broad categories: the mechanical and the organic. Many mechanical phenomena can be found within the organic category. Certain aspects of many organically based phenomena can be defined or accounted for entirely within the category that comprises all forms of mechanism. Frequency, and its measurement, often appears to bridge this mechanical–organic divide. Organically based frequencies have an operating range which itself varies, e.g., the length of the lunar year. On the one hand, purely mechanical frequencies also have an operating range, and this range can be set or otherwise manipulated up to a point, e.g., the resonant frequency at which a bridge structure may collapse in a sustained high wind. On the other hand, although organically based frequencies can be detected and measured, there is usually little or nothing, beyond a very definite window that must be determined by trial and error, which can be done to manipulate such frequencies.

Since Galileo's brilliant and successful deployment of an elaborate water clock as an organic-frequency device for measuring with some precision the differential rates of descent to earth of freely falling masses of different weights, all kinds of apparently natural clocks have been deployed to calibrate many things. This includes even universal standards of the metric system, e.g., the cesium atom clock at a Paris laboratory used for setting the standard length of the meter.

Problems arise when such frequency-based devices are treated as the generator of values for a variable that is treated as being independent in the sense that we take Newton's fictional time variable t to be varying "independently" of whatever phenomenon it is supposed to measuring/calibrating/counting. Outside of a tiny instantaneous range, e.g., the period in which Δt approaches zero, naturally sourced frequencies cannot be assumed to be independent in that way. This is a false assumption whose uncritical acceptance vitiates much of the eventual output of the measuring/calibration effort.

Such problem arises the moment one makes the phenomenal assumption that frequency is fixed. That's the idea behind the unit of "second" for time (solar orbit to cesium radiation frequency). New Science fixed the frequency (it is like fixing speed of light), then back-calculated time. No wonder, later on, time was made into a function of perception (relativity) thereby making the unique functionality schizophrenic. Not only is it the case that "such a problem arises the moment you make the phenomenal assumption that frequency is fixed." Even if you allow that t is not fixed and undergoes changes in value, i.e., that its frequency is not necessarily fixed, this problem persists if the subtler but still toxic assumption is accepted that the rate at which the variable t changes, Δt, is constant in some "continuous" interval over which the derivative $df(t)/dt$ may be taken. Here is where we uncover the truly toxic power of Newton's laws of motion over conscience-based consciousness. That is when they invoke "known" function, which itself is aphenomenal. The only function that is valid is with infinite order of periodicity (this is beyond chaotic).

6.7 REDEFINING FORCE AND ENERGY

All currently available fundamental definitions in New Science emerge from Newton's laws. Let's review the conventional definitions and present thereafter the scientific definition that emerges from the above section.

Force: Conventionally, a force is defined to be an influence that tends to change the motion of an object. The inherent assumption is, this "force" is external to the object. This is a false premise because the entire creation is internal and connected to each other, as presented by recent works of Islam et al. (2010a, 2012) and Khan and Islam (2012).

Currently it is believed there are four fundamental forces in the universe: the gravity force, the nuclear weak force, the electromagnetic force, and the nuclear strong force in ascending order of strength. In mechanics, forces are seen as the causes of linear motion, whereas the causes of rotational motion are called torques. The action of forces in causing motion is described by Newton's laws under ordinary conditions. Subsequently, forces are inherently vector quantities, requiring vector addition to combine them. This further characterization is yet another tactic to cover up for the false first premise.

Khan and Islam (2012) provide one with detailed deconstruction of Newton's laws. With the scientific theory of the previous section, one can redefine force as something that drives the universal movement. It is constant, absolute, and immutable. With this definition, there is no need to further characterize force in the above-mentioned categories. This replaces the notion of gravity in conventional sense. The source of this force is the absolute light that is omnipresent. This description answers the questions regarding what forces make the entire galactic system move—a question that has perplexed modern scientists (Cowen, 2012).

6.7.1 Energy

Energy is commonly defined as the capacity for doing work. One must have energy to accomplish work—it is like the "currency" for performing work. To do 100 J of work, one must expend 100 J of energy.

New Science postulates the purest form of energy is light that is comprised of photons. These photons are thought to have zero mass. As stated earlier in this chapter, this assertion disconnects mass from energy but invokes Einstein's formula, $E = mc^2$, which itself is based on Maxwell's formula that considers energy as a collection of solid, spherical, rigid balls (similar to atoms). The assertion of zero mass also invokes infinite speed (a notion that was promoted by Aristotle but discarded by Ibn Al-Haytham, some 900 years ago). This obvious gaffe is "remedied" by forcing speed of light, "c" to be constant and maximum attainable speed by any particle. Apart from the fact that a zero mass would not qualify to be called a "particle," this also poses the obvious spurious solution to the equation, $E = mc^2$ and renders it an absurd concept. This mathematical fallacy is "solved" with dogmatic assertion of quantum physics. As such, the product of 0 times infinity gives a value that is a function of the frequency of the photon. Furthermore, it is asserted that a photon can be converted to a particle and its antiparticle (a process called pair creation), so it does convert energy into mass. All equations, therefore, give an answer but are completely devoid of any physical significance.

Similar illogical attributes are assigned to Higgs boson, neutrino, and a number of other particles, some of which have zero mass. In addition, it is asserted that certain particles, such as neutrino, can travel through opaque material, albeit with a speed lower than that of light (photons). In order to compensate for the concept of gravitational force that is conventionally nonexistent in zero-mass conditions, it is asserted that the Higgs particle is a carrier of a force. This force mediated by the Higgs boson is considered to be universal as the Higgs boson interacts with all kinds of massive particles, no matter whether they are quarks, leptons, or even massive bosons (the electroweak bosons). Only photons and gluons do not interact with the Higgs boson. Neutrinos, the lightest particles with almost zero mass, barely interact with a Higgs boson. This description and assignment of special 'power' to certain particles is characteristic of pragmatic approach (Khan and Islam, 2012). In simple terms, they are stop gap tactics for covering up the fundamental flaws in basic premises.

Neutrinos are considered to be similar to electrons, with one crucial difference: neutrinos do not carry electric charge. Because neutrinos are electrically neutral, they are not affected by the electromagnetic forces that act on electrons. Neutrinos are affected only by a "weak" subatomic force of much shorter range than electromagnetism, and are therefore able to pass through great distances in matter without being affected by it. If neutrinos have mass, they also interact gravitationally with other massive particles, but gravity is by far the weakest of the four known forces. Such repeated characterization of matter and energy with contradicting traits has been the most prominent feature of New Science. The characterization offered by Islam et al. (2012) and Khan and Islam (2012) eliminates such needs.

When it comes to "heat energy," New Science is full of gaffes as well. The entire "heat engineering" is based on Lord Kelvin's work. Lord Kelvin, whose "laws" are a must for modern day engineering design believed that the earth is progressively moving to worse status and which would eventually lead to the "heat death." So, if Kelvin were to be correct, we are progressively moving to greater energy crisis and indeed we need to worry about how to fight this "natural" death of our planet. Kelvin also believed flying an airplane was an absurd idea, so absurd that he did not care to be a member of the aeronautical club. Anyone would agree, it is not unreasonable to question this assertion of Lord Kelvin, but the moment one talks about the nature progressively improving, if left alone (by humans, of course), many scientists break out in utter contempt and invoke all kinds of arguments of doctrinal fervor. How do these scientists explain then, if the earth is progressively dying, how it happened that life evolved from the nonbiological materials and eventually very sophisticated creature, called *homo sapiens* (thinking group) came to exist? Their only argument becomes the one that has worked for all religions, "you have to believe." All of a sudden, it becomes a matter of faith and all the contradictions that arise from that assertion of Lord Kelvin becomes paradoxes and we mere humans are not supposed to understand them. Today, the Internet is filled with claims that Kelvin is actually a god and there is even a society that worships him. This line of argument cannot be scientific (Islam et al., 2010).

Modern scientists claim to have moved away from doctrinal claims of Kelvin. However, no theory has challenged the original premise of Kelvin.

Even Nobel laureate winning works (review, for instance the work of Roy J. Glauber, John L. Hall, and Theodor W. Hänsch, along with their Nobel prize winning work on light theory), consider Kelvin's concept of absolute temperature a fact. Theoretically, at that point, there is zero energy, hence matter would not exist either. A matter is being rendered nonexistence because it does not move—an absurd state. Table 6.5 shows how everything in creation is in a state of motion, including time itself. However, instead of realizing this obviously spurious premise, New Science offers the following explanation:

> *"If you take away energy from an atom, you do so by lowering the energy level of its electrons, which emits a photon corresponding to the energy gab between the electron bands. Keep doing that until the electron is absorbed by the nucleus, and converts a proton to a neutron. Now you need to extract energy from the nucleus. How are you going to do that? How are you going to shield the resulting neutron from the influence of the rest of the universe, including radio waves, that penetrate everything?"*

Another explanation attempts to justify discontinuity between mass and energy, by saying, "All matter has energy, unless it is at absolute zero temperature, true. But that amount of energy is tiny compared to the energy you could get if the matter were totally converted to energy via Einstein's famous equation, $E = mc^2$. But there is no way for that to happen unless you are dealing with antimatter. Even the sun converts only a tiny percentage of the matter to energy, but that tiny percentage (because of the c^2 term) produces a lot of energy." In this, the notion of "antimatter" is invoked.

Natural light or heat is a measure of radiation from a system (called "material" in the above section). This radiation is continuous and accounts for change in mass within a system. In this, there is no difference between heat generation and light generation, nor there is any difference in radiation of different types of "radiation" (such as X-ray, gamma ray, visual light, infrared, etc.) other than they are of various frequencies. This can be reconciled with New Science for the limiting cases that say that there is an exponential relationship between reactants and products (Arrhenius equation) through the time function. Such relationship is continuous in time and space. For instance, as long as the assumption of continuity is valid, any substance is going to react with the media. The term "reaction" here implies formation of a new system that will have components of the reactants. This reaction has been explained by Khan et al. (2008) as a collection of snow flakes to form avalanche. Islam et al. (2014) developed

NASA, ESA, Z. Levay and R. van der Marel (STScI), T. Hallas, and A. Mellinger

Picture 6.2 It is reported that two galaxies are in a collision course. *Cowan, 2012.*

similar theory that also accounts for energy interactions and eliminates separate balance equations for mass and energy. This theory considers energy or mass transfer (chemical reaction or phase change) as merger of two galaxies. Before merger, the two galaxies have different sets of characteristic frequencies. However, after merger, a new galaxy is formed with an entirely new set of characteristic frequencies. Such phenomena is well understood in the context of cosmic physics. Picture 6.2 shows NASA picture of two galaxies that are in a collision course. Cowan (2012) reports the following:

> *"Four billion years from now, the Milky Way, as seen from Earth in this illustration, would be warped by a collision with the Andromeda galaxy. It's a definite hit. The Andromeda galaxy will collide with the Milky Way about 4 billion years from now, astronomers announced today. Although the sun and other stars will remain intact, the titanic tumult is likely to shove the Solar System to the outskirts of the merged galaxies."*

Such collision does not involve merger of two suns or any planets or moons. It simply means reorientation of the starts and planets within a new family. Note how conservation of mass is strictly maintained as long as an artificial boundary is not imposed. In New Science, such artificial boundary is imposed by confining a system within a boundary and imposing "no-leak" boundary conditions. Similarly, adiabatic conditions are imposed after creating artificial heat barriers.

With the galaxy model, physical or chemical changes can both be adequately described as change in overall characteristic frequency. So, how does heat or mass gets released or absorbed? As stated above, "the titanic tumult" would cause the stars to be "shoved" toward the outskirt of the newly formed galaxy. In case, they are indeed placed around the outskirt, this would translate into excess heat near the boundary. However, if those stars are "shoved" inside the new giant galaxy, for an outsider, it would appear to be a cooling process, hence, endothermic reaction. In this context, the "titanic tumult" is equivalent to the "spark" that lights up a flame or starts a chain reaction. It is also equivalent of onset of life or death as well as "big bang" in the universal sense. Even though these terms have been naturalized in New science vocabulary, they do not bear scientific meaning. Islam et al. (2012, 2014) recognized them to be unknown and unexplainable phenomena that cause onset of a phase change. They can be affected by heat, light, pressure that are direct results of changes within the confine of a certain system.

Source of heat is associated to "collisions" as represented above in the context of galaxies, be it in subatomic level (known as chemical reactions), in combustion within a flame, or in giant scale (such as solar radiation). For our system of interest, i.e., the earth, our primary source of heat is the sun that radiates mass in various wavelengths. New Science recognizes "the solar constant" as the amount of power that the sun deposits per unit area that is directly exposed to sunlight. The solar constant is equal to approximately 1368 W/m^2 at a distance of one astronomical unit (AU) from the sun (that is, on or near earth). Sunlight at the top of earth's atmosphere is composed (by total energy) of about 50% infrared light, 40% visible light, and 10% ultraviolet light. In another word, the heat source is inherently linked to light source. As discussed in previous sections, this transition between different forms of energy is continuous and should be considered to be part of the same phenomenon characterized here as "dynamic nature of everything in creation." These are not "mass-less" photons or "energy-less" waves, they are actually part of mass transfer that originates from radiation of the sun.

Before solar emissions enter the atmosphere of the earth, nearly one-third of the irradiative material are deflected through filtering actions of the atmospheric particles. How does it occur? It is similar to the same process described above as galactic collision. During this process, the composition of the atmospheric layer changes continuously and "new galaxies" form continuously in the "tumult" mode, while some of the material are deflected outside the atmosphere and the rest penetrating the atmosphere to trigger similar 'tumult" events through various layers of the atmosphere. These atmospheric layers are such that all the layers act similar to a stacked up filtering system. Following is a brief description of different layers of the atmosphere.

1. The exosphere is the thinnest (in terms of material concentration) layer. This is the upper limit of the earth atmosphere.
2. The thermosphere is a layer with auroras. This layer sees intensive ionic activities.
3. The next layer is mesosphere. This is the layer that burns up meteors or solid fragments. The word "solid" implies most passive levels of activities of the constitutive material. See Figure 6.8 with reference to "solid" representing collection of "dust specks" that exhibit the slowest characteristic speed. Meteors or rock fragments burn up in the mesosphere. Another way to characterize matter in terms of solid liquid and vapor state.

   Within earth, the following configuration applies. It is possible that such configuration of various states will apply to other celestial entities, but that is not the subject of interest in the current context. Figure 6.12 shows how the relationship between characteristic speed and physical state of matter is a continuous function. Note that natural state of matter is an important consideration, particularly in relation to human species and life. For instance, the most abundant matter on earth is water, most useful for human species in its liquid state. It turns out that water is also the most abundant in liquid state. In solid, clayey matter (SiO_2, see the

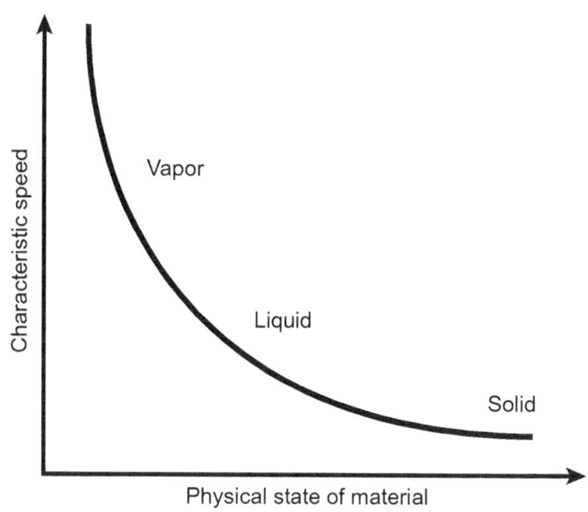

Figure 6.12 Characteristic speed (or frequency) can act as the unique function that defines the physical state of matter.

position of "dust speck" in Figure 6.8) is the most abundant solid and scientists are beginning to find out humans are also made out of such matter. Here is a quote from Daily mail (2013):

"The latest theory is that clay - which is at its most basic, a combination of minerals in the ground - acts as a breeding laboratory for tiny molecules and chemicals which it 'absorbs like a sponge'.

The process takes billions of years, during which the chemicals react to each other to form proteins, DNA and, eventually, living cells, scientists told the journal Scientific Reports.

Biological Engineers from Cornell University's department for Nanoscale Science in New York state believe clay 'might have been the birthplace of life on Earth'.

It is a theory dating back thousands of years in many cultures, though perhaps not using the same scientific explanation."

Clay also retain the most amount of water—the most essential ingredient of life and organic material. As would be seen in other chapters as well as later in this section, similar optima exist in terms of visible light being the most abundant of sunlight rays and earth being the most dense of all the planets in the solar system. Overall, all characteristic features for the earth makes it the most suitable as a "habitat for mankind" (Khan and Islam, 2012).

4. The next layer of the atmosphere, called stratosphere, is the most stable layer of the atmosphere. Many jet aircrafts fly in the stratosphere because it is very stable. Also, the ozone layer absorbs harmful rays from the sun. By the time, sun rays enter the final and fifth layer, almost 30% of the total irradiation have been removed. What energy (in form of light and heat) is ideal for rendering the earth system totally sustainable and ideal for human habitation? This layer is the most vulnerable to human intervention and is the cause of global warming (Islam et al., 2012). This aspect is elaborated below.

   The Intergovernmental Panel on Climate Change stated that there was a "discernible" human influence on climate and that the observed warming trend is "unlikely to be entirely natural in origin" (IPCC 2001). The Third Assessment Report of IPCC stated, "There is new and stronger evidence that most of the warming observed over the last 50 years is attributable to human activities." Khilyuk and Chilingar (2004) reported that the CO_2 concentration in the atmosphere between 1958 and 1978 was proportional to the CO_2 emission due to the burning of fossil fuel. In 1978, CO_2 emissions into the atmosphere due to fossil fuel burning stopped rising and were stable for nine years. They concluded that if burning fossil fuels was the main cause, then the atmospheric concentration should stop rising, and, thus, fossil fuel burning would not be the cause of the greenhouse effect. However, this assumption is extremely shortsighted and the global climate certainly does not work linearly, as envisioned by Khilyuk and Chilingar (2004). Moreover, the "Greenhouse Effect One-Layer Model," proposed by Khilyuk and Chilingar (2003, 2004), assumes there are adiabatic conditions in the atmosphere that do not and cannot exist. The authors have concluded that the human-induced emissions of carbon dioxide and other greenhouse gases have a very small effect on global warming. This is due to the limitation of the current linear computer models that cannot predict temperature effects on the atmosphere other than at low levels. Similar arguments were made while promoting dichlorodifluoromethane (CFC-12) in order to relieve environmental problems incurred by ammonia and other refrigerants after decades of use. CFC-12 was banned in USA in 1996 for its impacts on stratospheric ozone layer depletion and global warming. Khan and Islam (2012) presented detailed lists of technologies that were based on spurious promises. Zatzman and Islam (2007) complemented this list by providing a detailed list of economic models that are also counterproductive. Khilyuk and Chilingar (2004) explained the potential impact of microbial activities on the mass and content of gaseous mixtures in earth's atmosphere on a global scale. However, this study does not distinguish between biological sources of greenhouse gas emissions (microbial activities) and industrial sources (fossil fuel burning) of greenhouse gas emissions. Emissions from industrial sources possess different characteristics because they derive from diverse origins and travel different paths that, obviously, have significant impacts on atmospheric processes. Current climate models have several problems. Scientists have agreed on the likely rise in the global temperature over the next century. However, the current global climatic models can predict only global average temperatures. Projection of climate change in a particular region is considered to be beyond current human ability. Atmospheric Ocean General Circulation Models are used by the IPCC to model climatic features, but these models are not accurate enough to provide a reliable forecast on how climate may change. They are linear models and cannot forecast complex climatic features. Some climate models are based on CO_2 doubling and transient scenarios. However, the effect of climate in these models, while doubling the concentration of CO_2 in the atmosphere, cannot predict the climate in other scenarios. These models are insensitive to the difference between natural and industrial greenhouse gases. There are some simple models that use fewer dimensions than complex models and do not predict complex systems. The Earth System Models of Intermediate Complexity are used to bridge the gap between the complex and simple models, but these models are not able to assess the regional aspect of climate change (IPCC 2001).

Overall, any level of artificial products in the stratosphere will affect the final and the most important layer of the earth atmosphere.

5. The closest layer to the earth surface is troposphere. This layer contains half of the earth's atmosphere. All transient phenomena related to weather occur in this layer. This layer too contributes to attenuation of sunlight and at the end some 1000 W/m^2 falls on the earth when the sky is clear and the sun is near the zenith. The multiple filtering system of the atmosphere is such that it filters out 70% of solar ultraviolet, especially at the shorter wavelengths.

The immediate use of solar energy in terms of sustaining human life is photosynthesis—the process that allows plants to capture the energy (through mass transfer) of sunlight and convert it to "live" chemical form. The energy stored in petroleum and other fossil fuels was originally converted from sunlight by photosynthesis in the distant past.

The most significant is the photosynthetic mechanism. There are two classes of the photosynthetic cycle, the Calvin–Benson photosynthetic cycle and the Hatch–Slack photosynthetic cycle. The Calvin–Benson photosynthetic cycle is dominant in hard woods and conifers. The primary CO_2 fixation or carboxylation reaction involves the enzyme ribulose-1,5-diphosphate carboxylase and the first stable product is a three-carbon compound. This reaction is considered to be "light-independent." This series of reactions occur in the fluid-filled area of a chloroplast outside of the mytosis membranes. These reactions take the light-dependent reactions and perform further chemical processes on them. Various stages of this process are carbon fixation, reduction reactions, and ribulose 1,5-bisphosphate regeneration. In describing this cycle of reactions, the role of light energy is marginalized. This process occurs only when light is available. Plants do not carry out the Calvin cycle by night. They, instead, release sucrose into the phloem from their starch reserves. This process happens when light is available independent of the kind of photosynthesis (C_3 carbon fixation, C_4 carbon fixation, and Crassulacean acid metabolism. The exceptions are Crassulacean acid metabolism, also known as CAM photosynthesis, a carbon fixation pathway that is used by some plants as an adaptation to arid conditions. In a plant using full CAM, the stomata in the leaves remain shut during the day to reduce evapotranspiration, but open at night to collect carbon dioxide (CO_2). The CO_2 is stored as the four-carbon acid malate and then used during photosynthesis during the day. The precollected CO_2 is concentrated around the enzyme RuBisCO, increasing photosynthetic efficiency.

On the other hand, the Hatch–Slack photosynthetic cycle is the one used by tropical grasses, corn, and sugarcane. Phosphenolpyruvate carboxylase is responsible for the primary carboxylation reaction. The first stable carbon compound is a C-4 acid, which is subsequently decarboxylated. It is then refixed into a three-carbon compound. These three steps define the canonical C_4 photosynthetic pathway. Overall, the photosynthesis process shows how nature converts energy into mass, storing energy for long-term use. This must be understood in order to appreciate the role of natural processing in the context of petroleum usage.

The process of energy-to-mass conversion is greatly affected by temperature (Fink, 2013). Sometimes temperatures are used in connection with day length to manipulate the flowering of plants. Chrysanthemums will flower for a longer period of time if daylight temperatures are 50 °F. The Christmas cactus forms flowers as a result of short days and low temperatures. Also, temperatures alone also influence flowering. Daffodils are forced to flower by putting bulbs in cold storage in October at 35–40 °F. The cold temperature allows the bulb to mature. The bulbs are transferred to the greenhouse in midwinter where growth begins. The flowers are then ready for cutting in three–4 weeks.

Plants produce maximum growth when exposed to a day temperature that is about 10–15 °F higher than the night temperature. This allows the plant to photosynthesize (build up) and respire (break down) during an optimum daytime temperature and to curtail the rate of respiration during a cooler night. High temperatures cause increased respiration, sometimes above the rate of photosynthesis. This means that the products of photosynthesis are being used more rapidly than they are being produced. For growth to occur, photosynthesis must be greater than respiration. Temperature alone can affect this process.

Low temperatures can result in poor growth. Photosynthesis is slowed down at low temperatures. Since photosynthesis is slowed, growth is slowed, and this results in lower yields. Each plant has an optimum temperature that allows maximum growth. For example, snapdragons grow best when night time temperatures are 55 °F, while the poinsettia grows best at 62 °F. Florist cyclamen does well under very cool conditions, while many bedding plants grow best at a higher temperature.

Buds of many plants require exposure to a certain number of days below a critical temperature before they will resume growth in the spring. Peaches are a prime example; most cultivars require 700 to 1000 h below 45 °F and above 32 °F before they break their rest period and begin growth. This time period varies for different plants. The flower buds of forsythia require a relatively short rest period and will grow at the first sign of warm weather. During dormancy, buds can withstand very low temperatures, but after the rest period is satisfied, buds become more susceptible to weather conditions and can be damaged easily by cold temperatures or frost. This series of phenomena have immediate implications to seeds and future of the biomass.

Overall, temperature represents level of subatomic particle activities. Any rise in temperature increases movement of all particles of the system. For certain systems, this would suffice to trigger a chain reaction, while for others this temperature rise would simply facilitate dispersion of the mass. In terms of phase change, Figure 6.12 shows how any change in temperature can trigger phase change by altering the characteristic speed of a collection of particles.

Similar effects are expected with pressure. Photosynthesis offers an example of natural effect of pressure on organic reactions. Beer and Waisel (1982) studied photosynthetic responses to light and pressure (up to 4 atm) for two seagrass species abundant in the Gulf of Eilat (Red Sea). In *Halodule uninervis* (Forssk.) Aschers. pressure decreased net photosynthetic rates, while in *Halophila stipulacea* (Forssk.) Aschers. pressure had no effect on net photosynthetic rates. In both species, light saturation was reached at 300 μE (400–700 nm) m^2/s and the compensation point was at 20–40 μE (400–700 nm) m^2/s. Comparing these results to in situ light measurements, neither species should be light limited to a depth of about 15 m, and H. *stipulacea* should reach compensation light intensities at about 50 m. The latter depth corresponds well to the natural depth penetration of this species. H. *uninervis* is never found deeper than 5 m in the Gulf of Eilat, and it appears that pressure rather than light is one of the factors limiting the depth penetration of this species. The differential pressure response of the two species may be related to aspects of leaf morphology and gas diffusion.

Scientifically, confining pressure is responsible for creating a series of vibrations that are in conflict with natural frequencies of matter. Because of continuity of matter, the external vibrations cause reactions to matter that attempt to escape its confinement. Pressure, alone can cause a series of oscillatory events that prompt fundamental changes in the subatomic structure of matter.

6.8 WHAT IS A NATURAL ENERGY SOURCE?

Fossil fuel energy follows a very different pathway, even though fossil fuel is also being used to generate electricity. Fossil fuel is also the most abundant, only second to biomass (including wood).

There are three major forms of fossil fuels: coal, oil, and natural gas. All three were formed many hundreds of millions of years ago before the time of the dinosaurs—hence the name fossil fuels. The age they were formed is called the Carboniferous Period. It was part of the Paleozoic Era. "Carboniferous" gets its name from carbon, the basic element in coal and other fossil fuels. The Carboniferous Period occurred from about 360 to 286 million years ago. At the time, the land was covered with swamps filled with huge trees, ferns and other large leafy plants. The water and seas were filled with algae—the green stuff that forms on a stagnant pool of water. Algae is actually millions of very small plants.

Some deposits of coal can be found during the time of the dinosaurs. For example, thin carbon layers can be found during the late Cretaceous Period (65 million years ago)—the time of Tyrannosaurus Rex. But the main deposits of fossil fuels are from the Carboniferous Period. As the trees and plants died, they sank to the bottom of the swamps of oceans. They formed layers of a spongy material called peat. Over many hundreds of years, the peat was covered by sand and clay and other minerals, which turned into a type of rock called sedimentary.

More and more rock piled on top of more rock, and it weighed more and more. It began to press down on the peat. The peat was squeezed and squeezed until the water came out of it andeventually, over millions of years, it turned into coal, oil or petroleum, and natural gas. Conventional theories suggest that coal is the least processed in nature while natural gas is the longest processed. Table 6.6 shows the heat of combustion for various fuels. Note that data on crude oil are not available as crude oil is not used directly. On the other hand, natural gas shows very high heat of reaction. Even though hydrogen has the highest heat of reaction, it is practically

Table 6.6 Heat of Combustion for Some Common Fuels

Fuel	kJ/g
Hydrogen	141.9
Gasoline	47.0
Diesel	45.0
Ethanol	29.7
Propane	49.9
Butane	49.2
Wood	15.0
Coal (Lignite)	15.0
Coal (Anthracite)	27.0
Carbon	32.8
Natural Gas	54.0

impossible to find hydrogen gas in its elemental form. Natural gas, on the other hand is the most abundant gas from underground, below the vadose zone.

Table 6.7 shows the constituents and their volume concentrations of the atmosphere. Table 6.8 shows the composition of earth's crust. Table 6.9 shows the general composition of humans. These numbers are significant for demonstrating that fossil fuel combustion is the best way to generate energy to be utilized by humans.

Natural gas can be found in a variety of different underground formations, including shale formations, sandstone beds, and coal seams. Just like oil is the second most available liquid, natural gas is the most available gas in the earth crust beyond the vadose zone. Traditionally, the gas available in

Table 6.7 Constituents and Volume Concentration of Atmosphere

Constituent	Percent volume
Nitrogen	78.084
Oxygen	20.946
Argon	0.934
Carbon dioxide	0.036
Neon	0.00182
Helium	0.000524
Methane	0.00015
Krypton	0.000114
Hydrogen	0.00005

Table 6.8 Major Elements of Earth Crust

Element	Percent by Volume
Oxygen	46.60%
Silicon	27.72%
Aluminum	8.13%
Iron	5.00%
Calcium	3.63%
Sodium	2.83%
Potassium	2.59%
Magnesium	2.09%
Titanium	0.44%
Hydrogen	0.14%
Phosphorus	0.12%
Manganese	0.10%
Fluorine	0.08%
Barium	340 ppm
Carbon	0.03%
Strontium	370 ppm
Sulfur	0.05%
Zirconium	190 ppm
Tungsten	160 ppm
Vanadium	0.01%
Chlorine	0.05%
Rubidium	0.03%
Chromium	0.01%
Copper	0.01%
Nitrogen	0.005%
Nickel	Trace
Zinc	Trace

Table 6.9 Elemental Composition of a Human Body

Elements	Fraction (%)
Oxygen	65%
Carbon	18%
Hydrogen	10%
Nitrogen	3%
Calcium	1.5%
Phosphorus	1.0%
Potassium	0.35%
Sulfur	0.25%
Sodium	0.15%
Magnesium	0.05%
Copper, zinc, selenium, molybdenum, fluorine, chlorine, iodine, manganese, cobalt, iron	0.70%
Lithium, strontium, aluminum, silicon, lead, vanadium, arsenic, bromine	Trace amounts

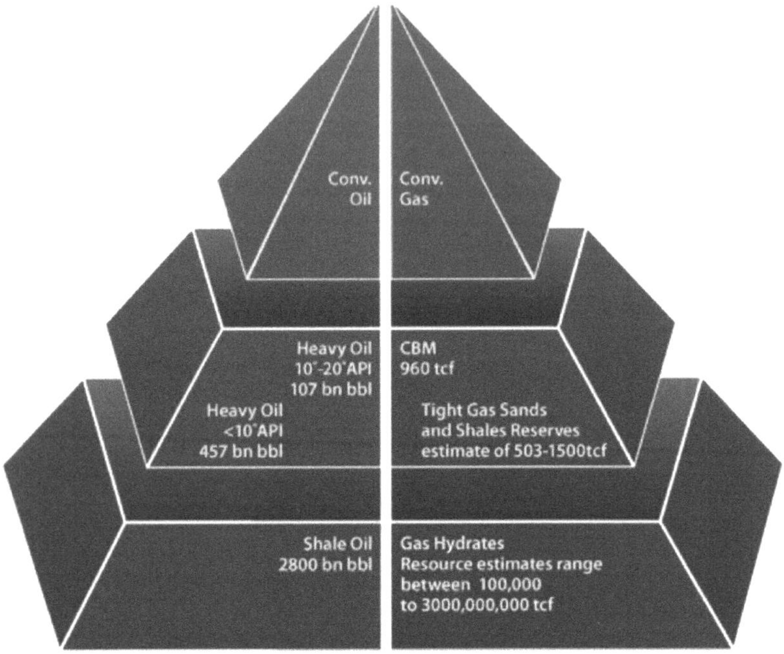

Figure 6.13 The amount of gas available in natural resources increases as one moves from conventional to unconventional resources.

the high permeability zone has been developed with conventional techniques. However, much more gas remains in reservoirs that are considered to be unconventional, mainly because they are not producible with conventional techniques. This natural gas is in fact cleaner than the conventional gas and has higher heating value. Figure 6.13 shows how the volume of natural gas increases exponentially when one moves away from conventional gas reserves. Similar trend is seen in oil reservoirs. However, oil from unconventional sources are increasingly more difficult to refine. Opposite trends exist for gas. For instance, cleanest and most readily combustible gas resides within gas hydrates (Picture 6.3).

Figure 6.14 shows the locations of unconventional gas around the world. Figure 6.15 shows coalbed methane and shale gas in Europe, whereas Figure 6.16 shows tight gas plays of the United States.

6.9 THE SCIENCE OF WATER AND PETROLEUM

Petroleum is the second most abundant fluid available on earth. Only water is more abundant than oil. While the role of water in creating and

Picture 6.3 Gas hydrate burning offers some of the most environment-friendly combustion available to mankind. *Picture from USGS.*

sustaining life is well recognized, the role of petroleum has been mischaracterized. Such mischaracterization is unique to the modern epoch and is paradoxical (Islam et al., 2010). This "bad name" comes from the original paradox, called "water–diamond paradox," first reported by Adam Smith, the father of modern economics. This paradox (also known as paradox of value) is the apparent contradiction that, although water is on the whole more useful, in terms of survival, than diamonds, diamonds command a higher price in the market. In a passage of Adam Smith's *An Inquiry into the Nature and Causes of the Wealth of Nations*, he discusses the concepts of value

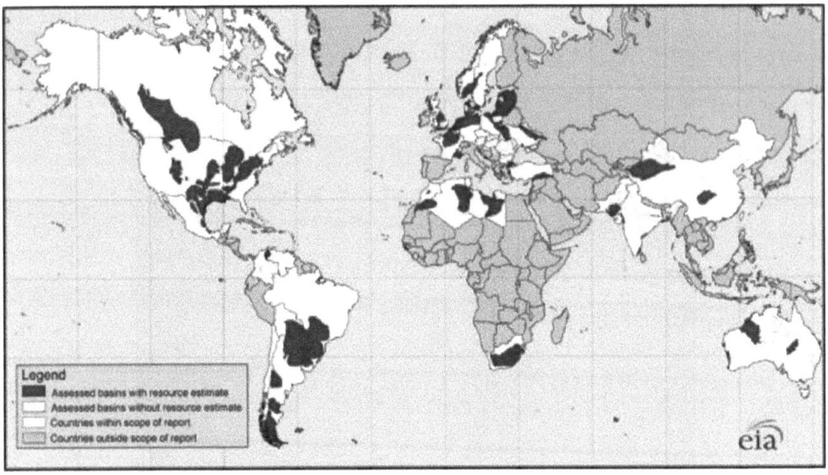

Figure 6.14 Unconventional gas around the world. *From EIA.*

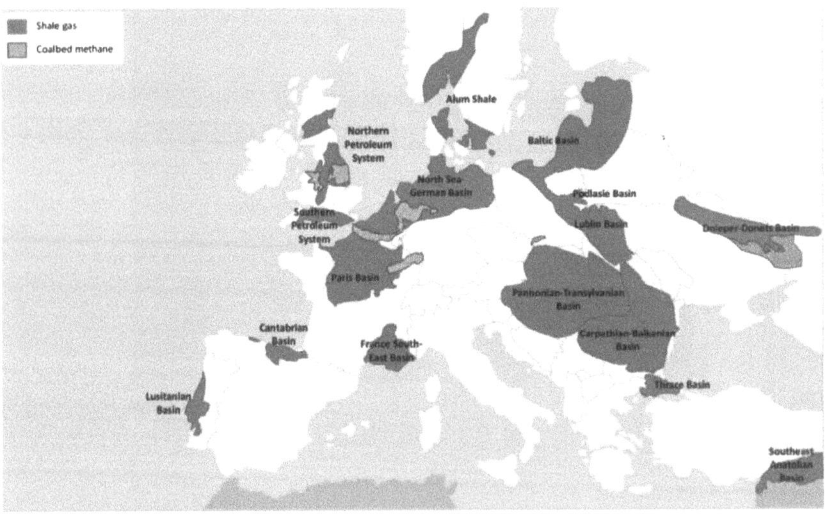

Figure 6.15 Shale gas and coalbed methane in Europe.

in use and value in exchange, setting stage for bifurcating trends in value in utility and value in exchange:

> *"What are the rules which men naturally observe in exchanging them [goods] for money or for one another, I shall now proceed to examine. These rules determine what may be called the relative or exchangeable value of goods. The word VALUE, it is to be observed, has two different meanings, and sometimes expresses*

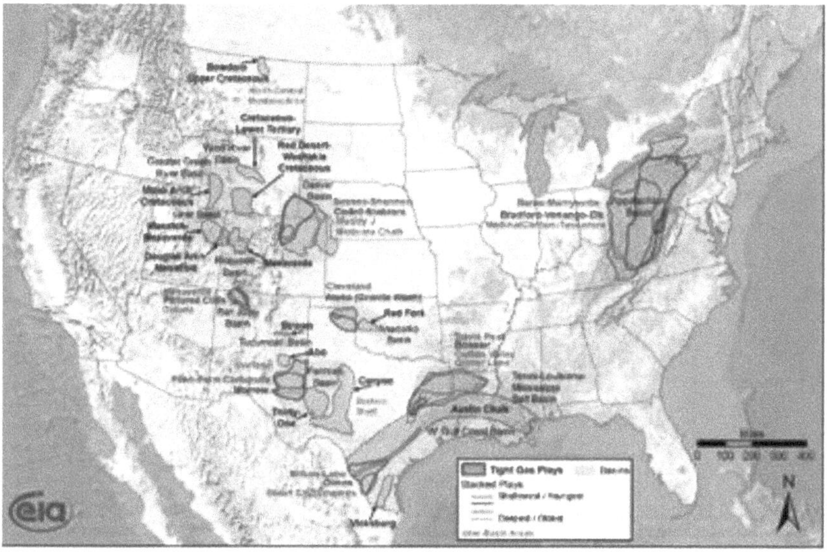

Figure 6.16 Tight gas plays in the United States.

the utility of some particular object, and sometimes the power of purchasing other goods which the possession of that object conveys. The one may be called "value in use;" the other, "value in exchange." The things which have the greatest value in use have frequently little or no value in exchange; on the contrary, those which have the greatest value in exchange have frequently little or no value in use. Nothing is more useful than water: but it will purchase scarce anything; scarce anything can be had in exchange for it. A diamond, on the contrary, has scarce any use-value; but a very great quantity of other goods may frequently be had in exchange for it."

He, then explained, "the real value." Furthermore, he explained the value in exchange as being determined by labor:

"The real price of every thing, what every thing really costs to the man who wants to acquire it, is the toil and trouble of acquiring it."

Instead of removing this paradox by finding a direct function that relates price with utility, pragmatic approach led to the resolution of this paradox by imposing price–production relationship and detaching consumers from the equation. In essence, this denomination of "value" created the basis for an inherently unsustainable pricing that in itself became the driver of technology development (Zatzman, 2012a; 2012b). Figure 6.17 shows how assigning artificial value has led to a preposterous functionality between real value and natural state.

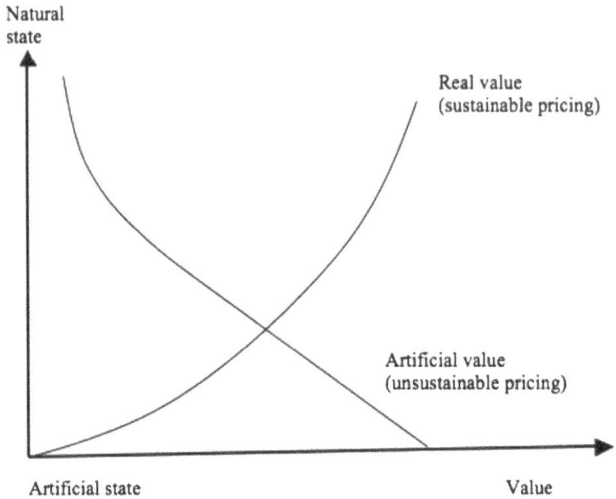

Figure 6.17 Natural state is synonymous with low value with current mode of energy pricing. *From Islam et al., 2010.*

In scientific term, the above manipulation amounts to removing time function from each processes. Only then, the utility of carbon in charcoal and carbon in diamond can be conflated (Picture 6.4).

A proper historical discourse, termed as delinearized history by Zatzman and Islam (2007a), can unravel the mysteries of sustainability. Figure 6.18 was produced by Khan and Islam (2007) shows how natural cycles are inherently sustainable. Note how the only source of energy is used to transform inorganic materials into organic ones. Such transformation

Picture 6.4 The difference between charcoal and diamond can be captured in the time function, which is either linearized or altogether eliminated in various economic models that drive modern technology.

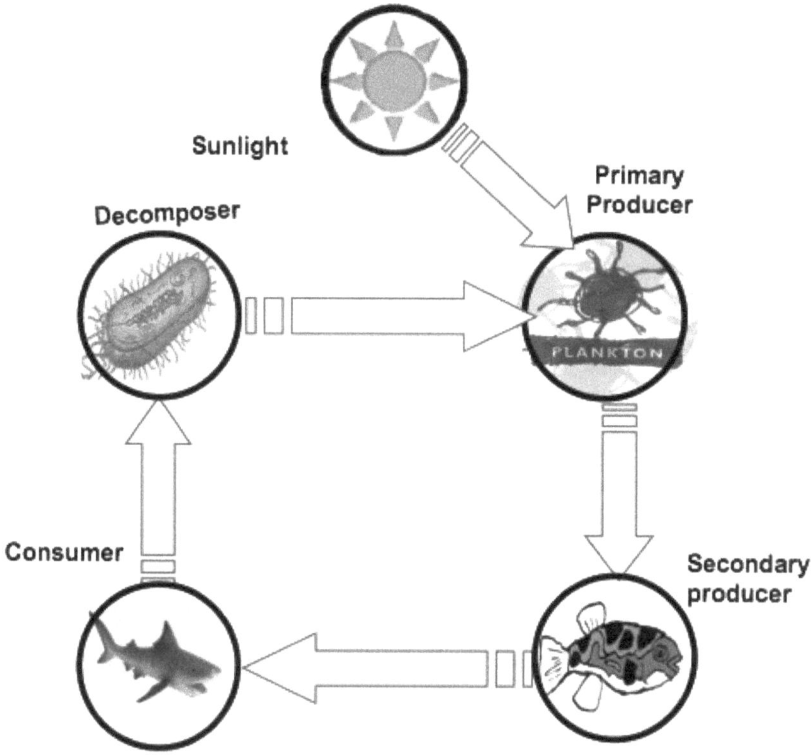

Figure 6.18 Scientific characterization is inherently sustainable. *From Khan and Islam, 2007.*

cannot take place in absence of water (H_2O) and carbon dioxide (CO_2). During this transformation, sunlight plays the role of a catalyst and its contribution is quantifiable with proper science (Khan and Islam, 2012). However, sunlight is not sufficient as the onset of life is the phenomenon that triggers conversion of inorganic matter into organic matter. Scientific description of this process is nonexistent or weak at best (Islam et al., 2014).

Scientifically, water represents the onset of life, whereas oil represents the end of life. Indeed, water and oil contain an array of contrasting, yet complimentary properties. Water is polar and is a good solvent due to its polarity. Oily materials are known to be hydrophobic. The ability of a substance to dissolve in water is determined by whether or not the substance can match or better the strong attractive forces that water molecules generate between other water molecules. If a substance has properties that do not allow it to overcome these strong intermolecular forces, the molecules are "pushed out" from the water and do not dissolve. Contrary to the common misconception, water and hydrophobic substances do not "repel," and the hydration of a hydrophobic surface is energetically favorable. The process of hydration can be best described by the process in which water molecules surround the molecule of another compound. Because, water molecules are relatively smaller, a number of water molecules typically surround the molecule of the other substance. Properties of water and oil are different and complementary. For instance, water and oil can form stable emulsions and eventually create soap. Life begins with water but ends with oil in its most stable and stabilized form. In fact, other than honey, oil is the most effective antibacterial natural liquid.

Petroleum is opposite to water in a complementary sense and form stable entities, such as soap that can act as a cleansing agent that works on both oil and water. Indeed, life begins with water and ends with oil. In molecular level, oil is hydrophobic but it is not water repellant. In fact, water molecules form very stable bonds around oil molecules. However, in broader scale, oil kills but water gives life. In microscale, they are opposite in every property but they are essential for life. This entire thing is like the Yin–Yang symbol that not only bonds together opposites (historically it meant fire, water; life, death; male, female; earth, sky; cold, hot; black, white) but also are embedded inside white background, while holding within each another circle that itself has similar Yin–Yang structures. The cycle continues all the down to Higgs boson (until 2013) and beyond (in future), never reaching the same trait as the homogenous, anisotropic, monochrome, boundary-less surrounding. At every stage, there is also another combination of opposite, i.e., intangible (time) and tangible (mass), which essentially is the program that defines the time function.

6.9.1 Comparison Between Water and Petroleum

Water is the source of life whereas petroleum is the end of a life cycle. These two form harmony in nature and coexist much like the Yin–Yang symbol. This fact was recognized throughout history and at no time petroleum products were considered harmful to the environment.

In its fundamental unit, snowflakes represent modules of water, whereas diatoms represent organic units of petroleum (Picture 6.5). In its original form, symmetry exists but only in broad sense. There is no local symmetry. Picture 6.6 shows various images of snow flakes. If diamonds are from

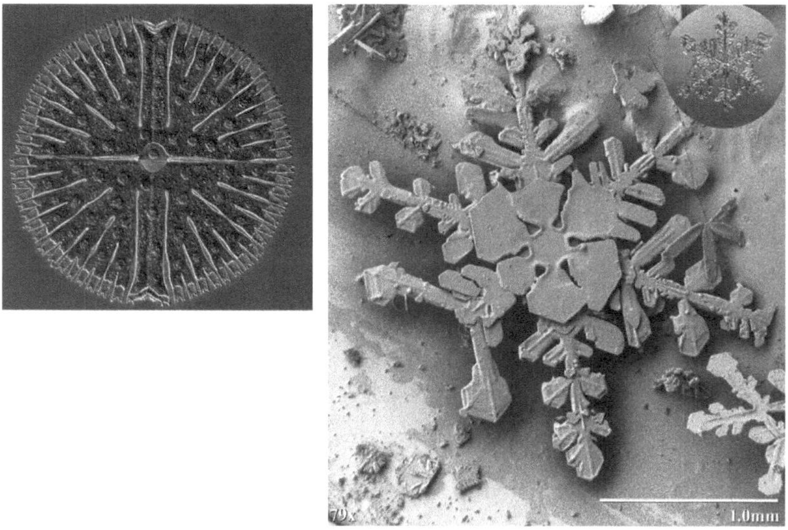

Picture 6.5 This single-celled green diatom won Rogelio Moreno Gill of Panama fifth place in the BioScapes Imaging Competition. Specimens for this composite image came from a lake (from National Geographic). Picture on the right is a snow flake. *From U.S. Dept. of Agriculture*

charcoal, petroleum is from diatoms (Picture 6.7). Table 6.10 shows various sources of water on earth.

Water and hydrocarbon are both essential to life, even though they play contrasting roles. Table 6.11 shows some of the unifying and contrasting features of water and petroleum.

The above opposites signal complimentary nature of water and petroleum. At a molecular level, the following reactions of opposites can be observed.

$$\text{Oxygen} + \text{Hydrogen} \rightarrow \text{Water} \tag{6.1}$$

The result is water vapor, with a standard enthalpy of reaction at 298.15 °K and 1 atm of −242 kJ/mol. While this equation is well known, it cannot be stated that original water or natural water is created this way. In fact, all evidence suggest that it is not and the suggestion that oxygen and hydrogen combined to form water as the basis of life bears the same first premise as the one imposed for the Big Bang theory. What we know, however, is if hydrogen burns in oxygen, it produces intense heat (around 2000 °C) as compared to heat of a natural flame (e.g., from candle) that is around 1000 °C. The above reaction does not take place unless there is a

Picture 6.6 Snow flakes are fundamental units of water.

presence of two other components, one tangible (catalyst) and one intangible (spark), that produce a flame. A discussion on what constitutes a flame and its consequences is presented later on in this chapter.

This reaction needs a spark that itself has catalysts (tangible) and energy (intangible). However, in nature water does not form by combining oxygen and hydrogen. One theory indicates water is the original matter as

Picture 6.7 Pictures of diatoms. *Picture from Colorado State geological survey, 2013.*

Table 6.10 Various Sources of Water on Earth

Sea water	
The oceans	97.2%
Inland seas and saline lakes	0.008%
Fresh water	
Freshwater lakes	0.009
All rivers (average levels)	0.0001
Arctic icecap	1.9
Arctic icecap and glaciers	0.21
Water in the atmosphere	0.001
Ground water within half a mile from surface	0.31
Deep-lying ground water	0.31

(data from USGS)

contrast to popular theory that puts hydrogen as the original mass (Islam et al., 2014b). Only recently this theory has gained ground as astrophysicists continue to find evidence of water in outer space (Farihi et al., 2011). Table 6.12 lists the fundamental properties of oxygen and hydrogen. Table 6.13 highlights qualities that unite and contrast oxygen and hydrogen.

$$\text{Carbon} + \text{Oxygen} \rightarrow \text{Carbon dioxide} \tag{6.2}$$

The above reaction takes place at all temperature (e.g., low-temperature oxidation). However, the most natural, yet rapid conversion takes place with

Table 6.11 Contrasting Features of Water and Petroleum

Water	Petroleum
Source of all organic matter	End product of all organic matter
Most abundant fluid on earth	Second most abundant fluid on earth
Oxygen 85.84; Sulfur 0.091 Hydrogen 10.82; Calcium 0.04 Chloride 1.94; Potassium 0.04 Sodium 1.08; Bromine 0.0067 Magnesium 0.1292; Carbon 0.0028	Carbon, 83–87% Hydrogen, 10–14% Nitrogen, 0.1–2% Oxygen, 0.05–1.5% Sulfur, 0.05–6.0% Metals < 0.1%
Mostly homogeneous	Hydrocarbon (15–60%), napthenes (30–60%), aromatics (3–30%), with asphaltics making up the remainder.
Reactivity of water towards metals: Alkali metals react with water readily. Contact of cesium metal with water causes immediate explosion, and the reactions become slower for potassium, sodium, and lithium. Reaction with barium, strontium, calcium are less well known, but they do react readily.	Non-reactive toward metal.
Nonmetals like Cl_2 and Si react with water $Cl_{2(g)} + H_2O_{(l)} \rightarrow HCl_{(aq)} + HOCl_{(aq)}$ $Si_{(s)} + 2H_2O_{(g)} \rightarrow SiO_{2(s)} + 2H_{2(g)}$ Some nonmetallic oxides react with water to form acids. These oxides are referred to as acid anhydrides.	Reaction with nonmetals is faster
High cohesion	Low cohesion
Unusually high surface tension; susceptible to thin film	Unusually low surface tension
Adhesive to inorganic	Adhesive to organic
Unusually high specific heat	Unusually low specific heat
Unusually high heat of vaporization	Unusually low heat of vaporization
Has a parabolic relationship between temperature and density	Has monotonous relationship between temperature and density
Unusually high latent heat of vaporization and freezing	Unusually low latent heat of vaporization and freezing
Versatile solvent	Very poor solvent
Unusually high dielectric constants	Unusually low dielectric constants
Has the ability to form colloidal solutions	Destabilizes colloids

Table 6.11 Contrasting Features of Water and Petroleum—cont'd

Water	Petroleum
Can form hydrogen bridges with other molecules, giving it the ability to transport minerals, carbon dioxide, and oxygen	Poor ability to transport oxygen and carbon dioxide
Unusually high melting point and boiling point	Unusually low melting point and boiling point
Unusually poor conductor of heat	Unusually good conductor of heat
Unusually high osmotic pressure	Unusually low osmotic pressure
Nonlinear viscosity pressure and temperature relationship (extreme nonlinearity at nanoscale, Hussain and Islam, 2010)	Mild nonlinearity in viscosity pressure and temperature relationship
Enables carbon dioxide to attach to carbonate	Absorbs carbon dioxide from carbonate
Allows unusually high sound travel	Allows unusually slow sound travel
Large bandwidth microwave signals propagating in dispersive media can result in pulses decaying according to a non-exponential law (Peraccini et al., 2009)	Faster than usual movement of microwave
Unusually high confinement of X-ray movement (Davis et al., 2005)	Unusually high facilitation of X-ray movement

From Hutchinson, 1957; Attwood, 1949; Handbook of Chemistry and Physics, 1981.

fire. Fire itself has tangible (mass of fire) and energy (heat of reaction, intangible). Table 6.14 lists the fundamental properties of oxygen and carbon. Table 6.15 highlights qualities that unite and contrast oxygen and carbon.

The above contrasting and complementary properties of hydrogen and oxygen and oxygen and carbon give rise to water and fire, respectively, creating a new set of contrasting and complementary components. Together, they form the basic ingredients of life on earth and exemplify natural sustainability.

Historically, water has always been recognized as the source matter of everything (Islam et al., 2010). As in early ancient Greek, ancient Chinese, and ancient Mesopotamia, water has been considered as the one that gives life while fire is the one that causes death. For fire to exist and complete the cycle of life, it must be accompanied with fuel, which is the essence of energy. The most efficient source of this fuel is natural gas.

Table 6.12 Fundamental Properties of Oxygen and Hydrogen

	Oxygen	Hydrogen
Atomic number	8	1
Atomic mass	15.999 g/mol	1.007825 g/mol
Electronegativity according to Pauling	3.5	2.1
Density	1.429 kg/m^3 at 20 °C	0.0899×10^{-3} g/cm^3 at 20 °C
Melting point	−219 °C	−259.2 °C
Boiling point	−183 °C	−252.8 °C
Van der waals radius	0.074 nm	0.12 nm
Ionic radius	0.14 nm (−2)	0.208 nm (−1)
Isotopes	4	3
Electronic shell	[He] $2s^2$ $2p^4$	$1s^1$
Energy of first ionization	1314 kJ/mol	1311 kJ/mol
Energy of second ionization	3388 kJ/mol	
Energy of third ionization	5300 kJ/mol	
Discovered by	Joseph Priestly in 1774	Henry Cavendish in 1766

The existence of water as a fundamental element is important. Ancient literature as well as Qur'an places the existence of water before anything else. In every culture water is synonymous with life and liveliness. Opposite to water is fire (number Two) at the lower left corner. The role of fire is opposite to water, yet it is essential to life. Without fire, there is no carbon dioxide, the essence of plant, and therefore, life. Fire represents transition from cold to hot, from life to death, from tangible (water or liquid) to intangible (vapor or gas). This phase change is typical of creation. In fact, the very fact that everything is moving (a function of time) makes it essential to go through this phase of tangible and intangible. Overall, this continues in an eternal circle.

Qur'an mentions water (*m'aa*) as the original creation. This "water" is not a combination of hydrogen and oxygen atoms as asserted in New Science, it is the essence of life in Arabic. The word, *m'aet* that stands for dehydration actually stands for moribund or dying. The following verse of Qur'an states:

> *And it is He who created the skies and the earth in six periods - and His Dominion (extends) upon water - that He might test you as to which of you is best in deed. But if you say, "Indeed, you are resurrected after death," those who rebel (against Allah) will surely say, "This is not but obvious magic." (11:7)*

Table 6.13 Common and Contrasting Features of Oxygen and Hydrogen

Oxygen	Hydrogen
Fundamental component of water (89% in mass and 33% in mole), which is ubiquitous on earth (70%).	Fundamental component of water (11% in mass and 67% in mole), which is ubiquitous on earth (70%)
Believed to be 3rd most abundant element in universe	Believed to be most abundant element in universe
If mass-energy discontinuity is removed, most abundant mass in universe	If mass-energy discontinuity is removed, second most abundant in universe
It is the essential element for respiratory processes for all living cells. It's the most abundant element in the Earth's crust. Nearly one-fifth (in volume) of the air is oxygen. Noncombined gaseous oxygen normally exists in form of diatomic molecules, O_2, but it also exists in triatomic form, O_3, ozone. Oxygen is reactive and will form oxides with all other elements except helium, neon, argon, and krypton. It is moderately soluble in water (30 cm^3/L of water dissolve) at 20 °C. Oxygen does not react with acids or bases under normal conditions.	Hydrogen is the most flammable of all the known substances. There are three hydrogen isotopes: protium, mass 1, found in more than 99,985% of the natural element; deuterium, mass 2, found in nature in 0.015% approximately; and tritium, mass 3, which appears in small quantities in nature. The dissociation energy of molecular hydrogen is 104 kcal/mol. Molecular hydrogen is not reactive. Atomic hydrogen is very reactive. It combines with most elements to form hydrides (e.g., sodium hydride, NaH), and it reduces metallic oxides, a reaction that produces the metal in its elemental state. The surfaces of metals that do not combine with hydrogen to form stable hydrides (e.g., platinum) catalyze the recombination of hydrogen atoms to form hydrogen molecules and are thereby heated to incandescence by the energy.
Strong bond with hydrogen (110 kcal/mol); slightly stronger bond with oxygen (119 kcal/mol).	Strong bond with oxygen; lesser strength bond with hydrogen (104 kcal/mol); lesser strength bond with carbon (98 kcal/mol).
The crust of earth is composed mainly of silicon-oxygen minerals, and many other elements are there as their oxides.	The earth crust has some 45 times less hydrogen than oxygen

Continued

Table 6.13 Common and Contrasting Features of Oxygen and Hydrogen—cont'd

Oxygen	Hydrogen
Oxygen gas makes up one-fifth of the atmosphere. The oxygen in the Earth's atmosphere comes from the photosynthesis of plants, and has built up in a long time as they utilized the abundant supply of carbon dioxide in the early atmosphere and released oxygen.	Only 0.000055% of earth atmosphere is hydrogen. Sunlight causes photosynthesis that utilizes hydrogen and releases oxygen, forming a closed loop.
Oxygen is fairly soluble in water (0.045 g/kg of water at 20 °C), which makes life in rivers, lakes, and oceans possible. The water in rivers and lakes needs to have a regular supply of oxygen, for when this gets depleted the water will no longer support fish and other aquatic species.	Low solubility in water (0.0016 g/kg of water at 20C).
Nearly every chemical, apart from the inert gasses, bind with oxygen to form compounds. Water, H_2O, and silica, SiO_2, main component of the sand, are among the more abundant binary oxygen compounds. Among the compounds which contain more than two elements, the most abundant are the silicates, that form most of the rocks and soils. Other compounds that are abundant in nature are calcium carbonate (limestone and marble), calcium sulfate (gypsum), aluminum oxide (bauxite), and various iron oxides that are used as source of the metal.	At normal temperature, hydrogen is a not very reactive substance, unless it has been activated somehow; for instance, by an appropriate catalyzer. At high temperatures it is highly reactive and a powerful reducing agent (anti-oxidant). It reacts with the oxides and chlorides of many metals, like silver, copper, lead, bismuth, and mercury, to produce free metals. It reduces some salts to their metallic state, like nitrates, nitrites, and sodium and potassium cyanide. It reacts with a number of elements, metals and nonmetals, to produce hydrides, like NAH, KH, H_2S, and PH_3. Atomic hydrogen produces hydrogen peroxide, H_2O_2, with oxygen.
Oxygen is essential for all forms of life since it is a constituent of DNA and almost all other biologically important compounds. Is it even more dramatically essential, in that animals must have minute by minute supply of the gas in order to survive. Oxygen	All compounds and elements produced through hydrogen reduction (see above) are potent toxins for all living organisms. However, organic form of the same toxin is necessary for living organisms. For instance, lack of

Table 6.13 Common and Contrasting Features of Oxygen and Hydrogen—cont'd

Oxygen	Hydrogen
in the lungs is picked up by the iron atom at the center of hemoglobin in the blood and thereby transported to where it is needed.	organic H_2S can trigger Alzheimer's disease.
Departure from normal atmospheric composition of oxygen (both too high or too low concentrations) causes lung damage	High concentrations of this gas can cause an oxygen-deficient environment. Individuals breathing such an atmosphere may experience symptoms that include headaches, ringing in ears, dizziness, drowsiness, unconsciousness, nausea, vomiting and depression of all the senses. Under some circumstances, death may occur.

Table 6.14 Fundamental Characteristics of Carbon

Atomic number	6
Atomic mass	12.011 g/mol
Electronegativity according to Pauling	2.5
Density	2.2 g/cm^3 at 20 °C
Melting point	3652 °C
Boiling point	4827 °C
Van der waals radius	0.091 nm
Ionic radius	0.26 nm (−4); 0.015 nm (+4)
Isotopes	3
Electronic shell	[He] $2s^2 2p^2$
Energy of first ionization	1086.1 kJ/mol
Energy of second ionization	2351.9 kJ/mol
Energy of third ionization	4618.8 kJ/mol
Discovered by	The ancients

Also, see the following hadith that compliments the above notion and clarifies the fundamental traits of both creator and creation, as well as the purpose of humans:

> *It was narrated that Ibn Mas'ood (R) said: Between the first heaven and the one above it is (a distance of) five hundred years. Between each of the skies is (a distance of) five hundred years. Between the seventh sky and the Throne is (a distance of) five hundred years. Between the Throne and the water is (a distance of) five hundred years, and the Throne is above the water, and Allah is above the Throne, and nothing whatsoever of your deeds is hidden from Him.*

Table 6.15 Contrasting and Unifying Features of Oxygen and Carbon

Oxygen	Carbon
Fundamental component of water (89% in mass and 33% in mole), which is ubiquitous on earth (70%). The most abundant in mass and numbers.	Fundamental component of living organisms, second most abundant in mass, and third most abundant in atomic numbers.
Most abundant in (65%) of a living body	Second most abundant (18%) of living body
Believed to be 3rd most abundant element in universe	Believed to be 4th most abundant element in universe
If mass-energy discontinuity is removed, most abundant mass in universe	If mass-energy discontinuity is removed, third most abundant (after oxygen and hydrogen) in universe.
Oxygen recycled through water cycle for sustenance of life	Carbon recycled through carbon cycle for sustenance of life
Oxygen burns hydrogen with the largest heat of reaction for any element (141.8 MJ/kg)	Oxygen burns carbon with the second largest heat of reaction for any element (32.8 MJ/kg)

Oxygen:

Photosynthesis (land)	16,500[a]
Photosynthesis (ocean)	13,500
Photolysis of N_2O	1.3
Photolysis of H_2O	0.03
Total gains	~ 30,000
Losses: Respiration and Decay	
Aerobic respiration	23,000
Microbial oxidation	5100
Combustion of fossil fuel (anthropogenic)	1200
Photochemical oxidation	600
Fixation of N_2 by lightning	12
Fixation of N_2 by industry (anthropogenic)	10
Oxidation of volcanic gases	5
Losses: Weathering	
Chemical weathering	50
Surface reaction of O_3	12
Total losses	~ 30,000

[a]In units of 10^{12} kg/year

Carbon:

Pool	Quantity (gigatons)
Atmosphere	720
Oceans (total)	38,400
Total inorganic	37,400
Total organic	1000
Surface layer	670
Deep layer	36,730
Lithosphere	
Sedimentary carbonates	> 60,000,000
Kerogens	15,000,000
Terrestrial biosphere (total)	2000
Living biomass	600–1000
Dead biomass	1200
Aquatic biosphere	1–2
Fossil fuels (total)	4130
Coal	3510
Oil	230
Gas	140
Other (peat)	250

able 6.15 Contrasting and Unifying Features of Oxygen and Carbon—cont'd

<table>
<tr><th>xygen</th><th>Carbon</th></tr>
<tr><td>is the essential element for respiratory processes
or all living cells.
's the most abundant element in the earth's crust.
learly one-fifth (in volume) of the air is oxygen.
Ioncombined gaseous oxygen normally exists in
rm of diatomic molecules, O_2, but it also exists in
iatomic form, O_3, ozone.</td><td>It is the second (second to hydrogen) most important fuel for living organism and sustenance of life.
Carbon is the 15th most abundant in earth's crust.</td></tr>
<tr><td>)xygen, major component of water, is essential for
fe. By far the largest reservoir of earth's oxygen is
ithin the silicate and oxide minerals of the crust
nd mantle (99.5%). Only a small portion has been
leased as free oxygen to the biosphere (0.01%) and
mosphere (0.36%). The main source of
mospheric free oxygen is photosynthesis, which
roduces sugars and free oxygen from carbon
ioxide and water.
<table>
<tr><th>Reservoir</th><th>Capacity (kg O_2)</th><th>Flux (kg/year)</th><th>Residence Time (years)</th></tr>
<tr><td>Atmosphere</td><td>$1.4.10^{18}$</td><td>3.10^{14}</td><td>4500</td></tr>
<tr><td>Biosphere</td><td>$1.6.10^{16}$</td><td>3.10^{14}</td><td>50</td></tr>
<tr><td>Lithosphere</td><td>$2.9.10^{20}$</td><td>6.10^{11}</td><td>500,000,000</td></tr>
</table></td><td>Carbon, major component of all organic matter
<table>
<tr><th>Reservoir</th><th>Size (gigatons of Carbon)</th></tr>
<tr><td>Atmosphere</td><td>750</td></tr>
<tr><td>Forests</td><td>610</td></tr>
<tr><td>Soils</td><td>1580</td></tr>
<tr><td>Surface ocean</td><td>1020</td></tr>
<tr><td>Deep ocean</td><td>38,100</td></tr>
<tr><td>Coal</td><td>4000</td></tr>
<tr><td>Oil</td><td>500</td></tr>
<tr><td>Natural gas</td><td>500</td></tr>
</table></td></tr>
<tr><td>he sun contributes to water mass through
otosynthesis and thereby contributes to carbon
cle.</td><td>A mass of about 7×10^{11} tons of carbon is in the atmosphere as CO_2 and about 4.5×10^{11} tons of carbon in vegetation as carbohydrate. The nominal percentage of CO_2 in the atmosphere is about 0.034%.</td></tr>
<tr><td>xygen is reactive and will form oxides with all
her elements except helium, neon, argon, and
ypton.
rong bond with hydrogen (110 kcal/mol); slightly
ronger bond with oxygen (119 kcal/mol).</td><td>Carbon's best reactant is oxygen that produces CO_2—the one needed for synthesis of carbohydrate.
The C–O bond strength is also larger than C–N or C–C. C–C = 83; C–O = 85.5; O-CO = 110; C=O = 192 (CO_2); C=O = 177 (aldehyde); C=O (ketone) = 178; C=O(ester) = 179; C=O(amide) = 179; C≡O = 258; C≡C = 200 (all values in kcal/mole)</td></tr>
</table>

Continued

Table 6.15 Contrasting and Unifying Features of Oxygen and Carbon—cont'd

Oxygen	Carbon
The crust of earth is composed mainly of silicon-oxygen minerals, and many other elements are there as their oxides.	Carbon is the major component of CO_2.
Oxygen gas makes up one-fifth of the atmosphere. The oxygen in the earth's atmosphere comes from the photosynthesis of plants, and has built up in a long time as they utilized the abundant supply of carbon dioxide in the early atmosphere and released oxygen.	After nitrogen, oxygen, and argon, carbon dioxide is the most abundant component of earth's atmosphere.
Oxygen is fairly soluble in water (0.045 g/kg of water at 20 °C), which makes life in rivers, lakes, and oceans possible.	Very low solubility in water
Nearly every chemical, apart from the inert gasses, bind with oxygen to form compounds. Oxygen is essential for all forms of life since it is a constituent of DNA and almost all other biologically important compounds.	The two most important characteristics of carbon, as a basis for the chemistry of life, are that it has four valence bonds and that the energy required to make or break a bond is just at an appropriate level for building molecules that are not only stable, but also reactive. The fact that carbon atoms bond readily to other carbon atoms allows for the building of arbitrarily long complex molecules and polymers. Uniquely suited for metabolism.
Departure from normal atmospheric composition of oxygen (both too high or too low concentrations) causes lung damage.	In its elemental form (graphite and diamond), is completely a benign and great fuel, only second to hydrogen as an elemental energy generator. Some simple carbon compound can be very toxic, such as carbon monoxide (CO) or cyanide (CN). Carbon 14 is one of the radionuclides involved in atmospheric testing of nuclear weapons. It is among the long-lived radionuclides that have produced and will continue to produce increased cancer risk for decades and centuries to come. It also can cross the placenta, become organically bound in developing cells and hence endanger fetuses.

Another verse of Qur'an has issued the following warning to humans:

Say: Have you considered if your water should go down then who would bring you back the flowing water? (67:30)

Water is the essence of life and it is to be noted that Allah calls Himself "*Al-Hayyu*" (The Alive), who never feels drowsiness or slumber (Qur'an 2:255). Death is associated with every entity of the creation and sleep is called synonymous to death (soul removed when someone sleeps). So, life and death are part of being creation.

"Have those who disbelieved not considered that the heavens and the earth were a joined entity, and We separated them and made from water every living thing? Then will they not believe?" (Qur'an 21:30)

It is of interest to note, New Science puts water as the first indicator of life. The search of water continues even in outside of our galactic reach. Only recently, such existence has been "confirmed" (Farihi et al., 2011).

According to Qur'anic narration, universe has no void and it is filled with water, which is the essence of life. However, at no time this means isolated water molecules and any other "fundamental" unit of water. Each particle is tagged with its own time function and forms integral part of the universal order.

6.9.2 Combustion and Oxidation

In a complete combustion reaction, a compound reacts with an oxidizing element, such as oxygen, and the products are compounds of each element in the fuel with the oxidizing element. The oxidation with oxygen is the most commonly occurring phenomena in nature. It is because of the abundance of oxygen as well as the ability of oxygen to react at all temperatures. In terms of generating energy, most notably heat generation, is through oxidation of hydrogen. Even though, in nature it is rarely the case, the oxidation of hydrogen produces the most intense heat in presence of a flame (2000 °C). This is the principle used in rocket engines. The second most intense heat is with carbon (1000 °C). This is the principle used in all forms of fossil fuel burning. Unlike hydrogen and oxygen, this reaction is natural and takes place at all temperatures, albeit as a strong function of temperature. The low-temperature oxidation is continuous and follows Arrhenius equation, which is an exponential relationship with temperature. However, oxidation of elemental carbon (e.g., graphite and diamond) are both rare because of rarity of those elements, compared to compound form of carbon. For instance, diamond and graphite both burn at 800 °C in presence of oxygen but in absence of oxygen they melt at very high temperature (3600 °C for graphite and 3800 °C for diamond). The next most heat generating combustion is with methane. This reaction is written as follows

$$CH_4(g) + 2O_2(g) \rightarrow CO_2(g) + 2H_2O(g) + \Sigma \tag{6.3}$$

The standard enthalpy of reaction for methane combustion at 298.15 °K and 1 atm is −802 kJ/mol. The symbol Σ signifies the time function that stores information regarding intangibles (Islam et al., 2010a), such as the history of methane (organic or otherwise), history of oxygen (organic or mechanical, as well as the collection of all elements that are present in nonmeasurable quantities. The usefulness of Σ is in its ability to track the history in order to chart the future pathway in terms of harm and beneficial quality. For instance, if the oxygen supply is restricted, the following reaction will take place, in stead of Eqn (6.2).

$$2C(s) + O_2(g) \rightarrow 2CO(g) + \Sigma \tag{6.4}$$

This reaction is typical of industry-standard producer gas that is produced by injecting oxygen through hot coke. The resulting gas is a mixture of carbon monoxide (25%), carbon dioxide (4%), nitrogen (70%), and traces of hydrogen (H_2), methane (CH_4), and oxygen (O_2). In addition to this information, Σ will also contain information regarding any other trace elements that can be present due to use of catalyst, heating mechanism, existence of flames, etc. In essence, Σ is the tracker of intangibles.

Any combustion reaction is known to be accelerated dramatically in presence of a flame. A flame is a mixture of reacting gases and solids emitting visible, infrared, and sometimes ultraviolet light, the frequency spectrum of which depends on the chemical composition of the burning material and intermediate reaction products. A standard and beneficial flame is fire, arising from burning wood. This process of heat and light generation is entirely sustainable (Chhetri and Islam, 2008) and produces no harmful or by-product, therefore, it is waste-free (Khan and Islam, 2012). The fundamental characteristic of this wood flame is that combustion is incomplete, thereby generating incandescent solid particles, called soot. It comes with red-orange glow of fire. This light has continuous spectrum, similar to sunlight spectrum. Even though it is rarely talked about, the orange glow of wood fire is also similar to the glow of sun. See Picture 6.8.

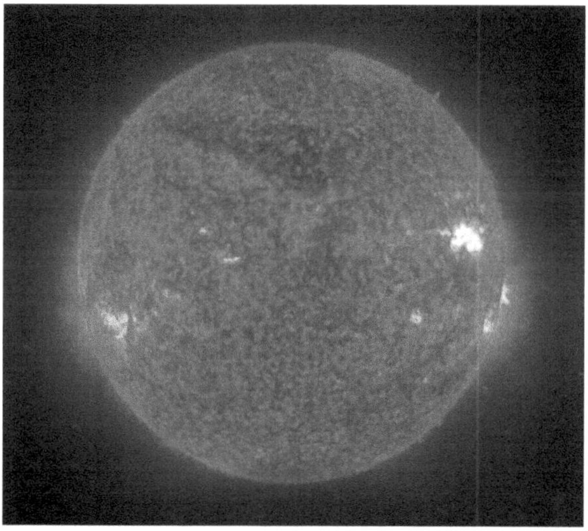

Picture 6.8 Sun picture taken at 9:19 a.m. EST on Nov. 10, 2004, by the SOHO (Solar and Heliospheric Observatory) spacecraft. *NASA/European Space Agency, 2004.*

6.9.3 Natural Energy versus Artificial Energy

The sun, a natural source of light is an essential element of the ecosystem. One of the benefits of the sun is day light and night light via the moon. The sun does not produce waste since all its resulting particles and effects are used by nature. The sun light service life is infinite. The sun consists of heterogeneous materials and particles. Then, this type of light sources is natural, heterogeneous, clean, vital, and efficient. Figure 6.19 shows the natural light pathway.

Light intensity or energy, efficiency, and quality are functions of the light source composition. The light source is composed of infinite particles with different sizes, d_i, masses, m_i, and temperature, T_i. The light source mass equals:

$$M = \sum_{i=1}^{\infty} m_i \tag{6.5}$$

A particle energy function equals:

$$E_i = a_i f_i \tag{6.6}$$

where a_i is a constant, and f_i is the frequency for the particle i.

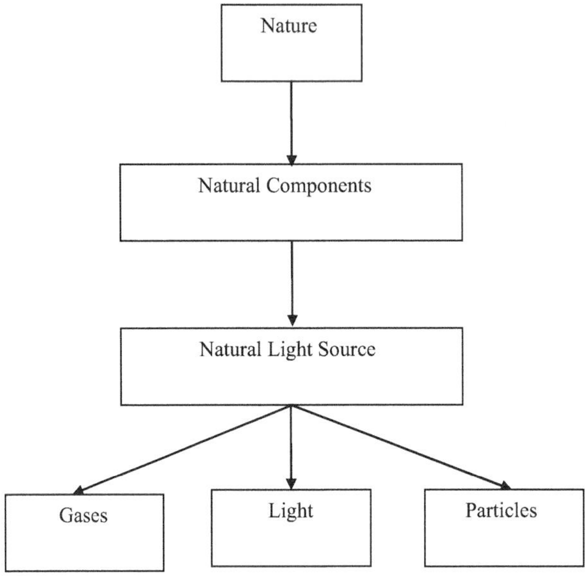

Figure 6.19 Natural light pathway.

The light energy of a particle i is also defined as follows:

$$E_i = b_i m_i^{p_i} v_i^{q_i} \tag{6.7}$$

where v_i is the speed of the particle i.

Combining Eqns (6.6) and (6.7) yields:

$$a_i f_i = b_i m_i^{p_i} v_i^{q_i} \tag{6.8}$$

Then, the frequency f_i for the particle i comes to:

$$f_i = \frac{b_i}{a_i} m_i^{p_i} v_i^{q_i} \tag{6.9}$$

where b_i, p_i, q_i are the constants defining the particle composition and properties.

As a result, the particle speed v_i amounts to:

$$v_i = \left(\frac{a_i f_i}{b_i m_i^{p_i}} \right)^{1/q_i} \tag{6.10}$$

The total light energy is the sum of all particle energy values.

$$E = \sum_{i=1}^{\infty} E_i \tag{6.11}$$

The wavelength is the inverse of the frequency:

$$\lambda_i = v_i / f_i \tag{6.12}$$

where v_i is the speed of the particle i:

$$v_i = l_i / t_i \tag{6.13}$$

l_i is the distance traveled by the particle i and t_i the travel time.

The distance traveled by a particle i is a function of its size, d_i, mass, m_i, and temperature, T_i. The particle mass, m_i, depends on the particle composition. Since this particle i consists of the smallest particle in the universe, its composition is unique and corresponds to one material.

The density of the particle i is:

$$\rho_i = m_i / V_i \tag{6.14}$$

where V_i is the particle volume:

$$V_i = \alpha_i d_i^{\beta_i} \tag{6.15}$$

α_i and β_i are the particle size constants.

The distance traveled by light particle is described by:

$$l_i = v_i t_i \tag{6.16}$$

which is equivalent to:

$$l_i = \left(\frac{a_i f_i}{b_i m_i^{p_i}} \right)^{1/q_i} t_i \tag{6.17}$$

The solar light spectrum is shown in Figure 6.20. Sunlight as the source of energy on earth must be understood in the context of photosynthesis reaction that creates vegetation on earth. Table 6.16 shows the composition of the sun. Considering some 8000 tons of loss of mass per second from the sun, it is reasonable to assume most of the mass loss involves hydrogen. Consequently, this hydrogen must constitute the most active role in photosynthesis. It is indeed the case. Compare the picture with the

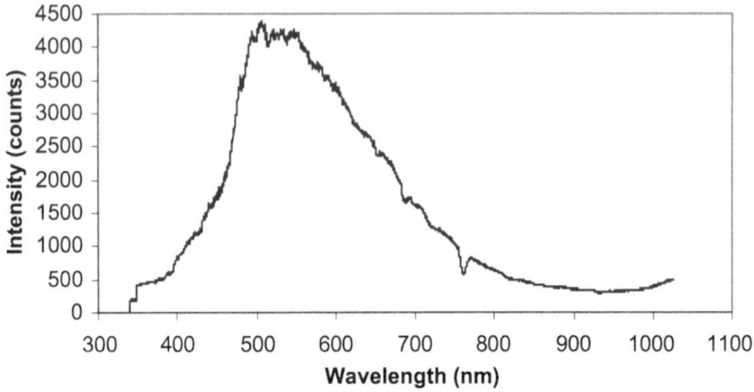

Figure 6.20 Wavelength spectrum of sunlight.

Table 6.16 Sun Composition

Element	Abundance (percentage of total number of atoms)	Abundance (percentage of total mass)
Hydrogen	91.2	71.0
Helium	8.7	27.1
Oxygen	0.078	0.97
Carbon	0.043	0.40
Nitrogen	0.0088	0.096
Silicon	0.0045	0.099
Magnesium	0.0038	0.076
Neon	0.0035	0.058
Iron	0.0030	0.14
Sulfur	0.0015	0.040

Chaisson and McMillan, 1997.

following picture of wood burning fire (Picture 6.9). Furthermore, this composition is important in terms of overall elemental balance of the ecosystem. It is also important for consideration of beneficial energy. If nature is taken to be perfect and beneficial, solar energy as well as the elements present in the sun must be in beneficial form and should be considered to be the standard of energy.

All vegetation on earth starts off with solar energy. If the artificial barrier between energy and mass is removed, immediate consequence of solar irradiation would be manifested in the light spectrum of sunlight. Interestingly, the most abundant section of the solar light spectrum is the section

Picture 6.9 Wood fire is natural, beneficial, and sustainable.

Table 6.17 Wavelengths of Various Visible Colors

Wavelength (nm)	Color
<400	Ultraviolet (invisible)
400–450	Violet
450–490	Blue
490–560	Green
560–590	Yellow
590–630	Orange
630–670	Bright red
670–750	Dark red
>750	Infrared (invisible)

that produces visible light (wavelength range of 400–750 nm). Table 6.17 lists wavelengths of various visible color lights.

All wavelengths beyond these wavelengths of visible light are inherently harmful. The premise that nature is perfect leads to the conclusion that other rays are also necessary but their intensity must be very low, in line

Table 6.18 Wavelengths of Known Waves

Type of rays	Wave length	
Gamma ray	10^{-2}–10^{-6} nm	
X-ray	10–10^{-1} nm	
Ultraviolet	10–400 nm	
Visible (by humans) light	Violet	400–450 nm
	Blue	450–490 nm
	Green	490–560 nm
	Yellow	560–590 nm
	Orange	590–630 nm
	Bright red	630–670 nm
	Dark red	670–750 nm
Infrared	800–1000 nm	
Microwave	0.001–0.3 m	
Radio wave	1–1000 m	

with the corresponding low intensities Table 6.18 lists wavelengths of various known waves.

It is important to identify the sources of nonvisible rays. While we know all of them are emitted from the sun, Table 6.19 shows artificial sources of the same waves. Because artificial sources render these rays inherently unnatural, they make natural materials vulnerable to harm.

For every natural ray, there is an artificial version. While each of the natural rays is essential and beneficial, the artificial counterpart is harmful to natural objects. Khan et al. (2008) demonstrated the nature of such artificial mass or energy by eliminating the assumption that transition from mass to energy is discrete and nonreactive.

6.9.4 From Natural Energy to Natural Mass

In nature, we have the most spectacular example of conversion of energy into mass. The process is called photosynthesis. For most plants, photosynthesis occurs within chlorophyll bodies. Chlorophylls are arranged in something called "photosystems," which are in the thylakoid membranes of chloroplasts. The main function of chlorophyll is to absorb light energy and transfer it to the reaction center chlorophyll of the photosystem.

- *Chlorophyll a* has an approximate absorption peak of 665 nm and 465 nm.
- *Chlorophyll b* has an approximate absorption peak of 640 nm and 450 nm. In addition, there are accessory pigments that are able to absorb light. Chlorophyll a & b are green and are able to best absorb light in the 450 nm (violet-blue) and 650 nm (red) area of the light spectrum. That

Table 6.19 Artificial Sources of Various Waves

Type of rays	Artificial sources
Gamma ray	Co-60 or Cs-137 isotopes. When an unstable (radioactive) atomic nucleus decays into a more stable nucleus, the "daughter" nucleus is sometimes produced in an excited state. The subsequent relaxation of the daughter nucleus to a lower-energy state results in the emission of a gamma-ray photon.
X-ray	30–150 kV with tungsten, molybdenum, or copper. X-rays are produced when electrons strike a metal target. The electrons are liberated from the heated filament and accelerated by a high voltage towards the metal target. The X-rays are produced when the electrons collide with the atoms and nuclei of the metal target.
Ultraviolet	UV rays can be made artificially by passing an electric current through a gas or vapor, such as mercury vapor.
Infrared	Tungsten, kanthal filaments, sapphire, calcium fluoride, zinc selenide, silicon nitride, laser, etc.
Microwave	Klystron (high power amplifiers), and reflex klystron (low power oscillators). Magnetron (high power pulsed oscillator) Semiconductors Specialized transistors and integrated amplifiers, especially using gallium arsenide instead of silicon. Often found in wireless networking devices, GPS receivers, etc.
Radio wave	When a direct electrical current is applied to a wire the current flow builds an electromagnetic field around the wire. This field sends a wave outward from the wire. When the current is removed, the field collapses which again sends a wave. If the current is applied and removed over and over for a period of time, a series of waves is propagated at a discrete frequency. If the current changes polarity, or direction repeatedly, that could make waves, too. This phenomenon is the basis of electromagnetivity and basically describes how radio waves are created within transmitters.

leaves the green, yellow, and orange parts of the spectrum unusable. This is why plants have extra pigments (colors), in order to take in light from different wavelengths that chlorophyll is not good at absorbing.

- *Carotene* is an orange pigment capable of photosynthesis. This pigment transmits light energy to chlorophyll. As well as photosynthesis, these pigments also help protect against too much light, photoinhibition.
- *Phaeophytin a* are gray-brown in colour.

- *Phaeophytin b* are yellow-brown.
- *Xanthophyll* are yellow pigments in the carotenoid group. These pigments seem to absorb best at 400–530 nm. These are involved with photosynthesis with chlorophyll. Chlorophyll is often much more abundant than xanthophylls, and this is why the leaves are still a green color. When fall arrives in many countries and the leaves change color, the chlorophyll "dies back" and the xanthophylls are more apparent in the yellow color you see (like a maple tree)
  - *The xanthophyll cycle* is a wonderful skill a plant has. In order to protect itself from absorbing too much light, and thus causing photoinhibition, xanthophyll cycle converts pigments that do not quench energy into ones that do. When a plant receives too much light, the xanthophyll cycle changes violoxanthin to antheraxanthin and zeaxanthin, which are photoprotective pigments.
- *Anthocyanin* pigments are often red, purple, or blue. These pigments have been said to help a plant against light stress and act to help protect a plant from blue-green and UV light. Cacti do not have these, they have betalain instead.
- *Betalain* are pigments found in caryophyllales (cacti and beets for example). They are often a red–yellow–purple color that is often found in flower color, but it can also be found in leaves, stems, fruits, and roots of these plants as well. It is not really known what the exact purpose of these pigments are.
  - *Betacyanins* are reddish to violet betalain pigments. They absorb light best at 535 nm.
  - *Betaxanthins* are yellow to orange betalain pigments. They absorb light best at 480 nm.

Given the various pigments, and the areas they are most abundant, that Chlorophyll a & b, and to a lesser extent, the various carotenoids (such as carotene and xanthophyll) would be the most productive in the absorption of light for photosynthesis. When applying this to cultivation and artificial lights, it would seem logical to choose lights that peak in the 430–470 nm and 640–680 nm range, to allow the two main chlorophyll types to gather the most energy. Light in the blue spectrum may also be a little stronger to allow the carotenes and xanthophylls to absorb more light as well. Figure 6.21 shows the existence of these wavelengths in visible light.

If the fundamental premise that natural is beneficial and artificial is harmful (Khan and Islam, 2012) is invoked, the picture depicted by Figure 6.22 emerges.

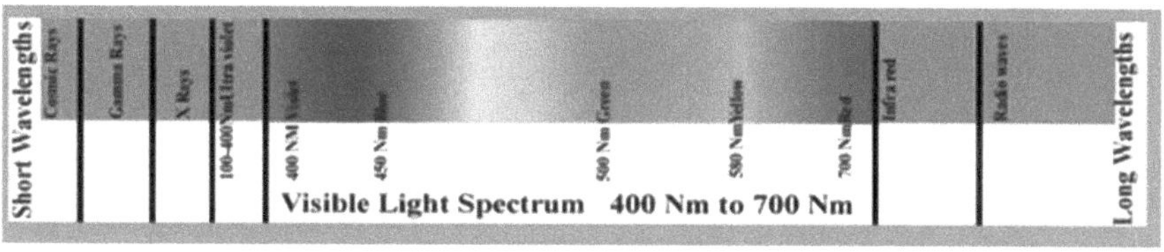

Figure 6.21 Colors and wave lengths of visible light.

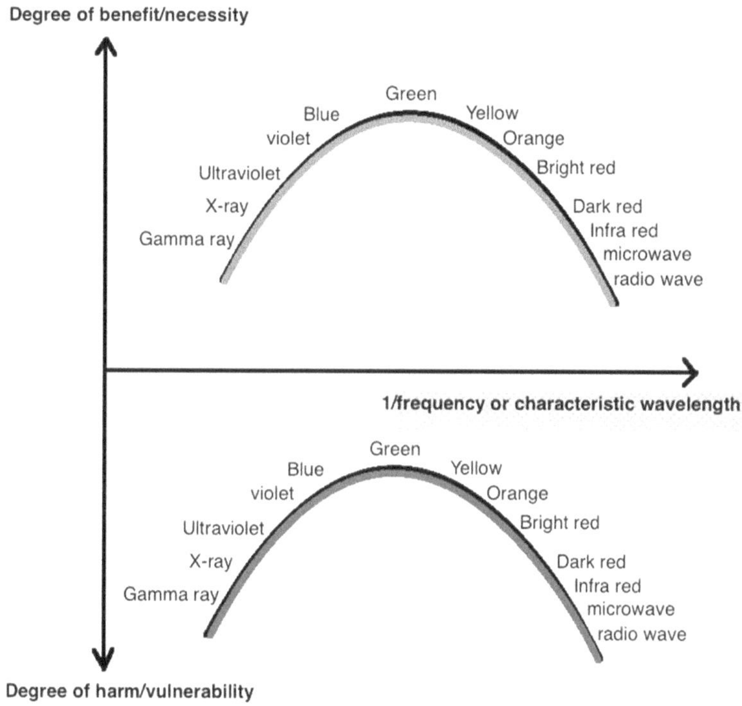

Figure 6.22 Artificial and natural lights affect natural material differently.

Of importance in the above graph is the notion that artificial rays are harmful at all times. As the exposure is increased, the harm is accentuated. For the short-term, artificial visible light is less harmful than artificial nonvisible rays (e.g., gamma ray, X-ray, etc.) on both sides of the spectrum (both long wavelengths and short ones). The reason for such behavior has been discussed by Khan and Islam (2012) and will be discussed later in this section. The above graph follows the same form as the wavelength spectrum of visible sunlight (Figure 6.23).

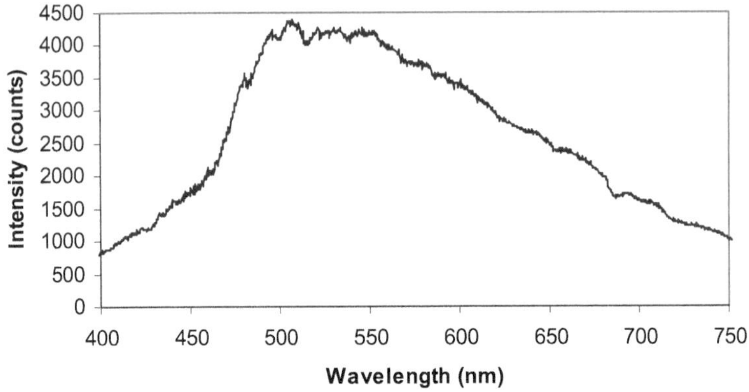

Figure 6.23 Wavelength spectrum of visible part of sunlight.

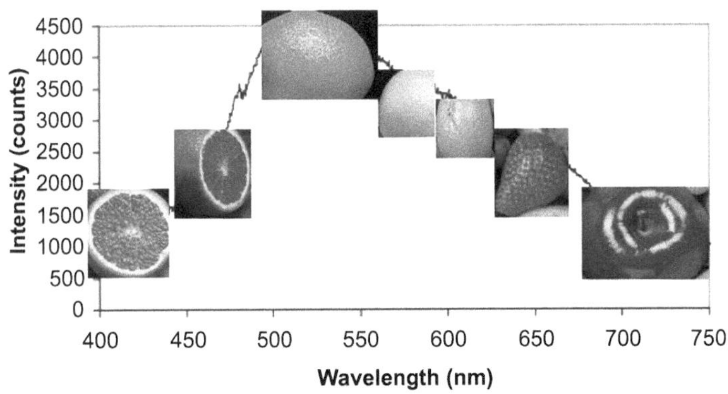

Figure 6.24 Visible natural colors as a function of various wavelengths and intensity of sunlight.

Figure 6.24 recasts visible colors on intensity of solar radiation for the visible light section. This figure confirms that green vegetation should be the most abundant color on earth for which the sun is the only natural source of energy. This figure also shows that the area under the intensity–wavelength curve is the greatest for green materials. Red has longer wavelength but their intensity in sunlight is much smaller than green lights.

Figure 6.25 plots radiance values for various wavelengths observed in forest fire as compared to grass and warm ground. For the visible light range, forest fire follows the same trend as grass very closely. Also, comparable is warm ground. For the invisible range, however, forest fire

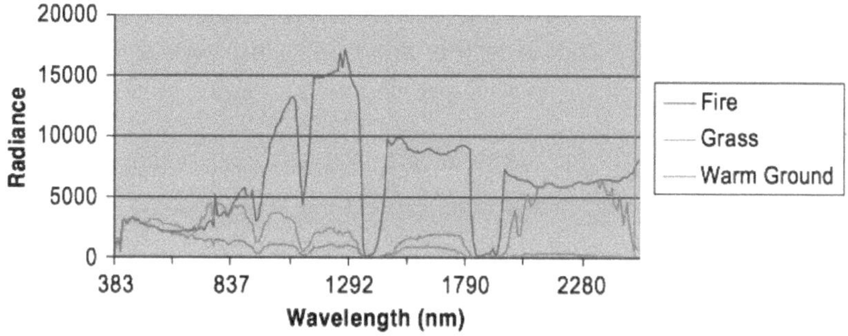

Figure 6.25 Wavelength and radiance for forest fire, grass, and warm ground. *From Li et al., 2005.*

produces high radiance values for larger (than infrared) values. For wavelengths larger than 2 mm, both fire and warm ground produce similar radiance, whereas grass does not show any radiation.

Oxidation of butane creates a blue flame. Typically, the separation of one particular component of a natural material skews the balance that a whole natural material would have. The burning of butane is, therefore, a skewed version of forest fire. Figure 6.26 shows how the butane flame produces spikes in the wavelength versus irradiance graph. This light, even

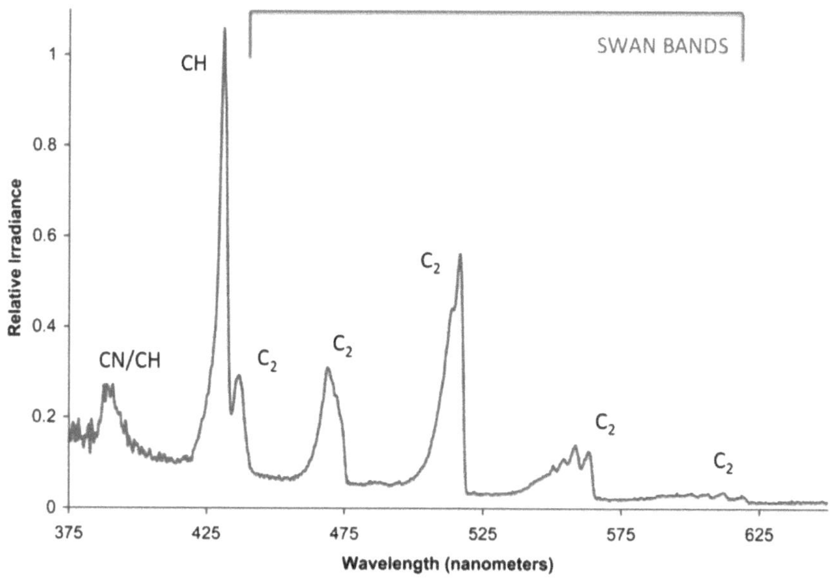

Figure 6.26 Blue flame radiance for butane.

though they are from a natural source, lacks balance—the likes of which persisted with sunlight and forest fire. Such imbalance would lead to harm of organic bodies, similar to the one shown in Figure 6.39 (bifurcation graph above) However, modern engineering typically ignores this fact and manufactures artificial material (energy or matter) that are similar to the natural counterpart only in the external features. For instance, for the case of electronic books, the main feature is to produce writings/pictures on a white background. All colors are artificial but white background is the most toxic because of its deviation from natural light spectrum. Figure 6.27 shows the light spectrum for Kindle Fire HD, Nexus 7, and new iPad. Compare these spectra with that of sunlight and further consider irradiation from a white page compared to irradiation from an electronic device. It becomes clear that the artificial device is both imbalanced and will create long-term harm to humans as well as the environment.

Figure 6.28 shows sunlight along with light produced from a paraffin candle, incandescent light, and other light sources. Note how red LED is the most skewed from sunlight spectrum. The deviation is the most in visible light zone (wavelength of 400–750 nm). With the exception of two spikes at 600 nm and 700 nm, red LED produces very little irradiation in the visible light zone, whereas it produces much higher irradiation in the infrared zone and beyond. Fluorescent light produces similar spikes at 600 nm and 700 nm points but with less intensity than red LED. Overall,

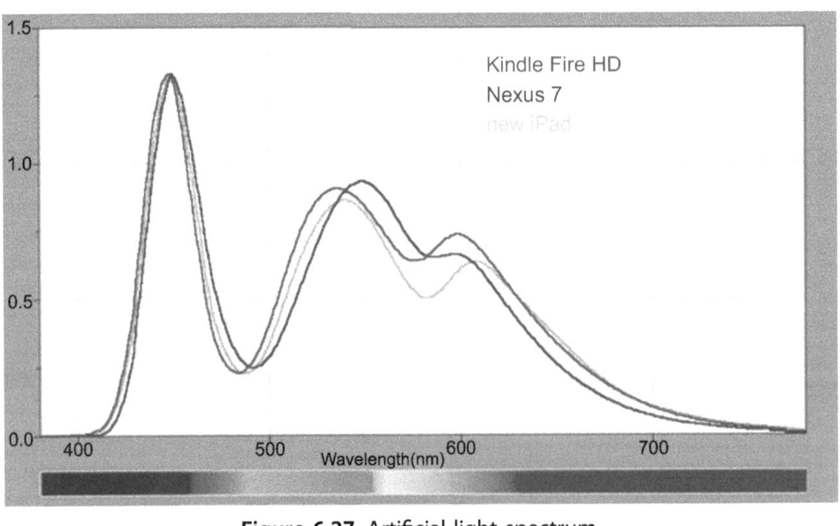

Figure 6.27 Artificial light spectrum.

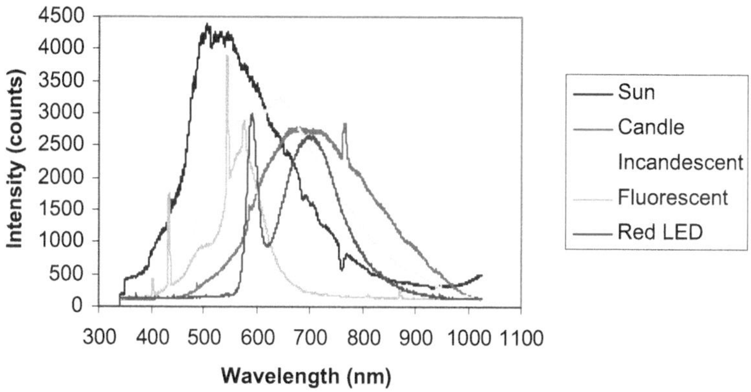

Figure 6.28 Comparison of various artificial light sources with sunlight.

candle is the only one among artificial light that produces a broad band of wavelengths. In terms of harm to the environment, red LED is the worst offender, followed by fluorescent, then incandescent, and finally candle light. This vulnerability ranking is done by comparing the area under the curve within the visible light zone (Figure 6.29).

If sunlight represents the original and the most beneficial energy source, any natural process emerging from sunlight will become beneficial. Let us consider forest fire. It comes from a flame that trees or vegetation as the most important ingredient. All vegetations are indeed a product of natural processing of sunlight, air, water, and carbon components.

When a flame is visible, oxidation of wood is rapid. As oxidation takes place, movement of each particle within the system is greatly enhanced,

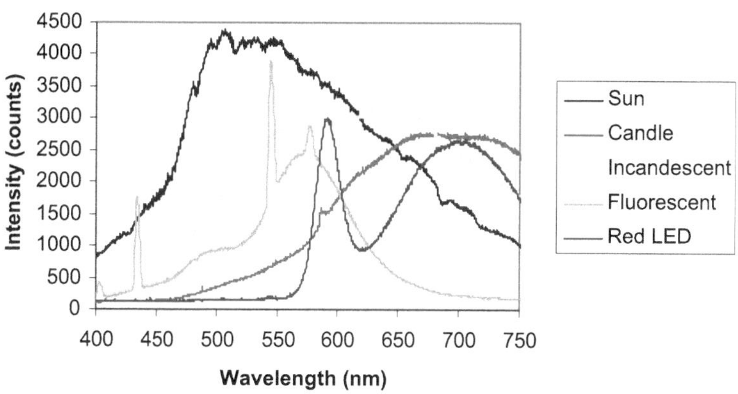

Figure 6.29 Comparing within the visible light zone will enable one to rank various artificial light sources.

creating a sharp increase in natural frequencies of every particle. For instance, a solid can burn into gases unleashing natural frequency change for each particle. The recent model developed by Islam et al. (2014a) describes this process as equivalent to merger of two galaxies in which each of them has numerous components with respective natural frequencies. However, after the reaction occurs (oxidation in this case), the resulting products have a frequency that is different from previous ones. If each particle is tagged, this model can help track a natural process apart from an artificial process. Figure 6.30 shows how this model casts the number of particles with their respective numbers in a natural system. Here, no distinction is made between light and mass particles as imposing such a distinction is contrary to natural order and renders the model aphenomenal.

Figure 6.30 shows how any natural flame will have a smooth spectrum as shown in the spectrum of the sunlight. Any alteration of light source would create a spectrum that is not natural, hence harmful. The above figure also indicates that photon emission is similar to any other radiation from a body of mass. This emission within the visible wavelengths is related to the existence of a flame. Even though a flame is typical of visible light emission, most recent theories indicate the presence of continuous emission throughout the entire spectrum.

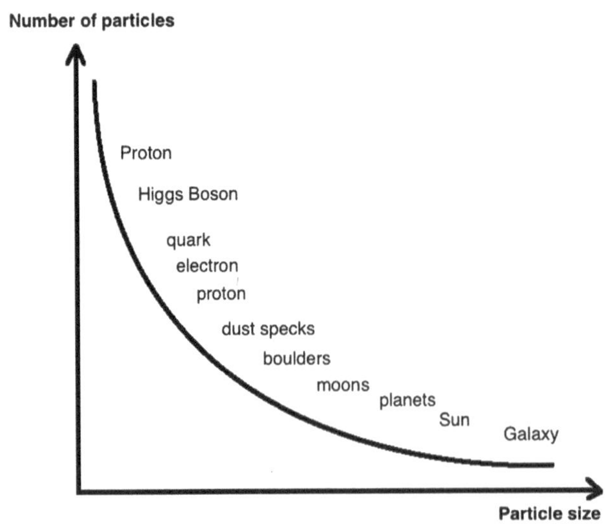

Figure 6.30 Number of particles versus particle size (not to scale). *Modified from Khan and Islam, 2012.*

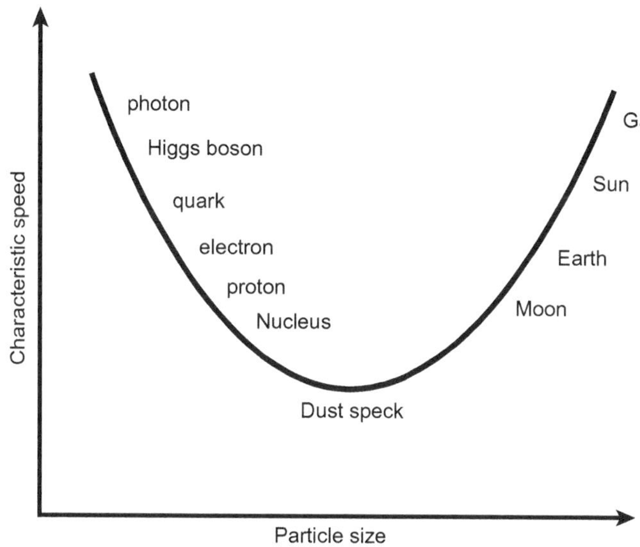

Figure 6.31 Particle size versus characteristic speed of matter (not to scale).

As a flame burns, the characteristic features of each particle changes drastically. Figure 6.31 shows how dust specks (similar to pulverized graphite) present an optimum case in terms of stability. This state is typical of a solid state. This state represents the most stable as well as most nonreactive or conservative state of matter. At subatomic level, a reversal in characteristic versus particle size trend line takes place and the speed increases as the particle size becomes smaller.

Such transition from matter to energy (light) can explain the existence of a flame. In addition, this treatment of matter and energy enables one to track the source of light pollution. The onset of flame is invariably associated with a temperature rise, which in turn triggers vigorous changes in particles, leading to the formation of different structures that are similar to the galaxy in mega scale.

Because of the change in characteristic speed due to the onset of a flame that invariably follows changes in temperature, heat being the result of particle motion triggers radiation. Such connection of radiation with particle movement and heat of reaction is new (Islam et al., 2014a).

The rate of emission is a strong function of temperature and is responsible for changing color of the flame. As stated earlier, radiation takes place in all values of spectrum. Much of the radiation is emitted in the visible and infrared bands, as seen earlier in the context of forest fire. The

Picture 6.10 Forest fire in Canada is an excellent example of natural flame.

color of a flame depends on temperature (for the black-body radiation), and on composition of the emission spectra. The photo of the forest fire in Canada is an excellent example of this variation (Picture 6.10).

Let us review the colors of a flame (with carbon particles emitting light) for various temperatures. Table 6.20 shows the temperature for various colors of flame. With these colors, one can analyze the above forest fire.

Near the ground, where most burning is occurring, the fire is white, the hottest color possible for organic material in general, or yellow. Above the yellow region, the color changes to orange, which is cooler, then red, which is cooler still. Above the red region, the flame is no longer visible. The black smoke that is visible is essentially pulverized carbon particles. These particles form the soot. The common distribution of a flame under normal gravity conditions depends on convection, as soot tends to rise to the top of a general flame, as in a candle in normal gravity conditions, making it yellow. In micro gravity, such as an environment in outer space, convection slows down significantly, leading to a more symmetric shape of the black smoke. This is almost a spherical flame with a blue center. While the presence of blue indicates perfect combustion, such flame cannot be sustained as the produced CO_2 tend to smother the flame, especially around the connection between carbon matter and the flame. There are several possible explanations for this difference, of which the most likely is that the temperature is sufficiently evenly distributed that soot is not formed and complete combustion occurs. Experiments by NASA reveal that diffusion flames in micro gravity allow more soot to be completely oxidized after they are produced than diffusion flames on earth, because of a series of mechanisms that behave differently in micro gravity when compared to

Table 6.20 Various Colors versus Temperature for an Organic Flame

Color	Temperature (°C)
Red	
Just visible	525
Dull	700
Cherry, dull	800
Cherry, full	900
Cherry, clear	1000
Orange	
Deep	1100
Clear	1200
White	
Whitish	1300
Bright	1400
Dazzling	1500

normal gravity conditions. Existing theories cannot account for such dependence of gravity on color of a region within the flame. This is because zero mass is assigned for both photon and Higgs boson. If that spurious assumption is removed, flames in any location can be explained with the emerges of a flame as the trigger event.

Picture 6.11 show the color of burning cars. They essentially represent burning of artificial carbon material (e.g., plastic, refined oil, etc.) The color yellow and red are dispersed throughout the flame body and there is no segregation between red and yellow colors.

This is not unlike the existence of a trigger that onsets life within inorganic bodies. The Picture 6.12 shows a depiction of onset of fire as well as of life.

Consider what happens with life, a living plant and a dead plant have little tangible difference for some time period. The reason the exact time of death cannot be identified is it is an intangible. Similar to what was discussed in terms Yin–Yang duality, both life and death have itself tangible and intangible components within them. When a seed becomes alive, no tangible change occurs in the seed or the surrounding. Similarly, when death occurs in plant, there is no tangible change. It is not until a few cycles have passed that one notices tangible changes. This cycle is characteristic of a living object.

Picture 6.11 Burning vehicles are examples of artificial flame.

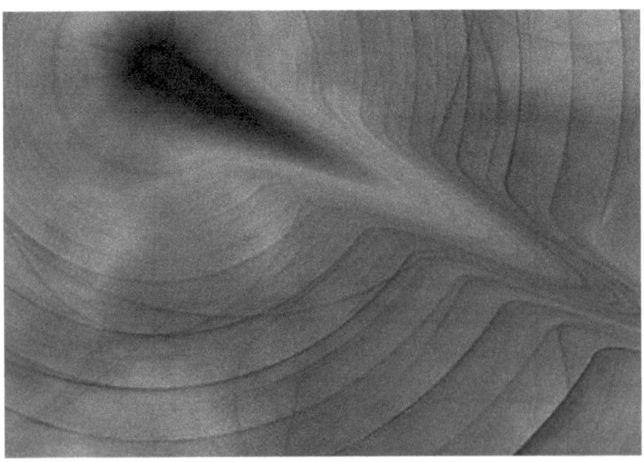

Picture 6.12 Depicture of a flame.

Similarly, extinction or onset of a flame involves an intangible. When a flame is onset, there is no tangible change, for instance, in terms of temperature, ingredient. When a flame is extinguished, the only change that is visible is the disappearance of flame's glow.

While it is true, heat alone can act as spark for a flame, the fact that a spark triggers a flame cannot be explained with conventional science. It is because New Science is grossly deficient of details of factors that are not amenable to linearization (Zatzman et al., 2007b).

Following are some of the typical temperatures for various types of flames and fires.

1. Oxyhydrogen flame: 2000 °C
2. Bunsen burner flame: 1300 to 1600 °C
3. Blowtorch flame: 1300 °C
4. Candle flame: 1000 °C
5. Smoldering cigarette: Always hotter in the middle.
   a. Temperature without drawing: side of the lit portion; 400 °C; middle of the lit portion: 585 °C
   b. Temperature during drawing: middle of the lit portion: 700 °C

This confirms that the minimum temperature associated with a flame is 1000 °C. The highest temperature is recorded for the case of oxyhydrogen flame. However, this flame is not natural because no such reaction takes place on earth under natural conditions. Bunsen burner, on the other hand, represents natural gas burning. Natural gas is such that it does not oxidize in any substantial amount if there is no flame. However, if there is a flame, ambient conditions offer the best condition. When the exposure to air is reduced, the completeness of oxidation reaction is affected. Less air yields an incomplete and thus cooler reaction, while a gas stream well mixed with air provides oxygen in an equimolar amount and thus a complete and hotter reaction. The hottest flame emerges with a blue color when air is mixed freely with the fuel. If the mixing is reduced by choking the inlet of air, the flame will be less hot, however, the brightness of the flame will be increased. The yellow flame is called "luminous flame." In contrast, when the burner is regulated to produce a hot, blue flame it can be nearly invisible against some backgrounds. The hottest part of the flame is the tip of the inner flame, while the coolest is the whole inner flame. Increasing the amount of fuel gas flow through the tube by opening the needle valve will increase the size of the flame. In brief, the Bunsen burner offers a contradictory behavior between heat and light generation, higher light

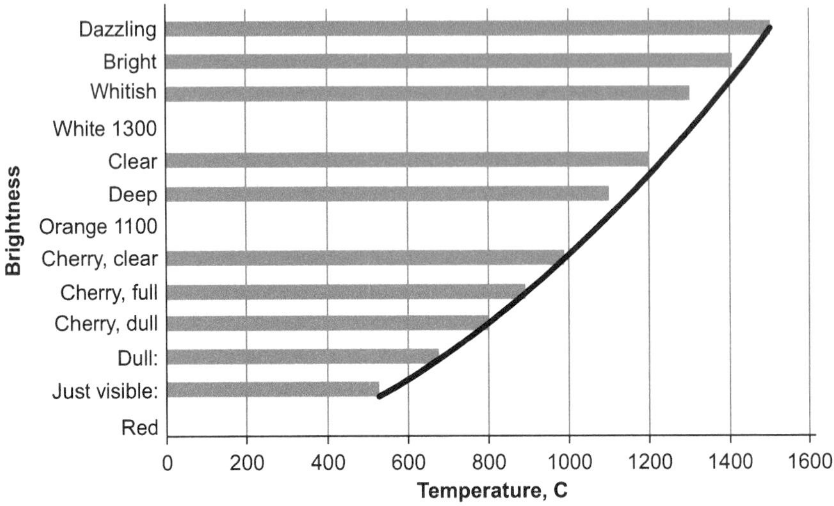

Figure 6.32 Natural flame colors and temperature.

leading to less efficient burning. This is in sharp contrast to the trend observed in natural flame (Figure 6.32)

Bunsen burner produces luminosity by decreasing air supply. In another word, there is a reverse relationship between yield (or efficiency) and luminosity. In the simplest case, the yellow flame is luminous due to small soot particles in the flame which are heated to incandescence. The flame is yellow because of its temperature. To produce enough soot to be luminous, the flame is operated at a lower temperature than its efficient heating flame. The color of simple incandescence is due to black-body radiation. This phenomenon is captured with Planck's law that models black-body radiation as an inverse function of temperature, going from blue to yellow. Luminosity is similarly affected by pressure. These factors are captured in designing artificial lights.

Such behavior is typical of artificial light that employs chemical alteration. Such is typical of a pyrotechnic colorant that triggers chemical reaction to "burn" into a certain color. These colorants are used to create the colors in pyrotechnic compositions like fireworks and colored fires. The color-producing species are usually created from other chemicals during the reaction. Metal salts are commonly used; elemental metals are used rarely (e.g., copper for blue flames).

The color of the flame is dependent on the metal cation, while the anion of the salt has very little direct influence. The anions however influence the flame temperature, both by increasing it (e.g., nitrates, chlorates) and decreasing it (e.g., carbonates, oxalates), indirectly influencing the flame brightness and brilliancy. For temperature-decreasing additives, the limit of colorant may be about 10–20 wt% of the composition. Table 6.21 shows how various colors can be produced with artificial flames.

Picture 6.13 shows fire from wood (top left) is part of the organic cycle whereas smoke from a tungsten bulb (bottom right) is that of mechanical (hence implosive and non-sustainable) cycle. While these extremes are well known, confusion arises as to how to characterize plastic fire (top right) and smoke from a cigarette (bottom left) that have very similar CO_2 emission as in natural wood burning.

The visible particulate matter in such smokes is most commonly composed of carbon (soot). This is the most tangible part. Other particulates may be composed of drops of condensed tar, or solid particles of ash. The presence of metals in the fuel yields particles of metal oxides. Particles of inorganic salts may also be formed, e.g., ammonium sulfate, ammonium nitrate, or sodium chloride. Inorganic salts present on the surface of the soot particles may make them hydrophilic. Many organic compounds, typically the aromatic hydrocarbons, may be also adsorbed on the surface of the solid particles. Metal oxides can be present when metal-containing fuels are burned, e.g., solid rocket fuels containing aluminum. Depleted uranium projectiles after impacting the target ignite, producing particles of uranium oxides. Magnetic particles, spherules of magnetite-like ferrous ferric oxide, are present in coal smoke. New Science does not have any means of characterizing these emissions based on artificiality, thereby failing to distinguish between organic and nonorganic emissions (Islam et al., 2010a; 2012a; Khan and Islam, 2012).

The following table (Table 6.22) shows the relative amount of various elements in the earth crust as well as the lithosphere. It shows oxygen as the most prevalent in the earth crust, followed by silicon, aluminum, iron, and others in lesser quantity. Hydrogen, the component of water, is 10th in the list. The essential component of living organism, viz., carbon is a distant 15th.

In order to determine the overall mass balance of the ecosystem, one should look into the source of carbon as well as hydrogen. It is known that the atmosphere is composed of approximately 78% nitrogen, 21% oxygen, and less than 1% argon. Theoretically, all other elements in the earth crust should also appear in the atmosphere. This composition remains fairly constant throughout the atmosphere. However, as the altitude goes up, the

Table 6.21 Colors and Sources of Artificial Flames

Color	Compound name	Chemical formula	Notes
Red	Strontium nitrate	$Sr(NO_3)_2$	Common. Used with chlorine donors. Excellent red, especially with metal fuels. Used in many compositions including road flares.
Red	Strontium carbonate	$SrCO_3$	Common. Produces good red, slows burning of compositions, decomposes yielding carbon dioxide. Fire retardant in gunpowders. Inexpensive, nonhygroscopic; neutralizes acids. Superior over strontium oxalate in absence of magnesium.
Red	Strontium oxalate	SrC_2O_4	Decomposes yielding carbon dioxide and carbon monoxide. In presence of magnesium fuel, carbon monoxide reduces particles of magnesium oxide, yielding gaseous magnesium and eliminating the black-body radiation of the MgO particles, resulting in clearer color.
Red	Strontium sulfate	$SrSO_4$	Common. High-temperature oxidizer. Used in strobe mixtures and some metal-based red compositions.
Red	Strontium chloride	$SrCl_2$	Common. Produces bright red flame.
Orange	Calcium carbonate	$CaCO_3$	Produces orange flame. Yields carbon dioxide on decomposition. Often

Table 6.21 Colors and Sources of Artificial Flames—cont'd

Color	Compound name	Chemical formula	Notes
			used in toy fireworks as a substitute for strontium.
Orange	Calcium chloride	$CaCl_2$	
Orange	Calcium sulfate	$CaSO_4$	High-temperature oxidizer. Excellent orange source in strobe compositions.
Orange	Hydrated calcium sulfate	$CaSO_4(H_2O)_x$*	
Gold/Yellow	Charcoal powder	C	
Gold/Yellow	Iron powder with oxygen based carbon OC12	Fe + C	
Yellow	Sodium bicarbonate	$NaHCO_3$	Compatible with potassium chlorate. Less burning rate than sodium carbonate. Incompatible with magnesium and aluminum, but reacts evolving hydrogen gas.
Yellow	Sodium carbonate	Na_2CO_3	Hygroscopic. Significantly decreases burning rate, decomposes evolving carbon dioxide. Strongly alkaline. Very effective colorant, can be used in small amounts. Corrodes magnesium and aluminum, incompatible with them.
Yellow	Sodium chloride	NaCl	Loses hygroscopicity on heating. Corrodes metals.
Yellow	Sodium oxalate	$Na_2C_2O_4$	Nonhygroscopic. Slightly reacts with magnesium, no reaction with aluminum.
Yellow		$NaNO_3$	

Continued

Table 6.21 Colors and Sources of Artificial Flames—cont'd

Color	Compound name	Chemical formula	Notes
	Sodium nitrate		Also acts as oxidizer. Bright flame used for illumination.
Yellow	Cryolite	Na_3AlF_6	One of the few sodium salts that is nonhygroscopic and insoluble in water.
Green	Barium chloride	$BaCl_2$	
Green	Barium chlorate	$Ba(ClO_3)_2$	Classic exhibition green with shellac fuel. Sensitive to shock and friction. Oxidizer.
Green	Barium carbonate	$BaCO_3$	Pretty color when ammonium perchlorate is used as oxidizer.
Green	Barium nitrate	$Ba(NO_3)_2$	Not too strong effect. With chlorine donors yields green color, without chlorine burns white. In green compositions usually used with perchlorates.
Green	Barium oxalate	BaC_2O_4	
Blue	Copper(I) chloride	CuCl	Richest blue flame. Almost insoluble in water.
Blue	Copper(I) oxide	Cu_2O	Lowest cost blue colorant.
Blue	Copper(II) oxide	CuO	Used with chlorine donors. Excellent in composite stars.
Blue	Copper carbonate	$CuCO_3$	Best when used with ammonium perchlorate.
Blue	Basic copper carbonate	$CuCO_3 \cdot Cu(OH)_2$, 2 $CuCO_3 \cdot Cu(OH)_2$	Occurs naturally as malachite and azurite. Good with ammonium perchlorate and for high-temperature flames with

Table 6.21 Colors and Sources of Artificial Flames—cont'd

Color	Compound name	Chemical formula	Notes
			presence of hydrogen chloride. Not easily airborne, less poisonous than Paris Green.
Blue	Copper oxychloride	$3CuO \cdot CuCl_2$	Good blue colorant with suitable chlorine donor.
Blue	Paris Green	$Cu(CH_3COO)_2 \cdot 3Cu(AsO_2)_2$	Copper acetoarsenite; Emerald green; toxic. With potassium perchlorate produces the best blue colors. Nonhygroscopic. Fine powder readily becomes airborne; toxic inhalation hazard. Used in majority of Japanese blue compositions as it gives very pretty color.
Blue	Copper arsenite	$CuHAsO_3$	Almost nonhygroscopic; Almost as good colorant as copper acetoarsenite; Toxic; Can be used with chlorate oxidizers.
Blue	Copper sulfate	$CuSO_4 \cdot 5H_2O$	Can be used with nitrates and perchlorates. Acidic, incompatible with chlorates. With red phosphorus in presence of moisture liberates heat; may spontaneously ignite. Less expensive than copper acetoarsenite. Anhydrous copper sulfate is hygroscopic, can be used as a desiccant. With ammonium perchlorate produces almost as pretty blue color as achievable

Continued

Table 6.21 Colors and Sources of Artificial Flames—cont'd

Color	Compound name	Chemical formula	Notes
			with copper acetoarsenite.
Blue	Copper metal	Cu	Rarely used, other compounds are easier to work with. Yields pretty blue color in ammonium perchlorate based compositions; but reacts with ammonium perchlorate and liberates ammonia in presence of moisture. The composition must be kept dry.
Purple	Combination of red and blue compounds	Sr + Cu	
Purple	Rubidium compounds	Rb	Rarely used
Silver/ White	Aluminium powder	Al	
Silver/ White	Magnesium powder	Mg	
Silver/ White	Titanium powder	Ti	
Silver/ White	Antimony (III) sulfide	Sb_2S_3	
Infrared	Caesium nitrate	$CsNO_3$	Two powerful spectral lines at 852.113 nm and 894.347 nm
Infrared	Rubidium nitrate	$RbNO_3$	

density is decreased, leading to "thinning" of the air. This thinning leads to the formation of various degrees of ozone within the stratosphere. This ozone layer acts as shield against some of the nonvisible emission of the sunlight. The high distribution of visible light, as reported earlier in this chapter, is possible in part due to the presence of this shield. Figure 6.33 shows how such protection is done with a clear and dark lens.

Picture 6.13 Fire from wood (top left) is part of the organic cycle, whereas smoke from a tungsten bulb (bottom right) is that of mechanical (hence implosive and non-sustainable) cycle.

This figure shows that how the presence of an even "transparent" lens can alter the wavelength spectrum significantly.

Above the mesosphere, the composition changes significantly, both in content and form. The overall composition is still dominated by nitrogen and oxygen, gases are highly ionized, and bond between oxygen atoms are broken. Conventional theories cannot explain these phenomena, but it is

Table 6.22 Various Elements in Earth Crust and Lithosphere

N	Element	Symbol	Lithosphere (ppm)	Crust (ppm)
8	Oxygen	O	460,000	460,000
14	Silicon [A]	Si	277,200	270,000
13	Aluminium	Al	81,300	82,000
26	Iron	Fe	50,000	63,000
20	Calcium	Ca	36,300	50,000
11	Sodium	Na	28,300	23,000
19	Potassium	K	25,900	15,000
12	Magnesium	Mg	20,900	29,000
22	Titanium	Ti	4400	6600
1	Hydrogen	H	1400	1500
15	Phosphorus	P	1200	1000
25	Manganese	Mn	1000	1100
9	Fluorine	F	800	540
56	Barium	Ba		340
6	Carbon [B]	C	300	1800
38	Strontium	Sr		360
16	Sulfur	S	500	420
40	Zirconium	Zr		130
74	Tungsten	W		1.1
23	Vanadium	V	100	190
17	Chlorine	Cl	500	170
24	Chromium	Cr	100	140
37	Rubidium	Rb	300	60
28	Nickel	Ni		90
30	Zinc	Zn		79
29	Copper	Cu	100	68
58	Cerium	Ce		60
60	Neodymium	Nd		33
57	Lanthanum	La		34
39	Yttrium	Y		29
7	Nitrogen	N	50	20
27	Cobalt	Co		30
3	Lithium	Li		17
41	Niobium	Nb		17
31	Gallium	Ga		19
21	Scandium	Sc		26
82	Lead	Pb		10
62	Samarium	Sm		6
90	Thorium	Th		6
59	Praseodymium	Pr		8.7
5	Boron	B		8.7
64	Gadolinium	Gd		5.2

Table 6.22 Various Elements in Earth Crust and Lithosphere—cont'd

N	Element	Symbol	Lithosphere (ppm)	Crust (ppm)
66	Dysprosium	Dy		6.2
72	Hafnium	Hf		3.3
68	Erbium	Er		3.0
70	Ytterbium	Yb		2.8
55	Caesium	Cs		1.9
4	Beryllium	Be		1.9
50	Tin	Sn	0	2.2
63	Europium	Eu		1.8
92	Uranium	U		1.8
73	Tantalum	Ta		1.7
32	Germanium	Ge		1.4
42	Molybdenum	Mo		1.1
33	Arsenic	As		2.1
67	Holmium	Ho		1.2
65	Terbium	Tb		0.94
69	Thulium	Tm		0.45
35	Bromine	Br		3
81	Thallium	Tl		0.530
71	Lutetium[7]	Lu		
51	Antimony	Sb		0.2
53	Iodine	I		0.490
48	Cadmium	Cd		0.15
47	Silver	Ag		0.080
80	Mercury	Hg		0.067
34	Selenium	Se		0.05
49	Indium	In		0.160
83	Bismuth	Bi		0.025
52	Tellurium	Te		0.001
78	Platinum	Pt		0.0037
79	Gold	Au		0.0031
44	Ruthenium	Ru		0.001
46	Palladium	Pd		0.0063
75	Rhenium	Re		0.0026
77	Iridium	Ir		0.0004
45	Rhodium	Rh		0.0007
76	Osmium	Os		0.0018

considered to be essential for earth's sustainability. In the exosphere, the outer layer of Earth's atmosphere, air molecules can easily escape the earth's gravity and float into space. This process is similar to atomic radiation, which can be captured as long as the artificial boundary between mass and energy is removed (Islam et al., 2014).

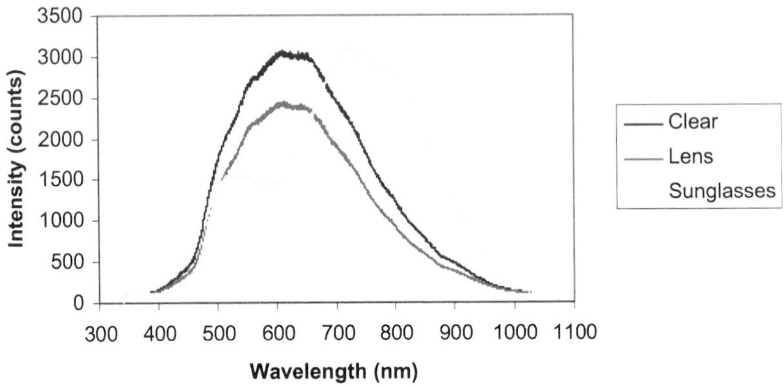

Figure 6.33 Formation of a shield with dark and clear lenses.

In this context, the composition of human body is important. Table 6.23 presents the elemental composition of a typical human body (70 kg). This table does not contain some trace elements. Through continuity, all elements of the earth crust should also be present in a human body. Interestingly, carbon is the 2nd most important component of a human body, followed by hydrogen, nitrogen, and calcium, etc.

Obliviously, human needs for various chemicals are met through breathing and consumption of food. The composition of the atmosphere shows that breathing alone would provide very little carbon, which has to be taken from plants. In this regard, the composition of plant is of utmost importance. The exact chemical composition of plants varies from plant to plant, and within different parts of the same plant. Chemical composition also varies within plants from different geographic locations, ages, climate, and soil conditions (Reimann et al., 2001; Shtangeeva, 1994). However, the most abundant chemical in plants as well as other living bodies is cellulose. The basic component of this chemical is sugar or carbohydrate. This also forms the basis for all petroleum products, irrespective of their physical state. Also, plants are known to show variable compositions in terms of Cd, V, Co, Pb, Ba, and Y, while maintain a surprisingly similar levels in all plants in some other elements, e.g., Rb, S, Cu, K, Ca, P, and Mg (Reimann et al., 2001).

Even though no evidence exists in nature that hydrogen combined with oxygen in their elemental form to produce water, it is commonly accepted that elemental balance in oxygen and hydrogen exists independently. This connection comes from the Big Bang theory that assumes

Table 6.23 Table of elements in the human body by mass

Element	Mass
Oxygen	43 kg (61%, 2700 mol)
Carbon	16 kg (23%, 1300 mol)
Hydrogen	7 kg (10%, 6900 mol)
Nitrogen	1.8 kg (2.5%, 129 mol)
Calcium	1.0 kg (1.4%, 25 mol)
Phosphorus	780 g (1.1%, 25 mol)
Potassium	140 g (0.20%, 3.6 mol)
Sulfur	140 g (0.20%, 4.4 mol)
Sodium	100 g (0.14%, 4.3 mol)
Chlorine	95 g (0.14%, 2.7 mol)
Magnesium	19 g (0.03%, 0.78 mol)
Iron	4.2 g
Fluorine	2.6 g
Zinc	2.3 g
Silicon	1.0 g
Rubidium	0.68 g
Strontium	0.32 g
Bromine	0.26 g
Lead	0.12 g
Copper	72 mg
Aluminum	60 mg
Cadmium	50 mg
Cerium	40 mg
Barium	22 mg
Iodine	20 mg
Tin	20 mg
Titanium	20 mg
Boron	18 mg
Nickel	15 mg
Selenium	15 mg
Chromium	14 mg
Manganese	12 mg
Arsenic	7 mg
Lithium	7 mg
Cesium	6 mg
Mercury	6 mg
Germanium	5 mg
Molybdenum	5 mg
Cobalt	3 mg
Antimony	2 mg
Silver	2 mg

Continued

Table 6.23 Table of elements in the human body by mass—cont'd

Element	Mass
Niobium	1.5 mg
Zirconium	1 mg
Lanthanum	0.8 mg
Gallium	0.7 mg
Tellurium	0.7 mg
Yttrium	0.6 mg
Bismuth	0.5 mg
Thallium	0.5 mg
Indium	0.4 mg
Gold	0.2 mg
Scandium	0.2 mg
Tantalum	0.2 mg
Vanadium	0.11 mg
Thorium	0.1 mg
Uranium	0.1 mg
Samarium	50 μg
Beryllium	36 μg
Tungsten	negligible

From Emsley, 1998.

that the original mass was hydrogen. This new version of atomism has been challenged by several researchers and remains a subject of ongoing debate (Islam et al., 2014a). In every cycle, however, there are components that cannot be accounted for with conventional scientific analysis. Figure 6.35 shows how oxygen cycle is complete within the echo system. In every step, there is involvement of living organism. That itself is a matter of intangible as "life" cannot be quantified or even qualified and is inherently intangible. The first reaction identified in the following figure is photolysis. This is a term coined to include the role of sunlight in sustaining the terrestrial ecosystem. Photolysis is part of the light-dependent reactions of photosynthesis. The general reaction of photosynthetic photolysis can be given as

$$H_2A + 2\ \text{photons(light)} \rightarrow 2e^- + 2^+ + A + \Sigma \tag{6.18}$$

The chemical nature of "A" depends on the type of organism. For instance, in purple sulfur bacteria, hydrogen sulfide (H_2S) is oxidized to sulfur (S). In oxygenic photosynthesis, water (H_2O) serves as a substrate for photolysis resulting in the generation of diatomic oxygen (O_2). The Σ

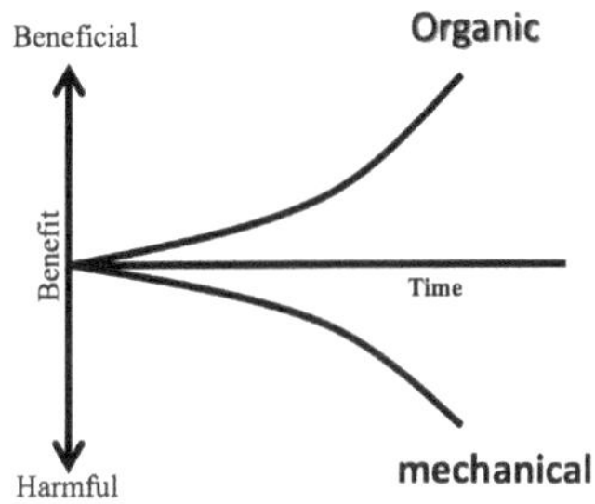

Figure 6.34 Benefit to environment depends entirely on the organic nature of energy and mass.

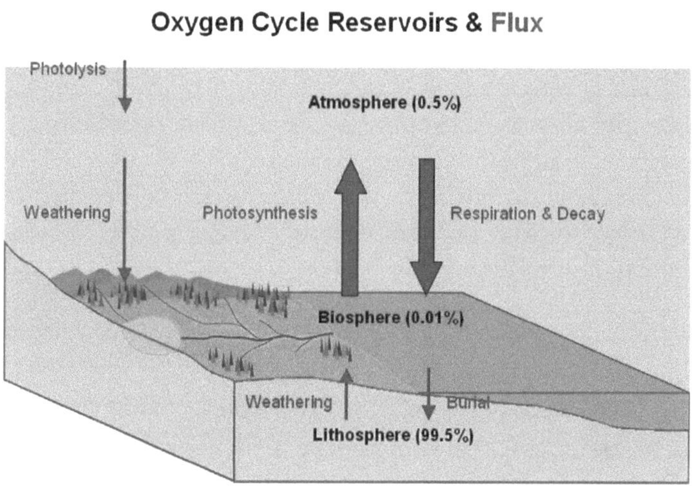

Figure 6.35 Oxygen cycle in nature involving the earth.

symbol includes information about the pathway, *f*(t), for the photons. For instance, for sunlight it would be something intangible that is beneficial in the long term and for artificial light, it would be something intangible that is harmful in the long term (Figure 6.34). This is the process that returns oxygen to earth's atmosphere (Figure 6.35). Photolysis of water occurs in the thylakoids of cyanobacteria and the chloroplasts of green algae and plants.

Photosynthesis is the next process that involves the sunlight. Similar to photolysis, photosynthesis also involves living organisms. Photosynthetic organisms are photoautotrophs, which means that they are able to synthesize food directly from carbon dioxide and water using energy from light. However, not all organisms that use light as a source of energy carry out photosynthesis, since photoheterotrophs use organic compounds, rather than carbon dioxide, as a source of carbon. In plants, algae and cyanobacteria, photosynthesis releases oxygen. This is called oxygenic photosynthesis. However, there are some types of bacteria that carry out an oxygenic photosynthesis, which consumes carbon dioxide but does not release oxygen. In that case, they act as an oxygen sink.

Carbon dioxide is converted into "sugars" in a process called carbon fixation. Carbon fixation is an endothermic redox reaction, so photosynthesis needs to supply both a source of energy to drive this process and the electrons needed to convert carbon dioxide into a carbohydrate. This addition of the electrons is a reduction reaction. In general outline and in effect, photosynthesis is the opposite of cellular respiration, in which "glucose" and other compounds are oxidized to produce carbon dioxide and water, and to release exothermic chemical energy to drive the organism's metabolism. In this process, the intangibles are captured by another set of Σ_2. This symbol contains two sets of information, one regarding the source of carbon dioxide and the other regarding the source of light. The general equation for photosynthesis is:

$$\begin{aligned}&\text{Carbon dioxide} + \text{electron donor} + \text{light energy} \rightarrow \text{carbohydrate}\\&\quad + \text{oxidized electron donor} + \Sigma_2\\&\text{or,}\\&2n\ CO_2 + 2n\ DH_2 + \text{photons} \rightarrow 2(CH_2O)_n + 2n\ DO + \Sigma_2\end{aligned} \tag{6.19}$$

In *oxygenic* photosynthesis water is the electron donor and, since its hydrolysis releases oxygen, the equation for this process is:

$$\begin{aligned}&\text{Carbon dioxide} + \text{water} + \text{light energy} \rightarrow \text{carbohydrate} + \text{oxygen}\\&\quad + \text{water} + \Sigma_2\\&\text{Or,}\\&2n\ CO_2 + 2n\ H_2O + \text{photons} \rightarrow 2(CH_2O)_n + 2n\ O_2 + \Sigma_2\end{aligned} \tag{6.20}$$

Figure 6.36 shows the cycle involving hydrogen balance. Figure 6.37 shows the overall water balance. Scientifically, water balance is the only natural balance.

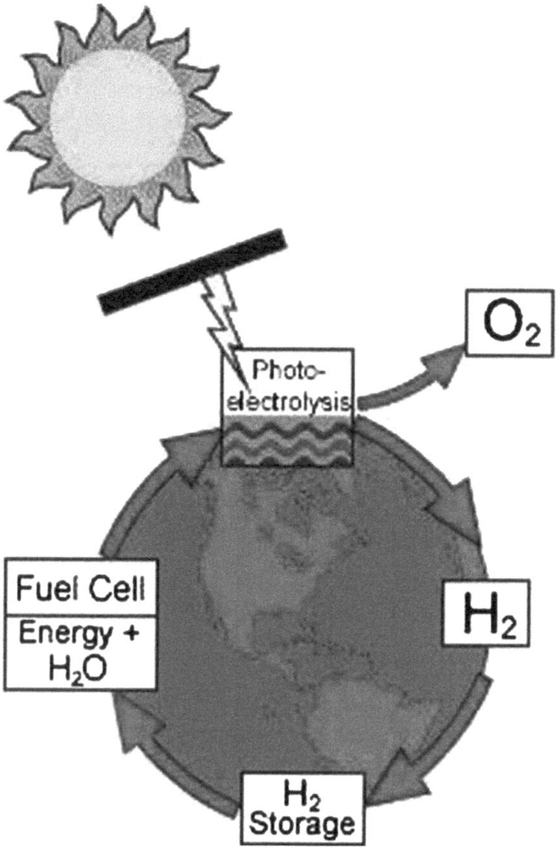

Figure 6.36 Hydrogen cycle in nature involving the earth.

6.10 ORGANIC ORIGIN OF PETROLEUM

The most notable groups of chemicals used in the processes of living organisms include:

Proteins, which are the building blocks from which the structures of living organisms are constructed (this includes almost all enzymes, which catalyze organic chemical reactions)

Nucleic acids, which carry genetic information

Carbohydrates, which store energy in a form that can be used by living cells

Lipids, which also store energy, but in a more concentrated form, and which may be stored for extended periods in the bodies of animals.

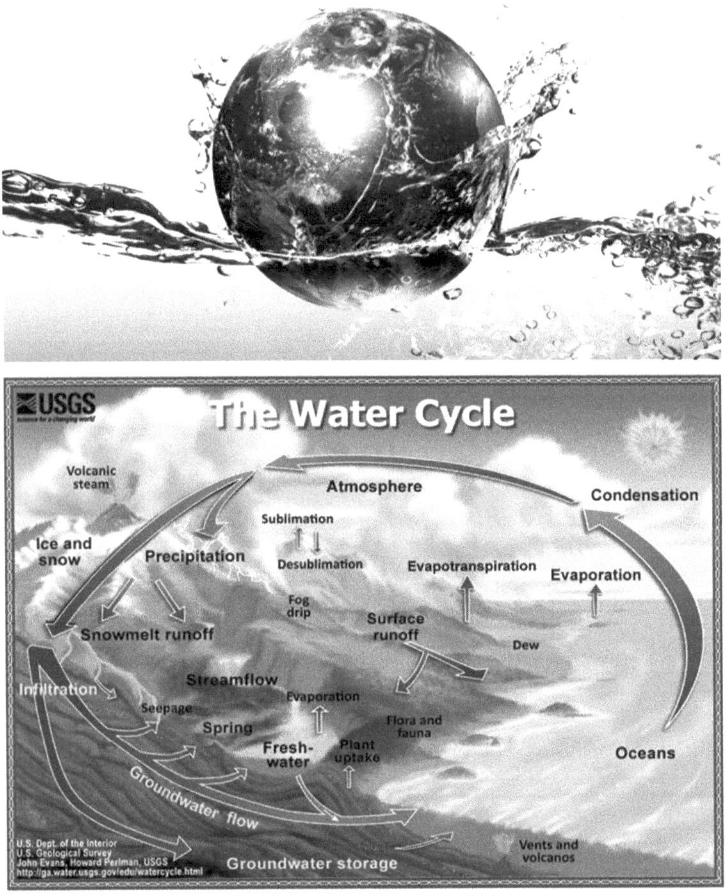

Figure 6.37 Water cycle, involving energy and mass.

Silicon has been a theme of noncarbon-based life since it also has four bonding sites and is just below carbon on the periodic table of the elements. This means silicon is very similar to carbon in its chemical characteristics. In cinematic and literary science fiction, when artificial machines cross from nonliving to living, this new form would be an example of noncarbon-based life. Since the advent of the microprocessor in the late 1960s, these machines are often classed as "silicon-based life." Another example of "silicon based-life" is the episode "The Devil in the Dark" from Star Trek: The Original Series, where a living rock creature's biochemistry is based on silicon.

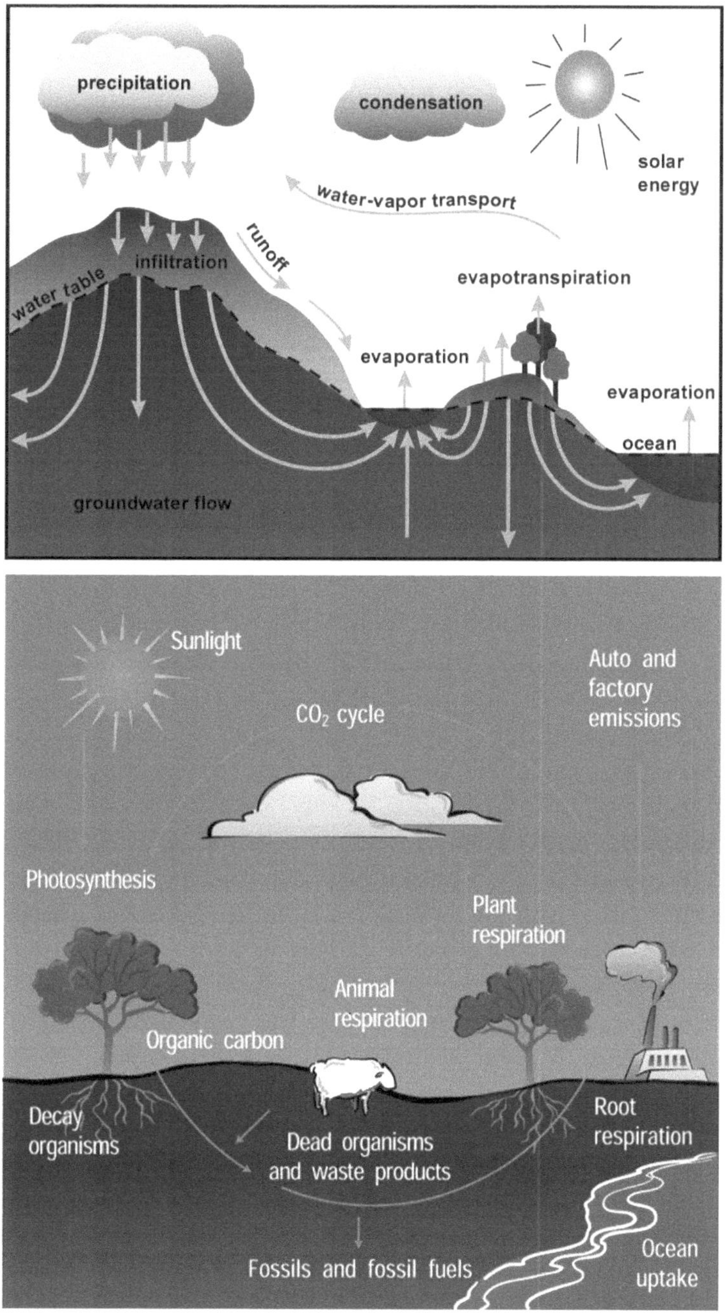

Figure 6.37 Cont'd.

Scientifically, natural processing of organic materials into petroleum products is entirely environment-friendly as each stage of processing is well balanced through proper use of characteristic frequencies (Islam et al., 2014). It is commonly agreed that petroleum has an organic origin. Organic materials are transformed by heat and pressure into a complex mixture, known as kerogen. Depending on the initial ingredients and the geologic conditions, kerogen can produce either coal (a solid carbon-rich fuel derived mostly from woody plants) or hydrocarbons (a relatively hydrogen-rich substance that comes from algae and various lipid-containing plant parts). Oil is formed from kerogen, a mixture of organic compounds in sedimentary rocks. It is most abundant in shales. Shales are analyzed and characterized as potential "source rocks" for oil based largely on their total organic content.

For kerogen to be transformed into oil, it must be buried to a depth where the temperature and pressure are sufficiently high to convert the kerogen into oil. The place where the depth is sufficient to achieve this, is called the "oil window." Oil shale which contains only kerogen was never buried deeply enough to generate oil. Therefore, humans must artificially convert the kerogen in oil shale into oil at high temperatures. The processing time of petroleum is longer than a million years. Such long processing time makes petroleum fluids the most stabilized and environmentally benign fluid other than water. Petroleum is equivalent to natural batteries that pack solar energy in the most efficient and environment-friendly way known to mankind. In contrast, pyrolysis is inherently artificial as it uses artificial sources of energy (heat and pressure) and often contains catalysts (artificial chemicals). As a consequence, pyrolysis leaves behind a trail of heavy foot prints and sets off negative impact on the environment, even though the final products of pyrolysis and natural processing are similar (e.g., methane, ethane, propane, and others). In terms of environmental impact, naturally processed petroleum products are completely benign to the environment. History tells us that the usage of petroleum products in their natural state does not endanger the environment. At the onset of modern age, starting with the mechanization of natural science and engineering, a bifurcation between economic index (driver of the modern civilization) and environmental index has taken place. Using 2000 as the reference point, Vassilis et al. (2013) presented the following graph that shows the nature of this bifurcation (Figure 6.38).

It is, therefore, unfathomable that such modus operandi had been in place in previous civilizations that were no less glamorous than the modern

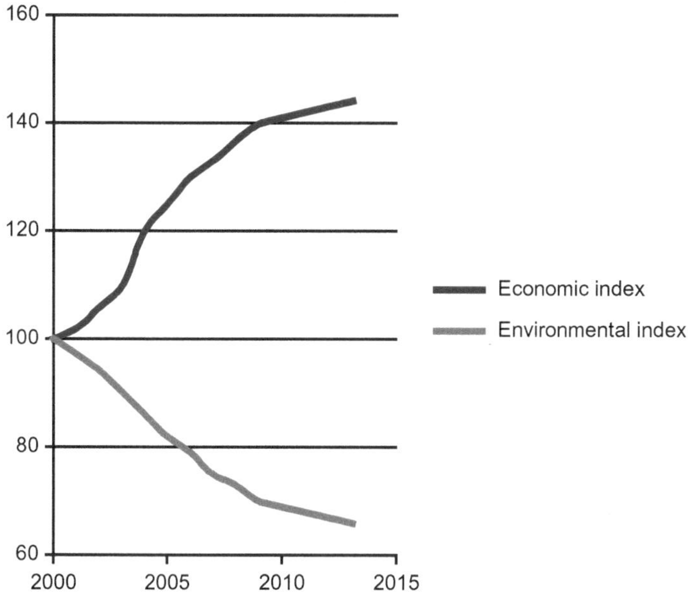

Figure 6.38 Even is short-term, modern age is synonymous with decoupling of economic index from environmental index.

age. The most remarkable distinction is in the use of petroleum as a fuel. Today, 85% of petroleum use in the United States is as a fuel. Nonfuel applications include lubricants for cars, asphalt for roads, tars for roofing, waxes for food wrapping, as well as solvents for paints, cosmetics, and dry cleaning products. Petrochemicals also provide the building blocks for a vast panoply of plastics and foams. Pharmaceutical industry also uses petroleum plastics as well as chemicals derived from petroleum. In each application, however, numerous stages of refinement and processing are involved.

Originally, petroleum played an entirely different role. There were oil pits near Ardericca (near Babylon). the Chinese drilled for "rock oil" to provide heating and lighting, and the Byzantines sprayed "Greek fire" as an incendiary weapon.

Many cultures have also employed petroleum as a medicinal cure, giving it names such as "St Quirinus oil," "Barbados tar," and "Seneca oil." Our modern society continues to use petroleum jelly as a skin ointment.

Petroleum became a significantly valuable commodity in the mid-ninteenth century, when modern techniques of chemical distillation were developed to separate kerosene from crude oil. The refined kerosene could be used in lamps, replacing the more expensive whale oil.

Natural state of petroleum fluids has numerous qualities that make it an ideal antidote for many applications, including medicinal. It is difficult to characterize these qualities within a single index because the only commonality they have is natural processing. Figure 6.39 shows wide range of variability in crude oil characteristics.

The use of petroleum products in its natural state has been there for thousands of years. Petroleum has been used for thousands of years. The ancient Babylonians built walls and towers with asphalt. City walls were constructed to include gates and watchtowers and usually a ditch running around the outer perimeter of the wall, which could be filled with water. King Hammurabi surrounded his city of Babylon with more impressive walls than usually seen shortly after he assumed the throne in 1792 BC, but the credit for transforming the city of Babylon into an awe-inspiring wonder belongs to King Nebuchadnezzar II. Nebuchadnezzar built three walls around Babylon at heights of 40 feet and so broad at the top that chariots could race around them. The Ishtar Gate in the wall of Nebuchadnezzar II's Babylon was claimed by some to be greater than any of the listed Wonders of the Ancient World. Heavy petroleum components are

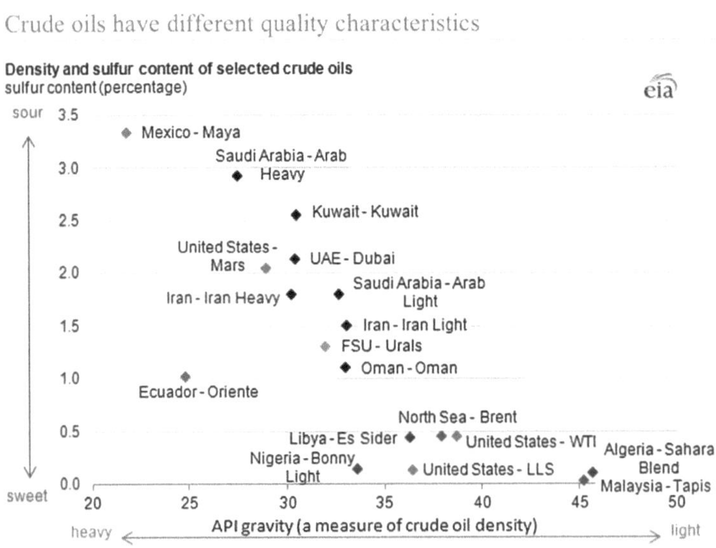

Figure 6.39 Crude oil characteristics vary widely, making it difficult to characterize with a single criterion.

effective cementing materials and this property of crude petroleum is rarely exploited today.

Archeologists have uncovered tools used by Neanderthals dating back some 40,000 years ago. Neanderthals used oil from naturally occurring oil seeps to adhere stones to sticks.

As human society developed, so did the use of oil and natural gas. Around 3000 BC, man was using oil for the construction of buildings and waterproofing canoes, baskets, baths, and drains.

In ancient Mesopotamia, natural asphalt was used in the construction of the walls and towers of Babylon.

Around 1272 BC, Marco Polo, while traveling through the city of Baku (Azerbaijan), witnessed oil seeps being worked in hand-dug wells.

Natural gas was being used too.

Around 1000 BC, a goat herdsman in Greece came across what looked like a burning stream at the top of Mount Parnassus. It was a flame rising from a crack in the rock. Most likely, a lightening strike ignited the natural gas.

The ancient Greeks, believing the flame to be of divine origin, built a temple around it. This temple housed a priestess who was known as the Oracle of Delphi, giving out prophecies she claimed were inspired by the flickering flame.

Fast-forward 2000 years to 1821, when the first natural gas well was drilled in the United States.

William Hart drilled a 27-foot well in Fredonia, New York. How did he know where to drill? He asked the Native Americans who worshipped the naturally occurring gas flowing out of "burning creek."

Hart's success lead to the first-ever utility company in America, called the Fredonia Gas Light Company.

And ever since, oil and gas companies have used the observation of naturally occurring seeps to find massive oilfields.

That's because typically where you see a seep, you find a highly-pressurized reservoir below. The high quality oil is literally being pushed out of the ground.

How did Captain Drake know where to drill in Titusville? Well it was actually quite easy for him. Since Drake did not have the technological advantages of satellite images or 3D seismic studies, he took the easiest path for discovering oil: He looked down at the ground.

Titusville is a small town that sits near a creek. The name of the creek? Oil Creek.

For centuries, the Seneca tribe used the oil that seeped into the creek for many things, including cures for sickness.

For the next 40 years after Drake struck oil, Pennsylvania would supply the world one-third of its oil.

Then came Spindletop in 1901.

Before the legendary Spindletop field was drilled in 1901, Texas residents had for years witnessed thousands of "tar balls" and miles-long oil slicks right off shore in the Gulf of Mexico.

Spindletop quickly became the largest oilfield in America, larger than all of the previous oil fields in the United States, combined.

And it occurs all around the world. The second-largest oilfield in the world—Kuwait's Burgan—was discovered by British geologists who observed hundreds of oil seeps at the surface in 1912.

To this day, oil seeps are spewing crude in deserts, mountains, rivers, and oceans.

6.11 PLACEMENT OF UNCONVENTIONAL GAS IN THE ENERGY PICTURE

This section is aimed at developing a protocol for scientific characterization of unconventional gas. Instead of following conventional protocol of delineating reservoirs based on permeability, a new approach is taken. In this, US reserves are used as a prototype and USGS report (2008) on US reserve growth is used as starting point. Both siliciclastic (largely sandstone) and carbonate (limestone and dolomite) reservoirs in North America are considered. Initial division is made based on geological environment, namely:

1. Eolian environments (Jurassic Norphlet Formation of the Gulf Coast and Pennsylvanian–Permian Minnelusa Formation of the Powder River Basin);
2. Interconnected fluvial, deltaic, and shallow marine environments (Oligocene Frio Formation of the Gulf Coast and the Pennsylvanian Morrow Formation of the Anadarko and Denver basins);
3. Deeper marine environments (Mississippian Barnett Shale of the Fort Worth Basin and Devonian-Mississippian Bakken Formation of the Williston Basin);
4. Marine carbonate environments (Ordovician Ellenburger Group of the Permian Basin and Jurassic Smackover Formation of the Gulf of Mexico Basin);

5. Submarine fan environment (Permian Spraberry Formation of the Midland Basin);
6. A fluvial environment (Paleocene-Eocene Wasatch Formation of the Uinta-Piceance Basin).

Reservoirs in each formation were further subdivided into categories, as appropriate, where the reservoirs had sufficiently different geological attributes to warrant separate treatment. For each category, the potential of unconventional gas and oil is considered. Important variables considered are source rock, depositional setting and then postdepositional alteration. Type of traps or seals is left under a different category that is responsible for engineering decisions, rather than scientific characterization.

As discussed in previous chapters, origin of fluid as well as rock must be tracked in order to characterize gas resources. In order to determine origin of fluids, the connection between petroleum history and geological background is highlighted. This analysis also helped refine basis for reservoir heterogeneities that have different impacts on gas trapping as well as gas quality. Production heterogeneity itself serves as an indicator to the origin of fluid and helps one determine protocol for developing delinearized history of petroleum.

Conventionally, reserve growth has been attributed to homogeneity of production history as well as reservoir rocks. It is perceived that reserve growth in existing fields is most predictable for those in which reservoir heterogeneity is low and thus production differs little between wells, probably owing to relatively homogeneous fluid flow. In fields in which reservoirs are highly heterogeneous, prediction of future growth from infill drilling is notably more difficult. If the same criterion is used to determine potential reserve of unconventional gas or to characterize such reservoirs, one is faced with inherent flaws. Unconventional resources thrive in heterogeneous formations as well as fluids of previously undefined origins. As discussed in previous chapters and in previous sections of this chapter, scientific characterization is all about paradigm shift and the conventional paradigm places unconventional reserves at a disadvantage and falsifies predictions. This section is aimed at lifting that barrier and determining true potential of unconventional resources.

6.11.1 Reserve Growth Potential of Unconventional Gas

Historically, the majority of additions to oil and gas reserves are attributed to growth of existing fields and reservoirs. In fact, from 1978 to 1990, growth of known fields in the United States accounted for more than 85% of known additions to proven reserves (Root and Attanasi, 1993; McCabe, 1998). Thus, field growth and reserve growth are essentially synonymous for discussions of domestic resources. The exception occurred when unconventional resources became exploitable. However, it would be illogical to apply the principle of conventional reserve growth to unconventional reserves that have little in common with the conventional reserves in terms of technology. Even though, it is promoted that unconventional gas and oil are of lesser quality and are in need of more expensive technology for future development, previous chapters demonstrated that notion to be false. In fact, the quality of gas improves as the quantity of the resource increases, the case in point being hydrate. It is also true, vast amount of unconventional gas can be exploited with little or no exploration. In addition, many existing conventional reservoirs have links to unconventional resources and even the abandoned fields can be accessed for tapping the unconventional resources.

While evaluating the nature of growth in fields requires understanding of both geologic and nongeologic factors that affect growth estimations, geology is the pivotal factor in determining accumulation, quality, and producibility of petroleum resource. Fields may grow when

1. additional geologic data on existing reservoirs become available and are used to identify new reservoirs or to guide infill drilling;
2. there are annual updates of reserves data;
3. field boundaries are extended;
4. recovery technology is improved; and/or
5. nongeologic factors such as economics, reporting policies, or politics favor expanded production and development.

Past reserve estimates and characterization both used mathematics and not science as the basis. Attanasi et al. (1999) wrote:

> *"...the modeling approach used by the USGS (U.S. Geological Survey) to characterize this phenomenon is statistical rather than geologic in nature."*

Islam et al. (2010) discussed in details the short-comings of these statistical models and characterized them as unscientific. Yet, volumetric estimates of reserve growth are calculated by using these mathematical approaches and large data bases that record field reserves through time. Crovelli and Schmoker (2001), Verma (2003), and Klett (2003) present details of various methods used to estimate reserve growth. Application of these models to unconventional gas is of particular concern. It is because one of the most important basis for statistical models is the past history, which is practically nonexistence for unconventional reserves.

Geologists tend to think in terms of entire reservoirs, in some cases down to facies level, whereas reservoir engineers deal with measurements at the wellbore. To fully understand the geologic factors that affect growth in reserves, this gap in investigative approaches must be bridged. It has become increasingly important to integrate different scales and different observational techniques as secondary and tertiary recovery methods are applied more frequently in mature petroleum provinces.

Fluid-flow pathways, governed predominantly by rock porosity and permeability, are a reflection of heterogeneities of varying scales within a reservoir. Because these reservoir heterogeneities are fundamentally geologic in 2nature (Hamilton et al., 1998; Dyman et al., 2000), an adequate understanding of the reservoir architecture, obtained through evaluation of geologic, engineering, or production data, or a combination of these data sets, can provide the basis for scientific characterization of unconventional gas reservoirs.

Proper estimates and growth potential of unconventional gas depends on geological classification and linking of conventional resources with unconventional ones through scientific characterization. This is done in the following sections.

6.11.2 Reservoir Categories in the United States

US Geological Survey (2008) evaluated the geologic factors that affect reserve growth in both siliciclastic (largely sandstone) and carbonate (limestone and dolomite) reservoirs. This study included 10 formations in the United States (one of which extends into southern Canada) that represent various depositional environments in both siliciclastic and carbonate settings. This is shown in Table 6.24.

Reservoirs were then categorized based on geological criteria, such as source rock, depositional setting, and postdepositional alteration. The following environments were studied:

1. Eolian environments—Norphlet Formation of the Gulf of Mexico Basin (Figure 6.40) and Minnelusa Formation of the Powder River Basin (Figure 6.41);
2. Interconnected fluvial, deltaic, and shallow marine environments—Frio Formation of the Gulf of Mexico Basin (Figure 6.42) and Morrow Formation of the Anadarko and Denver Basins (Figure 6.41);

Table 6.24 Depositional Environments and Rock Units Selected for Study of Reserve Growth, and Geologic Age and General Location of Units

Depositional environment and formation studied	Age	General location
Eolian sandstone Norphlet Formation Minnelusa Formation	Upper Jurassic Pennsylvanian-Permian	Gulf of Mexico Basin Powder River Basin
Fluvial or deltaic—shallow marine Frio Formation Morrow Formation	Tertiary (Oligocene) Pennsylvanian (Morrowan)	Gulf of Mexico Basin Anadarko and Denver Basin
Marine shale Barnett Shale Bakken Formation	Mississippian (Chesterian) Devonian-Mississippian	Fort Worth Basin Williston Basin
Marine carbonates Ellenburgcr Group Smackovcr Formation	Ordovician (Early Ordovician) Upper Jurassic (late Oxfordian)	Permian Basin
Submarine sands Spraberry Formation	Permian (Leonardian)	Gulf of Mexico Basin Permian Basin
Nonmarine fluvial—deltaic Wasatch Formation	Tertiary (Paleocene—Eocene)	Unita-Piceance Basin

3. Deeper marine environments—Barnett Shale of the Fort Worth Basin (Figure 6.42) and Bakken Formation of the Williston Basin (Figure 6.41);
4. Marine carbonate environments—Ellenburger Group of the Permian Basin (Figure 6.42) and Smackover Formation of the Gulf of Mexico Basin (Figure 6.40);
5. Submarine fan environment—Spraberry Formation of the Midland Basin (Figure 6.42); and
6. Fluvial environment—Wasatch Formation of the Uinta-Piceance Basin (figure 6.41).

6.11.2.1 Eolian Reservoirs

6.11.2.1.1 Norphlet Formation

The Middle to Upper Jurassic Norphlet Formation of the Gulf of Mexico Basin consists largely of eolian sandstones, with minor black shale,

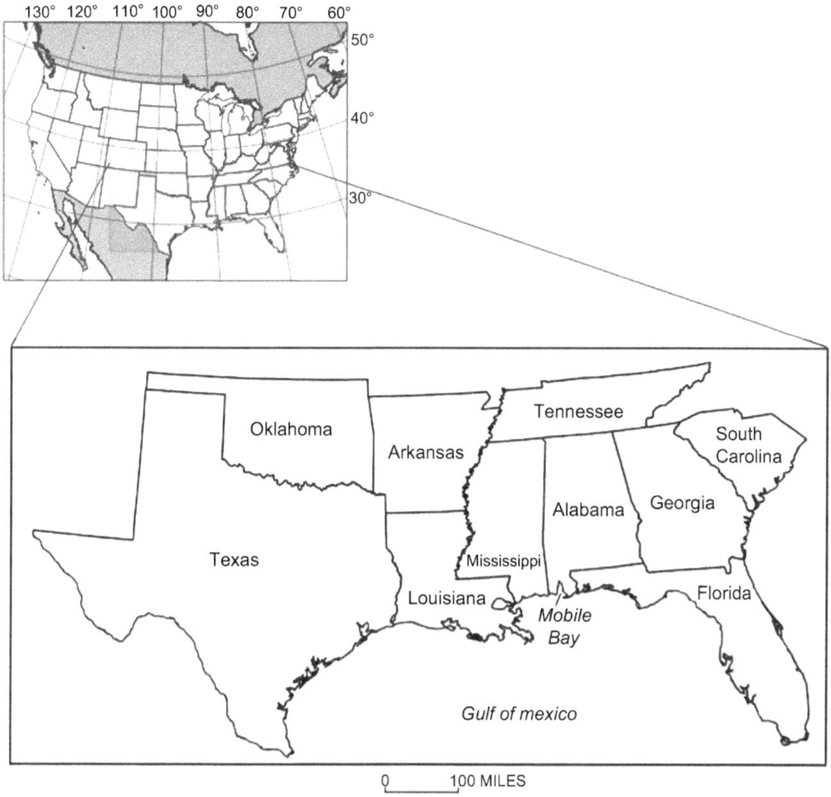

Figure 6.40 Gulf of Mexico Basin region, the petroleum-producing region of the Norphlet and Smackover Formations. Both formations produce in both onshore and offshore locations; the Norphlet produces from Mobile Bay. *From USGS, 2008.*

conglomerate, and red beds; thicknesses are as much as 100 ft. The Norphlet produces oil and gas largely in Alabama, offshore in Mobile Bay, and in Mississippi (Figure 6.40). Principal reservoirs in the Norphlet are eolian sandstones (Table 6.25), which are known to have excellent porosity (as much as 20%) and permeability (as much as 500 mD).

As can be seen from Table 6.25, broad similarities in reservoir characteristics throughout the area of production suggest that only a single reservoir category is warranted. This reservoir was considered to be a single one with homogeneous rock and fluid properties. While such reservoir is not considered to be a source of unconventional gas, this reservoir offers the least expensive access to vast resource that is not conventionally included in all analyses. For instance, any cutoff point of shale in shale breaks or

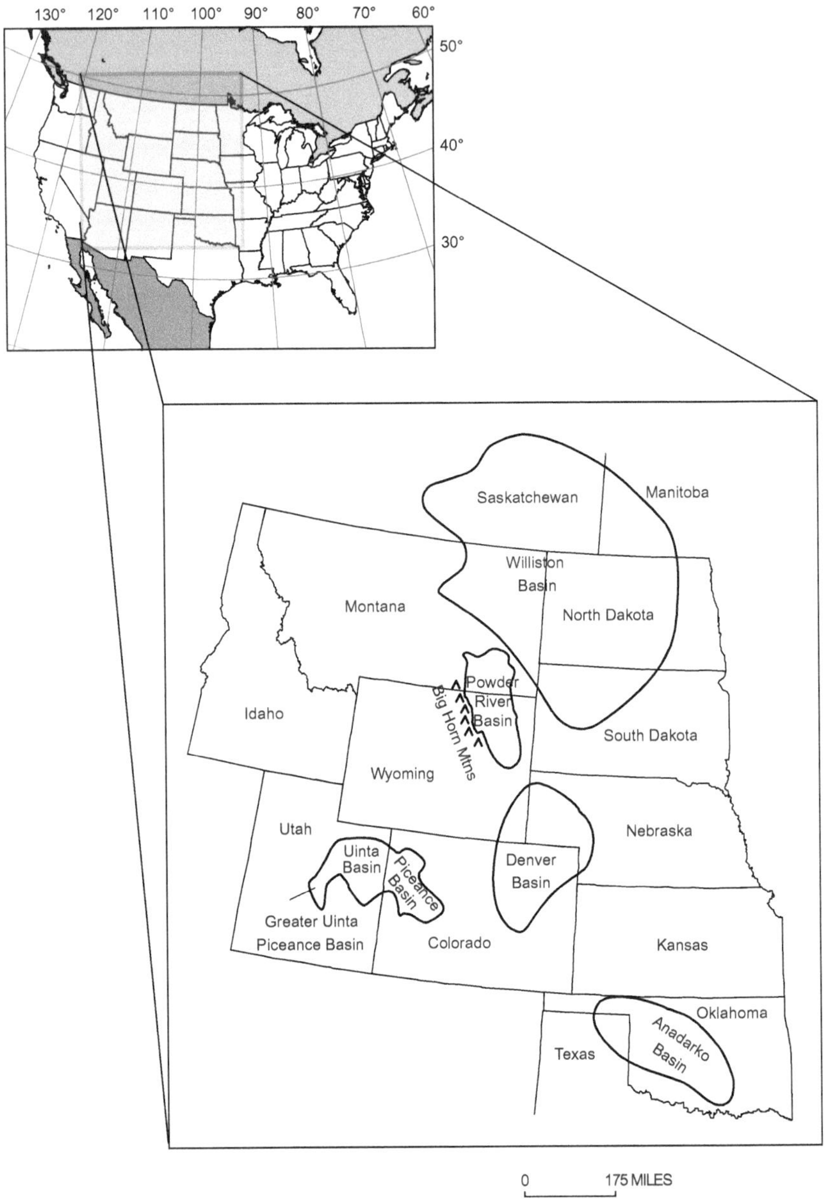

Figure 6.41 General region from which petroleum is produced from formations discussed in this paper, including the Minnelusa (Powder River Basin), Morrow (Anadarko and Denver Basins), Bakken (Williston Basin), and Wasatch (Uinta and Piceance Basins) Formations. *From USGS, 2008.*

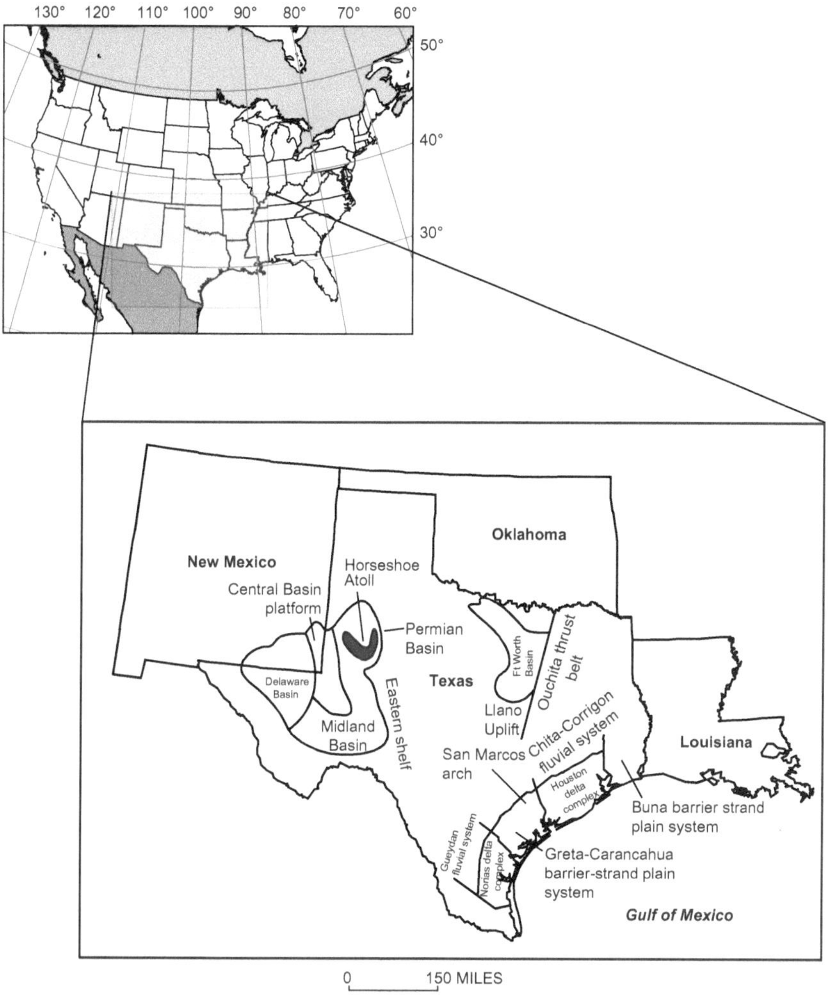

Figure 6.42 Area from which petroleum is produced from the Frio Formation, Barnett Shale, Ellenburger Group, and Spraberry Formation. Extent of depositional environments in the Frio (such as the Norias delta complex or the Buna barrier–strandplain) *(From Galloway et al. (1982))*. For the Barnett, the locations of the Llano uplift and Ouachita thrust belt mark the southern and eastern limits of the Fort Worth Basin, respectively. Horseshoe Atoll is a Pennsylvanian structure that effectively separates productive rocks of the Spraberry Formation (to the south) from nonproductive rocks (to the north). *From USGS, 2008.*

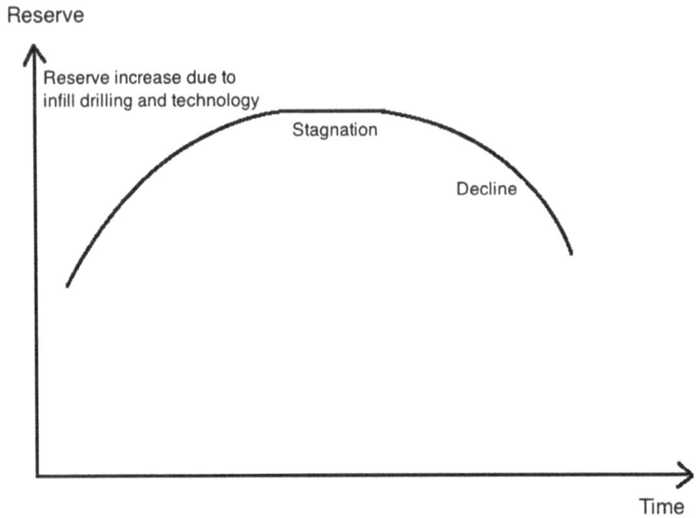

Figure 6.43 Three phases of conventional reserve.

caprocks eliminate as much as 10% of the reserve that lies within low-permeability, low effective-porosity formations. As conventional reserves become insensitive to infill drilling or other means (including enhanced oil/gas recovery), the potential of unconventional gas increases. Figure 6.43 shows general trend in reserve for conventional reservoirs. Previously, extensive exploration has kept the growth part active. In United States, however, main contributor to reserve growth has been infill drilling and enhanced oil recovery. In the infill drilling scheme, one must acknowledge 90% of the new wells are horizontal. This means the role of horizontal well technology has been embedded in this reserve growth. However, despite best efforts the reserve will decline, particularly for the "homogeneous" formations.

For such reservoirs, a vast amount of unconventional reservoirs are associated and are accessible readily. This has been the case for shale gas as well as tight gas, as has been evidenced in recent gas boom in the United States. However, the use of horizontal wells and hydraulic fracturing as the sole mode of reservoir development will lead to similar stagnation as in conventional reservoirs (Figure 6.44). This can be given a boost by accessing reserve that was previously considered to be part of conventional reservoirs and was excluded through the use of "cutoff" points. The dotted line in Figure 6.44 shows how immediate boost can be invoked. This can be followed by subsequent use of enhanced gas recovery schemes, as

Table 6.25 Norphlet Formation, Gulf of Mexico Basin—Summary of Geological Characteristics and Reserve Growth Potential of Reservoirs

	Depositional characteristics			Reservoir characteristics				
					Porosity (bulk rock)			
Reservoir category	Environment	Reservoir facies	Nonreservoir facies	Lithology	Principal pore space	Diagenetic enhancement	Diagenetic occlusion	Porosity
Norphlet	Sand sea	Eolian sands	Overlying and interbedded marine shale and interdune sediments	Sandstone	Primary intergranular and secondary intergranular and moldic	Dissolution of early authigenic cements and authigenic chlorite	Local quartz, anhydrite, halite, illite. Intense quartz cementation may seal some accumulations	As much as 20% in onshore reservoirs and 12% in deeper offshore reservoirs

Reservoir characteristics—Continued			Stratigraphic controls		Structural controls		
Permeability	Fractures	Source rock	Reservoir location	Traps or seals	Reservoir location	Traps or seals	Oil or gas
Generally high; as much as 500 mD	May be complexity faulted	Overlying marine shale of Smackover Formation; interbedded or interfingering organic-rich shale in Norphlet Formation	Updip pinchout against basement complex	Overflying shale and interbedded interdune, sabkha, or playa units	Reservoir rocks thicken in basement controlled grabens and are absent or thin over basement controlled highs	Anticlines, faulted anticlines, faults associated with basement structures, and halokinesis of Louann Salt	Dominantly nonassociated gas (cracked) and minor oil

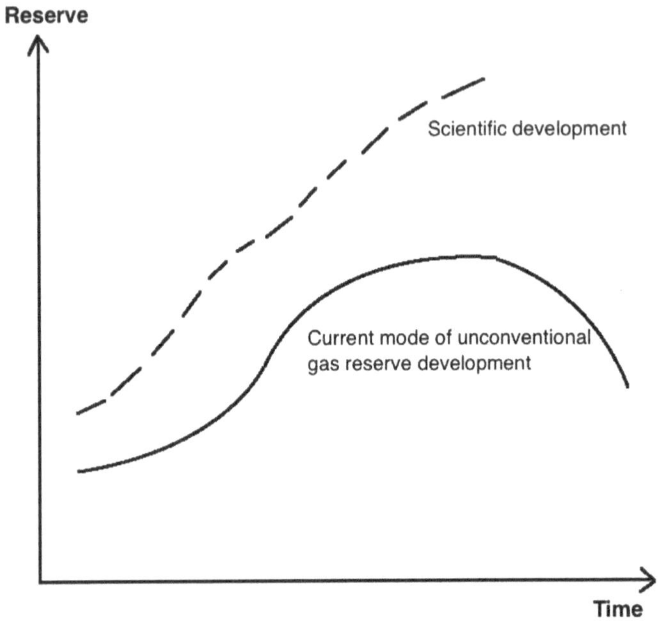

Figure 6.44 Unconventional reserve growth can be given a boost with scientific characterization.

outlined in previous chapters of this book. Ironically, the greatest potential for most economic recovery of unconventional gas lies within high productivity, "homogeneous" formations as exemplified in this category of Norphlet Formation.

6.11.2.1.2 Minnelusa Formation

The Pennsylvanian to Early Permian Minnelusa Formation of the Powder River Basin, northeastern Wyoming, consists largely of eolian sandstones, with minor shale and carbonate; thicknesses are as much as 1200 ft. Most production is in the north–central and northeastern parts of the basin; lesser production is in the southerly and southeastern parts (Figure 6.41). Principal reservoirs are the eolian sandstones (Table 6.26), which can have excellent porosity (as much as 47%) and permeability (as much as 830 mD).

Reservoirs in the Minnelusa Formation are placed into two categories, Minnelusa and Leo (Table 6.26). This twofold division was necessary because of differences in stratigraphic position, depositional environment, and geographic distribution of producing wells; in addition, reservoirs in the two categories may have different source rocks. Reservoir rocks of the Leo category have been variously referred to by previous workers as the

Table 6.26 Minnelusa Formation, Powder River Basin—Summary of Geological Characteristics and Reserve Growth Potential of Reservoirs

	Depositional characteristics			Reservoir characteristics				
					Porosity (bulk rock)			
Reservoir category	**Environment**	**Reservoir facies**	**Nonreservoir facies**	**Lithology**	**Principal pore space**	**Diagenetic enhancement**	**Diagenetic occlusion**	**Porosity**
Minnelusa	Coastal sand sea	Eolian dunes	Overlying shallow marine shales, anhydrite, and carbonate rocks	Sandstone, quartz arenite, sublitharenite	Primary and secondary intergranular; moldic	Dissolution of early authigenic cements and of some unstable denial grains	Quartz, carbonates minerals, and anhydrite/gypsum where not dissolved. Cemented zones may act as seals	Averages 12–24% but may be as high as 47%
Leo	Coastal dunes	Eolian dunes	Overlying shallow marine shales, anhydrite, and carbonate rocks	Sandstone, quartz arenite, sublitharenite	Primary and secondary intergranular; moldic	Dissolution of early authigenic cements and of some unstable denial grains	Quartz, carbonates minerals, and anhydrite/gypsum where not dissolved. Cemented zones may act as seals	Averages 12–24%

"Leo sandstone" (Hunt, 1938), "Leo section" (Desmond et al., 1984), "Leo Formation" (Morel et al., 1986), or the "Leo sandstone of the Minnelusa Formation" (Dolton and Fox, 1995).

6.11.2.2 Interconnected Fluvial, Deltaic, and Shallow Marine Reservoirs

6.11.2.2.1 Frio Formation

The Oligocene Frio Formation of the Gulf of Mexico Basin consists largely of sandstone and shale deposited in various environments; it is as much as 15,000-ft thick. The Frio produces largely from onshore and offshore locations in Texas. Principal reservoirs in the Frio are sandstones (Table 6.27), which are known to have good to excellent porosity (as much as 35%) and variable permeability (as much as 3500 mD).

Reservoir categories defined in the Frio Formation are fluvial, deltaic, strandplain-barrier, and shelf sandstones (Table 6.27). These four categories were selected principally because reservoirs within them differ in terms of their broad depositional and geographic settings, structural setting, proximity to structures and potential source rocks, and reservoir characteristics.

These reservoirs contain thick shale deposits that are always excluded in conventional reserve calculations. However, these shales are the source of unconventional gas and are accessible through current development schemes. As these conventional resources are subject to enhanced oil recovery, unconventional resources should be considered because often such resources would contain higher saturation of oil and gas than the conventional reserve, particularly for matured reservoirs.

6.11.2.2.2 Morrow Formation

The Lower Pennsylvanian Morrow Formation of the Anadarko and Denver Basins consists largely of sandstone and shale; it is as much as 1500-ft thick. The Morrow produces oil and gas in Oklahoma, Texas, Kansas, and Colorado (Figure 6.41). Principal reservoirs in the Morrow are sandstones (Table 6.28), which are known to have good porosity (as much as 22%) and permeability (as much as several darcies).

Petroleum reservoirs in the Morrow Formation were placed into three categories—incised valley-fill, deltaic, and shallow marine (Table 6.28). These categories were selected because reservoirs within them differ in terms of their broad geographic and depositional setting. The differing depositional settings of the reservoir categories have led to differing

	Depositional characteristics			Reservoir characteristics				
					Porosity (bulk rock)			
Reservoir category	Environment	Reservoir facies	Nonreservoir facies	Lithology	Principal pore space	Diagenetic enhancement	Diagenetic occlusion	Porosity
Fluvial, chiefly the Gueydan and Chita/Corrigan fluvial systems	Chiefly fluvial with associated channel fill, point bar, crevasse splay, and floodplain sediments	Channel sands, point bars, and crevasse splay sands	Floodplain and lacustrine muds	Feldspathic litharenite, litharenite, and sublitharenite sandstone	Intergranular and moldic	Dissolution of unstable detrital grains and earlier formed cements, resulting in secondary pore space	Quartz, calcite, and clay cements; mechanical compaction	15–35%
Deltaic; chiefly the Norias and Houston delta complexes	Delta-plain, delta-front, and delta-flank environments of a prograding continental margin in the Gulf Basin. Norias contains more sediment and more sand, and was less influenced by marine processes than Houston	Distributary channel, delta-front and delta-flank, and channel-mouth bar sands	Prodelta and shelf shales	Feldspathic litharenite, litharenite, and sublitharenite sandstone	Intergranular and moldic	Dissolution of unstable detrital grains and earlier formed cements, resulting in secondary pore space	Quartz, calcite, and clay cements; mechanical compaction	10–35%
Strandplain-barrier; chiefly the Buna and Greta/Carancahua barrier strandplains	Shoreface, beach, barrier, and lagoonal deposits adjacent to deltaic depocenters	Shoreface, beach, and barrier sands	Marsh and lagoonal muds	Feldspathic Litharenite, litharenite, and sublitharenite sandstones	Intergranular and moldic	Dissolution of unstable detrital grains and earlier formed cements, resulting in secondary pore space	Quartz, calcite, and clay cements; mechanical compaction	20–35%

Continued

Table 6.27 Frio Formation, Gulf of Mexico Basin—Summary of Geological Characteristics and Reserve Growth Potential of Reservoirs—cont'd

	Depositional characteristics			Reservoir characteristics				
					Porosity (bulk rock)			
Reservoir category	**Environment**	**Reservoir facies**	**Nonreservoir facies**	**Lithology**	**Principal pore space**	**Diagenetic enhancement**	**Diagenetic occlusion**	**Porosity**
Shelf: off-shore Gulf Coast Basin	Shelf, slope, and perhaps submarine fan environments in deeper parts of the Gulf Coast Basin	Shelf, slope, and possibly fan sand-stones	Marine shales and siltstones	Feldspathic Litharenite, litharenite, and sublitharenite sandstones	Intergranular and moldic	Dissolution of unstable detrital grains and earlier formed cements, resulting in secondary pore space	Quartz, calcite, and clay cements; mechanical compaction	As much as 30%

Reservoir characteristics—Continued			Stratigraphic controls		Structural controls		
Permeability	**Fractures**	**Source rock**	**Reservoir location**	**Traps or seals**	**Reservoir location**	**Traps or seals**	**Oil or gas**
20–1500 mD	Important in hydrocarbon migration from source to reservoir	Shales that underlie reservoirs	Gueydan system largely a single drainage; leads to stacked channels and lateral amalgamation of channels. Chita Corrigan largely multiple channels with somewhat less stacking of sands	Stratigraphic component of trap is the interval where facies change to mud-rich floodplain rocks; mud-rich rocks are seals	Production best where fluvial and splay sands cross anticlines, faulted anticlines, or growth-fault trends, and faults served as conduits for upward petroleum migration	Rollover anticlines, particularly on downdip side of Vicksburg growth fault	Oil and gas

10–2400 mD	Important in hydrocarbon migration from source to reservoir; also juxtapose reservoirs and seals	Shales that underlie or are basinward facies of reservoirs	Abundant sediment supply and single fluvial system input lead to vertically stacked sandy deltaic lobes (Norias), whereas Houston delta fed by several smaller fluvial systems that led to numerous small dispersed lobes with less continuous sands	Stratigraphic component of trap is at abrupt facies changes from reservoir to fine-grained rocks; mud-rich rocks are seals	Syndepositional movement on growth faults and salt diapirs but no thickening of deltaic sediments, including reservoir rocks	Anticlines and faulted anticlines, some of which are associated with growth faults (Noria and Houston) or salt diapirism (Houston); also growth faults juxtapose reservoirs with seals or compartmentalize reservoirs	Associated gas and oil from more proximal parts, and nonassociated gas from more distal parts
8–3500 mD	Important in hydrocarbon migration from source to reservoir; also juxtapose reservoirs and seals	Shales that underlie or are basinward facies of reservoirs	Greater marine influence on Houston delta led to greater redistribution of sands into strandplain systems than on sands that originated in Norias delta	Stratigraphic component of trap is the interval where facies change to mud-rich floodplain rocks; mud-rich rocks are seals	Vertical stacking of sands and strike-parallel orientation of sands greatly influenced by orientation and movement of growth faults	Anticlines, rollover anticlines, and faulted anticlines	Associated gas and oil
As much as 1500 mD	Important in hydrocarbon migration from source to reservoir; also juxtapose reservoirs and seals	Shales that interbed with or underlie reservoir rocks	Stratigraphic controls on reservoir location unclear	Stratigraphic component of trap is at abrupt change from reservoir to fine-grained rocks; fine-grained rocks serve as seals	Sediment accumulation in submarine canyons or intraslope basins that formed from active faulting or salt diapirs (or both)	Faulted anticlines and salt-related structures. Seals formed by fault-related juxtaposition of reservoirs with impermeable rocks	Largely gas

Table 6.28 Morrow Formation, Anadarko and Denver Basins—Summary of Geological Characteristics and Reserve Growth Potential of Reservoirs

	Depositional characteristics			Reservoir characteristics				
					Porosity (bulk rock)			
Reservoir category	Environment	Reservoir facies	Nonreservoir facies	Lithology	Principal pore space	Diagenetic enhancement	Diagenetic occlusion	Porosity
Incised valley fill	Braided streams that grade upward into meandering and estuarine environments	Dominantly in coarser grained fluvial sands that fill incised valleys	Floodplain, estuarine, and marine mud-stone	Sandstone; varies from quartz arenite to litharenite or arkosie	Intergranular; variable volume of moldic porosity due to dissolution of detrital grains	Secondary pore space from dissolution of early formed authigenic cements and some unstable detrital grains	Extensive cement in lower parts of channel sands with calcite or iron carbonate minerals, or both	12–21%
Deltaic	Lower delta plain	Point bar, meander channel, stream-mouth bar, and distributary channel sands	Overbank, backswamp marsh, prodelta, and marine mudstone	Sandstone; varies from quartz arenite to litharenite or arkosie		Secondary pore space from dissolution of early formed authigenic cements and some unstable detrital grains	Late-stage calcite or iron carbonate minerals, or both	12–22%

Shallow marine	Near-shore and marginal marine	Beach, barrier island, and shoreline parallel sandbar sands	Marine shale and siltstone	Sandstone; varies from quartz arenite to litharenite or arkosie; locally fossiliferous		Secondary pore space from dissolution of early formed authigenic cements and some unstable detrial grains	Late-stage calcite or iron carbonate minerals, or both; mechanical compaction	4–20%

Reservoir characteristics—Continued			Stratigraphic controls		Structural controls		
Permeability	Fractures	Source rock	Reservoir location	Traps or seal	Reservoir location	Traps or seals	Oil or gas
As much as several darcies	Could have helped hydrocarbons to migrate from any overlying or underlying sources	Possibly marine muds of the Morrow Formation, where mature in Anadarko Basin; other organic-bearing formations outside the Morrow	Downcutting and formation of palcovalleys localized fluvial channel-reservoirs, dominantly in upper part of Morrow	Underlying marine limestone or shale and overlying floodplain muds	Palcostructures and perhaps subsidence from dissolution of underlying evaporates may have localized areas of downcutting and incision	Anticlines may influence but are secondary to stratigraphic controls	Associated gas and oil

Continued

Table 6.28 Morrow Formation, Anadarko and Denver Basins—Summary of Geological Characteristics and Reserve Growth Potential of Reservoirs—cont'd

Reservoir characteristics—Continued			Stratigraphic controls		Structural controls		
Permeability	**Fractures**	**Source rock**	**Reservoir location**	**Traps or seal**	**Reservoir location**	**Traps or seals**	**Oil or gas**
1–100 mD	Could have helped hydrocarbons to migrate from any overlying or underlying sources	Possibly marine muds of the Morrow Formation, where mature in Anadarko Basin; other organic-bearing formations outside the Morrow	Unclear	Lateral pinchout of sands into fine-grained marine muds	Unclear	Anticlines may influence but are secondary to stratigraphic controls	Dominantly gas
<1–200 mD	Could have helped hydrocarbons to migrate from any overlying or underlying sources	Possibly marine muds of the Morrow Formation, where mature in Anadarko Basin; other organic-bearing formations outside the Morrow	Location of sands in part a function of longshore currents, dominantly in lower part of Morrow	Lateral pinchout of sands into fine-grained marine muds	Unclear	Anticlines may influence but are secondary to stratigraphic controls	Dominantly nonassociated gas

reservoir-rock characteristics, such as porosity and permeability, which bear directly on the reservoir properties and contained resources. The shallow marine category offers the greatest potential for unconventional gas reserves. However, caprock of high-porosity reservoirs also contain large volume of natural gas.

6.11.2.3 Deeper Marine Shales

6.11.2.3.1 Barnett Shale

The Middle to Late Mississippian Barnett Shale of the Fort Worth Basin, Texas, consists largely of black marine shales with some limestone; it is as much as 650-ft thick. Most production is of nonassociated gas, principally in the northeastern part of the basin (Figure 6.42). Reservoirs of this self-sourced unit are marine shales in the Barnett (Table 6.29), which have very low porosity (less than 6%) and extremely low permeability (a few nanodarcies).

Reservoirs in the Barnett Shale are grouped in a single category termed the shale (unconventional) category (Table 6.29). Until recently, the lower shale member has been the more productive, although considerable production is now being realized from the upper shale member as well (Bowker, 2002). Both members characteristically have a high content of organic material, which is largely Type–II (Jarvie et al., 2001; Hill et al., 2007). In general, the current average content of organic material in both members is four–5% (Jarvie et al., 2007), although in places the Barnett is thought to have contained as much as 20% total organic carbon when it was deposited (Bowker, 2002). The organic material serves as the source of the gas, thereby defining these reservoirs as self-sourced and unconventional.

This formation is characteristically categorized as unconventional gas reserve. However, this formation makes up for only a fraction of total unconventional reserve that would be evident through scientific characterization.

6.11.2.3.2 Bakken Formation

The Late Devonian to Early Mississippian Bakken Formation (of the Williston Basin of North Dakota, Montana, and the Canadian provinces of Saskatchewan and Manitoba (Figure 6.41)) consists largely of marine shale with minor sandstone; it is as much as 140-ft thick. The Bakken produces mostly oil, principally in North Dakota and Montana and lesser amounts in Saskatchewan and Manitoba. Reservoirs in the Bakken are principally marine shales, although smaller reservoirs are found in interbedded near-shore to shoreface sandstones (Table 6.30). Porosity of the shales is very

Table 6.29 Barnett Shale, Fort Worth Basin—Summary of Geological Characteristics and Reserve-Growth Potential of Reservoirs

	Depositional characteristics				Reservoir characteristics			
					Porosity (bulk rock)			
Reservoir category	Environment	Reservoir facies	Nonreservoir facies	Lithology	Principal pore space	Diagenetic enhancement	Diagenetic occlusion	Porosity
Shale (unconventional)	Offshore marine	Marine shale	Dense limestone	Organic-rich shale	Matrix, but very low	Uncertain	Calcite along fractures	Very low, typically <6%

Reservoir characteristics—Continued			Stratigraphic controls		Structural controls		
Permeability	Fractures	Source rock	Reservoir location	Traps or seals	Reservoir location	Traps or seals	Oil or gas
Very low, typically in the range of nanodarcies	Naturally fractured in deeper parts of basin and over structures; fractures reduce productivity	Organic-rich shale in the Barnett that also serves as reservoir rock	Uncertain	Gas trapped by fine grained nature of shale reservoir	Best production away from fractured areas	Open faults tended to leak gas out of formation, whereas calcite filled faults prevented gas migration	Non-associated gas

Table 6.30 Bakken Formation, Williston Basin—Summary of Geological Characteristics and Reserve-Growth Potential of Reservoirs

	Depositional characteristics			Reservoir characteristics				
					Porosity (bulk rock)			
Reservoir category	**Environment**	**Reservoir facies**	**Nonreservoir facies**	**Lithology**	**Principal pore space**	**Diagenetic enhancement**	**Diagenetic occlusion**	**Porosity**
Shale (unconventional)	Deep marine, below wave base	Black, organic-rich mudstone	Overlying shallow marine carbonates and shales	Black mudstone	Fracture	Little or none	Little or none	Very low, typically <5%
Siltstone-sandstone (unconventional)	Near-shore and shoreface	Siltstone and very fine to medium-grained sandstone	Enclosing black mudstone	Dolomitic siltstone and sandstone	Fracture	Dissolution of carbonate cement	Carbonate cement	Can be >10% but typically 3–10%

Reservoir characteristics—Continued			Stratigraphic controls		Structural controls		
Permeability	**Fractures**	**Source rock**	**Reservoir location**	**Traps or seals**	**Reservoir location**	**Traps or seals**	**Oil or gas**
<0.01–60 mD	Critical for production	Black, organic-rich mudstone; is also the reservoir rock	Apparently not important	Apparently not important	Fracture zones overlying anticlinal or monoclinal folds and solution fronts in underlying salts	Minimal; reservoirs unconventional	Oil

Continued

Table 6.30 Bakken Formation, Williston Basin—Summary of Geological Characteristics and Reserve-Growth Potential of Reservoirs—cont'd

Reservoir characteristics—Continued			Stratigraphic controls		Structural controls		Oil or gas
Permeability	**Fractures**	**Source rock**	**Reservoir location**	**Traps or seals**	**Reservoir location**	**Traps or seals**	
<0.01–109 mD	Critical for production	Organic-rich mud in Bakken, interbedded with or perhaps downdip from reservoirs	Local thickening owing to subsidence associated with dissolution of underlying salts	Overlying shales of the Bakken	Fracture zones overlying anticlinal or monoclinal folds and solution fronts in underlying salts	Updip against enclosing mudstone strata	Oil

low (typically less than 5%) as is their permeability (<0.01–60 mD). Porosity of sandstone reservoirs is higher (as much as 10%) as is permeability (<0.01–109 mD).

Two categories of reservoirs were defined in the Bakken Formation—shale (unconventional) and siltstone sandstone (unconventional) (Table 6.30). These two categories were selected because they have different characteristics, stratigraphic positions, and geographic distributions. In each, however, the petroleum is thought to be generated within the Bakken, so both categories are considered to be unconventional, similar to those in the Barnett Shale. Recent success in producing from these unconventional sources with conventional technology (e.g., horizontal well and fracturing) point out that there is tremendous potential for expanding this resource base.

6.11.2.4 Marine Carbonate Reservoirs

6.11.2.4.1 Ellenburger Group

The Early Ordovician Ellenburger Group of the Permian Basin (Figure 6.43) consists largely of marine carbonate rocks; the group is as much as 1500-ft thick. Units in the Ellenburger produce oil and gas chiefly in Texas. Principal reservoirs in the Ellenburger are in karstified parts of a carbonate platform and in dolomitized carbonate muds (Table 6.31). Reservoirs in karstified rocks have low but variable porosity (2–7%) and moderate but variable permeability (2–750 mD). Reservoirs in dolomitized muds have higher porosity (2–14%) but lower permeability (1–44 mD) than karstified reservoirs.

Reservoirs in the Ellenburger Group are placed into three categories (Table 6.31)—karstified, platform, and tectonically fractured—based primarily on differences in the nature and volume of porosity and permeability, geographic distribution, produced petroleum, and the degree to which structure influenced reservoir development. This threefold division is similar to that presented by others (Kerans et al., 1989; Kosters et al., 1989c; Holtz and Kerans, 1992) and is also consistent with that presented by Ball (1995). This is the category rarely connected to unconventional gas. However, significant amount of oil and gas exists within such reservoirs that fit the description of unconventional oil and gas, including in low-permeability patches, caprock, and volcanic rocks. Some of these resources are in high pressure and temperature conditions. They form a special candidate for reverse thermal recovery. It involves injection of cold water in order to induce thermal fracturing owing to large temperature

Table 6.31 Ellenburger Group, Permian Basin—Summary of Geological Characteristics and Reserve-Growth Potential of Reservoirs

	Depositional characteristics			Reservoir characteristics				
					Porosity (bulk rock)			
Reservoir category	**Environment**	**Reservoir facies**	**Nonreservoir facies**	**Lithology**	**Principal pore space**	**Diagenetic enhancement**	**Diagenetic occlusion**	**Porosity**
Karstified, principally in Central Basin platform and Midland Basin	Shallow aggrading marine carbonate platform	Inner platform	Reef, forereef, supratidal	Dolo mitized mudstone	Interbreccia fragment and within fractures	Dissolution of lime mud leading to karstification and brecciation; intercrystalline owing to dolomitization of muds	Late-stage saddle dolomite	Average, 3% Range, 2–7%
Platform, dominantly in southern and eastern pans of Midland Basin	Shallow aggrading marine carbonate platform	Middle to outer platform	Reef, forereef, Supratidal	Dolomitized packstone and muds tone	Intercrystalline	Intercrystalline porosity owing to dolomitization	Late-stage saddle dolomite	Average, 14% Range, 2–14%

Tectonically fractured, dominantly in the eastern Delaware Basin	Shallow aggrading marine carbonate platform	Inner platform	Reef, forereef, supratidal	Dolomitized mudstone	Fracture (tectonic)	Dissolution of lime mud leading to karstification and brecciation; intercrystalline owing to dolomitization of muds	Late-stage saddle dolomite	Average, 4% Range, 1–8%

Reservoir characteristics—Continued			Stratigraphic controls		Structural controls		
Permeability	Fractures	Source rock	Reservoir location	Traps or seals	Reservoir location	Traps or seals	Oil or gas
Mean, 32-mD Range, 2–750 mD	Channeled pore fluids that allowed vertical infiltration of dissolving waters into various stratigraphic horizons to promote karstification	Overlying Ordo vician Simpson Group	Lime muds remaining after early dolomitization, which became horizons subject to dissolution leading to karstification	Traps and seals include overlying Simpson Group and unkarsted Ellenburger dolomite. Seals also include impermeable cave-fill sediments and collapse zone adjacent to reservoirs	Anticlines, faulted anticlines, and fault-bounded anticlines	Uncertain	Principally oil with some associated gas and gas condensate

Continued

Table 6.31 Ellenburger Group, Permian Basin—Summary of Geological Characteristics and Reserve-Growth Potential of Reservoirs—cont'd

Reservoir characteristics—Continued			Stratigraphic controls		Structural controls		
Permeability	**Fractures**	**Source rock**	**Reservoir location**	**Traps or seals**	**Reservoir location**	**Traps or seals**	**Oil or gas**
Average, 12-mD range, <1–44 mD	Focused early dolomitizing fluids, which resulted in intercrystal-line porosity and permeability	Overlying Devonian Woodford Shale?	Lime muds that were dolomitized	Traps and seals include overlying Simpson Group	Anticlines, faulted anticlines	Uncertain	Largely oil
Average, 4-mD range, 1-100 mD	Early fracturing promoted karstification, whereas later fracturing improved porosity and permeability of the reservoir	Overlying Ordovician Simpson Group	Lime muds that were dolomitized	Traps and seals include overlying Simpson Group	Fractured anticlines and faults critical	Uncertain	Nonassociated gas

gradient. In most reservoirs, this is easy to accomplish. These formations typically are not suitable for hydraulic fracturing for reasons described in previous chapters.

6.11.2.4.2 Smackover Formation

The Upper Jurassic Smackover Formation in onshore parts of Texas, Arkansas, Louisiana, Mississippi, Alabama, and Florida, as well as offshore in the Gulf of Mexico Basin, consists largely of carbonate rocks with minor black shale and siltstones; it is as much as 1000-ft thick. Most oil and gas is produced from onshore locations in the above-listed states (Figure 6.40). Principal reservoirs in the Smackover are in carbonate rocks deposited in a ramp setting (Table 6.32) that have good to excellent porosity (as much as 35%) and variable permeability (<1–4100 mD). Reservoir categories in the Smackover Formation are salt structure, basement structure, graben, stratigraphic, and updip fault (Table 6.32). These categories, which were defined or later refined through regional studies by other workers (for example, Bishop, 1973; Collins, 1980; Moore, 1984; Mancini et al., 1990; Kopaska-Merkel and Mann, 1993; Tew et al., 1993) were selected because of differences in their geographic extent and in the role that structures played in both source-rock deposition and petroleum trapping.

This type of formations is not typically considered to generate unconventional gas. However, significant amount of gas is present in these formations in caprcoks, salt domes, and other locations. Salt domes have not been included in this book as a source of potential unconventional gas. However, they do contain natural gas in many instances and each case should be investigated in order to access unconventional gas with minimal cost (see for instance, Dronkert and Remmelts, 1996).

6.11.2.5 Submarine Fan Reservoir

6.11.2.5.1 Spraberry Formation

The Early Permian Spraberry Formation of the Midland Basin consists largely of turbiditic sandstones, with minor black shales, silty dolostones, and argillaceous siltstones; it is as much as 1000-ft thick. Most production of oil is in west-central Texas, in the Midland Basin (Figure 6.42). Principal reservoirs in the Spraberry are the tubiditic sandstones (Table 6.33), which have good porosity (as much as 18%) but relatively low permeability (maximum, 10 mD). A single reservoir category, submarine sand, was defined for the Spraberry Formation.

Table 6.32 Smackover Formation, Gulf Coast region—Summary of Geological Characteristics and Reserve-Growth Potential of Reservoirs

	Depositional characteristics			Reservoir characteristics				
					Porosity (bulk rock)			
Reservoir category	Environment	Reservoir facies	Nonreservoir facies	Lithology	Principal pore space	Diagenetic enhancement	Diagenetic occlusion	Porosity
Salt structure. dominantly in southern and eastern Texas, southern Arkansas, southern and central Mississippi, southwestern Alabama, and northern Louisiana	Slow regressive to stillstand marine carbonate ramp	Ramp, higher energy shoaling facies	Subtidal mudstone, wackestone, supratidal units, and outer ramp dolostones	Largely dolomitic oolitic grainstones and packstones	Dominantly intercrystalline where dolomitized, oomoldic in, updip regions, intergranular in basinal regions	Intercrystalline owing to dolomitization; ooid dissolution; late calcite dissolution; diagenesis most pronounced on structural highs	Late-stage saddle dolomite, anhydrite, and calcite	2–35%
Basement structure, primarily in eastern Texas, central Mississippi, southern Arkansas, and southwestern Alabama	Slow regressive to stillstand marine carbonate ramp	Ramp, higher energy shoaling facies	Subtidal mudstone, wackestone, supratidal units, and outer ramp dolostones	Largely dolomitic oolitic grainstones and packstones	Principally oomoldic; minor primary interparticle and intercrystalline where dolomitized	Principally oomoldic; minor intercrystalline owing to minor dolomitization; diagenesis pronounced on structural highs	Late-stage calcite and dolomite	As much as 20%

Graben, principally along Arkansas–Louisiana border	Slow regressive to stillstand marine carbonate ramp	Ramp, higher energy-shoaling facies	Subtidal mudstone, wackestone, supratidal units, and outer ramp dolostones	Oolitic limestone, locally dolomitic	Considerable interparticle pore space preserved; also oomoldic	Some interparticle and intercrystalline owing to dolomitization; some oomoldic	Partial cementation by calcite	4–19%
Stratigraphic, principally in southern Arkansas	Slow regressive to stillstand marine carbonate ramp	Ramp, higher energy shoaling facies	Subtidal mudstone, pelloid packstone, wackestone, supratidal units, and outer ramp dolostones	Oolitic, oncolitic, or skeletal, grainstone limestone minimally dolomitized	Considerable interparticle; some oomoldic and intercrystalline where dolomitized	Some inter particle and intercrystalline owing to dolomitization; considerable early- and late-stage dissolution of particles and late-stage cement	Cements such as early and late stage calcite and anhydrite; some compaction	3–30%
Updip fault, principally in eastern Texas, southern Arkansas, central Mississippi, southwestern Alabama and Florida Panhandle	Slow regressive to stillstand marine carbonate ramp	Ramp, higher energy shoaling facies	Subtidal mudstone, wackestone, supratidal units, and outer ramp dolostones	Oolitic limestone, locally dolomitic	Principally oomoldic	Ooid dissolution common; some dolomitization	Early calcite cement	10–20%

Reservoir characteristics—Continued			Stratigraphic controls		Structural controls		
Permeability	Fractures	Source rock	Reservoir location	Traps or seals	Reservoir location	Traps or seals	Oil or gas
<1–4100 mD	Large-scale open fractures not now widespread: however, fractures probably served as conduits for hydrocarbon migration	Organic-rich units in lower part of Smackover Formation	Shoaling sequences best developed on positive features formed by salt diapirism during deposition	Fine-grained beds in overlying Buckner Formation acted as seals	Salt anticlines, faulted salt anticlines, faulted salt-pierced anticlines	Faults seal some reservoirs	Dominantly oil and associated gas with minor condensate
60–350 mD	Faults now act as seals owing to impermeability of fault zones but earlier probably served as conduits for hydrocarbon migration	Organic-rich units in lower part of Smackover Formation	Facies changes up on basement highs; shoaling on positive basement highs during deposition. Little evidence of halokinesis	Stratigraphic and structural trap with overlying Buckner Formation; pinchouts on basement highs serve as seals	Regional fault zones, anticlines, faulted anticlines	Downdip fault zone served as reservoir seal	Dominantly oil in updip areas; associated gas or gas condensate in basinal areas
<1–1000 mD	Faults now act as seals owing to impermeability of fault zones but earlier probably served as conduits for hydrocarbon migration	Organic-rich units in lower part of Smackover Formation	Shoaling sequences best developed on horst blocks adjacent to grabens	Structural and stratigraphic trap; overlying Buckner Formation serves as seal	Fault zones and faulted anticlines	Faults seal some reservoirs	Dominantly oil

1–250 mD	Faults probably-served as conduits for hydrocarbon migration	Organic-rich units in lower part of Smackover Formation	Facies changes and regressive units overlying reservoirs	Structural and stratigraphic trap; overlying Buckner Formation serves as seal	Likely; structures limited deposition of reservoir rocks or facilitated pinchouts	Faults seal some reservoirs	Dominantly oil; some associated gas
3–280 mD	Faults now act as seals owing to impermeability of fault zones but earlier probably served as conduits for hydrocarbon migration	Organic-rich units in lower part of Smackover Formation	Near updip limit of Smackover deposition	Dominantly structural trap; fault systems serve as seals	Uplift on faults juxtaposed reservoirs and impermeable beds	Fault zones	Dominantly oil; some gas or gas condensate

Table 6.33 Spraberry Formation, Midland Basin—Summary of Geological Characteristics and Reserve-Growth Potential of Reservoirs

	Depositional characteristics			Reservoir characteristics				
					Porosity (bulk rock)			
Reservoir category	Environment	Reservoir facies	Nonreservoir facies	Lithology	Principal pore space	Diagenetic enhancement	Diagenetic occlusion	Porosity
Submarine sand	Deep water submarine basin and fan	Submarine fan and turbidite sandstone	Silty dolostone, organic-rich shale, and argillaceous sandstone	Sandstone	Largely intergranular but some minor moldic	Dissolution of preexisting authigenic cements and unstable detrital grains	Mechanical compaction and authigenic cements such as illite, chlorite, quartz, and dolomite	Matrix porosity usually 5–15% but may be as high as 18%

Reservoir characteristics—Continued			Stratigraphic controls		Structural controls		
Permeability	Fractures	Source rock	Reservoir location	Traps or seals	Reservoir location	Traps or seals	Oil or gas
Average matrix permeability low, <1 mD, but may be as high as 10 mD	Very common; multiple orientations observed; fractures cemented to various degrees	Interbedded organic-rich shales	Most reservoirs downdip from the ancient Horseshoe Atoll at mouth of sub-marine canyons or where facies change from channel to interchannel deposits	Pinchouts of reservoir rocks updip and downdip into fine-grained rocks serve as traps. Shales seal reservoirs	Uncertain	Mostly stratigraphic traps; one small field on an anticline	Largely oil

Even though not explicitly recognized, these reservoirs form an excellent candidate for unconventional gas. These are thick formations that can have very high reserve once the cutoff point for porosity is removed.

6.11.2.6 Fluvial Reservoir

6.11.2.6.1 Wasatch Formation

The Paleocene–Eocene Wasatch Formation of the Uinta-Piceance Basin of Utah and Colorado consists largely of overbank and lacustrine mudstones with some fluvial and fluvial-dominated deltaic sandstones; it is as much as 5000-ft thick. The Wasatch produces oil and associated gas mostly in the Uinta Basin of northeastern Utah, although minor gas is also produced in the Piceance Basin of Colorado (Figure 6.41). Principal reservoirs in the Wasatch are the fluvial sandstones (Table 6.34), which are known to have good porosity (maximum, 15%) but low permeability (maximum, 40 mD).

Reservoirs in the Wasatch Formation are categorized as Green River source and Mesaverde source (Table 6.34). The two categories are distinguished by (1) the source of the petroleum produced from each, (2) the nature of the petroleum produced from each, and (3) the geographic distribution of production. This division is important because it recognizes that petroleum produced from the Wasatch comes from two different source rocks; hence, two petroleum systems generated economic amounts of petroleum within the greater Uinta-Piceance Basin.

In addition to unconfirmed patches of unconventional plays, these formations contain three types of continuous-type unconventional sources. They are (Spencer, 1995)

- oil in fractured Upper Cretaceous marine shale;
- gas in tight sandstone;
- coalbed methane.

Other unconventional gas may be present in heavy oil and tar sand formations.

6.11.3 Quantitative Measures of Well Production Variability

Production history offers a powerful tool for scientific characterization of a formation. It is not because it offers refinement of statistical tools, but because any history is evidence that can be used to refine a scientific model. USGS (2008) produced a detailed analysis of production history. This analysis was aimed at identifying origin of both fluid and rock systems. This report (USGS, 2008) compared historical well production data of the five formations by use of proprietary information. In addition, it

Table 6.34 Wasatch Formation, Uinta-Piceance Basin—Summary of Geological Characteristics and Reserve-Growth Potential of Reservoirs

	Depositional characteristics			Reservoir characteristics				
					Porosity (bulk rock)			
Reservoir category	Environment	Reservoir facies	Nonreservoir facies	Lithology	Principal pore space	Diagenetic enhancement	Diagenetic occlusion	Porosity
Green River source	Fluvial, deltaic, and lacustrine	Fluvial, channel sandstone, and sands deposited in lacustrine deltas	Overlying and interbedded overbank, floodplain, delta plain, and lacustrine mudstone and claystone	Sandstones, lithic arkoses, or feldspathic litharenites	Intergranular, principally secondary; some minor moldic	Dissolution of early authigenic cements and unstable detrital grains	Some quartz and carbonate cements and authigenic clays	Ranges up to 15% at shallow (<4000 ft) depths but <10% at greater depths (>8500 ft)
Mesaverde source	Fluvial, deltaic, and lacustrine	Fluvial, channel sandstone, and sands deposited in lacustrine deltas	Overlying and interbedded overbank, floodplain, delta plain, and lacustrine mudstone and claystone	Sandstones, lithic arkoses, or feldspathic litharenites	Intergranular, principally secondary; some minor moldic	Dissolution of early authigenic cements and unstable detrital grains	Some quartz and carbonate cements and authigenic clays	Ranges up to 15% at shallow (<4000 ft) depths but <10% at greater depths (>8500 ft)

Reservoir characteristics—continued			Stratigraphics controls		Structural controls		
Permeability	Fractures	Source rock	Reservoir location	Traps or seals	Reservoir location	Traps or seals	Oil or gas
Generally low; as much as 40 mD but commonly <0.1 mD	Reservoirs may be complexly faulted; faults allow production	Organic-rich lacustrine mad-stones of Green River Formation, which largely interfingers with the Wasatch	Reservoir rocks deposited adjacent to and in deltas within ancient Lake Uinta	Overlying and interbedded shales, mudstones, and claystones trap and seal reservoirs	Uncertain	Secondary to stratigraphic traps or seals	Dominantly oil; some associated gas
Generally low; as much as 40 mD but commonly <0.1 mD	Reservoirs may be complexly faulted; faults allow production; migration along fractures	Coals and organic-rich shale of the Mesaverde Group, which underlies the Wasatch	Reservoir rocks deposited adjacent to and in deltas within ancient Lake Uinta	Overlying and interbedded shales, mudstones, and claystones trap and seal reservoirs	In areas where gas could migrate up fractures that cut from source to reservoir rocks	Secondary to stratigraphic traps or seals	Nonassociated gas

considered data from two specific reservoir categories in the Ellenburger Group (karst and platform, Table 6.35), which are based on gross geologic differences, to evaluate the possible intraformational variability in production within that formation. Because in most wells production declines exponentially or hyperbolically as a function of time, cumulative production from older wells (those for which current monthly production is less than 10% of initial monthly production) asymptotically begins to approximate ultimate recovery. This is typical of conventional reserve analysis. In such wells, variations in cumulative production reflect variations in the volume of reservoir rocks accessed by the wellbore. The slopes of the probability distributions for cumulative production (Figure 6.45) are direct indicators of the variability as shown by the data set. For example, steeper slopes reflect greater production heterogeneity (Figure 6.45), whereas a horizontal line represents uniform production characteristics. A dimensionless parameter that is proportional to the slopes of the four probability distributions of Figure 6.45 would provide a quantitative numerical representation of production heterogeneity. Such a parameter, referred to here as a variation coefficient (VC), can be calculated by using a measure of the dispersion (range) of the data set divided by a measure of central tendency such as the mean or the median (Stell and Brown, 1992; Dyman et al., 1996; Schmoker, 1966; Dyman and Schmoker, USGS Report, 2008).

The different parts of the distribution behave differently—that is, a single straight-line fit does not adequately describe the behavior of the entire distribution of production data. Of interest from conventional perspective is the central part of the distribution because it represents production from the vast majority of wells. Extreme production behavior, categorized by wells in the upper 5 and lower 20 percent of the production distribution, forms excellent candidates for unconventional recovery. They either include old wells that would invariably have vast gas content that is deemed uneconomical with conventional analysis or they include low-permeability formation, suitable for unconvetional gas development.

Figure 6.46 lists wells that were sorted by production from lowest to highest and subdivided into two size classes: a central class representing a productive range of 20–60% along the distribution and an upper class representing a productive range of 80–95%.

Figure 6.46 shows fields producing oil from the reservoirs representing the (1) fluvial category of the Frio Formation, (2) incised valley-fill category of the Morrow Formation, (3) Green River–source category of the

Table 6.35 Location of, Number of Fields and Wells in, Cumulative Production of, and Largest Fields in Each Reservoir Category Analyzed in This Study

Reservoir category	Location	No. fields	No. wells	Cum. oil (MMBO)	Cum. gas (BCF)	Largest oil fields	Cum. oil (MMBO)	Cum. gas (BCF)
Frio	Texas[a]	272	6301	534.4	11,193.0	Seeligson	238.1	2306.0
						Tijerina-Canales-Blucher	87.0	753.8
						Stratton	41.8	1841.5
Morrow	Colorado[b]	38	386	74.7	116.8	Arapahoe	23.2	35.7
						Mt. Pearl	13.6	41.2
						Sorrento	12.6	8.5
Wasatch	Utah[c]	24	436	89.9	139.2	Altamont	48.6	74.5
						Bluebell	34.9	48.6
						Cedar Rim	4.4	6.6
Ellenburger (karst)	Texas[d]	141	2784	1155.8	1042.5	Andector	178.4	70.2
						TXL	129.3	29.8
						Pegasus	96.3	361.2
Ellenburger (ramp)	Texas[e]	134	928	65.0	29.0	Barnhart	16.7	11.9
						Swenson-Barron	5.8	1.2
						Swenson-Garza	4.2	0.7
Minnelusa	Wyoming[f]	315	1936	586.8	14.9	Raven Creek	44.2	0.03
						Timber Creek	16.2	1.2
						Dillinger Ranch	16.2	1.6

[a]All or parts of Starr, Hidalgo, Brooks, Jim Hills, and Kleburg Counties, Texas.
[b]Morrow Formation producing wells in Colorado.
[c]Wasatch Formation producing wells in Utah.
[d]All or parts of Andrews, Winkler, Ector, Midland, Upton, and Crane Counties, Texas.
[e]All or parts of Border, Garza, Scurry, Coke, Mitchell, Irion, Reagan, and Crockett Counties, Texas.
[f]All of Campbell, Crook, and Johnson Counties, Wyoming.

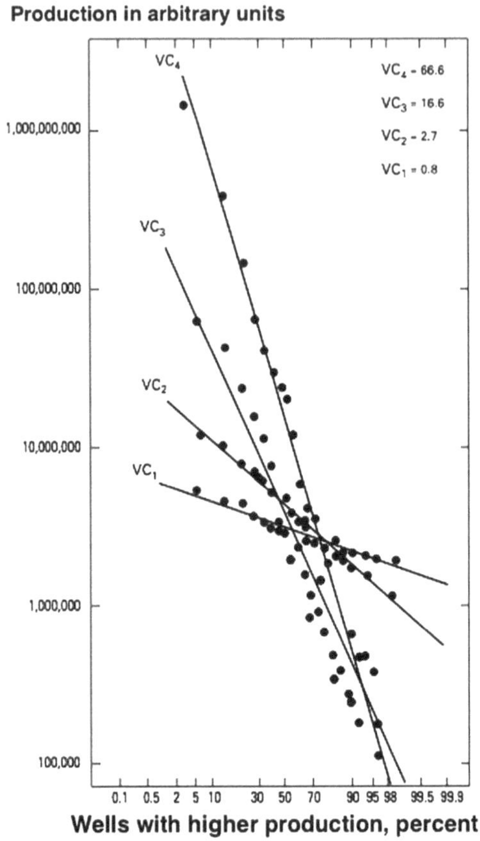

Figure 6.45 Probability distributions for production from wells of an oil or gas field (distributions based on hypothetical data—peak monthly production, peak yearly production, or cumulative production). Each point represents a well, and four fields (VC1–VC4) are depicted. In this type of plot log normal distributions plot as straight lines, and steeper slopes of lines correspond with a greater range of production and thereby greater production variability. The variation coefficient VC = (F5–F95)/F50 provides a dimensionless numerical value for the variability of each data set, and its value increases as slope increases. *From USGS, 2008.*

Wasatch Formation, (4) Minnelusa category of the Minnelusa Formation, and both the (5) platform and karst categories of the Ellenburger Group. Table 6.35 contains the basic data used in calculating production variability for each reservoir category. A minimum of 35 producing wells were used to describe the production behavior for each category and to calculate upper class and central class rates of recovery and slope ratios for each. We also identified a well productive life of at least 10 years on the basis of data in the IHS Energy Group production file. For example, 6301 wells were selected

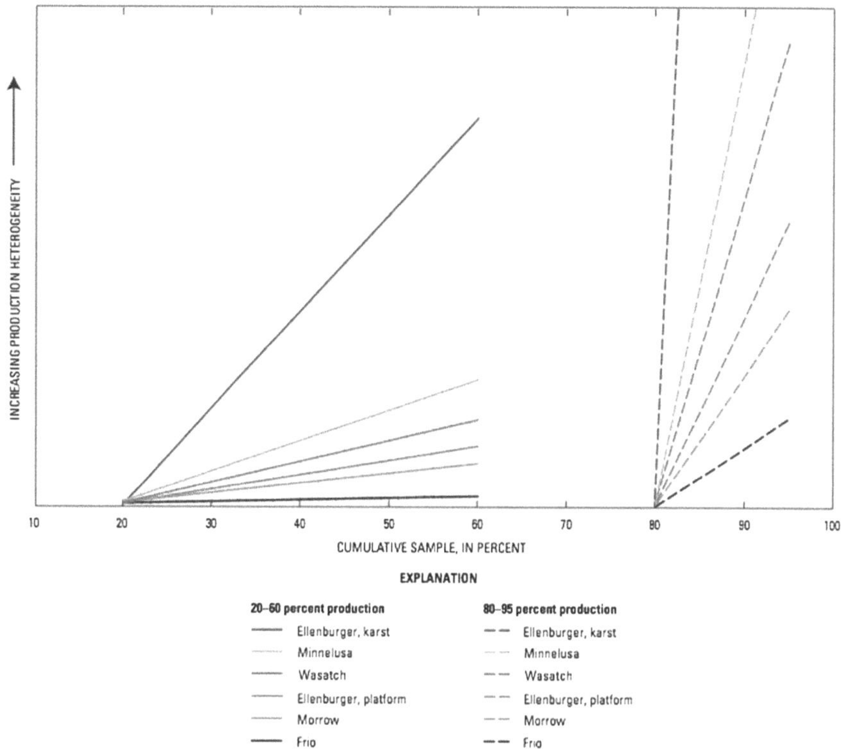

Figure 6.46 Production data of gas wells in fields in the Ellenburger Group karst and platform categories, Frio Formation fluvial category, Morrow Formation incised-valley category, Minnelusa Formation Minnelusa category, and Wasatch Formation Green River–source category. *From USGS, 2008.*

from IHS data as Frio Formation producers in all or parts of Starr, Hidalgo, Brooks, Jim Hills, and Kleburg Counties, Texas (Table 6.35). A computer program then calculated upper and central class rates of recovery and slope ratios on the basis of a subset of these wells that met our selection criteria. The six reservoirs analyzed have produced more than 2 billion bbl of oil and 12 trillion cubic feet of gas from nearly 13,000 producing wells. The results are plotted in Figure 6.46. Of interest is the role of depositional environment, diagenesis, and lithology on reservoir productivity. Comparing the slope ratios and variation coefficients of reservoirs with different geologic characteristics may provide insight into productivity analysis and ultimately into estimating field growth through time. An opposite trend will be followed by unconventional oil and gas. Whenever conventional resources hit stagnation, unconventional resource potentials increase.

CHAPTER 7

Overview of Reservoir Simulation of Unconventional Reservoirs

7.1 INTRODUCTION

In reservoir simulation, the principle of garbage in and garbage out is well known (Rose, 2000). This principle implies that the input data have to be accurate for the simulation results to be acceptable. Petroleum industry has established itself as the pioneer of subsurface data collection (Abou-Kassem et al., 2006). Historically, no other discipline has taken so much care in making sure input data are as accurate as the latest technology would allow. The recent superflux of technologies dealing with subsurface mapping, real-time monitoring, and high speed data transfer is an evidence of the fact that input data in reservoir simulation are not the weak link of reservoir modeling.

However, for a modeling process to be meaningful, it must fulfill two criteria, namely, the source has to be true (or real) and the subsequent processing has to be true (Islam et al., 2010). For unconventional reservoirs, they are both of concern. For instance, most commonly used well log data do not register useful information for unconventional reservoirs, which do not allow invasion of sonic signals, mud filtrate, etc., invalidating some of the well logging techniques. Processing data is also of concern. Unconventional reservoirs require special filtering tools that are not readily available. Some of them are presented in previous chapters of this book.

For reservoir simulation to yield meaningful results, the following logical steps have to be taken.

1. Collection of data with constant improvement of the data acquisition technique. The data set to be collected is dictated by the objective function, which is an integral part of the decision-making process. Decision-making, however, should not take place without the abstraction process. The connection between objective function and data needs constant refinement. This area of research is one of the biggest strength of the petroleum industry, particularly in the information age. Each of the data acquisition tools must be reevaluated for applicability in unconventional reservoirs.

Unconventional Gas Reservoirs
ISBN 978-0-12-800390-9
http://dx.doi.org/10.1016/B978-0-12-800390-9.00007-4

2. The gathered data should be transformed into information so that they become useful. With today's technology, the amount of raw data is so huge that the need for a filter is more important than ever before. However, it is important to select a filter that does not skew data set toward a certain decision. Mathematically, these filters have to be nonlinearized (Islam et al., 2010). While the concept of nonlinear filtering is not new, the existence of nonlinearized models is only beginning to be recognized (Hossain and Islam, 2009).
3. Information should be further processed into "knowledge" that is free from preconceived ideas or a "preferred decision." Scientifically, this process must be free from information lobbying, environmental activism, and other forms of bias. Most current models include these factors as an integral part of the decision-making process (Eisenack et al., 2007), whereas a scientific knowledge model must be free from those interferences as they distort the abstraction process and inherently prejudice the decision-making. Knowledge gathering essentially puts information into a big picture. For this picture to be distortion free, it must be free from nonscientific maneuvering.
4. Final decision-making is knowledge based, only if the abstraction from step 1 through step 3 has been followed without interference. Final decision is a matter of Yes or No (or True or False, or 1 or 0) and this decision will be either knowledge based or prejudice based. Figure 7.1 shows the essence of the knowledge-based decision-making.

The process of aphenomenal or prejudice-based decision-making is illustrated by the inverted triangle, proceeding from the top down (Figure 7.2). The inverted representation stresses the inherent instability and unsustainability of the model. The source data from which a decision eventually emerges already incorporate their own justifications, which are then massaged by layers of opacity and disinformation.

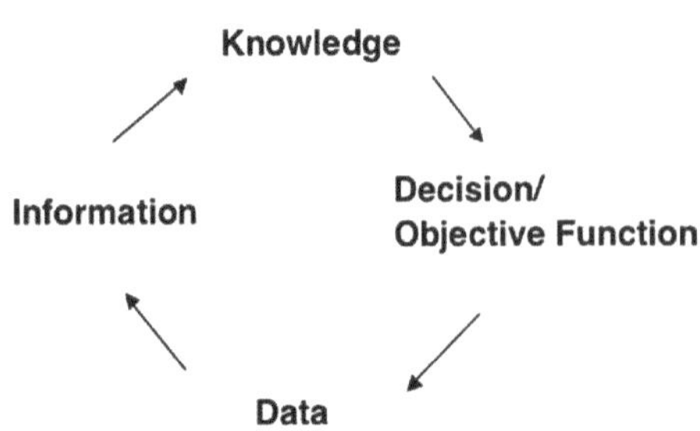

Figure 7.1 The knowledge model.

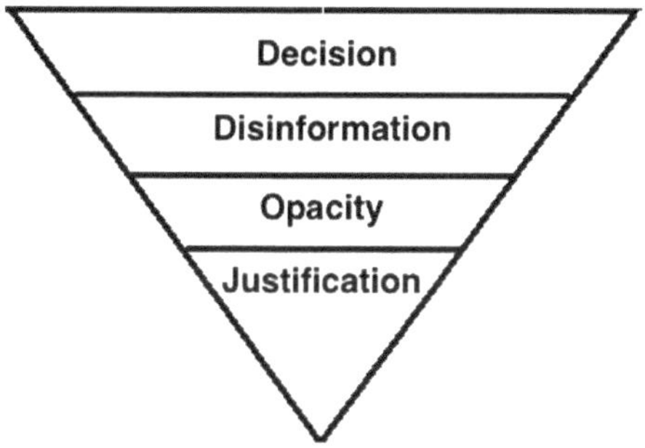

Figure 7.2 The aphenomenal model. *After Zatzman, 2012.*

7.2 ESSENCE OF RESERVOIR SIMULATION

Today, practically all aspects of reservoir engineering problems are solved with reservoir simulators, ranging from well testing to prediction of enhanced oil recovery. For every application, however, there is a custom-designed simulator. Even though, quite often, "comprehensive," "all purpose," and other denominations are used to describe a company simulator, every simulation study is a unique process, starting from the reservoir description to the final analysis of results. Simulation is the art of combining physics, mathematics, reservoir engineering, and computer programming to develop a tool for predicting hydrocarbon reservoir performance under various operating strategies. The first step in the development of a reservoir simulator is expressing the physical situation in terms of mathematical description. This is called the *formulation* step. The *formulation* step outlines the basic assumptions inherent to the simulator, states these assumptions in precise mathematical terms, and applies them to a control volume in the reservoir. Newton's approximation is used to render these control volume equations into a set of coupled, nonlinear partial differential equations (PDEs) that describe fluid flow through porous medium. These PDEs are then discretized, giving rise to a set of nonlinear algebraic equations. *Discretization* is the process of converting PDEs into algebraic equations. The discretization can be done using the finite difference methods or the approaches based on the variation methods such as finite element methods. However, the most common approach in the oil industry is the finite difference method. To carry out discretization, a PDE is written for a given point in space at a given time level using the Taylor series expansion to convert it in finite difference forms. The choice of time level (old, current, or the intermediate time level) leads to the explicit, implicit, or Crank–Nicolson formulation method. The discretization process results in a system of nonlinear algebraic equations. These equations generally cannot be solved with linear equation solvers and linearization of such equations becomes a necessary step before solutions can be obtained. *Well representation* is used to incorporate fluid production and injection into the nonlinear algebraic equations. *Linearization* involves approximating nonlinear terms in both space and time. Linearization results in a set of linear algebraic equations. Any one of several linear equation solvers can then be used to obtain the *solution*. The solution comprises of pressure and fluid saturation distributions in the reservoir and well flow rates. *Validation* of a reservoir simulator is the last step in developing a simulator, after which the simulator can be used for practical field applications. The validation step is necessary to make sure that no errors were introduced in the various steps of development and in computer programming.

It is possible to bypass the step of formulation in the form of PDEs and directly express the fluid flow equation in the form of nonlinear algebraic equation as pointed out in Abou-Kassem et al. (2006). In fact, by setting up the algebraic equations directly, one can make the process simple and yet maintain accuracy. This approach is termed the "engineering approach" because it is closer to the engineer's thinking and to the physical meaning of the terms in the flow equations. Both the engineering and mathematical approaches treat boundary conditions with the same accuracy, if the mathematical approach uses second order approximations. The engineering approach is simple and yet general and rigorous.

7.2.1 Assumptions behind Various Modeling Approaches

Reservoir performance is traditionally predicted using three methods, namely, (1) analogical, (2) experimental, and (3) mathematical. The analogical method consists of using properties of a mature reservoir that are similar to that of the target reservoir to predict the behavior of the reservoir. This method is especially useful when there is a limited available data. The data from the reservoir in the same geologic basin or province may be applied to predict the performance of the target reservoir. Experimental methods measure the reservoir characteristics in the laboratory models and scale these results to the entire hydrocarbon accumulation. The mathematical method applied basic conservation laws and constitutive equations to formulate the behavior of the flow inside the reservoir and the other characteristics in mathematical notations and formulations. The two basic equations are the material balance equation or continuity equation and the equation of motion or momentum equation. These two equations are expressed for different phases of the flow in the reservoir and are combined to obtain single equation for each phase of the flow. However, it is necessary to apply other equations or laws for enhanced oil recovery. As an example, the energy balance equation is necessary to analyze the reservoir behavior for the steam injection or in situ combustion reservoirs.

The mathematical model traditionally includes material balance equation, decline curve, statistical approaches, and also analytical methods. The Darcy's law is almost used in all of the available reservoir simulators to model the fluid motion. The numerical computations of the derived mathematical model are mostly based on the finite difference method. All these models and approaches are based on several assumptions and approximations that may cause to produce erroneous results and predictions.

7.2.1.1 Material Balance Equation

Material balance equation is known to be the classical mathematical representation of the reservoir. According to this principle, the amount of material remaining in the reservoir after a production time interval is equal to the amount of material originally present in the reservoir minus the amount of material removed from the reservoir due to production plus the amount of material added to the reservoir due to injection. This equation describes the fundamental physics of the production scheme of the reservoir. Listed below are several assumptions in the material balance equation.

- Rock and fluid properties do not change in space.
- Hydrodynamics of the fluid flow in the porous media is adequately described by Darcy's law.
- Fluid segregation is spontaneous and complete.
- Geometrical configuration of the reservoir is known and exact.
- Pressure–volume–temperature data obtained in the laboratory with the same gas-liberation process (flash vs differential) are valid in the field.
- Material balance equation is sensitive to inaccuracies in measured reservoir pressure. The model breaks down when no appreciable decline occurs in reservoir pressure, as in pressure maintenance operations.

The material balance equation is directly linked to original gas in place. In order to consider production of an unconventional gas reservoir, a sufficient amount of gas in place within the reservoir is a concern. For shale, this usually means it is a hydrocarbon source that generated large volumes of either thermal or biogenic gas. To have generated such large quantities of gas, shale needs to be rich in organic matter, relatively thick, and should have been exposed to source of heat in excess of usual global geothermal gradients. The presence of adsorbed, trapped, and free gases in fractures contributes to the complexity of the problem.

A general equation for calculating gas in place is:

$$G = A^* h^* \phi^* S_g \tag{7.1}$$

where,

G = Gas in Place
A = Area of the entire trap
h = Average thickness
S_g = Gas saturation
ϕ = Porosity

Note that conventionally, thickness is considered to be net thickness, from which shale breaks are deducted. For analysis of unconventional gas, the total thickness must be considered. Similarly, there cannot be any cutoff point for porosity. For unconventional reservoirs, most porosity occurs within very tight, shale formations and removing those components would seriously falsify total reserve estimate and subsequent prediction.

Total thickness can be calculated from geological estimate of the trap as well as geophysical information. This can later be refined with capillary pressure data. The best estimate of gas saturation is through core analysis.

Organic carbon concentration in the shales is important in deciding its productive potential. Among all shale basins studied here, it appears that organic carbon concentration in productive shales range anywhere between 1% and 10% or more. In some unconventional reservoirs, intervals with high carbon concentration exhibit higher gas in place and generally, the highest matrix porosity and the lowest clay content. It is difficult to come up with a deterministic value of total organic content (TOC, %) as it often differs from basin to basin.

Thermal maturity of the shales is an important parameter in deciding the oil and gas window. Prospective shale must be within thermal maturity window for shale gas production. It is said that shales act as semipermeable membrane and allow only smaller molecules to pass through the sieves, whereas larger molecules choke pore throat and cannot pass through. So it is important to locate on the transition from gas to oil window, as wells in the oil window are subjected to poor performance if developed as gas wells. Thermal maturity is generally represented by vitrinite reflectance (Ro, %). Vitrinite reflectance is measured in the core analysis. Vitrinite reflectance is one of the organic geochemical indicators of petroleum maturation. The principal maceral groups in coals provide the basic Van Krevelen diagram, which depicts path of their evolution during carbonization. Paths progressively approach origin depicting 100% carbon. The macerals are distinguished by their plan precursors. Vitrinite is one of the macerals that includes both telinite, in which woody structures are present and collinite, essentially structureless matrix, cement, and cavity infilling. Vitrinite is not fluorescent. It is primarily a humic organic material.

Such analysis also applies to other unconventional reservoirs for which the reservoir rock is the same as the source rock. This includes several shale plays within sandstone and even carbonate formations, volcanic reservoirs, and gas hydrates.

Water saturation is very difficult to measure in shale gas reservoirs. The reason being all water saturation equations developed till date are designed for nonshale lithology and concept of net pay based on nonshaly zones in a reservoir. So water saturation could be measured from the core analysis more reliably in shale reservoirs.

In the material balance process, gas storage mechanism is part of the history of the reservoir. It is important to know how the gas is stored before producing it. For shale gas, there are three possibilities:

1. Most of the gas (>50%) is adsorbed on the shale matrix and remaining is stored in the matrix.
2. Most of the gas is stored in the matrix and fractures, and adsorption is not so an important phenomenon.
3. Most of the gas is stored in fractures. Matrix storage is not possible due to absolute absence of porosity and permeability.

Out of these, item 3 is rarely seen but it is perceived as a possibility. When the gas is adsorbed on the matrix, shale gas reservoirs can be treated as special case of coalbed methane (CBM) reservoirs. Dewatering of the shales is required before actual gas production. So wells are treated at low operating pressures. In other Devonian shales, where both phenomena are present, i.e., adsorption and matrix storage both exist, production mechanism is decided on well to well basis. Reservoirs where matrix storage is major, it is seen that these wells have high initial decline but produce for longer time; so payout period for these wells is long but produce for 30–50 years, as matrix gas diffuses into fractures slowly.

7.2.1.2 Decline Curve

The rate of oil production decline generally follows one of the following mathematical forms: exponential, hyperbolic, and harmonic. The following assumptions apply to the decline curve analysis.

- The past processes continue to occur in the future;
- Operation practices are assumed to remain same.

Typically, it is assumed that there is a pseudostate for which the reservoir pressure behaves linearly with time. The necessary condition for decline curve analysis is that there should be no interference from neighboring wells. The decline curve analysis gives not only information about the extent of the reservoir, but also the future production rate versus time; hence this plot is often used to give a forecast of cumulative gas production. The reservoir is assumed to act as being infinite in extent till the pressure diffusion hits the boundaries. The duration until the reservoir acts as infinite in extent is called transient behavior. The analysis of this transient behavior gives us information about the permeability and skin of the well. Since well test analysis analyzes transient behavior, the resolution of pressure gauge should be as high as possible and time duration should be in seconds preferably. The pressure diffusion in the reservoir depends upon the permeability and porosity of the reservoir. Since both the permeability and porosity are very low for tight matrix such as coalbed reservoirs, it takes very long time for the pressure to diffuse deeper and further away from the wellbore. Well test analysis can be used to detect not only the boundaries, but also the type of boundaries. For special cases such as dual porosity reservoirs, well test analysis can reveal the fracture and matrix permeability. However, to date, no reservoir simulator or analytical tool exists that can discern between fracture and matrix permeability.

Recovery, at any point in time, can be defined as the percentage of original hydrocarbons that have been produced. Recovery from any natural gas reservoir can be expressed as:

$$G_{pD} = \frac{G_p}{G} = \frac{G - G_a}{G} \tag{7.2}$$

where,

G_{pD} = recovery factor, dimensionless
G_p = cumulative volume of gas produced, Mscf
G = original volume of gas in place, Mscf
G_a = volume of gas in place at abandonment, Mscf

If the real gas law is invoked, for n moles of gas, Eqn (7.2) can be turned into:

$$G_{pD} = 1 - \frac{P_a V_a z_i}{P_i V_i z_a} \tag{7.3}$$

where,

P_i = initial reservoir pressure, psia
V_i = initial hydrocarbon pore volume (HCPV)
P_a = reservoir pressure at abandonment
V_a = HCPV at abandonment
Z = real gas compressibility factor, dimensionless

For decline curve analysis, pressure conditions prevalent in an unconventional reservoir are of paramount importance. Powers (1967) pointed out that there are two types of water in clays: normal pore water and structured water that is bonded to the layer of montmorillonite clays (smectites). When illitic or kaolinitic clays are buried, a single phase of water emission occurs because of compaction in the first 2 km of burial. When montmorillonite rich-muds are buried, however, two periods of water emission occur: an early phase and a second quite distinct phase when the structured water is expelled during the collapse of the montmorillonite lattice as it changes to illite. Further work by Burst (1969) detailed the transformation of montmorillonite to illite and showed that this change occurred at an average temperature of some 100–110 °C, right in the middle of oil generation window. The actual depth at which this point is reached varies with the geothermal gradient, but Burst (1969) was able to show a normal distribution of productive depth at some 600 m above the clay dehydration level. By integrating geothermal gradient, depth, and the clay change point, it was possible to produce a fluid distribution model for the Gulf Coast area. Barker (1972) has pursued this idea, showing that not only water, but also hydrocarbons may be attached to the clay lattice.

Obviously, the hydrocarbons will be detached from the clay surface when dehydration occurs. The exact physical and chemical process whereby oil is expelled from the source rock is not clear. Clay dehydration is only one of the several causes of supernormal pressure. Inhibition of normal compaction due to rapid sedimentation, and the formation of pore-filling cements, can also cause high pore pressures. Furthermore, some major hydrocarbon provinces do not have supernormal pressures. In some present normally pressurized basins, and some instances, such as the Wessex

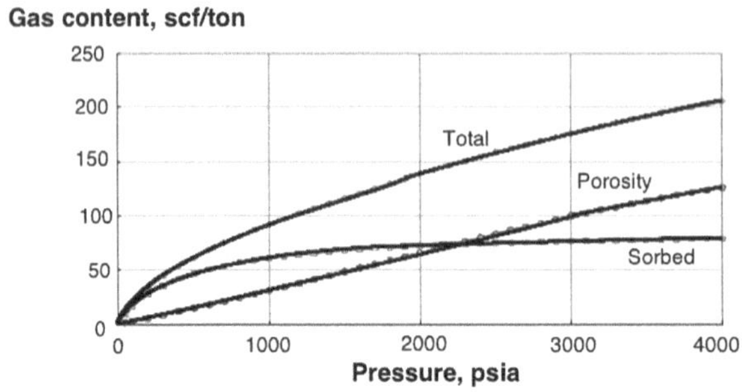

Figure 7.3 Gas content in free and absorbed gas. *From Frantz et al., 2005.*

Basin of southern England, the presence of fibrous calcite along veins shows traces of petroleum (Stonely, 1982).

Figure 7.3 shows how total gas content can be much higher than free gas, which is the only one accounted for in a conventional reservoir. This change in gas content with pressure in the Barnett shale indicates that for low-permeability formations, pseudo-radial flow can take over 100 years to be established. Thus, most gas flow in the reservoir is a linear flow from the near fracture area toward the nearest fracture face. This graph shows the complication in estimating performance of unconventional gas reservoir.

Similarly, pressure and drainage in CBM reservoirs are different from conventional reservoirs. Methane drainage is the process of removing the methane gas contained in the coal seam and surrounding strata through pipelines. Control techniques for use in coal mines can be divided into three categories: (1) dilution ventilation, (2) blocking or diverting gas flow in the coalbed by means of seals, and (3) removing relatively pure or diluted methane through boreholes. The drained gas is mostly methane. However, the main mechanism of methane production is dewatering or water production in contrast to conventional oil or gas production, which relies on depressurizing the pore pressure. There are four main methods used for methane recovery:

- the vertical degas;
- vertical gob degas system;
- horizontal borehole degas system; and
- the cross-measure borehole system.

Overall, unconventional gas wells show decline depicted in Figure 7.4. This figure shows how pressure decline can be arrested with hydraulic fracturing. However, this is an optimistic scenario. In previous chapters, it has been pointed out that the success of hydraulic fracturing depends

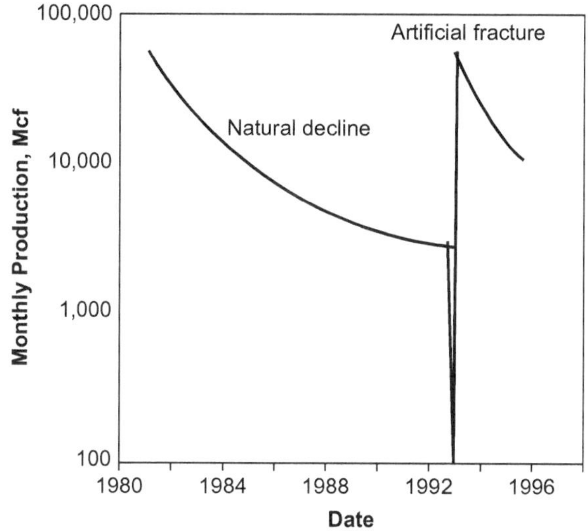

Figure 7.4 Typical decline curve for a tight gas reservoir. *Redrawn from Holditch, 2006.*

on specifics of a reservoir. Note that hydraulically fractured tight gas wells do not decline exponentially. In almost all cases, they will decline hyperbolically.

Gas production can be analyzed using log–log plots of rate or dimensionless rate versus time or dimensionless time, plots of inverse of normalized rates versus square root of time (called square root plots), following material balance plots, dimensionless plots, rate derivative plots, rate integral plots, and rate integral derivative plots, among others. Anderson et al. (2010) point out that the first three are particularly well-suited for tight gas and shale gas production analyses.

If one plots square root plots, log–log rate (q) versus time (t) plots, and log–log plots of dimensionless rate (q_D) versus dimensionless time (t_D), the plots show the different flow regimes exhibited by the reservoir, while the square root plot indicates the size of the "stimulated reservoir volume" during linear flow. The dimensionless variables are as defined below:

$$q_D = 141.2 \frac{q\mu}{kh(p_i - p_{wf})}$$

$$t_D = 0.0002637 \frac{kt}{\phi \mu c_t x_f^2}.$$

where,

μ = viscosity, cp

q = gas flow rate in reservoir, cu ft

k = the matrix permeability, mD

h = the reservoir thickness, ft

p_i = the initial reservoir pressure, psia

p_{wf} = the flowing bottomhole pressure, psia

t = time, h

ϕ = porosity as a fraction

c_t = total compressibility, psi^{-1}

x_f = the fracture half-length, ft

All decline curve analysis techniques depend on solving the diffusivity equation, which is given in the following form in the case of anisotropic reservoir (Eqn (7.4)):

$$k_x \frac{\partial^2 p}{\partial x^2} + k_y \frac{\partial^2 p}{\partial y^2} + k_z \frac{\partial^2 p}{\partial z^2} = \phi \mu c_t \frac{\partial p}{\partial t} \quad (7.4)$$

All decline curve analyses involve solution of the diffusivity equation with analytical models. Because the equation is inherently nonlinear, simplifications must be made in order to solve the equation analytically. Most common assumption involves a semi-infinite domain. Others have used infinite extent of the reservoir. Yet, others have also used closed rectangular system in order to solve the governing equation. The infinite model has no-flow boundaries at the top and the bottom. The semi-infinite reservoir model has three no-flow boundaries (top, bottom, and left). The closed reservoir model has all four no-flow boundaries. Placing the well within the boundary poses additional constraints due to the imposition of a boundary condition. Typically, unconventional reservoirs are modeled with horizontal wells. Such wells are simulated as a line drive. This modeling makes the additional assumption that there is no pressure drop along the wellbore. For most cases, of gas flow, this is a reasonable assumption. Ozkan et al. (1987) solved the equation with Laplace transform, using infinite domain for a horizontal well and closed rectangular box for the reservoir. Results are compared with numerical simulation results in Figure 7.5.

The presence of natural fracture is inherent to unconventional gas. As such, any modeling should include natural fractures. While it has been recognized that natural fractures should produce oscillatory surge in oil/gas production, no one has attempted to model natural fractures that would

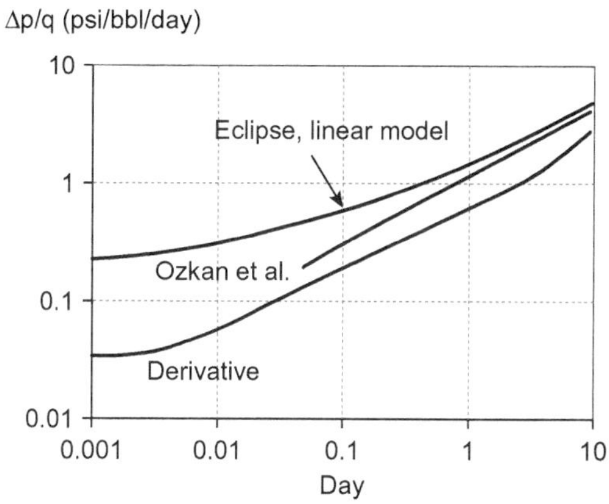

Figure 7.5 Comparison of linear model (homogeneous, zero skin, closed) with Fekete, ECLIPSE, and Ozkan Laplace solution. *Redrawn from Bello, 2009.*

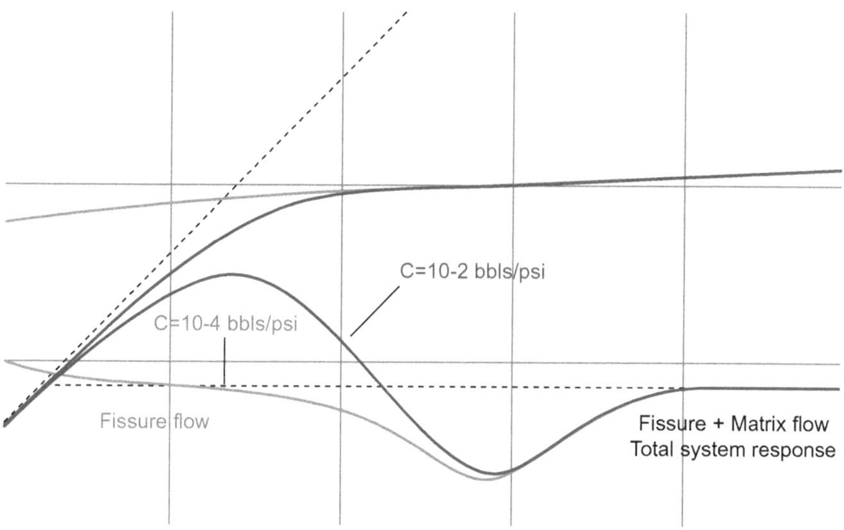

Figure 7.6 Pressure response in a fracture formation.

produce such a production regime. Figure 7.6 shows how a dual porosity, dual permeability regime would produce gas surges in periodic fashion. Such behavior can be explained with interactions between capillary flow, storage, and fracture flow. Figure 7.7 shows experimental results conducted by Delgadillo-Velazquez et al., (2008). Such behavior occurs because of difference in shear rates and capillary intrusion, which is typical in a porous medium with fractures. In addition, fluid rheology plays a role, which adds further complication in modeling (Hossain and Islam, 2008).

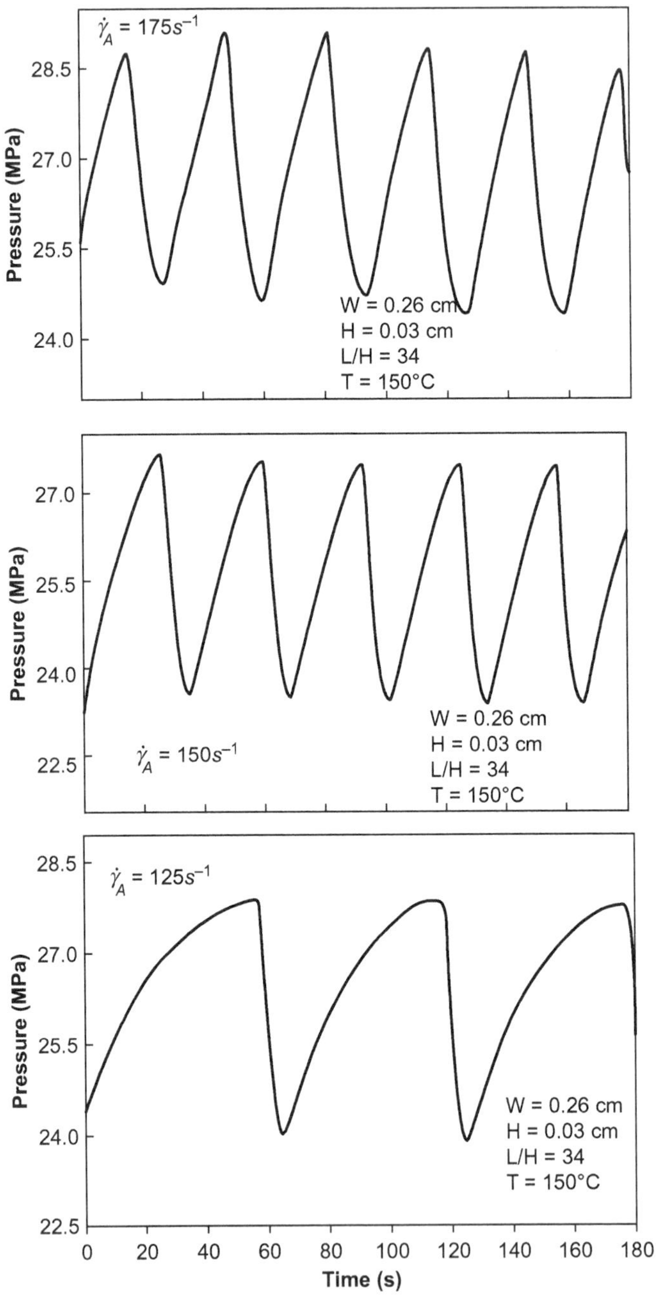

Figure 7.7 Typical pressure oscillations in slit flow at 150 °C. *From Delgadillo-Velazquez et al., 2008.*

In order to model shale gas and tight gas reservoirs, various analytical and semianalytical solutions have been proposed. Gringarten (Gringarten and Ramey, 1973; Gringarten, 1984) developed some of the early analytical models for flow through domains involving a single vertical fracture and a single horizontal fracture. A more complex series of semianalytical models for single vertical fractures were developed much later (Blasingame and Poe Jr., 1993). Prior to the development of models for multiple-fracture horizontal wells (Medeiros et al., 2006), it was common practice to represent these multiple fractures with an equivalent single fracture. In all these, however, the assumptions are the same and ultimate results are not improved from scientific perspective.

Several other analytical and semianalytical models have been developed since Bello and Wattenbarger (2008); Mattar (2008); Anderson et al., (2010). Although these models are much faster than numerical simulators, they generally cannot accurately handle the very high nonlinear aspects of shale gas and tight gas reservoirs because these analytical solutions address the nonlinearity in gas viscosity, compressibility, and compressibility factor with the use of pseudo-pressures (an integral function of pressure, viscosity, and compressibility factor) rather than solving the real gas flow equation. Other limitations include the difficulty in accurately capturing gas desorption from the matrix, multiphase flow, unconsolidation, and several nonideal and complex fracture networks (Houze et al., 2010).

The limitations of the analytical and semianalytical models have led to the use of numerical reservoir simulators to study the reservoir performance of these unconventional gas reservoirs. Miller et al., (2010) and Jayakumar et al., (2011) showed the application of numerical simulation to history-matching and forecasting production from two different shale gas fields, while Cipolla et al., (2009), Freeman et al., (2009), and Moridis et al. (2010) conducted numerical sensitivity studies to identify the most important mechanisms and factors that affect shale gas reservoir performance. These numerical studies show that the characteristics and properties of the fractured system play a dominant role in the reservoir performance. Hence, significant effort needs to be invested in the characterization and representation of the fractured system.

Carvalho and Rosa (1988) presented solutions for an infinite conductivity horizontal well in a semi-infinite reservoir. The reservoir was considered to be homogeneous and isotropic. The horizontal well is modeled as a line source. The solutions for the homogeneous case were then extended to the dual porosity case by substituting $s^*f(s)$ for s in Laplace space for the pressure derivative (homogeneous). Wellbore storage and skin are incorporated into their model using Laplace space.

Aguilera and Ng (1989) presented analytical equations for pressure transient analysis. Their model is a horizontal well in a semi-infinite, anisotropic, naturally fractured reservoir. Transient and pseudo-steady state interporosity flow is considered. Six flow periods are identified—first radial flow (at early times, from fractures), transition period, second radial flow in vertical plane, first linear flow, pseudo-radial flow, and late linear flow—with expressions for determining the skin provided.

Ng and Aguilera (1989) presented analytical solutions using a line source and then compute pressure drop on a point away from the well axis to account for the radius of the actual well. A method for determining the numerical Laplace transform is presented. This method was then used to compute the dual porosity response (pseudo-steady state).

Thompson et al. (1991) presented an algorithm for computing horizontal well response in a bounded dual porosity reservoir. Their model is a horizontal well in a closed rectangular reservoir. Their procedure involves converting a known analytic solution to Laplace space numerically point by point and then inverting using the Stehfest algorithm. This is similar to the procedure presented by Ohaeri and Vo (1991) who use a numerical Laplace space algorithm, but present alternative equations determined by parameter ranges that result in computational efficiency.

Du and Stewart (1992) described situations that can yield linear flow behavior: a multilayered reservoir (one layer has a very high permeability relative to the other); naturally fractured reservoir (flow from matrix into horizontal well intersecting fractures); and areal anisotropy (vertical fractures aligned predominantly in one direction). Their model is a horizontal well in a homogeneous, infinite acting reservoir. Three flow regimes are identified: radial vertical flow, linear flow opposite to completed section, and pseudo-radial flow at late time. A bilinear flow behavior was also identified.

7.2.1.3 Statistical Method

In this method, the past performance of numerous reservoirs is statistically accounted for to derive the empirical correlations, which are used for future predictions. It may be described as a "formal extension of the analogical method." The statistical methods have the following assumptions.

- Reservoir properties are within the limit of the database.
- Reservoir symmetry exists.
- Ultimate recovery is independent of the rate of production.

In addition, Islam et al. (2014) recently pointed out a more subtle, yet far more important shortcoming of the statistical methods. Practically, all statistical methods assume that two or more objects based on a limited number of tangible expressions make it legitimate to comment on the underlying science. It is equivalent to stating that if effects show a reasonable correlation, the causes can also be correlated. As Zatzman and Islam (2006) pointed out, this poses a serious problem as, in absence of time–space correlation (pathway rather than end result), anything can be correlated with anything, making the whole process of scientific investigation spurious. They make their point by showing the correlation between global warming (increases) with a decrease in the number of pirates (Figure 7.8). The absurdity of the statistical process becomes evident by drawing this analogy. Shapiro et al. (2006) pointed out another severe limitation of the statistical method. Even though they commented on the polling techniques used in various surveys, their comments are equally applicable in any statistical modeling. They wrote: "Frequently, opinion polls generalize their results to a U.S. population of 300 million or a Canadian population of 32 million on the basis of what 1,000 or 1,500 'randomly selected' people are recorded to have said or answered. In the absence of any further information to the contrary, the underlying theory of mathematical statistics and random variability assumes that the individual selected 'perfectly' randomly is no more nor less likely to have any one opinion over any other. How perfect the randomness may be determined from the 'confidence' level attached to a survey, expressed in the phrase that

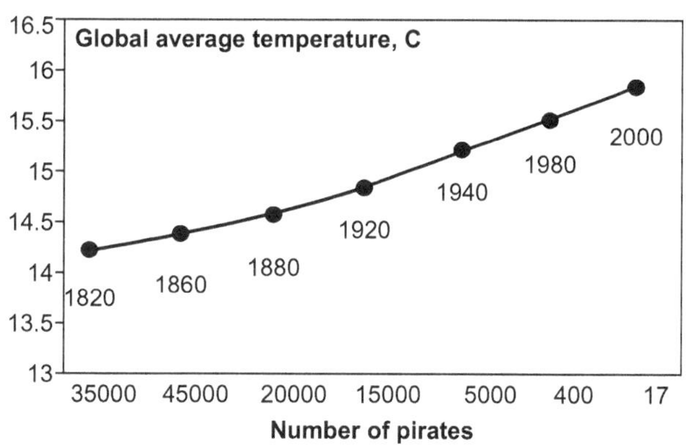

Figure 7.8 Using statistical data to develop a theoretical correlation that can make an aphenomenal model appealing, depending on which conclusion would appeal to the audience.

describes the margin of error of the poll sample lying plus or minus some low single-digit percentage 'nineteen times out of twenty', *i.e.*, a confidence level of 0.95. Clearly, however, assuming *in the absence of any knowledge otherwise* a certain state of affairs to be the case, *viz.*, that the sample is random and no one opinion is more likely than any other, seems more useful for projecting horoscopes than scientifically assessing public opinion."

Probability experiments have been some of the least controversial techniques for determining future course of actions. However, recently it has come to light that such assertion is not warranted (see Islam et al., 2014a, 2014b, 2015).

Consider the following example. If you know there are three white balls and two black balls in a bag, what is the probability of finding a white ball if you attempt to withdraw a ball from the bag? Of course, it is three out of five. The same way, probability of drawing a black ball is two out of five. However, if one asks what is the probability of drawing a pink ball, what would be the answer? If the answer is "zero," it would violate the quantum physics principle that nothing is absolutely improbable, except the existence of a creator. What other principle does a zero probability violate?

Scientific answer is the probability of finding a pink ball is zero unless someone truthful confirms that the premise that only black and white balls exist in the bag is untrue or the actual content of the bag is unknown as a function of time.

Say after a certain time, a draw was made and a pink ball was drawn at the first trial. This observation shows that the premise that there were only black and white balls present in the bag is untrue. This confusion could be avoided if there were some room to account for the time function or history of the bag in question in probability theories. There is none. Probability theory by definition assumes steady state in all matters and does not include the time function. How would it look like if the time function were included?

Consider the following example. A draw of head and tail is made repeatedly. Probability theory assumes that if the draw is made infinite number of times, the ratio of head and tail would be 1. In essence, it introduces a time function that is self-imposed and there is no way to verify the validity of this time function. It is because infinity is not achievable in a physical experiment. In Figure 7.9, an actual observation of coin toss is plotted. The y-axis represents the ratio of head and tail, while the x-axis represents the number of coin toss. It is clear that as the number of coin toss

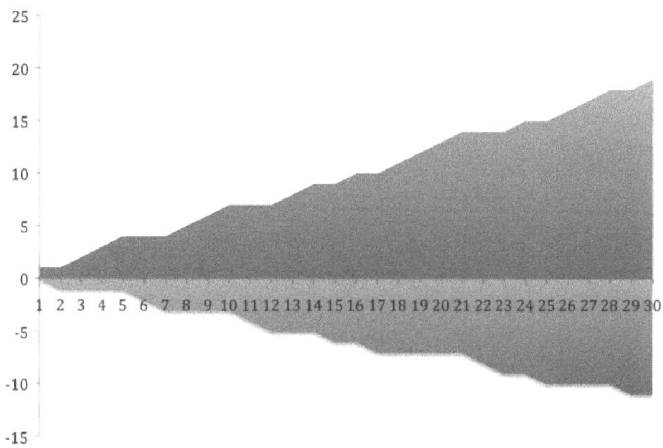

Figure 7.9 Number of coin toss vs head (+sign) and tail (−sign) in an actual experiment.

increases, the most probable occurrence of head or tail settles around a ratio of 2. This is in sharp contrast to the conventional notion of 1.

Figure 7.10 also shows how at no time the distribution of head and tail follows the 1:1 rule. As the number of coin toss is increased, the distribution actually diverges and settles toward a value outside of 1:1 distribution. If the number of coin toss is increased, the conventional approach says the number of head and tail ratio would be closer to 1. It means, if this number is increased to infinity, the ratio would converge to 1.

This simple experiment goes to demonstrate how a fundamentally flawed paradigm is introduced by selecting a probability model.

Theoretically, if the premise of "steady state" probability is removed, data gathered on certain population can be helpful only if the data volume

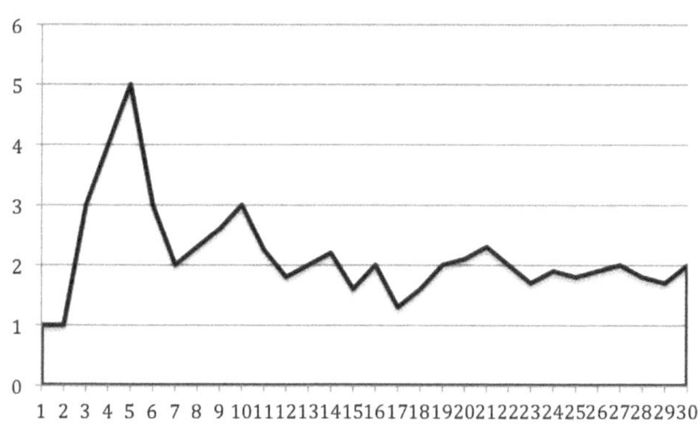

Figure 7.10 As the number of coin toss is increased, the head and tail toss ratio converges toward 2 and not 1.

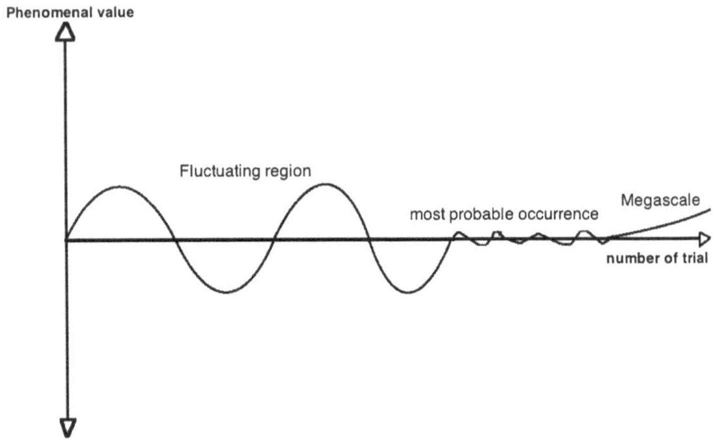

Figure 7.11 Number of trials is likely to yield different sets of phenomenal values that apply to different entities.

and time involved are representative. In engineering, the notion of representative elementary volume (REV) is well known. Here, we introduce the notion of representative elemental time, which is in essence characteristic frequency of a process. For every entity, there is a characteristic frequency, which itself is a function of time. This notion has been discussed by Islam et al. (2014a). Figure 7.11 represents the relationship between characteristic value and number of trials. Similar figure applies to the number of subjects in a survey.

One side effect of statistical rules is the so-called Simpson's paradox. This paradox states that for continuous data, a positive trend appears for two separate groups (blue and red) and a negative trend (black, dashed) appears when the data are combined (Figure 7.12). In probability and statistics, Simpson's paradox, or the Yule–Simpson effect, is a paradox in which a trend that appears in different groups of data disappears when these groups are combined, and the reverse trend appears for the aggregate data. This result is often encountered in social-science and medical-science statistics. Islam et al. (2010a) discussed this phenomenon as something that is embedded in Newtonian calculus that allows taking the infinitely small differential and turning that into any desired integrated value, while giving the impression that a scientific process has been followed. Furthermore, Khan and Islam (2012) showed that true trend line should contain all known parameters. The Simpson's paradox can be avoided by including full historical data, followed by scientifically true processing (Islam et al., 2014a; Pearl, 2009).

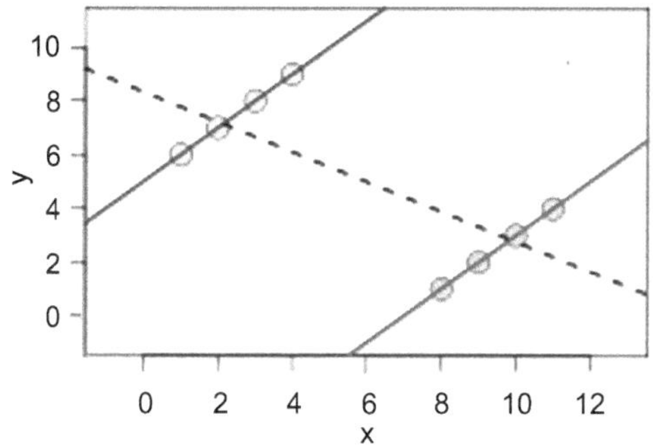

Figure 7.12 Simpson's paradox highlights the problem of targeted statistics.

The above difficulty with statistical processing of data was brought into highlight through the publication of the following correlation between number of pirates versus global temperature with the slogan: Join piracy, save the planet.

Similar paradox arises from the so-called ecological fallacies. It is best described in Wikipedia with the following examples:

> *Assume that at the individual level, being Protestant impacts negatively one's tendency to commit suicide but the probability that one's neighbor commits suicide increases one's tendency to become Protestant. Then, even if at the individual level there is negative correlation between suicidal tendencies and Protestantism, there can be a positive correlation at the aggregate level.*
>
> *Similarly, even if at the individual level, wealth is positively correlated to tendency to vote Republican, we observe that wealthier states tend to vote Democrat. For example, in 2004, the Republican candidate, George W. Bush, won the fifteen poorest states, and the Democratic candidate, John Kerry, won 9 of the 11 wealthiest states. Yet 62% of voters with annual incomes over $200,000 voted for Bush, but only 36% of voters with annual incomes of $15,000 or less voted for Bush.*

The prosecutor's fallacy is a fallacy of statistical reasoning, typically used by the prosecution to argue for the guilt of a defendant during a criminal trial. In its crudest form, it involves the assertion that the probability of defendant to be guilty is 90% because the perpetrator and the defendant are known to be sharing the blood type that has a probability of 10% in the general population. It is purported that a DNA match is not a fallacy because the probability of match is much greater. However, this is also an example of how New Science has refined techniques in favor of opacity instead of transparency. The use of DNA as the only match has the gravest risk of "creating" evidence in case there is no other evidence. This is rarely talked about. Such mind-set has promoted prosecution tactics that allowed torture as a means of extracting "evidence." In a broader sense, it has also allowed notorious conclusions, such as the nonconsideration of junk DNA, asserting probability of Big Bang as 97%, probability of life as "reasonable," probability of "intelligent life" as even "more reasonable," and others. All of them suffer from the fundamental basis that a "fact is not a matter of probability." For instance, the proclamation that orangutan is linked to humans because the DNA match is the greatest (Derbyshire, 2011), What can be said when a greater match of some other component of genome is found with certain plants?

7.2.1.4 Finite Difference Methods

Finite difference calculus is a mathematical technique, which is used to approximate values of functions and their derivatives at discrete points, where they are not known. The history of differential calculus dates back to the time of Leibnitz and Newton. In this concept, the derivative of a continuous function is related to the function itself. The Newton's formula is the core of differential calculus and suffers from the approximation that magnitude and direction change independently of one another. There is no problem in having separate derivatives for each component of the vector or in superimposing their effects separately and regardless of order. That is what mathematicians mean when they describe or discuss Newton's derivative being used as a "linear operator." Following this, comes Newton's difference quotient formula. When the value of a function is inadequate to solve a problem, the rate at which the function changes, sometimes, becomes useful. Therefore, the derivatives are also important in reservoir simulation. In Newton's difference quotient formula, the derivative of a continuous function is obtained. This method relies implicitly on the notion of approximating instantaneous moments of curvature, or infinitely small segments, by means of straight lines. This alone should have tipped everyone off that his derivative is a linear operator precisely because, and to the extent that, it examines change over time (or distance) within an already established function (Islam et al., 2010). This function is applicable to an infinitely small domain, making it nonexistent. When integration is performed, however, this nonexistent domain is assumed to be extended to finite and realistic domain, making the entire process questionable.

The finite difference methods are extensively applied in petroleum industry to simulate the fluid flow inside the porous medium. The following assumptions are inherent to the finite difference method.

1. The relationship between derivative and the finite difference operators, e.g., forward, backward, and central difference operators, is established through the Taylor series expansion. The Taylor series expansion is the base element in providing the differential form of a function. It converts a function into polynomial of infinite order. This provides an approximate description of a function by considering a finite number of terms and ignoring the higher order parts. In other words, it assumes that a relationship between the operators of discrete points and the operators of the continuous functions is acceptable.
2. The relationship involves truncation of the Taylor series of the unknown variables after few terms. Such truncation leads to accumulation of error. Mathematically, it can be shown that most of the error occurs in the lowest order terms.
   a. The forward and the backward difference approximations are the first order approximations to the first derivative.
   b. Although the approximation to the second derivative by the central difference operator increases accuracy because of a second order approximation, it still suffers from the truncation problem.
   c. As the spacing size reduces, the truncation error approaches to zero more rapidly. Therefore, a higher order approximation will eliminate the need of same number of measurements or discrete points. It might maintain the same level of accuracy; however, less information at discrete points might be risky as well.
3. The solutions of the finite difference equations are obtained only at the discrete points. These discrete points are defined either according to block-centered or point-distributed grid system. However, the boundary condition, to be specific, the constant pressure boundary, may appear important in selecting the grid system with inherent restrictions and higher order approximations.
4. The solutions obtained for grid points are in contrast to the solutions of the continuous equations.
5. In the finite difference scheme, the local truncation error or the local discretization error is not readily quantifiable because the calculation involves both continuous and discrete forms. Such difficulty can be overcome when the mesh size or the time step or both are decreased leading to minimization in local truncation error. However, at the same time

the computational operation increases, which eventually increases the round-off error.

Zatzman et al., 2007 identified the most significant contribution of Newton in mathematics: the famous definition of the derivative as the limit of a difference quotient involving changes in space or in time as small as anyone might like, but not zero, *viz.*

$$\frac{d}{dt}f(t) = \lim_{\Delta t \to 0} \frac{f(t+\Delta t) - f(t)}{\Delta t} \tag{7.5}$$

Without regards to further conditions being defined as to when and where differentiation would produce a meaningful result, it was entirely possible to arrive at "derivatives" that would generate values in the range of a function at points of the domain where the function was not defined or did not exist. Indeed, it took another century following Newton's death before mathematicians would work out the conditions—especially the requirements for continuity of the function to be differentiated within the domain of values—in which its derivative (the name given to the ratio quotient generated by the limit formula) could be applied and yield reliable results. Kline (1972) detailed the problems involving this breakthrough formulation of Newton. However, no one in the past did propose an alternative to this differential formulation, at least not explicitly. The following figure (Figure 7.13) illustrates this difficulty.

In this figure, economic index (it may be one of many indicators) is plotted as a function of time. In nature, all functions are very similar. They do have local trends as well as global trends (in time). One can imagine how the slope of this graph on a very small time frame would be quite arbitrary,

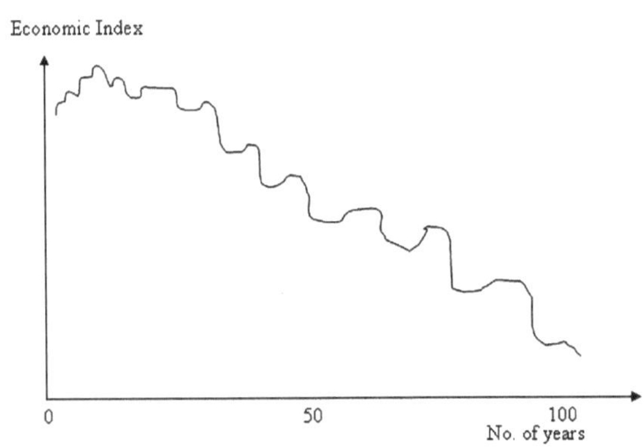

Figure 7.13 Economic well-being is known to fluctuate with time.

and how devastating it would be to take that slope to a long term. One can easily show that the trend emerging from Newton's differential quotient would be diametrically opposite to the real trend. Zatzman and Islam (2007b) provided a basis for determining real gradient, rather than local gradient that emerges from Newton's differential quotient. In that formulation, it is shown that the actual value of Δt over which the reliable gradient has to be observed needs to be several times greater than the characteristic time of a system. The notion of REV, as first promoted by Bear (1972), is useful in determining a reasonable value for this characteristic time. The second principle is that at no time Δt be allowed to approach 0 (Newton's approximation), even when the characteristic value is very small (e.g., phenomena at nanoscale). With the engineering approach, it turns out such approximation is not necessary (Abou-Kassem, 2007) because this approach bypasses the recasting of governing equations into Taylor series expansion, instead relying on directly transforming governing equations into a set of algebraic equations. In fact, by setting up the algebraic equations directly, one can make the process simple and yet maintain accuracy (Mustafiz et al., 2008). Finally, initial analysis should involve the extension Δt to ∞ in order to determine the direction, which is related to sustainability of a process (Zatzman et al., 2008).

Figure 7.14 shows how formulation with the engineering approach ends up with the same linear algebraic equations if the inside steps are avoided. Even though the engineering approach was known for decades (known as the control volume approach), no one identified in the past the advantage of removing in-between steps.

7.2.1.4.1 Darcy's Law

Because practically all reservoir simulation studies involve the use of Darcy's law, it is important to understand the assumptions behind this momentum balance equation. The following assumptions are inherent to Darcy's law and its extension.

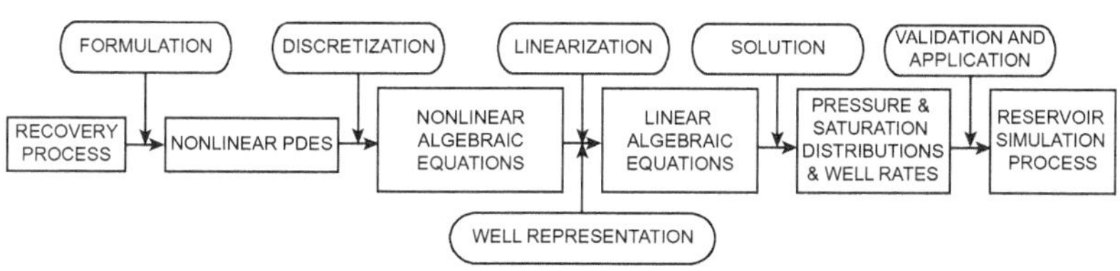

Figure 7.14 Major steps to develop reservoir simulators. *Redrawn from Abou-Kassem et al., 2006.*

- The fluid is homogenous, single phase and Newtonian.
- No chemical reaction takes place between the fluid and the porous medium.
- Laminar flow condition prevails.
- Permeability is a property of the porous medium, which is independent of pressure, temperature, and the flowing fluid.
- There is no slippage effect, e.g., Klinkenberg phenomenon.
- There is no electrokinetic effect.

Clearly, Darcy's law does not apply to unconventional reservoirs. The matrix permeability is too low, which means dispersive term should be included. The inclusion of Brinkman term can solve this problem. As for the fracture, inertial forces are high, thereby making Darcy's law invalid. This is particularly important for gas reservoirs, for which Forchheimer equation should be used. This aspect is discussed later in this chapter.

7.3 RECENT ADVANCES IN RESERVOIR SIMULATION

The recent advances in reservoir simulation may be viewed as:

- speed and accuracy,
- new fluid flow equations,
- coupled fluid flow and geomechanical stress model, and
- fluid flow modeling under thermal stress.

7.3.1 Speed and Accuracy

The need for new equations in oil reservoirs arises mainly for fractured reservoirs as they constitute the largest departure from Darcy's flow behavior. Advances have been made in many fronts. As the speed of computers increased following Moore's law (doubling every 12–18 months), the memory also increased. For reservoir simulation studies, this translated into the use of higher accuracy through inclusion of higher order terms in Taylor series approximation as well as great number of grid blocks, reaching as many as billion blocks. The greatest difficulty in this advancement is that the quality of input data did not improve at par with the speed and memory of the computers. As Figure 7.15 shows that the data gap remains possibly the biggest challenge in describing a reservoir. Note that the inclusion of large number of grid blocks makes the prediction more arbitrary than that predicted by fewer blocks, if the number of input data points is not increased proportionately. The problem is particularly acute when fractured formation is being modeled. The problem of reservoir cores being smaller than the

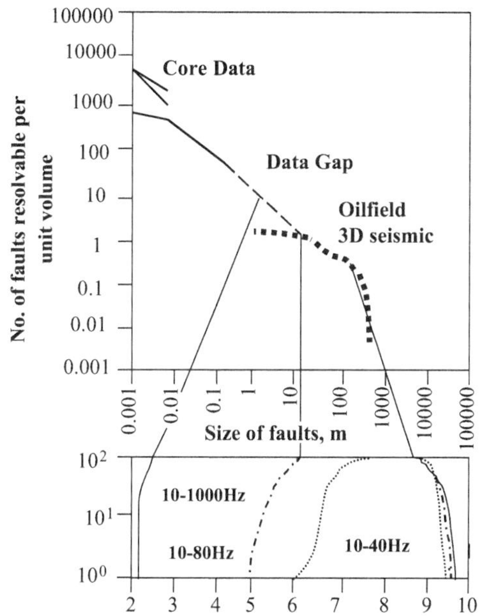

Figure 7.15 Data gap in geophysical modeling. *After Islam, 2001.*

REV is a difficult one, which is more accentuated for fractured formations that have a higher REV. For fractured formations, one is left with a narrow band of grid blocks, beyond which solutions are either meaningless (large grid blocks) or unstable (too small grid blocks).

The problem is particularly intense for unconventional reservoirs for which few logging techniques produce meaningful results and core data are mostly nonrepresentative. The reservoir characterization technique proposed in Chapter 5 outlines a way out of this dilemma. With proper reservoir characterization, the data gap can be filled with filtered data.

7.3.2 New Fluid Flow Equations

A porous medium can be defined as a multiphase material body (solid phase represented by solid grains of rock and void space represented by the pores between solid grains) characterized by two main features: that a REV can be determined for it, such that no matter where it is placed within a domain occupied by the porous medium; and it will always contain both a persistent solid phase and a void space. The size of the REV is such that parameters that represent the distributions of the void space and the solid matrix within it are statistically meaningful.

Theoretically, fluid flow in porous medium is understood as the flow of liquid or gas or both in a medium filled with small solid grains packed in homogeneous manner. The concept of heterogeneous porous medium is then introduced to indicate change in properties (mainly porosity and permeability) within that same solid grains packed system. An average estimation of properties in that system is an obvious solution, and the case is still simple.

Incorporating fluid flow model with a dynamic rock model during the depletion process with a satisfactory degree of accuracy is still difficult to attain from currently used reservoir simulators. Most conventional reservoir simulators do not couple stress changes and rock deformations with reservoir pressure during the course of production and do not include the effect of change of reservoir temperature during thermal or steam injection recoveries. The physical impact of these geomechanical aspects of reservoir behavior is neither trivial nor negligible. Pore reduction and/or pore collapse leads to abrupt compaction of reservoir rock, which in turn causes miscalculations of ultimate recoveries, damage to permeability, and reduction to flow rates and subsidence at the ground and well casing's damage. In addition, there are many reported environmental impacts due to the withdrawal of fluids from underground reservoirs. Using only Darcy's law to describe hydrocarbon fluid behavior in petroleum reservoirs when high gas flow rate is expected or when encountered highly fractured reservoir is totally misleading. Nguyen (1986) has showed that using standard Darcy flow analysis in some circumstances can overpredict the productivity by as much as 100%.

Fracture can be defined as any discontinuity in a solid material. In geological terms, a fracture is any planar or curvy planar discontinuity that has formed as a result of a process of brittle deformation in the earth's crust. Planes of weakness in rock respond to changing stresses in the earth's crust by fracturing in one or more different ways depending on the direction of the maximum stress and the rock type. A fracture can be said to consist of two rock surfaces, with irregular shapes, which are more or less in contact with each other. The volume between the surfaces is the fracture void. The fracture void geometry is related in various ways to several fracture properties. Fluid movement in a fractured rock depends on discontinuities, at a variety of scales ranging from microcracks to faults (in length and width). Fundamentally, describing flow through fractured rock involves describing physical attributes of the fractures: fracture spacing, fracture area, fracture aperture, and fracture orientation and whether these parameters allow percolation of fluid through the rock mass. Fracture parameters also influence the anisotropy and heterogeneity of flow through fractured rock. Thus the conductivity of a rock mass depends on the entire network within the particular rock mass and is thus governed by the connectivity of the network and the conductivity of the single fractures. The total conductivity of a rock mass depends also on the contribution of matrix conductivity at the same time.

A fractured porous medium is defined as a portion of space in which the void space is composed of two parts: an interconnected network of fractures and blocks of porous medium, and the entire space within the medium occupied by one or more fluids. Such a domain can be treated as a single continuum, provided an appropriate REV can be found for it.

The fundamental question to be answered in modeling fracture flow is the validity of the governing equations used. The conventional approach involves the use of dual porosity, dual permeability models for simulating flow through fractures. Choi et al. (1997) demonstrated that the conventional use of Darcy's law in both fracture and matrix of the fractured system is not adequate. Instead, they proposed the use of the Forchheimer model in the fracture while maintaining Darcy's law in the matrix. Their work, however, was limited to single-phase flow. In future, the present status of this work can be extended to a multiphase system. It is anticipated that gas reservoirs will be suitable candidates for using Forchheimer extension of the momentum balance equation, rather than the conventional Darcy's law. Similar to what was done for the liquid system (Cheema and Islam, 1995), opportunities exist in conducting experiments with gas as well as multiphase fluids in order to validate the numerical models. It may be noted that in recent years several dual porosity, dual permeability models have been proposed based on experimental observations (Tidwell and Robert, 1995; Saghir et al., 2001).

For unconventional reservoirs, Brinkman equation has to be used in the matrix. It is the case because of the low permeability of the matrix. This will be detailed in a later section of this chapter.

7.3.3 Coupled Fluid Flow and Geomechanical Stress Model

Coupling different flow equations has always been a challenge in reservoir simulators. In this context, Pedrosa et al. (1986) introduced the framework of hybrid grid modeling. Even though this work was related to coupling cylindrical and Cartesian grid blocks, it was used as a basis for coupling various fluid flow models (Islam and Chakma, 1990). Coupling flow equations in order to describe fluid flow in a setting, for which both pipe flow and porous media flow prevail, continues to be a challenge (Islam et al., 2010). For unconventional gas, horizontal wells are of great importance. Even though it is commonly perceived that pressure drop in the horizontal well within a gas reservoir is negligible, this is not the case in many scenarios, particularly the ones involving multiphase flow. For example, one can cite coalbed methane, condensate, and even some tight gas formations with oil flow.

Geomechanical stresses are very important in production schemes. However, due to strong seepage flow, disintegration of formation occurs and sand is carried toward the well opening. The most common practice to prevent accumulation as followed by the industry is to take filter measures, such as liners and gravel packs. Generally, such measures are very expensive to use and often, due to plugging of the liners, the cost increases to maintain the same level of production. In recent years, there have been studies in various categories of well completion including modeling of coupled fluid flow and mechanical deformation of medium (Vaziri et al., 2003). Vaziri et al. (2003) used a finite element analysis developing a modified form of the Mohr–Coulomb failure envelope to simulate both tensile and shear-induced failure around deep wellbores in oil and gas reservoirs. The coupled model was useful in predicting the onset and quantity of sanding. Nouri et al. (2006) highlighted the experimental part of it in addition to a numerical analysis and measured the severity of sanding in terms of rate and duration. It should be noted that these studies (Nouri et al., 2002; Vaziri et al., 2003 and Nouri et al., 2006) took into account the elastoplastic stress–strain relationship with strain softening to capture sand production in a more realistic manner. Although, at present these studies lack validation with field data, they offer significant insight into the mechanism of sanding and have potential in smart designing of well completions and operational conditions.

Settari et al. (2006) applied numerical techniques to calculate subsidence induced by gas production in the North Adriatic. Due to the complexity of the reservoir and compaction mechanisms, Settari et al. (2006) took a combined approach of reservoir and geomechanical simulators in modeling subsidence. As well, an extensive validation of the modeling techniques was undertaken, including the level of coupling between the fluid flow and geomechanical solution. The researchers found that a fully coupled solution had an impact only on the aquifer area and an explicitly coupled technique was good enough to give accurate results. On grid issues, the preferred approach was to use compatible grids in the reservoir domain and to extend that mesh to geomechanical modeling. However, it was also noted that the grids generated for reservoir simulation are often not suitable for coupled models and require modification.

In fields, on several instances, subsidence delay has been noticed and related to overconsolidation, which is also termed as the threshold effect (Merle et al., 1976; Hettema et al., 2002). Settari et al. (2006) used the numerical modeling techniques to explore the effects of small levels of overconsolidation in one of their studied fields on the onset of subsidence and the areal extent of the resulting subsidence bowl. The same framework that Settari et al. (2006) used can be introduced in coupling the multiphase, compositional simulator and the geomechanical simulator in future.

7.3.4 Fluid Flow Modeling under Thermal Stress

The temperature changes in the rock can induce thermoelastic stresses (Hojka et al., 1993), which can either create new fractures or can alter the shapes of existing fractures, changing the nature of primary mode of production. It can be noted that the thermal stress occurs as a result of the difference in temperature between injected and reservoir fluids or due to the Joule–Thomson effect. However, in the study with unconsolidated sand, the thermal stresses are reported to be negligible in comparison to the mechanical stresses (Chalaturnyk et al., 1995). Similar trend is noticeable in the work by Chen et al. (1995), which also ignored the effect of thermal stresses, even though a simultaneous modeling of fluid flow and geomechanics is proposed.

Most of the past research has been focused only on thermal recovery of heavy oil. Modeling subsidence under thermal recovery technique (Tortike and Farouq Ali, 1987) was one of the early attempts that considered both thermal and mechanical stresses in their formulation. There are only few investigations that attempted to capture the onset and propagation of fractures under thermal stress. Zekri et al. (2006) investigated the effects of thermal shock on fractured core permeability of carbonate formations of UAE reservoirs by conducting a series of experiments. Also, the stress–strain relationship due to thermal shocks was noted. Apart from experimental observations, there is also the scope to perform numerical simulations to determine the impact of thermal stress in various categories, such as water injection, gas injection/production, etc.

7.3.5 Challenges of Modeling Unconventional Gas Reservoirs

Most modeling approaches in unconventional gas reservoirs involve flow equations that are not valid for unconventional cases. In addition, complexities arise from complex geometry that prevails in unconventional reservoirs. For instance, the traditional approach of modeling fractured shale gas reservoirs with regular Cartesian grids could be limited in that it cannot efficiently represent complex geologies, including nonplanar and nonorthogonal fractures, and cannot adequately capture the elliptical flow geometries expected around the fracture tips in such fractured reservoirs. It also suffers from the fact that the number of grids can easily grow into millions because of the inability to change the orientation and shape of the grids away from the fracture tips, thus requiring extremely fine discretization in an attempt to describe all possible configurations. This problem is also aggravated by the large number (up to 60) of hydraulic fractures in a clustered fracture system (Jayakumar et al., 2011). The problem is more intense for natural fractures.

Moridis et al. (2010) identified the distinct fracture systems present in producing shale gas and tight gas reservoirs. Figure 7.16 shows a graphical illustration of the following four fracture systems.

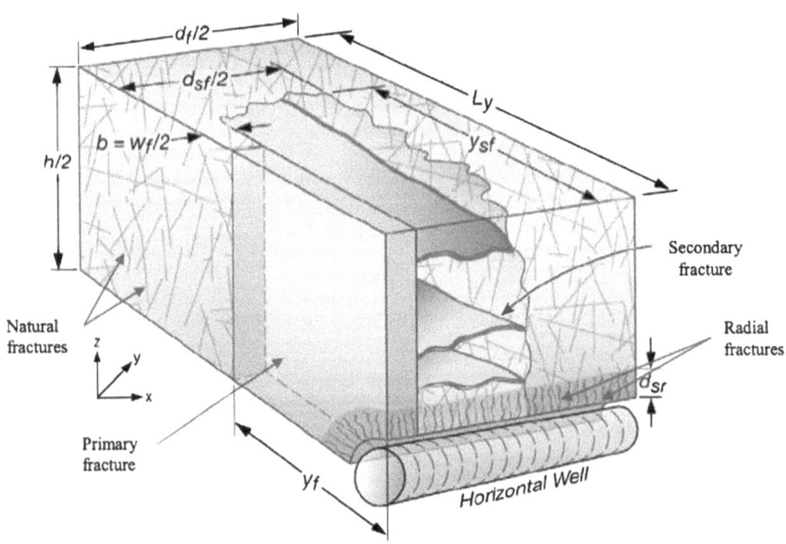

Figure 7.16 Identification of the four fractured systems. *From Moridis et al., 2010.*

- Primary or hydraulic fractures: These are fractures that are typically created by injecting hydrofracturing fluids (with or without proppants) into the formation. Proppants provide high-permeability flow paths that allow gas to flow more easily from the formation matrix into the well. Artificial fractures are predominantly orthogonal to natural fractures.
- Secondary fractures: These fractures are termed "secondary" because they are induced as a result of changes in the geomechanical status of a rock when the primary fractures are being created. Microseismic fracture mappings suggest that they generally intersect the primary fractures, either orthogonally or at an angle. Most prior studies assumed ideal configurations with orthogonal and planar fracture intersections so as to simplify the gridding; but in this work, we have developed an unstructured mesh maker that facilitates the gridding of nonorthogonal, nonplanar, and other nonideal fracture geometries.
- Natural fractures: As the name implies, these fractures are native to the formation in the original state, prior to any well completion or fracturing process. These fractures are result of faults that are part of broader tectonic of the reservoir.
- Radial fractures: These are fractures that are created as a result of stress releases in the immediate neighborhood of the horizontal well. This is integral part of the invasion due to drilling.

Efforts have been made in order to simplify the above geometry in a tractable form. Houze et al. (2010) recognized the importance of explicitly gridding secondary fractures so as to quantify the interaction between primary and secondary networks as distinct systems, using either a regular orthogonal pattern or a more random and complex system. In this work, we have identified two classes of possible fracture geometries/orientations:

- Regular or ideal fractures: These are idealized fracture geometries, which are usually planar and orthogonal. A perfectly planar (or orthogonal fracture) is the idealized geometry used in numerical studies using Cartesian grids. Figure 7.17a gives an illustration of this fracture geometry.
- Irregular or nonideal fractures: These are the kinds of fracture geometries we are likely to encounter in real life. They could be nonorthogonal, meaning that the fractures intersect either the well (for primary fractures) or primary fractures (for secondary fractures) at angles other than 90°, and they could be complex, meaning that the fractures are not restricted to a flat, nonundulating plane. Figures 7.17b and 7.17c give a diagrammatic illustration of two such scenarios.

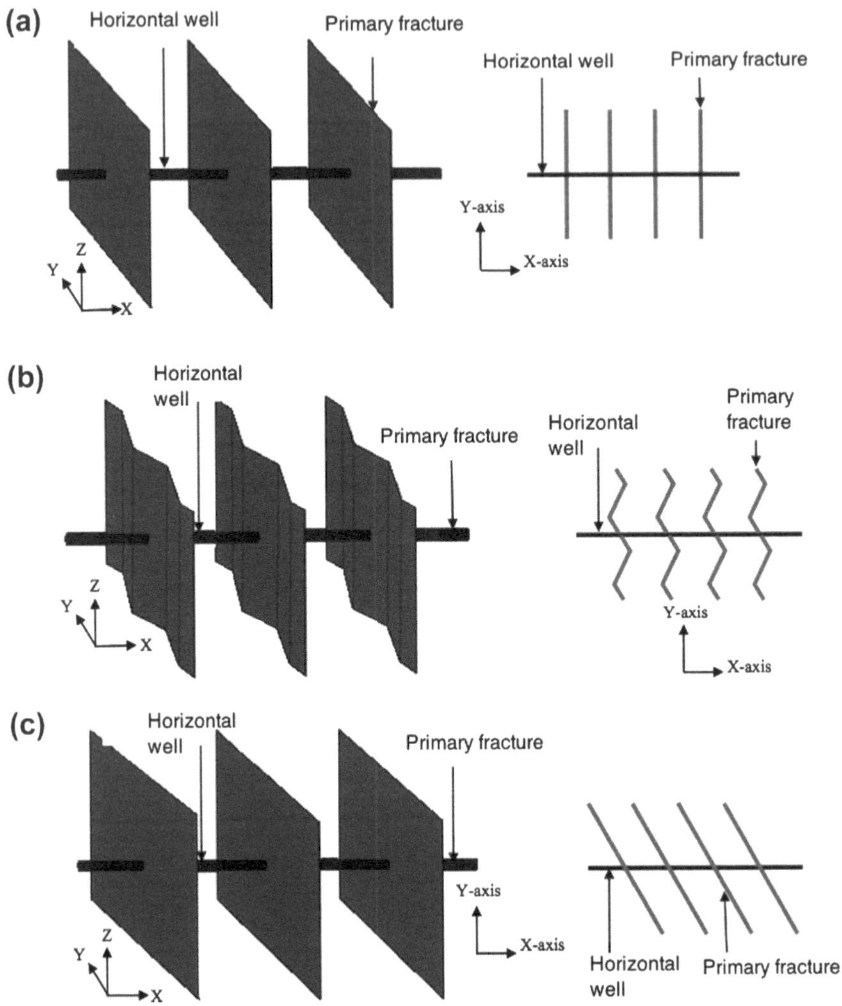

Figure 7.17 Simplification of fracture geometry.

Olorode (2011) reported modeling using nonorthogonal fractures (Figure 7.18). Figure 7.18 shows an almost identical rate profile for the nonplanar and nonorthogonal fracture systems at an angle of 60° to the horizontal. The fracture interference in these two cases becomes evident at the same time with the planar case that has a fracture half-length equal to the apparent length, l_a. The nonorthogonal fracture case with $\theta = 30°$ exhibits fracture interference earlier than the other cases, and this can be attributed to the fact that the apparent fracture half-length, l_a, is smaller in

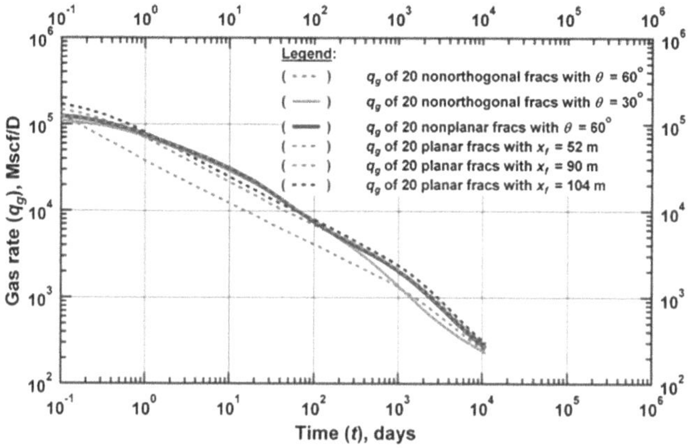

Figure 7.18 Flow profiles for nonplanar and nonorthogonal fractures. *From Olorode, 2011.*

comparison to the other nonideal cases. The cumulative production plot shows that the nonideal fractured systems have lower production than the planar cases with the same fracture half-length. This implies that to the extent possible, fractures should be designed such that their angle of inclination with the horizontal well should be as close to 90° as possible. This can be achieved by ensuring that the horizontal well is drilled in the direction of the minimum principal stress, as fractures usually propagate perpendicularly to the direction of the minimum principal stress.

Other challenges in modeling unconventional reservoirs lie within describing fracture flow and considering interactions between matrix and fracture. The original fracture flow model was developed in 1963 by Warren and Root. It assumed that the matrix has no permeability and the porosity is distributed in fractures as well as the matrix. Kazemi (1968) used a slab matrix model with horizontal fractures and unsteady state matrix–fracture flow to represent single-phase flow in the fractured reservoir. This was the first extension of the Warren–Root model. The assumptions included homogeneous behavior and isotropic matrix and fracture properties. He further assumed that the well is centrally located in a bounded radial reservoir. A numerical reservoir simulator was used. It was concluded that the results were similar to that of Warren–Root model when applied to a drawdown test in which the boundaries have not been detected. Two parallel straight lines were obtained on a semilog plot. The first straight line may be obscured by wellbore storage effects and the second straight line may lead to overestimating ω when boundary effects have been detected.

De Swaan (1975) presented a model that approximates the matrix blocks by regular solids (slab and spheres) and utilizes heat flow theory to describe the pressure distribution. It was assumed that the pressure in the fractures around the matrix blocks is variable and the source term is described through a convolution term. Approximate line-source solutions for early and late time are presented. The late time solutions are similar to those for early time except that modified hydraulic diffusivity terms dependent on fracture and matrix properties are included. The results are two parallel lines representing the early and late time approximations. The late time solution matches Kazemi (1968) for the slab case. De Swaan's model does not properly represent the transition period.

Najurieta (1980) presented a transient model for analyzing pressure transient data based on De Swaan's theory. Two types of fractured reservoir were studied: stratum (slabs) and blocks (approximated by spheres). The model predicted results similar to Kazemi (1968).

Serra et al. (1982) present methods for analyzing pressure transient data. The slab model used is similar to De Swaan (1975) and Najurieta (1980). The model considers unsteady state matrix–fracture transfer and is for an infinite reservoir. Three flow regimes were identified. Flow regime 1 and 3 are the Warren and Root's 18 early and late time semilog lines. A new flow regime 2 was also identified with half the slope of the late time semilog line.

Chen et al. (1985) presented methods for analyzing drawdown and buildup data for a constant rate producing well centrally located in a closed radial reservoir. The slab model similar to De Swaan (1975) and Kazemi (1968) is used. Five flow regimes are presented. Flow regimes 1, 2, and 3 are associated with an infinite reservoir and are described in Serra et al. (1982). Flow regime 1 occurs when there is a transient only in the fracture system. Flow regime 2 occurs when the transient occurs in the matrix and fractures. Flow regime 3 is a combination of transient flow in the fractures and "pseudo-steady state" in the matrix. Pseudo-steady state in the matrix occurs when the no-flow boundary represented by the symmetry center line in the matrix affects the response. Two new flow regimes associated with a bounded reservoir are also presented. Flow regime 4 reflects unsteady linear flow in the matrix system and pseudo-steady state in the fractures. Flow Regime 5 occurs when the response is affected by all the boundaries (pseudo-steady state).

Streltsova (1982) applied a "gradient model" (transient matrix–fracture transfer flow) with slab-shaped matrix blocks to an infinite reservoir. The model predicted results differ from the Warren–Root model in early time but converge to similar values in late time. The model also predicted a linear transitional response on a semilog plot between the early and late time pressure responses, which has a slope equal to half that of the early and late time lines. This linear transitional response was also shown to differ from the S-shaped inflection predicted by the Warren–Root model.

Cinco Ley and Samaniego (1982) utilized models similar to De Swaan and Najurieta and presented solutions for slab and sphere matrix cases. They utilize new dimensionless variables—dimensionless matrix hydraulic diffusivity, and dimensionless fracture area. They describe three flow regimes observed on a semilog plot: fracture storage dominated flow, "matrix transient linear"-dominated flow, and a matrix pseudo-steady state flow. The "matrix transient linear"-dominated flow period is observed as a line with one-half the slopes of the other two lines. It should be noted that the "matrix transient linear" period yields a straight line on a semilog plot indicating radial flow and might be a misnomer. The fracture storage dominated flow is due to fluid expansion in the fractures. The "matrix transient linear" period is due to fluid expansion in the matrix. The matrix pseudo-steady state period occurs when the matrix is under pseudo-steady state flow and the reservoir pressure is dominated by the total storativity of the system (matrix + fractures). It was concluded that matrix geometry might be identified with their methods, provided the pressure data is smooth.

Lai et al. (1983) utilize one-sixth of a cube matrix geometry transient model to develop well test equations for finite and infinite cases including wellbore storage and skin. Their model was verified with a numerical simulator employing the multiple interacting continua method.

Ozkan et al. (1987) presented an analysis of flow regimes associated with flow of a well at constant pressure in a closed radial reservoir. The rectangular slab model similar to De Swaan and Kazemi is used. Five flow regimes are presented: Flow regimes 1, 2, and 3 are described in Serra et al. (1982). Two new regimes are presented: Flow regime 4 reflects unsteady linear flow in the matrix system and occurs when the outer boundary influences the well response and the matrix boundary has no influence. Flow regime 5 occurs when the response is affected by all the boundaries.

Houze et al. (1988) presented type curves for analysis of pressure transient response in an infinite naturally fractured reservoir with an infinite conductivity vertical fracture.

Stewart and Ascharsobbi (1988) presented an equation for interporosity skin, which can be introduced into the pseudo-steady state and transient models. It should be noted that all the transient models previously described were developed for the radial reservoir cases (infinite or bounded).

El-Banbi (1998) was the first to present transient dual porosity solutions for the linear reservoir case. New solutions were presented for a naturally fractured reservoir using a dual porosity, linear reservoir model. Solutions are presented for a combination of different inner (constant pressure, constant rate, with or without skin and wellbore storage) and outer boundary conditions (infinite, closed, constant pressure).

Kucuk and Sawyer (1979, 1980) presented a model for transient matrix–fracture transfer for the gas case. Previous work had been concerned mainly with modeling slightly compressible (liquid) flow. They considered cylindrical and spherical matrix blocks cases. They also incorporate the pseudopressure definitions for gases. Techniques for analyzing buildup data are also presented for shale gas reservoirs. Their model results plotted on a dimensionless basis matched Warren and Root (1963) and Kazemi (1968) for very large matrix blocks at early time but differ at later times. They also conclude from their tests that naturally fractured reservoirs do not always exhibit the Warren–Root behavior (two parallel lines).

Carlson and Mercer (1989) coupled Fick's law for diffusion within the matrix and desorption in their transient radial reservoir model for shale gas. This is the closest to Brinkman formulation that any fluid flow model came in terms of unconventional reservoir modeling. Modifications included use of the pressure-squared forms valid for gas at low pressures to linearize the diffusivity equation. They provide a Laplace space equation for the cumulative gas production from their model and use it to history-match a sample well. They also show that semi-infinite behavior (portions of the matrix remain at initial pressure and is unaffected by production from the fractures) occurs in shale gas reservoirs regardless of matrix geometry. They present an equation for predicting the end of this semi-infinite behavior.

Gatens et al. (1987) analyzed production data from about 898 Devonian shale wells in four areas. They present three methods of analyzing production data: type curves, analytical model, and empirical equations. The empirical equation correlates cumulative production data at a certain time with cumulative production at other times. This avoids the need to determine reservoir properties. Reasonable matches with actual data were presented. The analytical model is used along with an automatic history-matching algorithm and a model selection procedure to determine statistically the best fit with actual data.

Watson et al. (1989) present a procedure that involves selection of the most appropriate production model from a list of models including the dual porosity model using statistics. The analytical slab matrix model presented by Serra et al. (1982) is utilized. Reservoir parameters are estimated through a history-matching procedure that involves minimizing an objective function comparing measured and estimated cumulative production. They incorporate the use of a normalized time in the analytical model to account for changing gas properties with pressure. Reasonable history matches were obtained with sample field cases but the forecast was slightly underestimated.

Spivey and Semmelbeck (1995) presented an iterative method for predicting production from dewatered coal and fractured shale gas reservoirs. The model used is a well producing at constant bottomhole pressure centered in a closed radial reservoir. A slab matrix is incorporated into these solutions. These solutions are extended to the gas case by using an adjusted time and pressure. Their method also uses a total compressibility term accounting for desorption.

None of these formulations can account for smooth transition between flow regimes. For gas hydrate as well as CBM, additional difficulties arise from the inability of conventional models to simulate phase transition and/or desorption. This is the only model that would be able to model different scenarios with the same model, as depicted in Figure 7.19.

A dynamic reservoir characterization tool is needed in order to introduce real-time monitoring. This tool can use the inversion technique to determine permeability data. At present, cuttings need to be collected before preparing petrophysical logs. The numerical inversion requires the

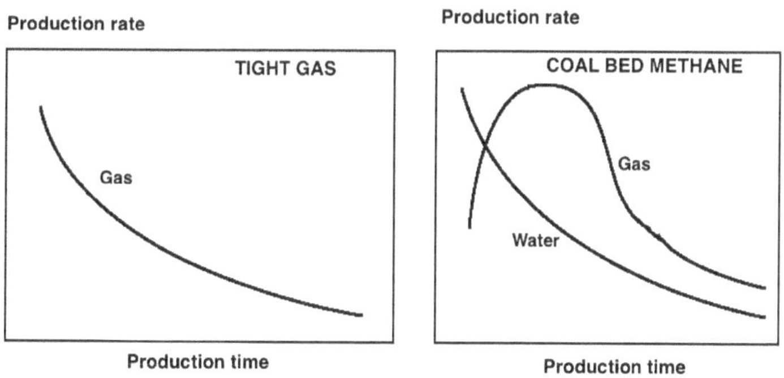

Figure 7.19 Conventional simulation techniques cannot simulate even fundamental features of various unconventional gas flow.

solution of a set of nonlinear PDEs. Conventional numerical methods require these equations to be linearized prior to solution (discussed early in the chapter). In this process, many of the routes to final solutions may be suppressed (Mustafiz et al., 2006a), while it is to be noted that a set of nonlinear equations should lead to the emergence of multiple solutions. Therefore, it is important that a nonlinear problem is investigated for multiple-value solutions of physical significance.

7.4 COMPREHENSIVE MODELING

This section introduces the governing PDEs that describe fluid flow in reservoir, where the matrix/fracture scheme controls the flow behavior. To represent flow behavior in porous media, suitable boundary and initial conditions must be considered and selected, and appropriate equations derived. Such governing equations describing the flow mechanism are the cornerstones of numerical modeling of an unconventional reservoir.

7.4.1 Governing Equations

In this section, the diffusivity equation of Forchheimer, the modified Brinkman Model, and the comprehensive model are discussed. Although modeled previously, Darcy's diffusivity equation is included because of its wide use throughout the oil industry. Here it will be used as a reference to compare its predictions with alternative models proposed herein. Forchheimer's model, popular in gas reservoirs' description, is used to derive a new diffusivity equation. What may be called modified Brinkman's model (MBM)—MBM's diffusivity equation incorporating both viscous and viscous/frictional terms from the original equation plus the inertial term from Forchheimer's basic equation—is also developed. The viscous terms of Brinkman's original equation are in fact Darcy's viscous term and another viscous/frictional term representing the forces generated by friction between layers of the fluid and its medium. The comprehensive is the ultimate proposed model; it has been called "comprehensive" because it includes all terms that might influence fluid flow in porous media in both Darcian and non-Darcian domains. It includes Darcy's viscous term, Forchheimer's inertial term, Brinkman's viscous/frictional term, and a Navier–Stokes convective term.

Although the situation of hydraulic fracturing, with the assumption of single horizontal fracture acting in the middle of a matrix system, can be modeled as a separate case, it is more appropriate to describe such a case using the models proposed in this section. Such models are valid for any hydraulic fracture aperture and length, any matrix size and extent; and also taking into account that inconsistent and heterogeneous matrix systems are highly preferable. Another unique model presented here deals with the coupled fluid flow/stress, and it should substantially enhance predictions of depletion while a reservoir is still in production. This model couples fluid flow conceptualization from this study with another stress model that updates petrophysical properties (porosity and permeability) according to effective stress variations arising from depletion, or other injection/flooding recovery techniques. Because, the model uses different flow equations in different sections of the reservoir, there is no need to make further assumptions.

7.4.1.1 Darcy's Model

Darcy's original equation (Eqn (7.6)) has been used to derive a diffusivity equation. This is the most commonly used equation in petroleum engineering analysis of well tests. This equation has also been used to predict reservoir pressure at any point of space and time within the reservoir. For single-phase linear flow, the pressure gradient in the x-coordinate is predicted by this equation:

$$\frac{\partial^2 P}{\partial x^2} = \frac{\phi \mu c}{k} \frac{\partial P}{\partial t} \tag{7.6}$$

where c, the total compressibility, is the sum of c_f (fluid compressibility) and c_r (rock compressibility).

In two dimensions, the diffusivity equation becomes:

$$\frac{\partial^2 P}{\partial x^2} + \frac{\partial^2 P}{\partial y^2} = \frac{\phi \mu c}{k} \frac{\partial P}{\partial t} \tag{7.7}$$

If the single phase considered flowing in the porous media is liquid, porosity (ϕ), permeability (k), viscosity (μ), and compressibility (c) are assumed to be constant. This assumption rests on the general and credible belief that liquids flowing in porous media are slightly compressible and the change in rock compressibility is negligible. Therefore, slight changes in these properties result in only negligible changes during flow. However, as will be seen in later section, this is not always the case. Many experimental investigations and field evidence show significant changes in rock properties during processes of depletion.

7.4.2 Forchheimer's Model

To develop a diffusivity equation similar to that of Darcy (Eqns (7.6) and (7.7)) for flow behavior replacing Darcy's linear equation by Forchheimer's nonlinear equation, consider an element of the reservoir through which a single phase (water, for example) is flowing in the x-direction and apply the conservation of mass balance equation (Figure 7.20):

Mass rate in − Mass rate out = Mass rate of accumulation.

In symbols:

$$(v_x\rho_x\Delta y\,\Delta z)-(v_{x+\Delta x}\rho_{x+\Delta x}\Delta y\Delta z)=(\Delta x\Delta y\Delta z)\phi\frac{(\rho_{t+\Delta t}-\rho_t)}{\Delta t} \tag{7.8}$$

Dividing Eqn (7.8) by $\Delta x\Delta y\Delta z$ *gives:*

$$-\frac{(v_{x+\Delta x}\rho_{x+\Delta x})-(v_x\rho_x)}{\Delta x}=\frac{\phi(\rho_{t+\Delta t}-\rho_t)}{\Delta t} \tag{7.9}$$

Take the limit as Δx and Δt go to zero simultaneously:

$$\frac{\partial(v\rho)}{\partial(x)}=-\phi\frac{\partial\rho}{\partial t} \tag{7.10}$$

This is the continuity equation in a linear system. We may therefore write this equation in the y and z coordinates:

$$\frac{\partial(v\rho)}{\partial(y)}=-\phi\frac{\partial\rho}{\partial t} \tag{7.11}$$

$$\frac{\partial(v\rho)}{\partial(z)}=-\phi\frac{\partial\rho}{\partial t} \tag{7.12}$$

Taking the derivative of the left-hand side of Eqn (7.10) in two dimensions, we obtain:

$$\rho\frac{\partial v}{\partial x}+v\frac{\partial\rho}{\partial x}+\rho\frac{\partial v}{\partial y}+v\frac{\partial\rho}{\partial y}=-\phi\frac{\partial\rho}{\partial t} \tag{7.13}$$

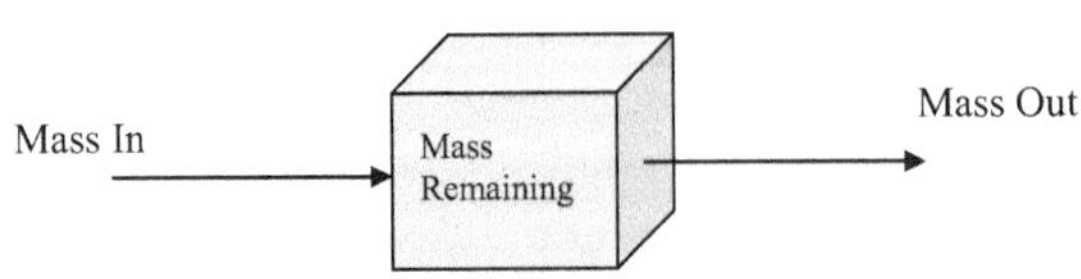

Figure 7.20 Schematic of the porous system considered for derivation of the diffusivity equation.

Now, Forchheimer's equation in the x-direction states:

$$-\frac{\partial P}{\partial x} = \frac{\mu}{k}v + \beta \rho v^2 \tag{7.14}$$

and in the y-direction:

$$-\frac{\partial P}{\partial y} = \frac{\mu}{k}v + \beta \rho v^2 \tag{7.15}$$

β is called the non-Darcy flow coefficient, Forchheimer's coefficient or turbulence coefficient, (1/ft). It is attributed to non-Darcy flow behavior. Several formulae are available for evaluating β, for example:

$$\beta = \frac{5.5 \times 10^9}{k^{5/4}\phi^{3/4}} \tag{7.16}$$

Taking the derivative with respect to x of both sides of Eqn (7.14) and with respect to y of both sides of Eqn (7.15):

$$-\frac{\partial^2 P}{\partial x^2} = \frac{\mu}{k}\frac{\partial v}{\partial x} + 2\beta \rho v \frac{\partial v}{\partial x} \tag{7.17}$$

$$-\frac{\partial^2 P}{\partial y^2} = \frac{\mu}{k}\frac{\partial v}{\partial y} + 2\beta \rho v \frac{\partial v}{\partial y} \tag{7.18}$$

Rearranging Eqns (7.17) and (7.18):

$$-\frac{\partial^2 P}{\partial x^2} = \rho\left(\frac{\mu}{k\rho} + 2\beta v\right)\frac{\partial v}{\partial x} \tag{7.19}$$

$$-\frac{\partial^2 P}{\partial y^2} = \rho\left(\frac{\mu}{k\rho} + 2\beta v\right)\frac{\partial v}{\partial y} \tag{7.20}$$

From Eqn (7.13):

$$\rho\left(\frac{\partial v}{\partial x}\right) = -\phi\frac{\partial \rho}{\partial t} - v\frac{\partial \rho}{\partial x} \tag{7.21}$$

Substituting the value of $\rho\left(\frac{\partial v}{\partial x}\right)$ in Eqn (7.19):

$$-\frac{\partial^2 P}{\partial x^2} = \left(\frac{\mu}{k\rho} + 2\beta v\right)\left(-\phi\frac{\partial \rho}{\partial t} - v\frac{\partial \rho}{\partial x}\right) \tag{7.22}$$

Dividing both sides of Eqn (7.22) by (−1):

$$\frac{\partial^2 P}{\partial x^2} = \left(\frac{\mu}{k\rho} + 2\beta v\right)\left(\phi\frac{\partial \rho}{\partial t} + v\frac{\partial \rho}{\partial x}\right) \tag{7.23}$$

By chain rule: $\frac{\partial \rho}{\partial x} = \frac{\partial \rho}{\partial P}\frac{\partial P}{\partial x}$ and $\frac{\partial \rho}{\partial t} = \frac{\partial \rho}{\partial P}\frac{\partial P}{\partial t}$ substituting in Eqn (7.23) we get:

$$\frac{\partial^2 P}{\partial x^2} = \left(\frac{\mu}{k\rho} + 2\beta v\right)\left(\phi\frac{\partial \rho}{\partial P}\frac{\partial P}{\partial t} + v\frac{\partial \rho}{\partial P}\frac{\partial P}{\partial x}\right) \tag{7.24}$$

Rearrange Eqn (7.24):

$$\frac{\partial^2 P}{\partial x^2} = \left(\frac{\mu}{k\rho} + 2\beta v\right)\left(\phi\frac{\partial P}{\partial t} + v\frac{\partial P}{\partial x}\right)\frac{\partial \rho}{\partial P} \tag{7.25}$$

$\frac{\partial \rho}{\partial P} = c\rho$, substituting in Eqn (7.25):

$$\frac{\partial^2 P}{\partial x^2} = \left(\frac{\mu}{k\rho} + 2\beta v\right)\left(\phi\frac{\partial P}{\partial t} + v\frac{\partial P}{\partial x}\right)c\rho \tag{7.26}$$

By rearrangement of Eqn (7.26):

$$\frac{\partial^2 P}{\partial x^2} = c\phi\left(\frac{\mu}{k} + 2\rho\beta v\right)\frac{\partial P}{\partial t} + cv\left(\frac{\mu}{k} + 2\rho\beta v\right)\frac{\partial P}{\partial x} \tag{7.27}$$

Equation (7.27) is the nonlinear diffusivity equation in one dimension for the Darcy/non-Darcy (Forchheimer Model). It can be expressed in two dimensions:

$$\frac{\partial^2 P}{\partial x^2} + \frac{\partial^2 P}{\partial y^2} = c\left(\frac{\mu}{k} + 2\beta\rho v\right)\left(\phi\frac{\partial P}{\partial t} + v\left[\frac{\partial P}{\partial x} + \frac{\partial P}{\partial y}\right]\right) \tag{7.28}$$

7.4.3 Modified Brinkman's Model

The common belief is that non-Darcian behavior can be caused by a number of factors and not the inertial term suggested by Forchheimer alone. Three aspects of non-Darcian behavior have been addressed in this section. The model presented herein is based on the two terms included in the previously derived Forchheimer's model (Darcy's viscous and Forchheimer's inertial) and the viscous/frictional term introduced in Brinkman's basic equation.

Having derived a diffusivity equation based on Forchheimer's fundamental equation, a diffusivity equation based on a combination of Darcy's, Forchheimer's, and Brinkman's terms may be derived as following:

Consider the flow in the x-direction for now. The pressure drop in the x-direction would be affected by Darcy's, Forchheimer's, and Brinkman's terms. The fundamental equation inferred from this setup takes the following form:

$$\frac{\partial P}{\partial X} = -\frac{\mu V}{K} - \rho\beta V^2 + \mu' \frac{\partial^2 V}{\partial X^2} \tag{7.29}$$

Rearrange:

$$\frac{\partial^2 V}{\partial X^2} = \frac{1}{\mu'}\left(\frac{\partial P}{\partial X} + \frac{\mu}{K}V + \rho\beta V^2\right) \tag{7.30}$$

From the same Eqn (7.10), we can obtain:

$$\rho\frac{\partial V}{\partial X} + V\frac{\partial \rho}{\partial X} = -\phi\frac{\partial \rho}{\partial t} \tag{7.31}$$

Take the derivative of both sides of Eqn (7.31) with respect to X:

$$\rho\frac{\partial^2 V}{\partial X^2} + \frac{\partial V}{\partial X}\frac{\partial \rho}{\partial X} + V\frac{\partial^2 \rho}{\partial X^2} + \frac{\partial \rho}{\partial X}\frac{\partial V}{\partial X} = -\phi\frac{\partial^2 \rho}{\partial t \partial X} \tag{7.32}$$

Substitute Eqn (7.30) into Eqn (7.32):

$$\frac{\rho}{\mu'}\left(\frac{\partial P}{\partial X} + \frac{\mu}{K}V + \rho\beta V^2\right) + \frac{\partial V}{\partial X}\frac{\partial \rho}{\partial X} + V\frac{\partial^2 \rho}{\partial X^2} + \frac{\partial V}{\partial X}\frac{\partial \rho}{\partial X} = -\phi\frac{\partial^2 \rho}{\partial t \partial X} \tag{7.33}$$

Rearrange:

$$\frac{\rho}{\mu'}\frac{\partial P}{\partial X} + \frac{\rho\mu}{\mu' K}V + \frac{\rho^2\beta}{\mu'}V^2 + 2\frac{\partial V}{\partial X}\frac{\partial \rho}{\partial X} + V\frac{\partial^2 \rho}{\partial X^2} = -\phi\frac{\partial^2 \rho}{\partial t \partial X} \tag{7.34}$$

Using chain rule for: $\frac{\partial^2 \rho}{\partial X^2}$ we obtain:

$$\frac{\partial^2 \rho}{\partial X^2} = \frac{\partial}{\partial X}\left(\frac{\partial \rho}{\partial X}\right) = \frac{\partial}{\partial X}\left(\frac{\partial \rho}{\partial P}\frac{\partial P}{\partial X}\right) = \frac{\partial}{\partial X}\left(\rho C\frac{\partial P}{\partial X}\right) = \rho C\frac{\partial^2 P}{\partial X^2} \tag{7.35}$$

Same way we have: for: $\frac{\partial^2 \rho}{\partial t \partial X}$ we obtain:

$$\frac{\partial^2 \rho}{\partial t \partial X} = \frac{\partial}{\partial t}\left(\frac{\partial \rho}{\partial X}\right) = \frac{\partial}{\partial t}\left(\frac{\partial \rho}{\partial P}\frac{\partial P}{\partial X}\right) = \frac{\partial}{\partial t}\left(\rho C \frac{\partial P}{\partial X}\right) = \rho C \frac{\partial^2 P}{\partial t \partial X} \quad (7.36)$$

and for: $\frac{\partial \rho}{\partial X}$ we obtain:

$$\frac{\partial \rho}{\partial X} = \frac{\partial \rho}{\partial P}\frac{\partial P}{\partial X} = \rho C \frac{\partial P}{\partial X} \quad (7.37)$$

$$C = \frac{1}{\rho}\frac{\partial \rho}{\partial P} \quad (7.38)$$

Substitute for $\frac{\partial^2 \rho}{\partial X^2}$, $\frac{\partial^2 \rho}{\partial t \partial X}$ and $\frac{\partial \rho}{\partial X}$ from Eqns (7.35), (7.36), and (7.37) respectively into Eqn (7.34):

$$\frac{\rho}{\mu'}\frac{\partial P}{\partial X} + \frac{\rho\mu}{\mu' K}V + \frac{\rho^2\beta}{\mu'}V^2 + 2\rho C \frac{\partial V}{\partial X}\frac{\partial P}{\partial X} + V\rho C \frac{\partial^2 P}{\partial X^2} = -\phi\rho C \frac{\partial^2 P}{\partial t \partial X} \quad (7.39)$$

Effective viscosity "μ'" is defined as:
$\mu' = \frac{\mu}{\phi}$, substitute into Eqn (7.39):

$$\frac{\rho\phi}{\mu}\frac{\partial P}{\partial X} + \frac{\rho\phi}{K}V + \frac{\rho^2\beta\phi}{\mu}V^2 + 2\rho C \frac{\partial V}{\partial X}\frac{\partial P}{\partial X} + V\rho C \frac{\partial^2 P}{\partial X^2} = -\phi\rho C \frac{\partial^2 P}{\partial t \partial X} \quad (7.40)$$

Divide both sides of Eqn (7.40) by $\phi\rho C$:

$$-\frac{1}{C\mu}\frac{\partial P}{\partial X} - \frac{1}{CK}V - \frac{\rho\beta}{C\mu}V^2 - \frac{2}{\phi}\frac{\partial V}{\partial X}\frac{\partial P}{\partial X} - \frac{V}{\phi}\frac{\partial^2 P}{\partial X^2} = \frac{\partial^2 P}{\partial t \partial X} \quad (7.41)$$

Again, using chain rule for $\frac{\partial V}{\partial X}$ we obtain:

$$\frac{\partial \mathrm{V}}{\partial \mathrm{X}} = \frac{\partial \mathrm{V}}{\partial \mathrm{P}}\frac{\partial \mathrm{P}}{\partial \mathrm{X}} \quad (7.42)$$

Substitute in Eqn (7.41):

$$-\frac{1}{C\mu}\frac{\partial P}{\partial X} - \frac{1}{CK}V - \frac{\rho\beta}{C\mu}V^2 - \frac{2}{\phi}\frac{\partial V}{\partial P}\left(\frac{\partial P}{\partial X}\right)^2 - \frac{V}{\phi}\frac{\partial^2 P}{\partial X^2} = \frac{\partial^2 P}{\partial t \partial X} \quad (7.43)$$

$\frac{\partial P}{\partial X}$ is a small quantity and $\left(\frac{\partial P}{\partial X}\right)^2$ is even smaller, therefore: $\left(\frac{\partial P}{\partial X}\right)^2 \approx 0$, substitute in Eqn (7.43):

$$-\frac{1}{C\mu}\frac{\partial P}{\partial X} - \frac{1}{CK}V - \frac{\rho\beta}{C\mu}V^2 - \frac{V}{\phi}\frac{\partial^2 P}{\partial X^2} = \frac{\partial^2 P}{\partial t \partial X} \tag{7.44}$$

Equation (7.44) is the diffusivity equation in the x-direction, for the y-direction we obtain:

$$-\frac{1}{C\mu}\frac{\partial P}{\partial Y} - \frac{1}{CK}V - \frac{\rho\beta}{C\mu}V^2 - \frac{V}{\phi}\frac{\partial^2 P}{\partial Y^2} = \frac{\partial^2 P}{\partial t \partial Y} \tag{7.45}$$

The summation of Eqns (7.44) and (7.45) gives the diffusivity equation in two dimensions:

$$-\frac{1}{C\mu}\left(\frac{\partial P}{\partial X} + \frac{\partial P}{\partial Y}\right) - \frac{2}{CK}V - \frac{2\rho\beta}{C\mu}V^2 - \frac{V}{\phi}\left(\frac{\partial^2 P}{\partial X^2} + \frac{\partial^2 P}{\partial Y^2}\right)$$
$$= \frac{\partial}{\partial t}\left(\frac{\partial P}{\partial X} + \frac{\partial P}{\partial Y}\right) \tag{7.46}$$

7.4.4. The Comprehensive Model

In this section, the principle idea underlying the comprehensive model is introduced. This is translated into a mathematical form that can be easily used to describe fluid flow in porous media.

The idea is very simple: the Navier–Stokes equations, representing fluid flow between two parallel plates, are applied. Figure 7.21 schematically describes such a system.

Fluid flow between two parallel plates or in a pipe as described by the Navier–Stokes equations is well documented. The representative equations give highly accurate results. They have been used in many applications and modified to suit many other circumstances. The pressure gradient in the x-direction, as described by the basic Navier–Stokes equations, takes this form:

$$-\frac{\partial P}{\partial x} = \rho\left(\frac{\partial U}{\partial t} + U\frac{\partial U}{\partial x} + V\frac{\partial U}{\partial y}\right) \tag{7.47}$$

In the y-direction, the pressure gradient takes the following form:

$$-\frac{\partial P}{\partial y} = \rho\left(\frac{\partial V}{\partial t} + U\frac{\partial V}{\partial x} + V\frac{\partial V}{\partial y}\right) \tag{7.48}$$

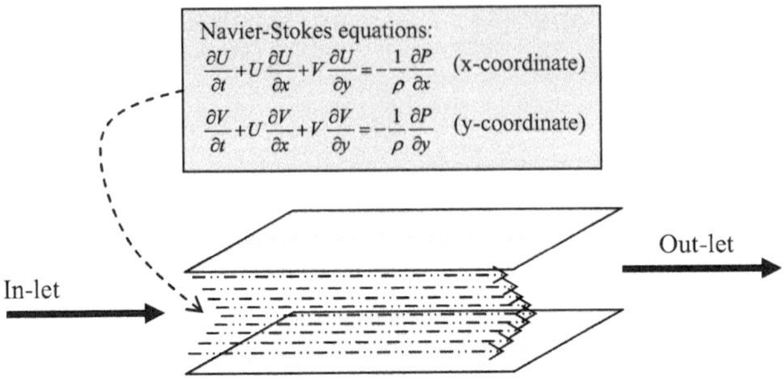

Figure 7.21 Schematic of fluid flow between two parallel plates described by the Navier–Stokes equations.

The summation of Eqns (7.47) and (7.48) is the total pressure gradient in two dimensions as described by the Navier–Stokes equations:

$$-\left(\frac{\partial P}{\partial x}+\frac{\partial P}{\partial y}\right)=\rho\left(\frac{\partial U}{\partial t}+\frac{\partial V}{\partial t}+U\frac{\partial U}{\partial x}+V\frac{\partial U}{\partial y}+U\frac{\partial V}{\partial x}+V\frac{\partial V}{\partial y}\right) \tag{7.49}$$

Introducing grain particles in the space between the two plates (Figure 7.22) would hinder the flow and cause extra resistance forces against Navier–Stokes prescribed fluid flow. Hence, a modification is required to account for this resistance.

The full boundary layer equations were analyzed to investigate forced convection from a horizontal flat porous medium filling the space between the two plates described in Figure 7.23. The porous material is assumed to be 100% saturated with the flowing fluid. The Navier–Stokes equations are adapted from the case described in Figure 7.23, where the

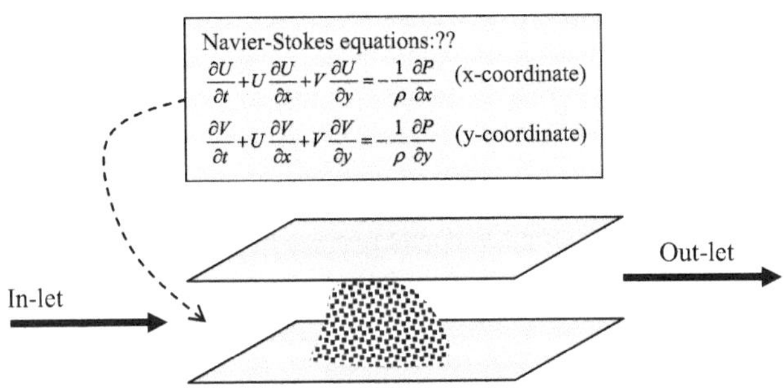

Figure 7.22 Schematic of fluid flow between two parallel plates with pile of grain particles in between causing extra resistance to flow.

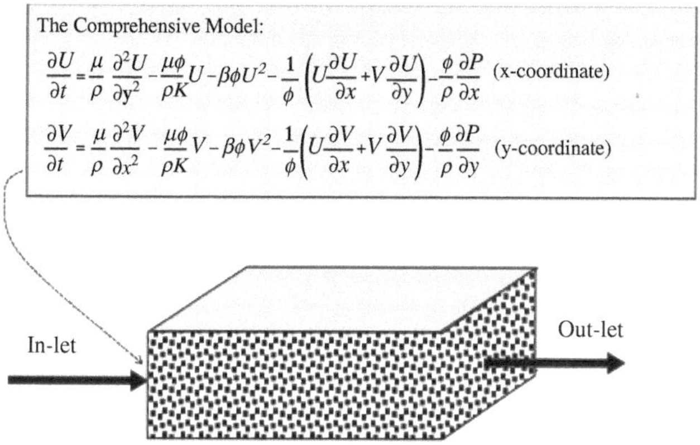

Figure 7.23 Schematic of fluid flow between two parallel plates filled with grain particles in between representing porous media flow.

space between the two plates is occupied by both the volume of grains and the space left between the grain particles through which the flowing fluid runs. Obviously, the fluid will be flowing only through that space (pore space) between grain particles opposed by convective forces in addition to the other viscous, frictional, and inertial forces usually associated with fluid flow in a porous system. Therefore, the term "porosity" denoted by "ϕ," become helpful in expressing the ratio of total pore space volume (space between grain particles) to the total volume of the system (total volume bounded by the two plates consists of both grain particles and pores in between):

$$\phi = \frac{(\text{Pore Space Volume})}{(\text{Total Volume})} \tag{7.50}$$

In fact, the pressure gradient mentioned in both Eqns (7.47) and (7.48) describes the situation where flow is subjected to the "total volume" between the plates. In the case considered in Figure 7.23, the convective forces are causing resistance to flow and contributing to a pressure gradient only within "pore space" through which the flow is actually taking place. We can define the pore space volume according to Eqn (7.50) then:

$$\text{Pore Space Volume} = \phi \times \text{Total Volume} \tag{7.51}$$

Hence, to express the effect of convective forces on the pressure gradient in the scheme represented by Figure 7.23, we should multiply by

"ϕ" to determine how much of the bulk volume between the plates is exposed to the flow. So Eqns (7.47) and (7.48) become:

$$-\phi\left(\frac{\partial P}{\partial x}+\rho\frac{\partial U}{\partial t}\right)=\rho\left(U\frac{\partial U}{\partial x}+V\frac{\partial U}{\partial y}\right) \tag{7.52}$$

$$-\phi\left(\frac{\partial P}{\partial y}+\rho\frac{\partial V}{\partial t}\right)=\rho\left(U\frac{\partial V}{\partial x}+V\frac{\partial V}{\partial y}\right) \tag{7.53}$$

However, the pressure gradient is affected not only by the convection term from Navier–Stokes, but also by viscous forces (Darcy's term), inertial forces (Forchheimer's term), and the viscous frictional drag forces (Brinkman's term). Considering the contribution of these terms, the above equations would take the following form, in which the x-direction component becomes:

$$-\frac{\partial P}{\partial x}=\rho\left(\frac{\partial U}{\partial t}+\frac{1}{\phi}\left[U\frac{\partial U}{\partial x}+V\frac{\partial U}{\partial y}\right]+\frac{\mu}{K}U+\rho\beta U^2-\frac{\mu}{\phi}\frac{\partial^2 U}{\partial y^2}\right) \tag{7.54}$$

The y-direction component becomes:

$$-\frac{\partial P}{\partial y}=\rho\left(\frac{\partial V}{\partial t}+\frac{1}{\phi}\left[U\frac{\partial V}{\partial x}+V\frac{\partial V}{\partial y}\right]+\frac{\mu}{K}V+\rho\beta V^2-\frac{\mu}{\phi}\frac{\partial^2 V}{\partial x^2}\right) \tag{7.55}$$

Rearrange Eqns (7.54) and (7.55) to evaluate velocity (U and V) at any time:

$$\frac{\partial U}{\partial t}=\frac{\mu}{\rho}\frac{\partial^2 U}{\partial y^2}-\frac{\mu\phi}{\rho K}U-\beta\phi U^2-\frac{1}{\phi}\left(U\frac{\partial U}{\partial x}+V\frac{\partial U}{\partial y}\right)-\frac{\phi}{\rho}\frac{\partial P}{\partial x} \tag{7.56}$$

$$\frac{\partial V}{\partial t}=\frac{\mu}{\rho}\frac{\partial^2 V}{\partial y^2}-\frac{\mu\phi}{\rho K}V-\beta\phi V^2-\frac{1}{\phi}\left(U\frac{\partial V}{\partial x}+V\frac{\partial V}{\partial y}\right)-\frac{\phi}{\rho}\frac{\partial P}{\partial y} \tag{7.57}$$

Equations (7.56) and (7.57) represent the final form of the comprehensive model that can be applied to porous media flow. It is capable of addressing a matrix–fracture scheme, including Darcian and non-Darcian domains.

7.5 TOWARDS SOLVING NONLINEAR EQUATIONS

Petroleum reservoir engineering problems are known to be inherently nonlinear. Consequently, solutions to the complete multiphase flow equations have been principally attempted with numerical methods. However, simplified forms of the problem have been solved some 60 years ago, when the Buckley–Leverett formulation was introduced. Ever since the pioneer work that neglected the capillary term, this formulation has been widely accepted in the petroleum industry. By using the method of characteristic, the multiphase one-dimensional fluid flow was solved. However, the resulting solution was a triple-valued one for a significant region. For decades, the existence of multiple solutions was considered to be the result of nonlinearity. Buckley and Leverett introduced shock utilizing the concept of material balance and two decades later, when numerical solutions were possible, it was discovered that the triple-value problem disappeared if the complete flow equation, including the capillary pressure form, is solved. Numerical methods, however, are not free from linearization. In fact, every numerical solution imposes linearization at some point of the solution scheme. Therefore, a numerical technique cannot be used to definitively state the origin of multiple solutions. In this section, a semianalytical technique, the Adomian decomposition method (ADM), capable of solving nonlinear PDEs without any linearizing assumptions, is used to unravel the true nature of the one-dimensional, two-phase flow. Results show that the Buckley–Leverett shock is neither necessary nor accurate portrayal of the displacement process. By using the ADM, the solution profile observed through numerous experimental studies was rediscovered. This paper opens up an opportunity to seek approximate but close to exact solutions to the multiphase flow problems in porous media.

7.5.1 Adomian Domain Decomposition Method

Since Adomian (1984) proposed his decomposition technique, the ADM has gained significant interests among researchers, particularly in the fields of physics and mathematics. They applied this method to many deterministic and stochastic problems (Adomian, 1986, 1994; Eugene, 1993). In this method, the governing equation is transformed into a recursive relationship and the solution appears in the form of a power series. The ADM has emerged as an alternative method to solve various mathematical models including algebraic, differential, integral, integro-differential, PDEs and systems, higher order ordinary differential equations, and others.

Guellal and Cherruault (1995) utilized the method to solve an elliptical boundary value problem with an auxiliary condition. Laffez and Abbaoui (1996) used it in modeling thermal exchanges associated to drilling wells. The application of this method is also noticeable in medical research, in which the decomposition technique is used to solve differential system of equations in modeling of HIV immune dynamics (Adjedj, 1999). Wazwaz (2001) and Wazwaz and El-Sayed (2001) reported that the ADM could be useful in solving problems without considering linearization, perturbation, or unjustified assumptions that may alter the nature of the problem under investigation. Also there have been suggestions that this method can be more advantageous over numerical methods by providing analytic, verifiable, rapidly convergent approximations that add insight into the character and the behavior of the solution as obtainable in the closed form solution (El-Sayed and Abdel-Aziz, 2003). In particular, they noticed the strength of the ADM in handling nonlinear problems in terms of rapid convergence. Biazar and Ebrahimi (2005) also expressed similar notion about rapid convergence and further added the advantage of the technique in terms of considerable savings in computation time when they attempted to solve hyperbolic equations.

Even though, petroleum problems are some of the most intrigue candidates for the ADM, there have been sparse attempts to utilize the method in the petroleum problems. Only recently, Mustafiz et al. (2005) reported the decomposition of the non-Darcy flow equations in porous media. In a different paper, Mustafiz and Islam (2005) transformed the diffusivity equations in well test applications into canonical forms. There is some success with ADM in modeling drilling activities, such as perforation by drilling (Rahman et al., in press) in specified drilling operations (Biazar and Islam, in press).

The Buckley–Leverett (1942) analysis is considered to be the first pioneer work in the study of linear displacement of a fluid by another fluid. The solution of their displacement study on two-phase fluid excluded the effect of capillary and gave multiple results for saturation at a given position. Realizing the fact that such coexistence of multiple saturation values is physically unrealistic, Buckley and Leverett (1942) used the fundamental concept of material balance to explain the shock. However, the lack of theoretical justification of shock was a major constraint in understanding the displacement phenomenon more vividly than before, and could not be reevaluated until the next significant work published by Holmgren and Morse (1951). They utilized the Buckley–Leverett theory to calculate the average water saturation at breakthrough and explained dispersion as a consequence of capillary effects. Welge (1952) continued the effort in finding the average saturation at water breakthrough in an oil reservoir. He also found that the method of *tangent construction* by Terwilliger et al. (1951) was equivalent to the shock proposed by Buckley and Leverett.

To solve the displacement equation including capillary as well as gravity, Fayers and Sheldon (1959) attempted a Lagrangian approach. They, however, did not succeed to determine the time required to obtain a particular saturation, which later was explained by Bentsen (1978) revealing the fact that the distance traveled by zero saturation is governed by a separate equation. Bentsen also noted that at slower injection rates, the input boundary condition of constant normalized saturation that Fayers and Sheldon used was incorrect in formulation. Also, there have been numerical investigations in the past to solve the displacement equation. Hovanessian and Fayers (1961) were able to avoid profiles of multiple valued saturations by considering the capillary pressure.

The ADM is applied to solve the nonlinear Buckley–Leverett equation. The solution for the water saturation is expressed in a series form. The base element of the series solution is obtained using the solution of the linear part of the Buckley–Leverett equation without the effect of the capillary pressure using the characteristic method. The other elements of the series solution are obtained recursively using the Adomian polynomial. The modification of the ADM in selection of the base element of the series solution makes the ADM feasible in solution of the nonlinear Buckley–Leverett equation. The computation is carried out for a reservoir of certain properties and initial conditions.

7.5.1.1 Governing Equations

The Buckley–Leverett equation is given by

$$\frac{\partial S_w}{\partial t} + \frac{q}{A\phi}\frac{\partial f_w}{\partial S_w}\frac{\partial S_w}{\partial x} = 0 \tag{7.58}$$

where f_w is expressed in the form

$$f_w = \left(\frac{1}{1 + \frac{k_{ro}\mu_w}{k_{rw}\mu_o}}\right)\left\{1 + \frac{Akk_{ro}}{q\mu_o}\left[\frac{\partial P_c}{\partial x} - (\rho_w - \rho_o)g\sin\alpha\right]\right\}. \tag{7.59}$$

This equation indicates that the fractional flow rate of water depends on reservoir characteristics, water injection rate, viscosity, and direction of flow. The effect of capillary pressure, P_c, which appears in the fractional flow equation, on saturation profiles is important, since these profiles affect the ultimate economic oil recovery (Bentsen, 1978). The ratio of effective permeability to viscosity is defined as the mobility that is shown for water and oil, respectively.

$$\lambda_w = \frac{kk_{rw}}{\mu_w} \quad \text{and} \quad \lambda_o = \frac{kk_{ro}}{\mu_o}. \tag{7.60}$$

The mobility ratio of oil to water is defined by

$$M = \frac{\lambda_o}{\lambda_w} = \frac{k_{ro}\mu_w}{k_{rw}\mu_o}. \tag{7.61}$$

The assumptions associated with the Buckley–Leverett equation, according to Ertekin et al. (2001) are:

- oil and water phases are assumed incompressible;
- the porous medium is assumed incompressible, which implies that porosity is constant;
- injection and production are taken care of by means of boundary conditions, which indicate that there are no external sink or source in the porous medium;
- the cross-sectional area that is open to flow is constant;
- the saturation-constraint equation for two-phase flow is valid; and
- the fractional flow of water is dependent on water saturation only.

The expression of the fractional flow rate of water in Eqn (7.59) suggests that Eqn (7.60) is a nonlinear differential equation and it is due to the effect of capillary pressure. In the simplest case of horizontal flow and neglecting the effects of capillary pressure variation along the reservoir, the expression for f_w in Eqn (7.59) is simplified to a linear differential equation, which is given by

$$f_w = \frac{1}{1 + M}. \tag{7.62}$$

The water saturation distribution along the reservoir can be found at different time steps by knowing the injection flow rate, the initial water saturation distribution, and the variation of fractional water flow rate. The water is normally the wetting fluid in the water–oil two-phase flow systems. The relative permeability of the water and oil for a specific reservoir are obtained with the drainage and imbibitions process on the core in the laboratory. However, in this paper, the variation of relative permeability to water and oil as a function of water saturation are obtained using following empirical relationships respectively.

$$k_{rw} = \alpha_1 S_{wn}^{n1} \tag{7.63}$$

$$k_{ro} = \alpha_2 (1 - S_{wn})^{n2} \tag{7.64}$$

where the normalized water saturation, S_{wn}, is defined as

$$S_{wn} = \frac{S_w - S_{wi}}{1 - S_{wi} - S_{or}} \tag{7.65}$$

If the effects of capillary pressure are included in a horizontal reservoir, Eqn (7.59) takes the form.

$$f_w = \frac{1}{1 + \frac{k_{ro}\mu_w}{k_{rw}\mu_o}} \left[1 + \frac{Akk_{ro}}{q\mu_o} \frac{\partial P_c}{\partial x} \right]. \tag{7.66}$$

Capillary pressure at any point is directly related to the mean curvature of the interface, which, in turn, is a function of saturation (Leverett, 1942). Therefore, it can be safely assumed that capillary pressure is only a function of water saturation. By applying the chain rule,

$$P_c = f(S_w) \rightarrow \frac{\partial P_c}{\partial x} = \frac{\partial P_c}{\partial S_w} \frac{\partial S_w}{\partial x} \tag{7.67}$$

and by incorporating Eqns (7.65) and (7.66) in Eqn (7.67), the following PDE is obtained

$$\begin{aligned}
&\frac{\partial S_w}{\partial t} + \frac{q}{A\phi} \frac{\partial}{\partial S_w} \left(\frac{1}{1 + \frac{k_{ro}\mu_w}{k_{rw}\mu_o}} \right) \frac{\partial S_w}{\partial x} \\
&+ \frac{kk_{ro}}{\mu_o\phi} \frac{\partial}{\partial S_w} \left(\frac{1}{1 + \frac{k_{ro}\mu_w}{k_{rw}\mu_o}} \right) \frac{\partial P_c}{\partial S_w} \left(\frac{\partial S_w}{\partial x} \right)^2 \\
&+ \frac{k}{\mu_o\phi} \left(\frac{1}{1 + \frac{k_{ro}\mu_w}{k_{rw}\mu_o}} \right) \frac{\partial k_{ro}}{\partial S_w} \frac{\partial P_c}{\partial S_w} \left(\frac{\partial S_w}{\partial x} \right)^2 \\
&+ \frac{kk_{ro}}{\mu_o\phi} \left(\frac{1}{1 + \frac{k_{ro}\mu_w}{k_{rw}\mu_o}} \right) \frac{\partial^2 P_c}{\partial S_w^2} \left(\frac{\partial S_w}{\partial x} \right)^2 = 0.
\end{aligned} \tag{7.68}$$

Equation (7.68) is a nonlinear PDE and the nonlinearity arises because of the inclusion of capillary pressure in it.

7.5.1.2 Adomian Decomposition of Buckley–Leverett equation

In the ADM, the solution of a given problem is considered as a series solution. Therefore, the water saturation is expressed as

$$S_w(x,t) = \sum_{n=0}^{\infty} S_{wn}. \tag{7.69}$$

If a functional equation is taken into account for water saturation, it can be written in the form

$$\sum_{n=0}^{\infty} S_{wn} = f(x,t) + \sum_{n=0}^{\infty} A_n. \tag{7.70}$$

where $f(x,t)$ is a given function and A_n (S_{wo}, S_{w1}, S_{w2},…, S_{wn}) or, simply, A_ns are called Adomian polynomials. The Adomian polynomials are expressed as

$$A_n = \frac{1}{n!}\frac{d^n}{d\lambda^n}\left[N\left(\sum_{i=0}^{\infty}\lambda^i S_{wn}(x,t)_i\right)\right]_{\lambda=0} \tag{7.71}$$

where N is an operator and λ is a parameter. The elements of the series solution for $S_{wn}(x,t)$ are obtained recursively by

$$S_{w0}(x,t) = f(x,t), \quad S_{w1}(x,t) = A_0, \quad \cdots \quad S_{wk}(x,t) = A_{k-1} \quad \cdots. \tag{7.72}$$

By this arrangement, the linear and nonlinear part of the functional Eqn (7.70) is replaced by a known function using the recursive Eqn (7.72). By integrating Eqn (7.68) with respect to t

$$\begin{aligned} S_w = S_w(x,0) &- \int_0^t \left[\frac{q}{A\phi}\frac{\partial}{\partial S_w}\left(\left(\frac{1}{1+\frac{k_{ro}\mu_w}{k_{rw}\mu_o}}\right)\frac{\partial S_w}{\partial x}\right)\right] dt \\ &- \int_0^t \left[\frac{kk_{ro}}{\mu_o\phi}\frac{\partial}{\partial S_w}\left(\frac{1}{1+\frac{k_{ro}\mu_w}{k_{rw}\mu_o}}\right)\frac{\partial P_c}{\partial S_w}\left(\frac{\partial S_w}{\partial x}\right)^2\right. \\ &+ \frac{k}{\mu_o\phi}\left(\frac{1}{1+\frac{k_{ro}\mu_w}{k_{rw}\mu_o}}\right)\frac{\partial k_{ro}}{\partial S_w}\frac{\partial P_c}{\partial S_w}\left(\frac{\partial S_w}{\partial x}\right)^2 \\ &\left.+ \frac{kk_{ro}}{\mu_o\phi}\left(\frac{1}{1+\frac{k_{ro}\mu_w}{k_{rw}\mu_o}}\right)\frac{\partial^2 P_c}{\partial S_w^2}\left(\frac{\partial S_w}{\partial x}\right)^2\right] dt. \end{aligned} \tag{7.73}$$

Comparing Eqn (7.73) with Eqn (7.70) and taking into account of Eqn (7.72), the elements of the water saturation series are obtained.

$$S_{w0}(x,t) = S_w(0,x) - \int_0^t \left[\frac{q}{A\phi} \frac{\partial}{\partial S_{w0}} \left(\frac{1}{1+\frac{k_{ro}\mu_w}{k_{rw}\mu_o}} \right) \frac{\partial S_{w0}}{\partial x} \right] dt$$

$$S_{w1}(x,t) = -\int_0^t \left[\frac{kk_{ro}}{\mu_o\phi} \frac{\partial}{\partial S_{w0}} \left(\frac{1}{1+\frac{k_{ro}\mu_w}{k_{rw}\mu_o}} \right) \frac{\partial P_c}{\partial S_{w0}} + \frac{k}{\mu_o\phi} \left(\frac{1}{1+\frac{k_{ro}\mu_w}{k_{rw}\mu_o}} \right) \right.$$
$$\left. \times \frac{\partial k_{ro}}{\partial S_{w0}} \frac{\partial P_c}{\partial S_{w0}} + \frac{kk_{ro}}{\mu_o\phi} \left(\frac{1}{1+\frac{k_{ro}\mu_w}{k_{rw}\mu_o}} \right) \frac{\partial^2 P_c}{\partial S_{w0}^2} \right] \left(\frac{\partial S_{w0}}{\partial x} \right)^2 dt$$

$$S_{w2}(x,t) = -\frac{d}{d\lambda} \int_0^t \left[\frac{kk_{ro}}{\mu_o\phi} \frac{\partial}{\partial(S_{w0}+S_{w1})} \left(\frac{1}{1+\frac{k_{ro}\mu_w}{k_{rw}\mu_o}} \right) \frac{\partial P_c}{\partial(S_{w0}+S_{w1})} \right.$$
$$+ \frac{k}{\mu_o\phi} \left(\frac{1}{1+\frac{k_{ro}\mu_w}{k_{rw}\mu_o}} \right) \frac{\partial k_{ro}}{\partial(S_{w0}+S_{w1})} \frac{\partial P_c}{\partial(S_{w0}+S_{w1})}$$
$$\left. + \frac{kk_{ro}}{\mu_o\phi} \left(\frac{1}{1+\frac{k_{ro}\mu_w}{k_{rw}\mu_o}} \right) \frac{\partial^2 P_c}{\partial(S_{w0}+S_{w1})^2} \right] \left(\frac{\partial(S_{w0}+\lambda S_{w1})}{\partial x} \right)^2 dt$$

$$S_{w3}(x,t) = \frac{1}{2!}\frac{d^2}{d\lambda^2} \int_0^t \left[\frac{kk_{ro}}{\mu_o\phi} \frac{\partial}{\partial(S_{w0}+S_{w1}+S_{w2})} \left(\frac{1}{1+\frac{k_{ro}\mu_w}{k_{rw}\mu_o}} \right) \frac{\partial P_c}{\partial(S_{w0}+S_{w1}+S_{w2})} \right.$$
$$+ \frac{k}{\mu_o\phi} \left(\frac{1}{1+\frac{k_{ro}\mu_w}{k_{rw}\mu_o}} \right) \frac{\partial k_{ro}}{\partial(S_{w0}+S_{w1}+S_{w2})^2} \frac{\partial P_c}{\partial(S_{w0}+S_{w1}+S_{w2})}$$
$$\left. + \frac{kk_{ro}}{\mu_o\phi} \left(\frac{1}{1+\frac{k_{ro}\mu_w}{k_{rw}\mu_o}} \right) \frac{\partial^2 P_c}{\partial(S_{w0}+S_{w1}+S_{w2})^2} \right]$$
$$\times \left(\frac{\partial\left(S_{w0}+\lambda S_{w1}+\lambda^2 S_{w2}\right)}{\partial x} \right)^2 dt$$

$$\cdots \quad \cdots \quad \cdots \quad \cdots \quad \cdots \quad \cdots \quad \cdots \quad \cdots \tag{7.74}$$

7.5.3 Discussion

The method is applied to a given reservoir with the initial water saturation of 0.18. This value is corresponding to the irreducible water saturation of the reservoir. Water is injected into the reservoir with a linear flow rate of 1 ft/day. The oil and water viscosities are 1.73 and 0.52 cp, respectively. The flow of the displaced phase (oil) ceases at $S_{oc} = 0.1$. The porosity of the medium is 0.25 with an absolute permeability of $k = 10$mD.

The normalized water saturation and the water and oil relative permeability are obtained using

$$S_{wn} = \frac{S_w - 0.18}{1 - 0.1 - 0.18}, \quad k_{rw} = 0.59439 S_{wn}^4, \quad k_{ro} = (1 - S_{wn})^2 \tag{7.75}$$

A typical plot of variation of relative permeability to water, k_{rw}, relative permeability to oil, k_{ro}, fractional flow curve, f_w and its derivative, df_w/dS_w are shown in Figure 7.24. The capillary pressure data are also shown in Table 7.1.

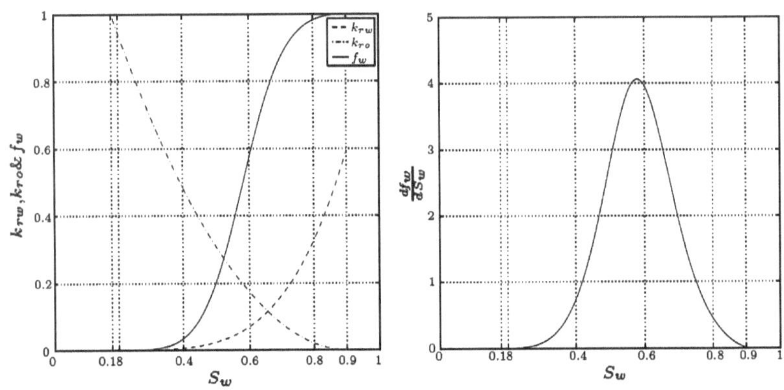

Figure 7.24 The variation of the water and oil relative permeability, the fractional water flow rate, and the differentiation of fractional water flow rate with the water saturation.

Table 7.1 Capillary Pressure Data

S_{wn}	P_c (atm)	S_{wn}	P_c (atm)	S_{wn}	P_c (atm)	S_{wn}	P_c (atm)
0.00	3.9921	0.11	1.2036	0.30	0.3600	0.65	0.0550
0.01	3.5853	0.15	0.8745	0.36	0.2699	0.72	0.0350
0.02	3.1987	0.18	0.7010	0.42	0.1980	0.87	0.0100
0.05	2.2577	0.21	0.5709	0.48	0.1450	0.95	0.0027
0.08	1.6209	0.25	0.4592	0.56	0.0920	1.00	0.0000

The solution for the first base element of the series solution of the water saturation, S_{w0}, is obtained through the solution of the first equation in Eqn (7.74). It can be written by integrating from the both sides of the first equation in Eqn (7.74) and using Eqn (7.61) that

$$\frac{\partial S_{w0}}{\partial t} + \frac{q}{A\phi}\frac{\partial}{\partial S_{w0}}\left(\frac{1}{1+M}\right)\frac{\partial S_{w0}}{\partial x} = 0. \tag{7.76}$$

This equation suggests that S_{w0} is constant along a direction that is called characteristic direction. The characteristic direction can be obtained by

$$\left(\frac{dx}{dt}\right)_{S_{w0}} = \frac{q}{A\phi}\frac{\partial}{\partial S_{w0}}\left(\frac{1}{1+M}\right)_t. \tag{7.77}$$

that is the Buckley–Leverett frontal advance equation. Integrating with respect to time, the distribution of S_{w0} is found in the form of

$$x\left(t, S_{w0}\right) = \frac{qt}{A\phi}\frac{\partial}{\partial S_{w0}}\left(\frac{1}{1+M}\right)_t. \tag{7.78}$$

The solution of Eqn (7.68) gives the variation of the S_{w0} along the reservoir at certain time. The distribution of S_{w0} along the reservoir is obtained based on the definition of the mobility ratio Eqn (7.61) and the relations for the relative permeability of water and oil in Eqn (7.75). It corresponds to the variation of the water saturation when the effect of the capillary pressure is ignored. The computations are carried out at different time of $t = 0.5$, 1, 2, 5,and 10 days. The solution for the water saturation without the capillary pressure effect shows the unrealistic physical situation that Buckley–Leverett mentioned in their pioneer paper with multiple saturations at each distance (x-position) as given partly in Figure 7.25. In order to avoid multiple saturation values at a particular distance, a saturation discontinuity at a distance, x_f, is generally created in such a way that the areas ahead of the front and below the curve are equal to each other.

The other elements of the series solution for the water are obtained recursively using Eqn (7.74) and the solution forS_{w0}. The solution for S_{w1} is obtained using the solution of S_{w0} at a certain point for a given time. To induce the nonlinear dependence of capillary on saturation, during decomposition, the capillary pressure and its derivatives are approximated by using the cubic spline, as shown in Figure 7.26. The interpolating splines are preferred over interpolating polynomials as they do not suffer from oscillations between the knots and are smoother and more realistic than

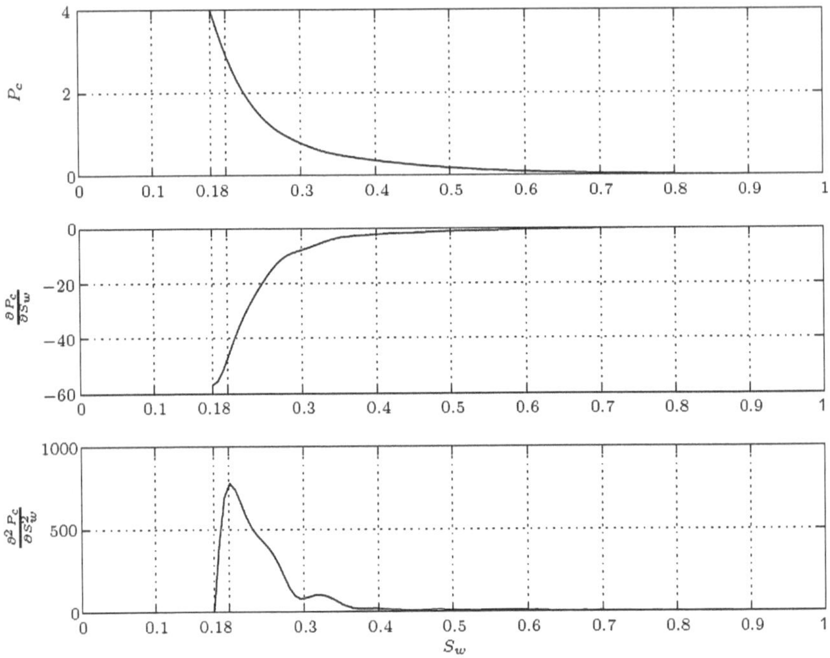

Figure 7.25 The water saturation distribution with and without the effect of capillary pressure using ADM.

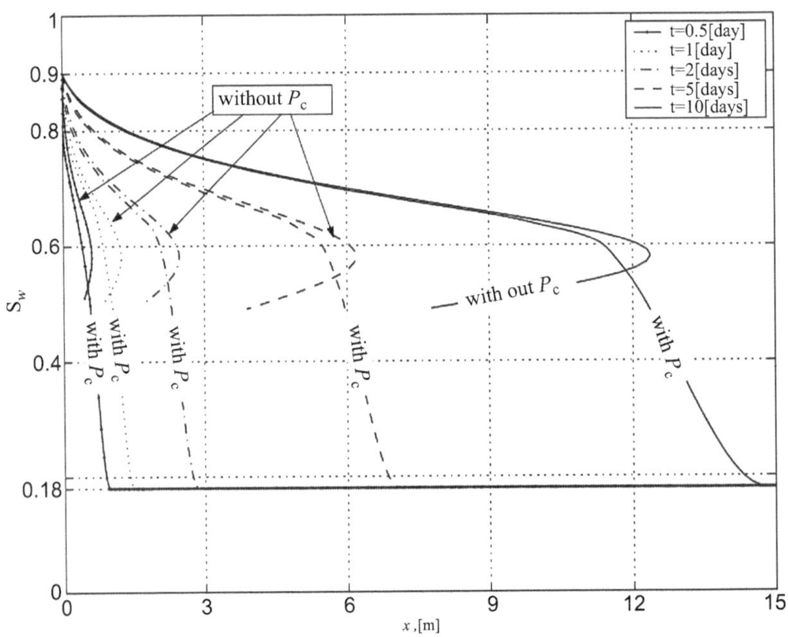

Figure 7.26 The capillary pressure variation and its first and second derivatives as a function of the water saturation.

linear splines (Ertekin et al., 2001). Such technique is also used to observe the effects of linearization in pressure in reservoir flow equations (Mustafiz et al., 2006). The solutions for S_{w2} and S_{w3} are obtained recursively using Eqn (7.74). The computations show that the series solution converges very fast and it is not necessary to go further thanS_{w3}.

The distribution of the water saturation along the reservoir is given in Figure 7.26. The results of the water saturation without the effect of the capillary pressure are also depicted in Figure 7.26 for the sake of comparison. This figure shows that by considering the effects of capillary pressure, it is possible to avoid unrealistic multiple saturation values. Moreover, the decomposition approach shows notable prediction of saturation profiles along the length. The gradual and mild changes observed in the saturation profile here, are perhaps less severe than the conventional prediction of shock-fronts.

CHAPTER 8

Economic Outlook and Conclusions

8.1 SUMMARY

Albert Einstein said, "One of nature's biggest forces is exponential growth." Information age has taught us that this exponential growth cannot be intangibles. Ever since the plastic revolution, there has been an exponential growth in GDP as well as nongenetic diabetes, cancer, Alzheimer, and numerous other maladies; so what? Success of the human race hinges upon how we manage our resources and how closely we emulate nature, because nature is the only one that can allow exponential growth. Unconventional gas has opened up opportunities that can revolutionize the energy outlook of the world.

There is this conventional analysis that would tell us how unconventional oil and gas will increase our disposable income, save in energy costs, and sustain economic growth for another century of new golden era of petroleum. Then there is this group of doubters, environmental Jihadists, who would remind us that such growth cannot be sustained and the human/environmental costs would add to the train wreck of the runaway "technological disaster" train. This book subscribes to neither of these two camps and puts unconventional gas in scientific perspective. It shows that unconventional gas has the potential of not only minimizing the environmental impact, but to improve the environment also. It not only saves in economic terms, but also creates sustainable growth in economy for both short term and long term.

This book deconstructs existing notions that mainly came from conventional analysis—the one that does not apply to unconventional reservoirs. It then shows the new synthesis that is entirely scientific and free from bias and prejudices of New Science that has been marred with contradictions and paradoxes. With the new synthesis presented is the energy outlook that indeed looks very rosy—something unprecedented in modern era.

This book neither invents any new theory nor introduces any new technology. It simply adds a new perspective that was not used in the context of unconventional gas. With this outlook, it shows how reservoirs

Unconventional Gas Reservoirs
ISBN 978-0-12-800390-9
http://dx.doi.org/10.1016/B978-0-12-800390-9.00008-6

should be characterized for proper assessment of reserve and strategies made for sustainable growth that is indeed exponential. It taps into existing technologies that are matured in the enhanced oil recovery (EOR) domain and shows that they are more easily applied for unconventional gas recovery. This can indeed lead to exponential growth with environmental appeal—a notion that is indeed unprecedented in the modern era.

8.2 ECONOMICS OF UNCONVENTIONAL AND ECONOMICS OF INTANGIBLES

The world looks at the United States for economic leadership. Energy being the driver of the current civilization, the role of energy economics cannot be overemphasized. At a current average global consumption growth rate of 2% annually (1995–2005), by 2025 the world will need 50% more oil (120 million bbl/day), and the International Energy Agency (IEA) admits that Saudi will have to double oil production to achieve this, which is not feasible in even the most optimistic scenario. And that is not even taking into account that 80% of the world is only just starting to use oil and gas. In recent years, energy demands from most emerging economies have increased dramatically in populous countries as their oil consumption per capita grows. The IEA estimates that 93% of all incremental demand comes from non-OECD countries. Therefore, in time, oil prices will continue to rise.

8.2.1 The Conventional Economic Analysis

In United States, the energy economics has been tremendously impacted by unconventional oil and gas production. The economic and employment contributions from US unconventional oil and gas production are now being felt throughout the US economy, increasing household incomes, boosting trade, and contributing to a new increase in US competitiveness in the world economy.

A recent report shows that unconventional oil and gas activity increased disposable income by an average of $1200 per US household in 2012, as savings from lower energy costs were passed along to consumers in the form of lower energy bills as well as lower costs for all other goods and services. This figure is expected to grow to just over $2000 in 2015 and reach more than $3500 in 2025 (HIS Report, 2013).

US trade position will continue to improve, owing to the significant reduction in energy imports and the increased global competitiveness of US-based energy-intensive industries. Driven by a rise in domestic production and manufacturing that will displace imports, as well as a favorable export position for these industries, the trade deficit will be reduced by more than US$ 164 billion in 2020—equivalent to one-third of the current US trade deficit. Unlike previous studies, which focused solely on upstream unconventional activities, this study considered overall activities and found that the sector of the energy value chain currently supports more than 1.7 million jobs and will grow to nearly 3 million by the end of the decade. The new study widens the breadth of the research to include the full energy value chain (upstream, midstream, and downstream energy and energy-related chemicals) and the overall macroeconomic contributions on the manufacturing sector and broader US economy. Midstream and downstream unconventional energy and energy-related chemicals activities currently support nearly 377,000 jobs throughout the economy, the study finds. Combined with upstream activity, the entire unconventional oil and gas value chain currently supports more than 2.1 million jobs. Total jobs supported by this value chain will rise to more than 3.3 million in 2020 and reach nearly 3.9 million by 2025.

"The unconventional oil and gas revolution is not only an energy story, it is also a very big economic story that flows throughout the U.S. economy in a way that is only now becoming apparent," said Daniel Yergin, IHS Vice Chairman and author of *The Quest: Energy, Security and the Remaking of the Modern World*. "In addition to significant job and economic impacts from energy production and its extensive supply chains, the growth of long-term, low-cost energy supplies is benefiting households and helping to revitalize U.S. manufacturing, creating a competitive advantage for U.S. industry and for the United States itself."

Energy-related chemicals and other energy-intensive industries such as petroleum refining, aluminum, glass, cement, and the food industry are some of the primary beneficiaries from secure supplies of low-cost energy from unconventional production, the study says. More than 70% of the cost of producing energy-related chemicals— which include major commodity petrochemicals such as olefins, methanol, and ammonia—is the cost of raw materials and energy feedstocks.

The chemical manufacturing industry accounted for 13% of all US merchandise exports ($198 billion) in 2012—compared to $ 152 billion in 2007.

This trend is expected to continue as energy-intensive industries benefit from lower energy and electricity prices and increased demand for their products, as growth in investment spurs domestic consumption. In addition to measuring jobs supported by the full unconventional value chain, the study also quantifies the additional manufacturing jobs attributed to the broader macroeconomic contributions that begin with unconventional oil and natural gas. More than 460,000 combined manufacturing jobs (3.7% of all manufacturing jobs) will be supported in 2020, rising to nearly 515,000 (4.2% of total manufacturing jobs) in 2025. The manufacturing sector will become increasingly connected to unconventional development as a primary source to create and sustain jobs over the course of the study period. Manufacturing jobs will represent 1 out of every 8 jobs supported by unconventional oil and gas development during that time.

The entire unconventional oil and gas value chain and energy-related chemicals will contribute $ 284 billion in value-added contributions to GDP in 2012, a figure that will increase to nearly $ 533 billion annually in 2025. The full value chain of industrial activity and employment associated with unconventional oil and natural gas contributed more than $ 74 billion in federal and state government revenues in 2012. Tax receipts will rise to more than $ 125 billion annually by 2020 and reach $ 138 billion by 2025. Workers' earnings from all unconventional energy and chemicals activities were nearly $ 150 billion in 2012. This total will rise to $ 207 billion in 2015 and will be nearly $ 269 billion in 2025. Industrial production increases directly resulting from lower feedstock prices and energy costs associated with the full value chain of unconventional activity will be $ 258 billion (3.5% increase) by 2020 and rise to $ 328 billion (3.9% increase) in 2025. Between 2012 and 2025, IHS (2013) projects a cumulative investment of nearly $ 346 billion across the midstream and downstream energy and energy-related chemical value chains. Close to $ 216 million of this will come in the midstream and downstream segments of the unconventional value chain, including 47,000 miles of new or modified pipeline infrastructure.

More than $ 31 billion in new capital investments will drive the addition of more than 16 million tons of chemical capacity by 2016. Cumulative investment will grow to more than $ 129 billion to support nearly 89 million tons of capacity by 2025.

Employment contributions from the midstream and downstream sector are at their greatest in the near term (currently supporting nearly 324,000 jobs), as expansions and other capital expenditures are made to increase capacity connecting the resource base with broader end users.

Energy-related chemicals (currently supporting more than 53,000 jobs) will support a growing number of jobs in the long term. By the end of the decade, energy-related chemicals will support more than 277,000 jobs—a fivefold increase—and rise to nearly 319,000 by 2025.

8.2.2 The New Synthesis

This book presents a new synthesis throughout. In scientific and technology terms, it meant material characterization, followed by the characterization of a process and energy source. In economic term, it should involve an approach that can be called 'economics of intangibles' (Zatzman and Islam, 2007). It includes all factors, rather than just one factor (often selected based on price index). Each of these factors has an index for environmental impact that is either negative or positive. The positive impacts are expected to keep an ecological balance. Most of the processes that are established to date disrupt ecological balance and produce enormous negative effects on all living beings. The new synthesis first adopted to petroleum engineering calculations by Islam et al. (2010) paved the way to accounting for environment-friendly and socially responsible technologies. Such approach indexes effects of:

1. Natural versus artificial: For instance, the use of Freon in a cooling system disrupts ozone layers and allows vulnerable rays of the sun to hit the earth and living beings. Similarly, chemically "purified" petroleum products pollute the environment by releasing harmful chemicals that add to global warming. Energy extraction from nuclear technology leaves harmful spent residues. All these can be accounted for if a negative index is assigned at every level of processing.
2. Quality of energy: The quality of energy is an important phenomenon. However, when it comes to energy, the talk of quality is largely absent. It is because of the disconnection between energy and mass analysis that itself is embedded in the definition of photon. With the scientific characterization tool introduced in this book, one can assign an index for quality for every type of energy source, by assigning a negative number for artificial sources and positive number for natural sources. Within a certain category, discussion of relative merit can be carried out.
3. The final value added products: During the value addition process, different indices should be attached to different processes depending on the contamination due to departure from natural processes. Consider, for instance, an operation with all natural chemicals and another with artificial chemicals. The former creates products that are part of environmental integrity, whereas the latter is part of the environmental insult. The new synthesis allows for a distinction between the two processes.

Table 8.1 shows the heating value of various fuels. The economic attractiveness of unconventional natural gas increases dramatically if the

Table 8.1 Heating Value of Various Fuel Types

Fuel type	Heating value
Premium wood pellets	13.6 million Btu/ton
Propane	71,000 Btu/gal
Fuel oil#2	115,000 Btu/gal
Fuel oil#6	124,000 Btu/gal
Seasoned firewood	15.3 million Btu/cord
Oven-dried switchgrass	14.4 million Btu/ton
Bituminous coal	26 million Btu/ton
Shelled corn @15% MC	314,000 Btu/bushel
Natural gas	1050 Btu/scf

From Islam et al., 2010.

development and operational costs are decreased or if there is a fundamental change invoked. This book shows how that is done through:

1. re-assessment of reserve based on conventional gas reserve (rather than separate unconventional gas reserve);
2. introducing cost-effective technologies for unlocking unconventional gas, rather than relying on expensive currently used technologies;
3. minimizing expensive data acquisition techniques that are not applicable to unconventional reservoirs;
4. properly characterizing (with dynamic refinement of data) reservoirs and identifying "sweet spots" within existing conventional reservoirs;
5. lower capital cost by selecting sites within developed regions.

In United States, since 2007, natural gas production has increased by 5931 Bcf. The overwhelming majority of this increase, i.e., 98.5%, came from unconventional oil and natural gas resources beneath onshore nonfederal land. This natural gas production accounted for approximately 82% of total US production in 2012 and grew by 40% between 2007 and 2012. Gas production from onshore federal lands also increased from 2007 to 2009, but has since returned to 2007 levels, representing a mere 3%. If new synthesis is applied to this unconventional outlook of the United States, a much brighter picture emerges, with all sustainable products to complement this production. As an example, the job creation with unconventional gas can be cited. Figure 8.1 combines published data with new synthesis outlook.

If this model is applied to the rest of the world, a total energy outlook emerges that gives great hopes for the future. Figure 8.2 plots the new energy outlook involving natural gas that includes unconventional gas. This figure shows, by the year 2030, natural gas alone can carry the bulk of the energy load of the world.

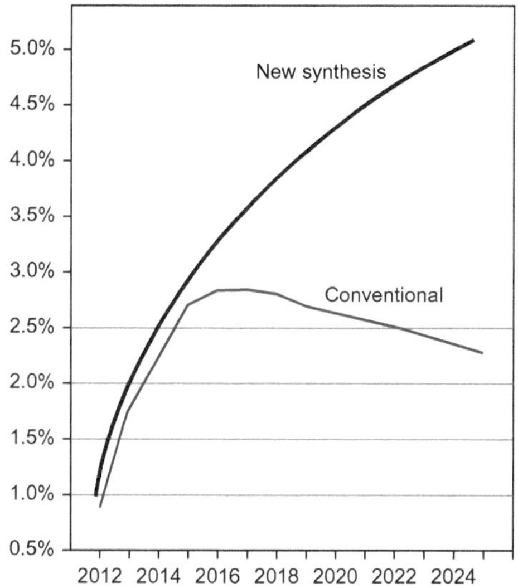

Figure 8.1 New synthesis versus conventional analysis.

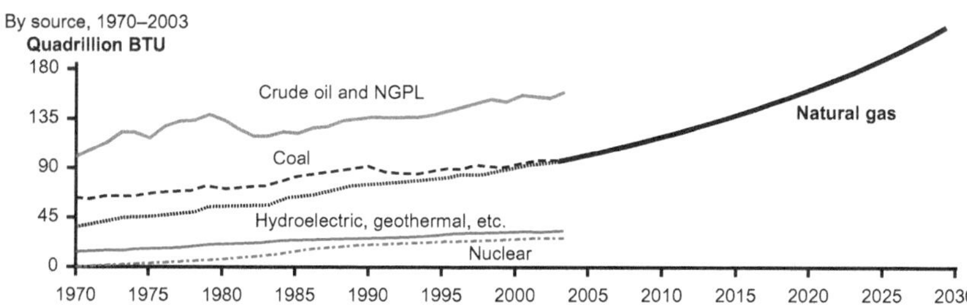

Figure 8.2 Energy outlook with natural gas under the new synthesis scenario.

8.3 CONCLUSIONS AND RECOMMENDATIONS

Ever since the oil embargo of 1972, the world has been gripped with the fear of "energy crisis." The U.S. President Jimmy Carter, in 1978, told the world in a televised speech that the world was in fact running out of oil at a rapid pace—a popular Peak Oil theory of the time—and that the US had to wean itself off of the commodity. Since the day of that speech, worldwide oil output has actually increased by more than 30%, and known available reserves are higher than they were at that time. This hysteria has survived the era of Reaganomics, President Clinton's cold war dividend, President G.W. Bush's post-9/11 era of "fearing everything but petroleum" and today even the most ardent supporters of petroleum industry have been convinced that there is an energy crisis looming and it is only a matter of time, we will be forced to switch no-petroleum energy source. This book can provide a delinearized history of energy and places unconventional gas in its proper place as the fuel for an energy revolution.

8.3.1 Unconventional Gas Reserve

Based on the discussion presented in the book, the following conclusions can be made.

1. The resource triangle is valid in its scope but the perceived difficulty associated to the development of unconventional reserve is unfounded.
2. Any resource base of unconventional gas should be conventional oil and gas that are already developed in order to minimize exploration and development costs. This includes abandoned or depleted reservoirs.
3. Conventional analysis with cutoff porosity or saturation points is not applicable to unconventional reservoirs.
4. In evaluating unconventional gas reserve, no cutoff point or net thickness calculations should be made, instead resorting to estimating the trap size from geological structural analysis, supported with seismic data and refined with core data.
5. Reserve estimates can increase as much as 100% in many instances with the new analytical tool presented in the book.

8.3.2 Unconventional Gas Designation

6. Natural gas contained in gas hydrate should be included in total reserve estimates.
7. Natural gas trapped in caprock, volcanic rocks, and other previously nondesignated formations should be considered.

8.3.3 Gas Reservoir Development

8. Systematic screening criteria presented in the book should be used prior to deciding on hydraulic fracturing or placement of horizontal wells.
9. Fracturing should be avoided in reservoirs that are deep and/or exhibit profuse natural fracture networks.
10. Existing EOR technologies should be considered to determine applicability to unconventional gas reserve. Most reservoirs are amenable to such EOR technologies and are likely to yield beneficial results.

8.3.4 Reservoir Characterization

11. Reservoir characterization should be performed with natural chronology, i.e., geology, followed by geophysics, drilling, coring, well test.
12. Well logging should not be automatic for known tight gas formations as they are not effective and often lead to erroneous conclusions.
13. Only a few geophysical tools function in an unconventional setting.
14. The origin of rock as well as fluid must be determined in order to characterize a reservoir.
15. Proper reservoir characterization can alter the reserve estimate; as well as screening criteria presented in the book should be used prior to deciding on hydraulic fracturing or placement of horizontal wells;

8.3.5 Reservoir Simulation

16. Most unconventional reservoirs are not represented with conventional simulation that uses Darcy's law.
17. While complex dual permeability, dual porosity models are difficult to implement, they do not improve accuracy of prediction over material balance calculations and simple rules of thumb.
18. Proper modeling should include Brinkman equation in the matrix and Forchheimer equation in the fracture.

8.3.6 Economic Outlook

19. Conventional economic analysis shows good performance of unconventional reservoirs but suffers from the short comings associated with centuries old accounting principles that do not apply in the information age.
20. Including salient features of long-term economic analysis and environment integrity, a much healthier energy outlook emerges. This analysis shows that unconventional gas reservoirs are the only ones that would support exponential growth in economics.

8.3.7 Overall Outlook

21. Unconventional gas has the potential of making natural gas, the most prominent and most environmentally appealing energy provider of the world.

REFERENCES

Aboudheir, A., Kocabas, I., Islam, M.R., 1999. Improvement of numerical methods in petroleum engineering problems. In: Proc. of the IASTED International Conference, Applied Modeling and Simulation, Cairns, Australia, September 1–3.

Abou-Kassem, J.H., Farouq Ali, S.M., Islam, M.R., 2006. A Basic Approach to Reservoir Simulation. Gulf Publishing Co., Houston, Texas, 455 pp.

Abou-Kassem, J.H., 2007. Engineering approach vs the mathematical approach in developing reservoir simulators. J. Nat. Sci. Sustain. Technol. 1 (1), 35–68.

Adams, W.L., 1964. Diagenetic aspects of lower Morrowan, Pennsylvanian, sandstones, northwestern Oklahoma. Am. Assoc. Pet. Geol. Bull. 48, 1568–1580.

Adams, C., 2010. Shale gas activity in British Columbia – exploration and development of BC's shale gas areas. In: Presentation to 4th Annual Unconventional Technical Forum, Victoria, BC.

Adams, C., 2011. Overview of shale gas operator activity in the Horn River Basin. In: Presentation to 5th Annual Unconventional Technical Forum, Victoria, BC.

Adjedj, B., 1999. Application of the decomposition method to the understanding of HIV immune dynamics. Kybernetes 28 (3), 271–283.

Adomian, G., Cherruault, Y., Abbaoui, K., 1996. A nonperturbative analytical solution of immune response with time-delays and possible generalization. Math. Comput. Model. 20 (10), 89–96.

Adomian, G., 1984. A new approach to nonlinear partial differential equations. J. Math. Anal. Appl. 102, 420–434.

Adomian, G., 1986. Nonlinear Stochastic Operator Equations. Academic Press, San Diego.

Adomian, G., 1994. Solving Frontier Problems of Physics: The Decomposition Method. Kluwer Academic Publishers, Dordrecht.

Agarwal, R.G., Carter, R.D., Pollock, C.B., March 1979. Evaluation and performance prediction of low-permeability gas wells stimulated by massive hydraulic fracturing. J. Pet. Technol. 362–372.

Aguilera, R., Ng, M.C., 1989. Transient pressure analysis of horizontal wells in anisotropic naturally fractured reservoirs. In: Paper SPE 19002 Presented at the 1989 Rocky Mountain Regional/Low Permeability Reservoir Symposium and Exhibition, Denver, Colorado, 6–8 March.

Aguilera, R., 2008. Role of natural fractures and slot porosity on tight gas sands. In: Paper SPE 114174 Presented at the 2008 SPE Unconventional Reservoirs Conference, Keystone, Colorado, USA, 10–12 February. http://dx.doi.org/10.2118/114174-MS.

Ahlbrandt, T.S., Taylor, D.J., 1993. Domestic conventional natural-gas reserves—can they be increased by the year 2010? In: Howell, D.G. (Ed.), Future Energy Gases: U.S. Geological Surv. Prof. Pap. 1570, pp. 527–546.

Ahr, W.M., 1973. The carbonate ramp—An alternative to the shelf model. Gulf Coast Assoc. Geol. Soc. Trans. 23, 221–225.

Aki, K., Chouet, B., 1975. Origin of coda waves, source attenuation, and scattering effects. J. Geophys. Res. 80, 3322–3342.

Aki, K., 1969. Analysis of seismic coda of local earthquakes as scattered waves. J. Geophys. Res. 74, 615–631.

Al-Dajani, A., Tsvankin, I., 1998. Nonhyperbolic reflection moveout for horizontal transverse isotropy. Geophysics 63, 1738–1753.

Allen, T.L., Fraser, T.A., Utting, J., 2009. Chapter 8 – upper Devonian to Carboniferous strata II – Tuttle formation play. In: Pyle, L.J., Jones, A.L. (Eds.), Regional Geoscience Studies and Petroleum Potential, Peel Plateau and Plain: Project Volume. Northwest Territories Geoscience Office and Yukon Geological Survey, NWT Open File 2009-02 and YGS Open File 2009-25, pp. 365–409.

Allen, T.L., Fraser, T.A., Lane, L.S., 2011. Preliminary results from a diamond drill hole study to assess shale gas potential of Devonian strata, Eagle Plain, Yukon. In: MacFarlane, K.E., Weston, L.H., Relf, C. (Eds.), Yukon Exploration and Geology 2010. Yukon Geological Survey, pp. 1–17.

Allen, T.L., 2000. An evaluation of coal-bearing strata at Division Mountain (115H/8 east-half, 105E/5 west-half), south-central Yukon. In: Emond, D.S., Weston, L.H. (Eds.), Yukon Exploration and Geology 1999. Exploration and Geological Services Division, Yukon, Indian and Northern Affairs Canada, pp. 177–198.

Allison, M., 1995. Increased Oil Production and Reserves from Improved Completion Techniques in the Bluebell Field, Uinta Basin, Utah. Utah Department of Natural Resources, Geological Survey Division. Annual Report, 123 pp.

Al-Maghrabi, I., Bin-Aqil, A.O., Chaalal, O., Islam, M.R., 1999. Use of thermophilic bacteria for bioremediation of petroleum contaminants. Energy Sources 21 (1/2), 17–30.

Alvarado, V., Manrique, E., 2010. Enhanced oil recovery: an update review. Energies 3, 1529–1575.

American Petroleum Institute (API), American Gas Association (AGA), and Canadian Petroleum Association (CPA), 1967–1980. Reserves of Crude Oil, Natural Gas Liquids, and Natural Gas in the United States and Canada as of December 31, vol. 25–34. American Petroleum Institute, Washington, D.C.

American Public Health Association, 1992. Standard Methods for the Examination of Water and Wastewater. American Public Health Association, Washington, DC.

Ammosov, I.I., Eremin, I.V., 1963. Fracturing in Coal. IZDAT Publishers, Moscow. Office of Technical Services, Washington, DC, 109 pp.

Anderson, D.M., Nobakht, M., Moghadam, S., Mattar, L. (Eds.), 2010. Analysis of Production Data from Fractured Shale Gas Wells. In: Paper SPE 131787 Presented at the SPE Unconventional Gas Conference, Pittsburgh, Pennsylvania, USA. http://dx.doi.org/10.2118/131787-MS.

Andrews, D.I., 1960. The Louann Salt and its relationship to Gulf Coast salt domes. Gulf Coast Assoc. Geol. Soc. Trans. 10, 215–240.

Aremu, O.J., Osisanya, S.O., 2008. Reduction of wellbore effects on gas inflow evaluation under underbalanced conditions. SPE J. 13 (2), 216–225. SPE 91586-PA. http://dx.doi.org/10.2118/91586-PA.

Arps, J.J., Roberts, T.G., 1958. Economics of drilling for cretaceous oil on the east flank of Denver-Julesburg Basin. Am. Assoc. Pet. Geol. Bull. 42 (11), 2549–2566.

Arps, J.J., 1945. Analysis of decline curves. Trans. AIME 160, 228–247.

Arps, J.J., 1956. Estimation of primary oil reserves. J. Pet. Technol. 182–191.

Arrington, J.R., February 29, 1960. Size of crude reserves is key to evaluating exploration programs. Oil Gas. J. 58 (9), 130–134.

Asquith, G., Krygowski, D., 2004. Basic well log analysis. – AAPG Methods in Exploration 16. AAPG, Tulsa, Oklahoma, 244 pp.

Ata, E., Michelena, R.J., 1995. Mapping distribution of fractures in a reservoir with P-S converted waves. Lead. Edge 12, 664–676.

Atkinson, B.K., Meredith, P.G., 1987. The theory of subcritical crack growth with applications to minerals and rocks. In: Atkinson, B.K.Z. (Ed.), Fracture Mechanics of Rock. Academic Press, Orlando, pp. 111–166.

Attanasi, E.D., Mast, R.F., Root, D.H., 1999. Oil, gas field growth projections—Wishful thinking or reality? Oil Gas. J. 97 (14), 79–81.

Ausubel, J.H., March/April 1996. Can Technology Spare the Earth? vol. 84. American Scientist, pp. 166–178.

Ayers, W.B., 2002. Coalbed gas systems, resources and production and a review of contrasting cases from the San Juan and Powder River Basins. Am. Assoc. Pet. Geol. Bull. 86, 1853–1890.

Baade, H.W., Lopez-Portillo, G.M., Guillermo, F., 1990. Tex. Int. Law J. 25 (3), 381–387.

Baker, R.A., Gehman, H.M., James, W.R., White, D.A., 1984. Geologic field number and size assessments of oil and gas plays. Am. Assoc. Pet. Geol. Bull. 68 (4), 426–437.

Ball, M.M., 1995. Permian Basin province (044). In: Gautier, D.L., Dolton, G.L., Takahashi, K.I., Varnes, K.L. (Eds.), 1995 National Assessment of United States Oil and Gas Resources. U.S. Geological Survey DDS–30, CD-ROM.

Banik, N.C., 1987. An effective anisotropy parameter in transversely isotropic media. Geophysics 52, 1654–1664.

Barbat, W.N., 1967. Crude-oil correlations and their role in exploration. Am. Assoc. Pet. Geol. Bull. 51, 1255–1292.

Barker, C., October 1972. Aquathermal Pressurizing: Role of Temperature in Development of Abnormal Pressure Zones. paper AAPG Bull. 65 (10), 2068–2071.

Barnes, V.E., Cloud Jr., P.E., Dixon, L.P., Folk, R.L., Jonas, E.C., Palmer, A.R., Tynan, E.J., 1959. Stratigraphy of Pre-Simpson Paleozoic Subsurface Rocks of Texas and Southeast New Mexico. University of Texas Bureau of Economic Geology Publication 5924, 836 pp.

Barnum, R.S., Brinkman, F.P., Richardson, T.W., Spillette, A.G., 1995. Gas condensate reservoir behavior: productivity and recovery reduction due to condensation. In: Presentation at SPE Annual Technical Conference and Exhibition, SPE 30767, Dallas, USA.

Barouch, E., Kaufman, G., 1977. Estimation of undiscovered oil and gas, in mathematical aspects of production and distribution of energy. In: Proceedings of Symposia in Applied Mathematics of the American Mathematical Society, vol. 21, pp. 77–91.

Barrett, M.L., 1986. Replacement Geometry and Fabrics of the Smackover (Jurassic) Dolomite, Southern Alabama. Gulf Coast Association of Geological Societies Transactions.

Bartberger, C.E., Dyman, T.S., Cook, S.A., 2001. Solution-subsidence origin, architecture, and hydrocarbon-trapping mechanisms of basal-Pennsylvanian Morrow fluvial valleys, southwest Kansas [abs.]. Abstracts with Program Am. Assoc. Pet. Geol. 10, A12–A13.

Bartley, J.T., Ruth, D.W., 2002. A look at break-through and end-point data from repeated water-flood experiments in glass bead-packs and sand-packs. SCA 2002-17. In: International Symposium Monterey, California, September 22–25, 2002, 12 pp.

Barton, C.A., Zoback, M.D., 2002. Discrimination of natural fractures from drilling-induced wellbore failures in wellbore image data – implication for reservoir permeability. SPE Res. Eval. Eng. 5 (3), 249–254. SPE 78599-PA. http://dx.doi.org/10.2118/78599-PA.

Barton, C.C., 1983. Systematic Jointing in the Cardium Sandstone along the Bow River, Alberta, Canada (Ph.D. dissert.). Yale Univ., 301 pp.

Basquet, R., Jeannin, L., Lange, A., Bourbiaux, B., Sarda, S., 2004. Gas-Flow simulation in discrete fracture-network models. SPE Res. Eval. Eng. 7 (5), 378–384. SPE 88985-PA. http://dx.doi.org/10.2118/88985-PA.

Batini, F., Dini, I., Certo, M., Cola, T., Cecconi, F., Soulier, L., Naville, C., Serbutoviez, S., 2001. Geothermal reservoir delineation by drillbit seismic while drilling. In: Proceedings of the Geothermal Resource Council 2001 – San Diego, USA, pp. 377–382.

Bean, D.W., Duchene, H.R., King, W.A., 1984. Edsel field (Minnelusa), powder river Basin, Wyoming. In: Goolsby, J., Morton, D. (Eds.), The Permian and Pennsylvanian Geology of Wyoming: Wyoming Geological Association, 35th Annual Field Conference, Guidebook, pp. 65–74.

Bear, J., 1972. Dynamics of Fluids in Porous Media. American Elsevier Publishing Co., New York, 761 pp.
Beaton, A.P., 2003. Production Potential of Coalbed Methane Resources in Alberta. Energy Utilities Board/Alberta Geological Survey Earth Sciences. Report 2003-03, 59 pp.
Beavers, W., 2009. Peace river coalbed gas project, NE british Columbia. In: Proceedings, 3rd Unconventional Gas Technical Forum – Petroleum Geology Information Circular 2009-02. B.C. Resource Development and Geoscience Branch, Oil and Gas Division, 1 CD-ROM.
Beer, S., Waisel, Y., 1982. Effects of light and pressure on photosynthesis in two seagrasses. Aquat. Bot. 13, 331–337.
Belhaj, H., Nouri, A., Vaziri, H., Islam, M.R., 2002a. Experimental Determination of the Effect of Pore Collapse on Permeability and Porosity of Sandstone Reservoirs. In: Proc. SPE Annual Technical Conference, Islamabad, November, 2002.
Belhaj, H., Nouri, A., Butt, S., Islam, M.R., 2002b. Experimental Modeling of Matrix/Fracture Flow in Sandstone Reservoirs. In: Proc. SPE Annual Technical Conference, Islamabad, November, 2002.
Belhaj, H., Agha, K., Islam, M.R., 2003a. Numerical modeling of Forchheimer's equation to describe Darcy and non-Darcy flow in porous medium system, SPE paper no. 81037. In: Proc. of SPE Latin American & Caribbean Petroleum Engineering Conference, Port of Spain, April 27–30.
Belhaj, H.A., Agha, K.R., Nouri, A.M., Butt, S.D., Islam, M.R., Vaziri, H.H., 2003b. Numerical modeling of Forchheimer's equation to describe Darcy and non-Darcy flow in porous medium system. In: Paper SPE 80440, Proc., SPE Asia Pacific Oil and Gas Conference and Exhibition, Jakarta, Indonesia, April 15–17.
Belhaj, H.A., Agha, K.R., Nouri, A.M., Butt, S.D., Vaziri, H.F., Islam, M.R., 2003c. Numerical simulation of non-Darcy flow utilizing the new Forchheimer's diffusivity equation. In: Paper SPE 81499, proc., Middle East Oil Show, Bahrain, April 5–8.
Belhaj, H.A., Biazar, J., Butt, S.D., Islam, M.R., 2004. Adomian solution of modified brinkman model to describe porous media flow. In: Paper SPE 89426 Presented at the SPE/DOE Fourteenth Symposium on Improved Oil Recovery, Tulsa, Oklahoma, USA, April 17–21.
Belhaj, H.A., Mousavizadegan, S.H., Ma, F., Islam, M.R., 2006. Three-Dimensional Permeability Utilizing a New Gas-Spot Permeameter. In: SPE 100428, SPE Gas Technology Symposium, Calgary, May 15–17.
Belhaj, H.A., Agha, K.R., Nouri, A.M., Butt, S.D., Phillips, W.J., Islam, M.R., April 2003a. Numerical and Experimental Modeling of Non-Darcy Flow in Porous Media. In: The SPE Latin American and Caribbean Petroleum Engineering Conference, Port-of-Spain, Trinidad, West Indies.
Belhaj, H., Agha, K.R., Nouri, A.M., Butt, S.D., Vaziri, H.H., Islam, M.R., April 2003b. Numerical Modeling of Forchheimer's Equation to Describe Darcy and Non-Darcy Flow in Porous Media. In: SPE Asia Pacific Oil and Gas Conference and Exhibition Held in Jakarta, Indonesia.
Belhaj, H., Agha, K.R., Nouri, A.M., Butt, S.D., Vaziri, H.H., Islam, M.R., April 2003c. Numerical Simulation of Non-Darcy Flow Utilizing the New Forchheimer's Diffusivity Equation. In: SPE 13th Middle East Oil Show & Conference, Bahrain.
Belhaj, H., Nouri, A., Vaziri, H., Islam, M.R., April, 2003d. Numerical modeling of forchheimer's equation to describe Darcy and non-Darcy flow in porous medium system. In: SPE Paper No. 80448, to Be Presented at SPE Asia Pacific Oil & Gas Conference and Exhibition (APOGCE).
Bello, R.O., Wattenbarger, R.A., 2008. Rate transient analysis in naturally fractured shale gas reservoirs. In: Paper SPE 114591 Presented at the CIPC/SPE Gas Technology Symposium 2008 Joint Conference, Calgary, Alberta, Canada. http://dx.doi.org/10.2118/114591-MS.

Bello, R.O., 2006. Rate Transient Analysis in Shale Gas Reservoirs with Transient Linear Behavior (Ph.D. dissertation). Texas A&M University.

Bennett, J.G., 1936. Broken coal. J. Inst. Fuel 10, 22–39.

Benson, D.J., Pultz, L.M., Bruner, D.D., 1996. The influence of paleotopography, sea level fluctuation, and carbonate productivity on deposition of the Smackover and Buckner Formations, Appleton field, Escambia County, Alabama. Gulf Coast Assoc. Geol. Soc. Trans. 46, 15–23.

Benson, D.J., Mancini, E.A., Groshong, R.H., Fang, J.H., Pultz, L.M., Carlson, E.S., 1997. Petroleum geology of Appleton field, Escambia County, Alabama. Gulf Coast Assoc. Geol. Soc. Trans. 47.

Bentsen, R.G., 1978. Conditions under which the capillary term may be neglected. J. Can. Pet. Technol. 17 (4), 25–30.

Beranek, L.P., Mortensen, J.K., Orchard, M.J., Ullrich, T., 2010. Provenance of North American Triassic strata from west-central and southeastern Yukon: correlations with coeval strata in the Western Canada Sedmentary Basin and Canadian Arctic Islands. Can. J. Earth Sci. 47, 53–73.

Bergin, T., April 26, 2012. Shell Bullish on China Shale Gas. Reuters.

Bertsekas, D., Tsitsiklis, J., 2000. Neuro-dynamic Programming. Athena Scientific.

Bethel, F.T., Calhoun, J.C., 1953. Capillary desaturation in unconsolidated beads. Pet. Trans. Am. Inst. Min. Eng. 198, 197–202.

Biazar, J., Ebrahimi, H., 2005. An approximation to the solution of hyperbolic equations by Adomian decomposition method and comparison with characteristics method. Appl. Math. Comput. 163, 633–638.

Biazar, M.R., Islam, 2004. "Solution of wave equation by Adomian decomposition method and the restrictions of the method". Applied Mathematics and Computation vol. 149, 807–814.

Bice, D., 2001. Modeling the Carbon Isotopes. Earthscape. www.earthscape.org/t1 /bid01/bid01f_03.html (cross referenced from Khan and Islam, 2007a).

Bienvenu, G., Kopp, L., 1980. Adaptivity to background noise spectral coherence for high resolution passive methods. In: Proc IEEE ICASSP (Denver, CO), pp. 307–310.

Bingham, T.L., Woodward, M., 1993. Morrow nearshore, Shallow-marine Sandstone, Arkoma Basin, Oklahoma and Arkansas—Atlas of Major Mid-continent Gas Reservoirs. Gas Research Institute Publication, 48–49.

Bingham, D.H., 1938. Developments in Arkansas-Louisiana Texas area, 1936–1937. Am. Assoc. Pet. Geol. Bull. 21, 1068–1073.

Bishop, W.F., 1968. Petrology of upper smackover limestone in North Haynesville field, Claiborne Parish, Louisiana. Am. Assoc. Pet. Geol. Bull. 52, 92–128.

Bishop, W.F., 1973. Late Jurassic contemporaneous faults in north Louisiana and south Arkansas. Am. Assoc. Pet. Geol. Bull. 57, 858–877.

Biswas, D., Suryanarayana, P.V., Frink, P.J., Rahman, S., 2003. An improved model to predict reservoir characteristics during underbalanced drilling. In: Paper SPE 84176 Presented at the 2003 SPE Annual Technical Conference and Exhibition, Denver, Colorado, USA, 5–8 October. http://dx.doi.org/10.2118/84176-MS.

Blangy, J.P., 1994. AVO in transversely isotropic media – an overview. Geophys. 59, 775–781.

Blanton, T.L. Propagation of hydraulically and dynamically induced fractures in naturally fractured reservoirs. Paper SPE 15261 Presented at the 1986 Unconventional Gas Technology Symposium, Louisville, May 18–21.

Blasingame, T.A., Poe Jr., B.D., 1993. Semianalytic solutions for a well with a single finite-conductivity vertical fracture. In: Paper SPE 26424 Presented at the SPE Annual Technical Conference and Exhibition, Houston, Texas, USA. http://dx.doi.org/10.2118/26424-MS.

Bloomberg, 2013. Bloomberg.com. May 23.

BMO Capital Markets Oil & Gas, 2010a. The Alberta Bakken: A New, Unconventional Tight Oil Resource Play. Analyst report for industry clients.

BMO Capital Markets Oil & Gas, 2010b. Not All of the Nikanassin is Created Equal. Analyst report for industry clients.

BMO Capital Markets Oil & Gas, 2011a. The Alberta Bakken Petroleum System (ABPS) is "Steady as She Goes" – An Emerging Unconventional Tight Oil Resource Play. Analyst report for industry clients. YGS MR 7-Scoping study of unconventional oil & gas potential, Yukon 89.

BMO Capital Markets Oil & Gas, 2011b. Swan Hills Platform of West Central Alberta: "The Sleeping Giant" – Opportunity Lies beyond the Reef Buildups. Analyst report for industry clients.

Bokhari, K., Islam, M.R., 2005. Improvement in the time accuracy of numerical methods in petroleum engineering problems – a new combination. Energy Sources 27 (1–2), 45–60.

Bokhari, K., 2003. Experiment and Numerical Modeling of Viscous Fingering – Liquid Liquid Miscible Displacement (MASc thesis). Dept. of Chemical Engineering, Dalhousie University, Halifax, Canada.

Boswell, R., Collett, T.S., 2011. Current perspectives on gas hydrate resources. Energy Environ. Sci. 4, 1206–1215.

Bowen, D.W., Weimer, Paul, 2003. Regional sequence stratigraphic setting and reservoir geology of Morrow incised-valley sandstones (Lower Pennsylvanian), eastern Colorado and western Kansas. Am. Assoc. Pet. Geol. Bull. 87, 781–815.

Bowker, K.A., 2002. Recent development of the barnett shale play, fort Worth Basin. In: Law, B.E., Wilson, Mike (Eds.), Conference on Innovative Gas Exploration Concepts: Rocky Mountain Association of Geologists, Proceedings, p. 16.

Bowker, K.A., 2007. Barnett Shale gas production, Fort Worth Basin: issues and discussion. Am. Assoc. Pet. Geol. Bull. 91, 523–533.

Boyd, D.R., Dyer, B.F., 1964. Frio barrier bar system of south Texas. Gulf Coast Assoc. Geol. Soc. Trans. 14, 309–322.

BP, June 2005. Putting Energy in the Spotlight. BP Statistical Review of World Energy p. 5.

Brady, B.T., Haramy, K.Y., 1994. High amplitude stress wave generation and damage induced by roof caving in mines. In: Nelson, P.P., Laubach, S.E. (Eds.), Proc. 1st North Am. Rock Mech. Symp. Balkema, Rotterdam, pp. 1033–1040.

Brekke, H., Kalheim, J.E., 1996. The Norwegian petroleum directorate's assessment of the undiscovered resources of the Norwegian continental shelf—background and methods. In: Dore, A.G., Sinding-Larsen, R. (Eds.), Quantification and Prediction of Petroleum Resources. Norwegian Petroleum Society Special Publication No. 6, pp. 91–103.

British Columbia Ministry of Energy and Mines and National Energy Board, 2011. Ultimate Potential for Unconventional Natural Gas in Northeastern British Columbia's Horn River Basin. Oil and Gas Reports 2011–1.

British Columbia Ministry of Energy, Mines, and Petroleum Resources, 2002. Coal in British Columbia. www.empr.gov.bc.ca/MINING/GEOSCIENCE/COAL/COALBC/Pages/default.aspx.

Brooks, R.O., 1993. Upper Tertiary/Quaternary detachment surface Gulf Coast—Texas and Louisiana. Gulf Coast Assoc. Geol. Soc. Trans. 43, 41–46.

Brown, D.S., Allison, J.D., 1992. MINTEQA2: An Equilibrium Metal Speciation Model. Environmental Research Laboratory, Office of Research and Development, U.S. Environmental Protection Agency, Athens, Georgia.

Brown, R.L., Parham, K.D., Campbell, J.A., 1993. Atlas of major midcontinent gas reservoirs. University of Texas, Bureau of Economic Geology, Austin, Tex, 44–47.

Brumfiel, G., September 23, 2011. Particles break light-speed limit. Nature News. http://dx.doi.org/10.1038/news 2011.554.

Brumfiel, G., 2012. Neutrinos not faster than light. Nature News. http://dx.doi.org/10.1038/nature2012.10249 (retrieved March 16, 2012).
Buckley, S.E., Leverett, M.C., 1942. Mechanism of fluid displacement in sands. Trans. AIME 146, 187–196.
Buckley, S.E., Leverett, M.C., 1942. Mechanisms of Fluid Displacement in Sands. Trans. AIME 146, 107–116.
Budd, D.A., Loucks, R.G., 1981. Smackover and Lower Buckner Formations, South Texas—Depositional Systems on a Jurassic Carbonate Ramp. Bureau of Economic Geology Report of Investigations 112, 38 pp.
Budiansky, B., O'Connel, R., 1976. Elastic moduli of a cracked solid. Int. J. Solids Struct. 12, 81–87.
Burchfiel, B.C., Royden, L.H., 1991. Tectonics of Asia 50 years after the death of Emile Argand. Eclogae Geol. Helv. 84, 599–629.
Burchfiel, B.C., Chen, Z., Liu, Y., Royden, L.H., 1996. Tectonics of the Longmen Shan and adjacent regions. Int. Geol. Rev. 37, 661–735.
Burnham, A.K., Sweeney, J.J., 1989. A chemical model of vitrinite maturation and reflectance. Geochim. Cosmochim. Acta 53, 2649–2657.
Burrus, J., Osadetz, K., Wolf, S., Doligez, W., Visser, K., Dearborn, D., 1996. A two-dimensional regional basin model of Williston Basin hydrocarbon systems. Am. Assoc. Pet. Geol. Bull. 80 (2), 265–291.
Burst, J.F., 1969. Diagenesis of Gulf Coast clayey sediments and its possible relation to petroleum migration. Am. Assoc. Petrol. Geol. Bull. 53, 73–93.
Bustin, R.M., Clarkson, C.R., 1998. Geological controls on coalbed methane reservoir capacity and gas content. Int. J. Coal Geol. 38, 3–26.
Bustin, R.M., February 2005. Gas shale tapped for big pay. AAPG Explor. 26 (2), 30.
Cameron, A.R., Beaton, A.P., 2000. Coal resources of northern Canada with emphasis on Whitehorse Trough, Bonnet Plume Basin and Brackett Basin. Int. J. Coal Geol. 43, 187–210.
Cameron, A.R., Goodarzi, F., Potter, J., 1994. Coal and oil shale of Early Carboniferous age in northern Canada: significance for paleoenvironmental and paleoclimate interpretations. Paleogeogr. Paleoclimatol. Paleoecol. 106, 135–155.
Cameron, A.R., 1993. Coal. Subchapter 6B. In: Stott, D.F., Aitken, J.D. (Eds.), Sedimentary Cover of the Craton in Canada. Geological Survey of Canada, Geology of Canada No. 5, pp. 563–598.
Campbell, J.D., 1979. Major cleat trends in Alberta Plains coals. Cim. Bull. 72, 69–75.
Cander, H., 2010. Granite to grass roots: understanding the geologic history of unconventional basins from bottom to top. Can. Soc. Pet. Geol. Reserv. 37 (3), 11.
Carlisle, Joe, 1991. The Bakken Formation of the Williston Basin—Deposition, maturation, and fracturing [abs.]. In: Hansen, W.B. (Ed.), Geology and Horizontal Drilling of the Bakken Formation: Montana Geological Society, 1991 Guidebook, p. 89.
Carlson, E.S., Mercer, J.C., 1989. Devonian shale gas production: mechanisms and simple models. In: paper SPE 19311 presented at the 1989 SPE Eastern Regional Meeting, Morgantown, West Virginia, 24–27 October.
Carter, R.D., October 1985. Type curves for finite radial and linear gas-flow systems: constant-terminal-pressure case. SPE J. 719–728.
de Carvalho, S.R., Rosa, A.J., 1988. Transient pressure behavior for horizontal wells in naturally fractured reservoir. In: Paper SPE 18302 Presented at the 1988 Annual Technical Conference and Exhibition, Houston, Texas, 2–5 October.
CBM Solutions, 2005. Gas shale Potential of Devonian Strata, Northeastern British Columbia. BC Ministry of Energy, Mines and Petroleum Resources, Resource Development and Geoscience Branch. Petroleum Geology Special Paper 2005-01, 110 pp.

Cengel, Y., Boles, A., 2008. Thermodynamics. An Engineering Approach, sixth ed. McGraw-Hill, New York, New York, USA, ISBN 9780073305370.
Chakravarti, S., Gupta, A., Hunek, B., 2001. Advanced technology for the capture of carbon dioxide. In: First National Conference on Carbon Sequestration, Washington, DC, USA, May 15–17, 2001.
Chalaturnyk, R., Scot, J.D., 1995. Geomechanics issues of steam assisted gravity drainage. In: Paper SPE 30280, Presented at the International Heavy Oil Symposium Held in Calgary, AB, Canada, June 19–21.
Chalmers, G.R.L., Bustin, R.M., 2008. Lower cretaceous gas shales in northeastern British Columbia, Part II: evaluation of regional potential gas resources. Bull. Can. Pet. Geol. 56, 22–61.
Chang, C.H., Gardner, G.H.F., 1993. Effects of vertically aligned fractures on reflection amplitudes. In: An Amplitudes-versus-offset Study, 63rd SEG Meeting Washington D.C. Expanded Abstracts, 769–771.
Charrete, E.E., 1991. Elastic Wave Scattering in Random Media (Ph.D. thesis). Massachusetts Institute of Technology, USA.
Chatterjee, S., Hadi, A.S., Price, B., 2000. Regression Analysis by Example, third ed. Wiley, New York. pp. 204–216.
Chatzis, I., Morrow, N.R., Lim, H.T., 1983. Magnitude and detailed structure of residual oil. SPE J. 23, 311–330.
Cheema, T.J., Islam, M.R., 1995. A new modeling approach for predicting flow in fractured formations. In: El-Kady (Ed.), Groundwater Models for Resources Analysis and Management. Lewis Publishers, Boca Raton, FL, pp. 327–338.
Chen, G., Harpalani, S., 1995. Study of Fracture Network in Coal. In: Myer, L.R., Cook, N.G.W., Goodman, C.F., Tsang (Eds.), Fractured and jointed rock masses, Rotterdam, Balkema, pp. 431–434.
Chen, C.-C., Serra, K., Reynolds, A.C., Raghavan, R., June 1985. Pressure transient analysis methods for bounded naturally fractured reservoirs. SPE J. 451–464.
Chen, H.Y., Teufel, L.W., Lee, R.L., 1993. Couple fluid flow and geomechanics in reservoir study-I, theory and governing equations. In: Paper SPE 30752, Paper Presented at the SPE Annual Technical Conference and Exhibition, Dallas, USA, October.
Chen, Z., Narayan, S.P., Yang, Z., Rahman, S.S., 2000. An experimental investigation of hydraulic behaviour of fractures and joints in granitic rock. Int. J. Rock Mech. Min. Sci. 37, 1061–1071.
Cheng, N., Cheng, C.H., Toksoz, M.N., 1995. Borehole wave propagation in three-dimensions. J. Acoust. Soc. Am. 97, 3483–3493.
Chernov, L.A., 1960. Wave Propagation in a Random Medium. McGraw-Hill, New York.
Chhetri, A.B., Islam, M.R., 2008. Inherently Sustainable Technology Developments. Nova Science Publishers, New York, 452 pp.
Choi, S.K., Wold, M.B., 1996. In: Aubertin, M. (Ed.), A Mechanistic Approach in the Numerical Modeling of the Cavity Completion Method of Coalbed Methane Stimulation.
Choi, I., Cheema, T., Islam, M.R., 1997. A new dual-porosity/dual permeability model with non-Darcian flow through fractures. J. Pet. Sci. Eng. 17, 331–344.
Choi, C.Q., October 13, 2011. Leading light: what would faster-than-light neutrinos mean for physics. Sci. Am. http://www.scientificamerican.com/article/ftl-neutrinos-new-physics-implications/.
Chopra, S., Castagna, J.P., Portniaguine, O., 2006. Seismic resolution and thin-bed reflectivity inversion. CSEG Recorder 31, 19–25.
Chopra, S., Sharma, R.K., Keay, J., Marfurt, K.J., 2012. Shale gas reservoir characterization work flow. In: 82nd Annual International Meeting, SEG. Expanded Abstracts. http://dx.doi.org/10.1190/segam2012-1344.1.

Chuoke, R.L., van Meurs, P., van der Poel, C., 1959. The instability of slow, immiscible, viscous liquid-liquid displacements in permeable media. Trans. AIME 216, 188–194.

CIBC World Markets, 2011. The Alberta Bakken – Improving Our Bias on a Prospective Early-stage Play. Analyst report for industry clients.

Cinco-Ley, H., Samaniego, V.-F., 1982. Pressure transient analysis for naturally fractured reservoirs. In: Paper SPE 11026 Presented at the 57th 1982 Annual Technical Conference and Exhibition of the Society of Petroleum Engineers of AIME, New Orleans, Louisiana, 26–29 September.

Cinco-Ley, H., Samaniego, V.F., 1982. Pressure transient analysis for naturally fractured reservoirs. In: Paper SPE 11026 Presented at the SPE 57th Annual Fall Technical Conference and Exhibition, New Orleans, LA, September 26–29.

Cipolla, C.L., Wood, M.C. A statistical approach to infill drilling studies: case history of the Ozona Canyon sands. Paper SPE 35628 Presented at the 1996 Gas Technology Symposium, Calgary, Alberta, Canada, 28 April–1 May.

Clayton, J.L., Ryder, R.T., 1984. Organic geochemistry of black shales and oils in the Minnelusa Formation (Permian and Pennsylvanian), Powder River Basin, Wyoming. In: Woodward, J., Meissner, F.F., Clayton, J.L. (Eds.), Hydrocarbon Source Rocks of the Greater Rocky Mountain Region. Rocky Mountain Association of Geologists, pp. 231–253.

Clemens, T., Wit, K., 2002. CO2 enhanced gas recovery studies for an example gas reservoir. In: SPE paper 77348, presented at SPE Annual Technical Conference and Exhibition held in San Antonio, Texas.

Clergeot, H., Tressens, S., Ouamri, A., 1989. Performance of high resolution frequencies estimation methods compared to the Cramer-Rao bounds. In: IEEE Transaction on Acoustics, Speech and Signal Processing, ASSP-37, pp. 1703–1720.

Close, J., Mavor, M., 1991. Influence of coal composition and rank on fracture development in Fruitland coal gas reservoirs of the San Juan Basin. In: Schwochow, S.D. (Ed.), Coalbed Methane of Western North America, Rk. Mt. Assoc. Geol. Field Conf, pp. 109–121.

Close, F., Conroy, D., Greig, A., Morin, A., Flint, G., Seale, R., 2005. Successful drilling of Basalt in a west of Shetland deepwater discovery. In: Paper SPE 96575, Presented at the SPE Offshore Europe Oil and Gas Conference and Exhibition, Aberdeen, 6–9 September.

Close, J., 1993. Natural fractures in coal. In: Law, B.E., Rice, D.D. (Eds.), Hydrocarbons from Coal, vol. 38. AAPG Studies in Geology, pp. 119–132.

Coates, R.T., Charrette, E.E., 1993. A comparison of signal scattering and finite difference synthetic seismograms in realizations of 2-D elastic random media. Geophys. J. Int. 113, 463–482.

Cobb, J.C., 1979. Timing and development of mineralized veins during diagenesis in coal beds. In: Cross, A.T. (Ed.), International Congress Carboniferous Stratigraphy and Geology, vol. 4. Southern Illinois Press, pp. 371–376.

Collins, S.E., 1980. Jurassic Cotton Valley and Smackover reservoir trends, east Texas, north Louisiana, and south Arkansas. Am. Assoc. Pet. Geol. Bull. 64, 1004–1013.

Colpron, M., 2010. Hydrocarbon Potential of Whitehorse Trough. Poster display.

Cook, A.E., 2010. Gas Hydrate-filled Fracture Reservoirs on Continental Margins (Ph.D. dissertation). Columbia University.

Cooke, L.W., Dellagiarino, G., 1990. Estimates of Undiscovered Oil and Gas Resources of the Outer Continental Shelf as of January 1987. Minerals Management Service OCS Report MMS 89-0090, 115 pp. plus 2 plates.

Cooke, L.W., 1985. Estimates of Undiscovered, Economically Recoverable Oil and Gas Resources for the Outer Continental Shelf as of July 1984. Minerals Management Service OCS Report MMS 85-0012, 45 pp.

Cooke, L.W., 1991. Estimates of Undiscovered, Economically Recoverable Oil and Gas Resources of the Outer Continental Shelf Revised as of January 1990. Minerals Management Service OCS Report MMS 91-0051, 30 pp.

Core Laboratories, 2012. Stratigraphic Correlation Chart for Western Canada. www.landman.ca/pdf/CORELAB.pdf.

Corrigan, D., Withers, R., Darnall, J., Skopinski, T., 1996. Fracture mapping from azimuthal velocity analysis using 3-D surface seismic data. 66th Ann. Intern. Mtg., Soc. Expl. Geophys. Expanded Abstracts, 1834–1837.

Cowen, R., May 31, 2012. Andromeda on collision course with the Milky Way. Nature. http://dx.doi.org/10.1038/nature.2012.10765, http://www.nature.com/news/andromeda-on-collision-course-with-the-milky-way-1.10765.

Cox, D.O., Kuuskraa, V.A., Hansen, J.T., 1996. Advanced type curve analysis for low permeability gas reservoirs. In: Paper SPE 35595 Presented at the 1996 Gas Technology Conference, Calgary, Alberta, Canada, 28 April–1 May. http://dx.doi.org/10.2118/35595-MS.

Cramer, D.D., 1991. Stimulation treatments in the Bakken Formation—Implications for horizontal drilling. In: Hansen, W.B. (Ed.), Geology and Horizontal Drilling of the Bakken Formation: Montana Geological Society, 1991 Guidebook, pp. 117–142.

Crampin, S., McGonigle, R., Ando, M., 1986. Extensive-dilatancy anisotropy beneath Mount Hood, origin and the effect of aspect ratio on seismic velocities through aligned cracks. J. Geophys. Res. 91, 12703–12710.

Crelling, J.C., Dutcher, R.R., Lange, R.V., 1982. Petrographic and fluorescent properties of resinite macerals from western U.S. coals. Utah Geol. Min. Surv. Bull. 118, 187–191.

Crovelli, R.A., Balay, R.H., 1986. FASP, an analytical resource appraisal program for petroleum play analysis. Comput. Geosci. 12 (4B), 423–475.

Crovelli, R.A., Schmoker, J.W., 2001. Probabilistic method for estimating future growth of oil and gas reserves. In: Dyman, T.S., Schmoker, J.W., Verma, Mahendra (Eds.), Geologic, Engineering, and Assessment Studies of Reserve Growth: U.S. Geological Survey Bulletin 2172–C. http://pubs.usgs.gov/bul/b2172-c/ (accessed 09.08.07.).

Crovelli, R.A., 1984. Procedures for petroleum resource assessment used by the U.S. Geological Survey—statistical and probabilistic methodology. In: Masters, C.D. (Ed.), Petroleum Resource Assessment, 17. International Union of Geological Sciences Publication, pp. 24–38.

Crovelli, R.A., 1986. A comparison of analytical and simulation methods for petroleum play analysis and aggregation. U.S. Geol. Surv. Open-File Report 86–97, 21 pp.

Crovelli, R.A., 1987. Probability theory versus simulation of petroleum potential in play analysis. In: Albin, S.L., Harris, C.M. (Eds.), Statistical and Computational Issues in Probability Modeling, Annual Operations Research, 8, pp. 363–381.

Crovelli, R.A., 1988. A microcomputer program for oil and gas resource appraisal. COGS 4 (3), 108–122.

Crovelli, R.A., 1990. FASPU English and Metric Version: Analytic Petroleum Resource Appraisal Microcomputer Programs for Play Analysis Using a Reservoir Engineering Model. U.S. Geological Survey Open-File Report 90-509, 23 pp.

Curtis, J.B., 2002. Fractured shale-gas systems. Am. Assoc. Pet. Geol. Bull. 86, 1921–1938.

Daily Oil Bulletin, 2011. Liard Piques Interest as Hedging Bolsters Lone Pine Income. Third quarter performance summary, released December 16, 2011. http://phx.corporate-ir.net/phoenix. zhtml?c=242054&p=irol-presentations, downloaded January 2012.

Daily Mail Online, November 6, 2013. "Was the Bible RIGHT about the origins of life? Scientists believe that we may have had our beginnings in CLAY". http://www.dailymail.co.uk/sciencetech/article-2488467/Scientists-believe-beginnings-CLAY.html.

Daley, P.T., Hron, F., 1977. Reflection and transmission coefficients for transversely isotropic media. Bull. Seism. Soc. Am. 67, 661–675.

Daniel, W., Graham, D.W., 2006. Explaining the Cosmos: The Ionian Tradition of Scientific Philosophy. Princeton, 368 pp, ISBN 9780691125404.

Daniels, E.J., Altaner, S.P., 1990. Clay mineral authigenesis in coal and shale from the anthracite region. Pa. Am. Min. 75, 825–839.

Daniels, E.J., Marshak, S., Altaner, S.P., 1996. Use of clay–mineral alteration patterns to define syntectonic permeability of joints (cleat) in Pennsylvanian anthracite coal. Tectonophysics 263, 123–136.

Das, S. Electro-magnetic heating in viscous oil reservoir (SPE-117693). In Proceedings of International Thermal Operations and Heavy Oil Symposium, Calgary, AB, Canada, October 20–23, 2008.

Davis, J.C., Chang, T., 1989. Estimating potential for small fields in a mature petroleum province. Am. Assoc. Pet. Geol. Bull. 73 (8), 967–976.

Davis, W., 2000. Air Pollution Engineering Manual. Wiley Interscience Publication, ISBN 978-0-471-33333-3.

Davis, G.R., 2005. Explicit control of image noise and error properties in cone-beam microtomography using dual concentric circular source loci. Nucl. Instrum. Methods Phys. Res. A 547, 679–685.

Dawson, F.M., Marchioni, D.L., Anderson, T.C., McDougall, W.J., 2000. An assessment of coalbed methane exploration projects in Canada. Geol. Surv. Can. Bull. 549, 217 pp.

De Bruin, R.H., 1993. Overview of oil and gas geology of Wyoming. In: Snoke, A.W., Steidtmann, J.R., Roberts, S.M. (Eds.), Geology of Wyoming: Geological Survey of Wyoming Memoir 5, pp. 837–873.

De Swaan, A., 1975. Analytic solutions for determining naturally fractured reservoir properties by well testing. In: Paper SPE 5346 Presented at the 1975 SPE-aime 45th Annual California Regional Meeting, Ventura, California, 2–4 April.

Dean, B., 1998. Keeping an eye on the geology improves well selection in the Frisco City/Norphlet play. Lead. Edge 17 (1), 24–28.

Deans, W.S., Scherer, L.R., Pulley, J., 1991. A study of horizontal drilling and the Bakken Formation, Williston Basin, North Dakota. In: Hansen, W.B. (Ed.), Geology and Horizontal Drilling of the Bakken Formation: Montana Geological Society, 1991 Guidebook, pp. 95–104.

DeBruin, R.H., Lyman, R.M., Jones, R.W., Cook, L.W., 2000. Coalbed Methane Development in Wyoming. Information Pamphlet Number 7. Wyoming State Geological Survey, Laramie, WY.

DeGraff, J.M., Aydin, A., 1993. Effect of thermal regime on growth increment and spacing of contraction joints in basaltic lava. J. Geophys. Res. 98, 6411–6430.

Delamaide, E., Corlay, P., Wang, D., 1994. Daqing oil field: the success of two pilots initiates first extension of polymer injection in a giant oil field (SPE-27819). In Proceedings of SPE/DOE Improved Oil Recovery Symposium, 17–20 April, Tulsa, Oklahoma.

Delgadillo-Velazquez, O., Georgiou, G., Sentmanat, M., Hatzikiriakos, S.G., 2008. Sharkskin and oscillating melt fracture: why in slit and capillary dies and not in annular dies? Polym. Eng. Sci. 405–414. http://dx.doi.org/10.1002/pen.

Demchuk, T., Strobl, R., 1989. Coal facies and in-seam profiling, Highvale no. 2 seam, Highvale Alberta. In: Advances in Western Canada Coal Geoscience – Forum Proceedings, Alberta Research Council Information Series No. 103, pp. 201–211.

DeMis, W.D., Milliken, J.V., 1993. Shongaloo field—a recent Smackover (Jurassic) discovery in the Arkan-sas-Louisiana state line graben. Gulf Coast Assoc. Geol. Soc. Trans. 43, 109–119.

Department Energy, Mines and Resources, 2011. Yukon Energy Prospectus Eagle Plain Basin. Government of Yukon, Department of Energy, Mines and Resources website. www.emr.gol.yk.ca/oilandgas/pdf/eagle_plain_prospectus_jan2011.pdf.

Department of Energy, Mines and Resources, 2011. Yukon Oil and Gas: A Northern Investment Opportunity. Government of Yukon, Department of Energy, Mines and Resources. Website. www.emr.gov.yk.ca/oilandgas/pdf/yukonoilandgasmarch2012.pdf.

DePriest, W., Van Laar, J., 1992. Engineering Evaluation of PRENFLO-Based Integrated-Gasification-Combined-Cycle (IGCC) Power Plant Designs. Chicago, Illinois, USA.

Derbyshire, D., January 27, 2011. How Humans are 97% the Same as Orangutans: New Research Shows how DNA Matches. The Daily Mail, UK.

Desmond, R.J., Steidtmann, J.R., Cardinal, D.F., 1984. In: Goolsby, J., Morton, D. (Eds.), Stratigraphy and Depositional Environments of the Middle Member of the Minnelusa Formation, Central Powder River Basin, Wyoming.

DeVries, A.J., 1958. Foam stability. Rubber Chem. Technol. 5, 1142–1205.

DeWitt Jr., W., Roen, J.B., Wallace, L.G., 1993. Stratigraphy of Devonian black shales and associated rocks in the Appalachian Basin. In: Petroleum Geology of the Devonian and Mississippian Black Shale of Eastern North America. U.S. Geological Survey Bulletin 1909, pp. B1–B57.

Dickinson, K.A., 1968. Upper Jurassic stratigraphy of some adjacent parts of Texas, Louisiana, and Arkansas. In: U.S. Geological Survey Professional Paper 5-E, 25 pp.

Dietrich, J., Dixon, J., 1996. Geology of the Beaufort Sea continental shelf. In: The Geology, Mineral and Hydrocarbon Potential of Northern Yukon Territory and Northwestern District of Mackenzie. Geological Survey of Canada Bulletin 422, pp. 319–322.

Dietrich, J., Lavoie, D., Hannigan, P., Pinet, N., Castonguay, S., Giles, P., Hamblin, T., 2011. Geological setting and resource potential of conventional petroleum plays in Paleozoic basins in eastern Canada. Bull. Can. Pet. Geol. 58, 54–84.

Dietze, P., 1997. Little Warming with New Global Carbon Cycle Model, vol. II. ESEF. http://www.john-daly.com/carbon.htm.

Dilgren, R.E., Deemer, A.R., Owens, K.B., 1982. Laboratory development and field testing of steam r non-condensable gas foams for mobility control in heavy oil recovery. In: SPE Paper 10774, California Regional Meeting, March 24–26.

Dixon, S.A., Summers, D.M., Surdam, R.C., 1989. Diagenesis and preservation of porosity in Norphlet Formation (Upper Jurassic), southern Alabama. Am. Assoc. Pet. Geol. Bull. 73, 707–728.

Dixon, J., Morrell, G.R., Dietrich, J., Taylor, G.C., Procter, R.M., Conn, R.E., Dallaire, S.M., Christie, J.A., 1994. Petroleum resources of the Mackenzie Delta and Beaufort sea. Geol. Surv. Can. Bull. 474, 52.

Dixon, J., 1992. Stratigraphy of Mesozoic strata, Eagle Plain area, northern Yukon. Geol. Surv. Can. Bull. 408, 58.

Dixon, J., 1996. Cretaceous and tertiary. In: The Geology, Mineral and Hydrocarbon Potential of Northern Yukon Territory and Northwestern District of Mackenzie. Geological Survey of Canada Bulletin, 422, pp. 301–317.

Djabbarah, N.F., Weber, S.L., Freeman, D.C., Muscatello, J.A., Ashbaugh, J.P., Covington, T.E., 1990. Laboratory design and field demonstration of steam diversion with foam. In: SPE Paper 20067, 60th California Regional Meeting, Ventura, April 4–6.

Djahanguiri, F., Stagg, M.S., Otterness, R.E., Kemeny, J.M., 1994. Rock fragmentation results for a simulated underground leaching stope. In: Nelson, P.P., Laubach, S.E. (Eds.), Proc. 1st North Am. Rock Mechanics Symp. Balkema, Rotterdam, pp. 293–300.

Dolton, G.L., Fox, J.E., 1995. In: Gautier, D.L. (Ed.), Powder River Basin Province (033).

Dolton, G.L., Carlson, K.H., Charpentier, R.R., Coury, A.B., Crovelli, R.A., Freon, S.E., Khan, A.S., Lister, J.H., McMullin, R.H., Pike, R.S., Powers, R.B., Scott, E.W.,

Varnes, K.L., 1981. Estimates of undiscovered recoverable conventional resources of oil and gas in the United States. U.S. Geol. Surv. Circ. 860, 87.
Dolton, G.L., Takahashi, K.I., Varnes, K.L. (Eds.), 1995. National Assessment of United States Oil and Gas Resources. U.S. Geological Survey DDS–30, CD-ROM.
Donaldson, A.B., 1997. Reflections on a downhole steam generator program (SPE-38276). In: Proceedings of SPE Western Regional Meeting, Long Beach, CA, USA, June 25–27, 1997.
Donnell, J.R., 1961. Tertiary geology and oil-shale resources of the Piceance Creek Basin between the Colorado and White Rivers, in contributions to economic geology, 1958. U.S. Geol. Surv. Bull. 1082-L L835–L891.
Doscher, T.M., Hammershaimb, E.C., 1981. Field demonstration of steamdrive with ancillary materials. In: SPE r DOE Paper 9777, SPE/DOE 2nd Symposium on Enhanced Oil Recovery, Tulsa, Oklahoma, April 5–8.
Dougherty, M.E., Stephen, R.A., 1987. Geoacoustic scattering from seafloor features in the ROSE area. J. Acoust. Soc. Am. 82, 238–256.
Drever, J.I., 1997. The Geochemistry of Natural Waters, Surface and Groundwater Environments. Prentice-Hall, Englewood Cliffs, NJ, pp. 41–58.
Drew, L.J., Lore, G.L., 1992. Field growth in the Gulf of Mexico—a progress report. In: Carter, L.M.H. (Ed.), USGS Research on Energy Resources, 1992, U.S. Geological Survey Circular, 1074, pp. 22–23.
Drew, L.J., Schuenemeyer, J.H., Baweic, W.J., 1982. Estimation of the future rates of oil and gas discoveries in the western Gulf of Mexico – a discovery-process model adapted to forecast future rates of oil and gas discoveries under the conditions of economic truncation. In: U.S. Geological Survey Professional Paper 1252, p. 7.
Dron, R.W., 1925. Notes on cleat in the Scottish coalfield. Trans. Inst. Min. Eng. 70, 115–117.
Dronkert, H., Remmelts, G., 1996. Influence of salt structures on reservoir rocks in Block L2, Dutch continental shelf. In: Rondeel, H.E., Batjes, D.A.J., Nieuwenhuijs, W.H. (Eds.), Geology of Gas and Oil under the Netherlands. Kluwer, Dordrecht, pp. 159–166.
Druyff, Lorraine, 1991. Reservoir properties of the Bakken Shale [abs.]. In: Hansen, W.B. (Ed.), Geology and Horizontal Drilling of the Bakken Formation: Montana Geological Society, 1991 Guidebook, p. 91.
Du, K.F., Stewart, G., 1992. Transient-pressure response of horizontal wells in layered and naturally fractured reservoirs with dual-porosity reservoirs. In: Paper SPE 24682 Presented at the 1992 Annual Technical Conference and Exhibition, Washington, DC, October 4–7.
Dubiel, R.F., 2003. Geology, Depositional Models, and Oil and Gas Assessment of the Green River Total Petroleum System, Uinta-Piceance Province, Eastern Utah and Western Colorado, Ch. 5, in USGS Uinta-Piceance Assessment Team, Compilers, Petroleum Systems and Geologic Assessment of Oil and Gas in the Uinta-Piceance Province, Utah and Colorado. U.S. Geological Survey Digital Data Series DDS– 69–B, version 1, CD-ROM.
Duff, B.A., Hall, D., 1996. A model-based approach to evaluation of exploration opportunities. In: Dore, A.G., Sinding-Larson, R. (Eds.), Quantification and Prediction of Petroleum Resources. Norwegian Petroleum Society Special Publication No. 6, pp. 183–198.
Duiveman, M.W.; Herwin, H.; Grivot, P. Integrated management of water, lean gas and air injection: the successful ingredients to IOR projects on the mature handil field (SPE-93858). In Proceedings of SPE Asia Pacific Oil and Gas Conference and Exhibition, Jakarta, Indonesia, April 5–7, 2005.
Dung, H.T., Piracha, A.L., 2000. J. Energy Dev. 25 (1), 47–70.
Dyer, J.R., 1983. Joints in sandstones, Arches National Park, Utah (Ph.D. dissertation). . Stanford Univ., CA, 202 pp.

Dyke, C.G., Wu, B., Milton-Tayler, D., 1995. Advances in characterizing natural-fracture permeability from mud-log data. SPE Form. Eval. 10 (3), 160–166. SPE 25022-PA. http://dx.doi.org/10.2118/25022-PA.

Dyman, T.S., Schmoker, J.W., 1998. The Use of Well-Production Data in Quantifying Gas-Reservoir Heterogeneity. U.S. Geological Survey Open-File Report 98–778, 7 pp.

Dyman, T.S., Schmoker, J.W., 2003. Well production data and gas reservoir heterogeneity—reserve growth applications. In: Dyman, T.S., Schmoker, J.W., Verma, M.K. (Eds.), Geologic, Engineering, and Assessment Studies of Reserve Growth, U.S. Geological Survey Bulletin 2172–E. http://pubs.usgs.gov/bul/b2172-e/ (accessed 07.08.07.).

Dyman, T.S., Schmoker, J.W., Quinn, J.C., 1996. Reservoir Heterogeneity as Measured by Production Characteristics of Wells—Preliminary Observations. U.S. Geological Survey Open-File Report 96–059, 14 pp.

Dyman, T.S., Schmoker, J.W., Quinn, J.C., 2000. Heterogeneity of morrow sandstone reservoirs measured by production characteristics. In: Johnson, K.S. (Ed.), Marine Clastics in the Southern Midcontinent, 1997 Symposium: Oklahoma Geological Survey Circular 103, pp. 119–127.

Editorial, 1996. Watching the world: tangible benefits of virtual reality. Oil Gas. J. 94 (13), 30.

Ehrenberg, S.N., Nadeau, P.H., Steen, Ø., October 2009. Petroleum reservoir porosity versus depth: influence of geological age. AAPG Bull. 93 (10), 1281–1296.

EIA, 2013. International Energy Outlook. July 25, EIA report.

Eisenack, K., Matthias, K.B., Lüdeke, M.K.B., Gerhard Petschel-Heldl, G., Scheffran, J., Kropp, J.P., 2007. Qualitative modelling techniques to assess patterns of global change. In: Kropp, J., Scheffran, J. (Eds.), Advanced Methods for Decision Making and Risk Management in Sustainability Science. Nova Science Publishers, pp. 83–127.

Ekstrom, M.P., Dahan, C.A., Chen, M.Y., Lloyd, P.M., Rossi, D.J., 1987. Formation imaging with microelectrical scanning arrays. Log Analyst 28, 294–306.

El-Banbi, A.H., May 1998. Analysis of Tight Gas Wells (Ph.D. dissertation). Texas A &M University.

Elliott, J.C., Bollet-Quivogne, F.R.G., Anderson, P., et al., 2005. Acidic demineralization of apatites studied by scanning X-ray microradiography and microtomography. Mineral. Mag. 69, 643–652.

El-Sayed, S.M., Abdel-Aziz, M.R., 2003. A comparison of Adomian's decomposition method and wavelet-Galerkin method for solving integro-differential equations. Appl. Math. Comput. 136, 151–159.

Energy Information Administration (EIA), 1990 (December). The Domestic Oil and Gas Recoverable Resource Base: Supporting Analysis for the National Energy Strategy. SR/NES/90–05. Washington, D.C.

Energy Information Administration, 1998. Natural Gas 1998. Issues and Trends, report.

Energy Resources Conservation Board, 2009. Development Entity Order No. DE 2006-2A. Calgary, Alberta.

Engelder, T., Fischer, M.P., 1996. Loading Configurations and Driving Mechanisms for Joints Based on the Griffith energy-balance concept: Tectonophysics, v. 256, 253–277.

Engelder, T., Lacazette, A., 1990. Natural hydraulic fracturing. In: Barton, C., Stephansson, O. (Eds.), Rock Joints. Balkema, Rotterdam, pp. 35–43.

Engelder, T., 1985. Loading paths to joint propagation during a tectonic cycle: an example from the Appalachian Plateau. U.S.A. J. Struct. Geol. 7, 459–476.

Engelder, T., 1987. Joints and shear fractures in rock. In: Atkinson, B.K. (Ed.), Fracture Mechanics of Rock. Academic Press, Orlando, pp. 27–69.

England, P., Houseman, G., 1989. Extension during the continental convergence, with application to the Tibetan Plateau. J. Geophys. Res. 98, 17561–17579.

English, J.M., Johannson, G.G., Johnston, S.T., Mihalynuk, M.G., Fowler, M., Wight, K.L., 2005. Structure, stratigraphy and petroleum resource potential of the Central Whitehorse Trough, northern Canadian Cordillera. Bull. Can. Pet. Geol. 53 (2), 130–153.

ERCB, 2008. ST98: Alberta's Energy Reserves 2007 and Supply/Demand Outlook 2008–2017. Alberta Energy Resources Conservation Board.

Erlend, H.V., Nygaard, G., Lorentzen, R.J., Fjelde, K.K., 2003. Reservoir characterization during UBD: methodology and active tests. In: Paper IADC/SPE 81634 Presented at the IADC/SPE Underbalanced Technology Conference and Exhibition, Houston, Texas, USA, 25–26 March. http://dx.doi.org/10.2118/81634-MS.

Ertekin, T., Abou-Kassem, J.H., King, G.R., 2001a. Basic Applied Reservoir Simulation. In: SPE Textbook Series, vol. 7. SPE, Richardson, TX, 406 pp.

Eson, R.L., 1982. Downhole steam generator—field tests (SPE-10745). In: Proceedings of SPE California Regional Meeting, San Francisco, CA, USA, March 24–26, 1982.

Esterhuizen, G.S., 1995. Rock engineering evaluation of jointing in South African coal seams and its potential effect on coal pillar strength. In: Rossmanith, H.-P. (Ed.), Mechanics of Faulted and Jointed Rock. Balkema, Rotterdam, pp. 807–812.

Eugene, Y., 1993. Application of the decomposition method to the solution of the reaction-convection-diffusion equation. Appl. Math. Comput. 56, 1–27. Expanded Abstracts, 723–730.

Falls, A.H., et al., August 1988. Development of a mechanistic foam simulator: the populations balance and generation by snap-off. SPE Reservoir Eng.

Falls, A.H., Musters, J.J., Ratulowski, J., 1989. The Apparent Viscosity of Foams in Homogeneous Bead Packs. SPE Res. Eng. 155–164. May.

Faraj, B., Williams, H., Addison, G., McKinstry, B., Winter 2004. Gas potential of selected shale formations in the western canadian sedimentary basin. Gas Tips 21–25.

Farihi, J., Brinkworth, C.S., et al., 2011. Possible signs of water and differentiation in a rocky exoplanetary body. Astrophys. J. Lett. 728 (1), L8.

Farouq Ali, S.M., 1972. "A current appraisal of in-situ combustion field tests". JPT 14 (No. 4), 477. April.

Farouq Ali, S.M., 1983. J. Pet. Tech. 35 (4). April.

Fayers, F.J., Sheldon, J.W., 1959. The effect of capillary pressure and gravity on two-phase fluid flow in a porous medium. Trans. AIME 216, 147–155.

Feature, 1997. Caution expressed over industry optimism on 4D seismic technology. First Break 15 (2), 51.

Feazel, C.T., 1985. Diagenesis of Jurassic grainstone reservoirs in the Smackover Formation, Chatom field, Alabama. In: Roehl, P.O., Choquette, P.W. (Eds.), Carbonate Petroleum Reservoirs. Springer-Verlag, New York, pp. 359–368.

Ferdous, H.S., Renaut, R.W., 1997. Dolomitization in the Bakken Formation in southeastern Saskatchewan, Canada [abs.]. Am. Assoc. Pet. Geol. Bull. 81, 685.

Fernández, E., Montalvo, S., 1998. Métodos económicos y ecológicamente viables para purificar gases contaminantes. Available from: http://www.bvsde.ops-oms.org/bvsaidis/caliaire/peru/cubpca004.pdf (accessed 26.04.11).

Ferri, F., Hickin, A.S., Huntley, D.H., 2011a. Besa River Formation, Western Liard Basin, British Columbia (NTS 094N): Geochemistry and Regional Correlations. In: Geoscience Reports 2011, BC Ministry of Energy and Mines, pp. 1–18.

Ferri, F., Hickin, A.S., Huntley, D.H., 2011b. Geochemistry and shale gas potential of the Garbutt formation, Liard Basin, British Columbia (Parts NTS 094N, O; 095B, C). In: Geoscience Reports 2011, BC Ministry of Energy and Mines, pp. 19–36.

Fetkovich, M.J., June 1980. Decline curve analysis using type curves. JPT 1066–1077.

Financial Post, June 30, 2011. Quebec moratorium leaves shale gas drillers staggering. Financ. Post.

Fink, J.K., 2013. Renewable Polymers. Wiley-Scrivener, 329 pp.
Finley, J.R., Pinter, J.D., Satish, M.G., 1998. Automatic model calibration applying global optimization techniques. J. Environ. Model. Assess. 3 (1–2), 117–126.
First Energy Capital, 2009. Nikanassin Visited. Analyst report for industry clients.
Forman, D.J., Hinde, A.L., 1985. Improved statistical methods for assessment of undiscovered petroleum resources. Am. Assoc. Pet. Geol. Bull. 69 (1), 106–118.
Forman, D.J., Hinde, A.L., 1986. Examination of the creaming methods of assessment applied to the Gippsland Basin, offshore Australia. In: Rice, D.D. (Ed.), Oil and Gas Assessment Methods and Applications. American Association of Petroleum Geologists Studies in Geology No. 21, pp. 101–110.
Fouch, T.D., Nuccio, V.F., Anders, D.E., Rice, D.D., Pitman, J.K., Mast, R.F., 1994. Green river (!) petroleum system, Uinta Basin, Utah, USA. In: Magoon, L.B., Dow, W.G. (Eds.), The Petroleum System—From Source to Trap. American Association of Petroleum Geologists Memoir 60, pp. 399–421.
Francois, D.K., 1995. Federal Offshore Statistics: 1994—Leasing, Exploration, Production, and Revenue as of December 31, 1994 (Minerals Management Service OCS Report).
Frankel, A., Clayton, R.W., 1986. Finite-difference simulations of seismic scattering: Implications for the propagation of short-period seismic waves in the crust and models of crustal heterogeneity. J. Geophys. Res. 91, 6465–6489.
Frantz, J.H., Waters, G.A., and Jochen, V.A. "Evaluating barnett shale production performance using an integrated approach," Paper SPE 96917 Presented at the 2005 SPE ATCE, Dallas, TX, 9–12 October.
Fraser, T.A., Allen, T.L., Lane, L.S., Reyes, J.C., 2012. Shale gas potential of Devonian shale in north Yukon: results from a diamond drillhole study in western Richardson Mountains. In: MacFarlane, K.E., Sack, P.J. (Eds.), Yukon Exploration and Geology 2011. Yukon Geological Survey, pp. 45–74.
Freedman, R., Minh, C.C., Gubelin, G., Freeman, J.J., McGinness, T., Terry, B., Rawlence, D., 1998. Combining NMR and density logs for petrophysical analysis in Gas-bearing formations. In: Paper II, SPAWLA 39th Annual Meeting, Colorado, May 26–29.
Freeman, C.M., Moridis, G.J., Ilk, D., et al., 2009. A numerical study of performance for tight gas and shale gas reservoir systems. In: Paper SPE 124961 Presented at the SPE Annual Technical Conference and Exhibition, New Orleans, Louisiana, USA. http://dx.doi.org/10.2118/124961-MS.
Friedel, T., Mtchedlishvili, G., Voigt, H.-D., Hafner, F., 2008. Simulation of inflow while underbalanced drilling with automatic identification of formation parameters and assessment of uncertainty. SPE Res. Eval. Eng. 11 (2), 292–297. SPE 93974-PA. http://dx.doi.org/10.2118/93974-PA.
Fryberger, S.G., 1984. The Permian upper Minnelusa Formation, Wyoming—Ancient example of an offshore prograding eolian sand sea with geomorphic facies and system-boundary traps for petroleum. In: Goolsby, Jim, Morton, D.M. (Eds.), The Permian and Pennsylvanian Geology of Wyoming: Wyoming Geological Association, 35th Annual Field Conference, Guidebook, pp. 241–271.
Fryberger, S.G., 1990. Bounding surfaces in eolian sediments. In: Fryberger, S.G., Krystinik, L.F., Schenk, C.J. (Eds.), Modern and Ancient Eolian Deposits—Petroleum Exploration and Production: Rocky Mountain Section S.E.P.M. Society for Sedimentary Geology, pp. 7-1–7-15.
Gal, L.P., Allen, T.L., Fraser, T.A., Hadlari, T., Lemieux, Y., Pyle, L.J., Zantvoort, W.G., 2007. Rock-Eval 6/TOC Analyses from Outcrop Samples in Northern Mackenzie Mountains, Eastern Richardson Mountains, and Southern Peel Plateau and Plain, Northwest Territories and Yukon, Canada. NWT Open Report 2007-002 and YGS Open File 2007-1, 11 page report, Microsoft Excel® spreadsheet and ESRI Arcview® files. Northwest Territories Geoscience Office and Yukon Geological Survey.

Gal, L.P., Allen, T.L., Hadlari, T., Zantvoort, W.G., 2009. Chapter 10 – petroleum systems elements. In: Pyle, L.J., Jones, A.L. (Eds.), Regional Geoscience Studies and Petroleum Potential, Peel Plateau and Plain, Northwest Territories and Yukon: Project Volume. Northwest Territories Geoscience Office and Yukon Geological Survey. NWT Open File 2009-02 and YGS Open File 2009-25, p. 477–549.

Gale, J.F.W., Reed, R.M., Holder, J., April 2007. Natural fractures in the Barnett Shale and their importance for hydraulic fracture treatments. AAPG Bull. 91 (4), 603–622.

Galloway, W.E., Hobday, D.K., Magara, K., 1982. Frio Formation of the Texas Gulf Coast Basin—Depositional Systems, Structural Framework, and Hydrocarbon Origin, Migration, Distribution, and Exploration Potential. University of Texas at Austin, Bureau of Economic Geology Report of Investigations 122, 78 pp.

Galloway, W.E., Ewing, T.E., Garrett, C.M., Tyler, N., Bebout, D.G., 1983. Atlas of Major Texas Oil Reservoirs. The University of Texas at Austin Bureau of Economic Geology, 139 pp.

Galloway, W.E., Jirik, L.A., Morton, R.A., DuBar, J.R., 1986. Lower Miocene (Fleming) Depositional Episode of the Texas Coastal Plain and Continental Shelf: Structural Framework, Facies, and Hydrocarbon Resources, 50 pp. The University of Texas at Austin Bureau of Economic Geology Report of Investigations No. 150.

Galloway, W.E., 1977. Catahoula Formation of the Texas Coastal Plain—Depositional Systems, Composition, Structural Development, Ground-Water Flow History, and Uranium Distribution. Report of Investigations 87, 59 pp. The University of Texas at Austin Bureau of Economic Geology.

Galloway, W.E., 1986. Depositional and Structural Framework of the Distal Frio Formation, Texas Coastal Zone and Shelf. The University of Texas at Austin, Bureau of Economic Geology Geological Circular 86–9, 16 pp.

Galloway, W.E., 1989a. Genetic stratigraphic sequences in basin analysis 1: architecture and genesis of flooding-surface bounded depositional units. Am. Assoc. Pet. Geol. Bull. 73 (2), 125–142.

Galloway, W.E., 1989b. Genetic stratigraphic sequences in basin analysis II—application to northwest Gulf of Mexico Cenozoic basin. Am. Assoc. Pet. Geol. Gists Bull. 73, 143–154.

Gamson, P.D., Beamish, B.B., Johnson, D.P., 1993. Coal microstructure and micro-permeability and their effects on natural gas recovery. Fuel 72, 87–99.

Gash, B.W., 1992. The effects of cleat orientation and confining pressure on cleat porosity, permeability, and relative permeability in coal. Log. Anal. 33, 176–177.

Gaskari, R., Mohaghegh, S.D., Jalali, J., 2006. An integrated technique for production data analysis with application to mature fields. In: Paper SPE 100562 Presented at the 2006 SPE Gas Technology Symposium, Calgary, Alberta, Canada, 15–17 May. http://dx.doi.org/10.2118/100562- MS.

Gaskari, R., Mohaghegh, S.D., Jalali, J., November 2007. An integrated technique for production data analysis (PDA) with application to mature Fields. SPE Prod. Oper. 403–416.

Gaspar, A.T.F.S., Suslick, S.B., Ferreira, D.F., Lima, G.A.C., March 7–9, 2005. Enhanced oil recovery with CO_2 sequestration: a feasibility study of a Brazilian mature oil field (SPE-94939). In: Proceedings of SPE/EPA/DOE Exploration and Production Environmental Conference, Galveston, TX, USA.

Gatens III, J.M., Lee, W.J., Lane, H.S., Watson, A.T., Stanley, D.K., Lancaster, D.E., 1987. Analysis of eastern Devonian gas shales production data. In: Paper SPE 17059 Presented at the 1987 SPE Eastern Regional Meeting, Pittsburgh, Pennsylvania, 21–23 October.

George, D.S., Islam, M.R., 1998. A micromodel study of acidization foam. paper no. 49489. In: Proc. SPE ADIPEC 98, Abu Dhabi, UAE.

Geosearch Logging Inc. Wellsite Geology and Gas Detection. http://www.geosearchlogging.com/employee/HydrocarbonWellLoggingHandout.pdf (accessed 20.10.11.).
Gibson, B.S., Levander, A.R., 1988. Modeling and processing of scattered waves in seismic reflection surveys. Geophysics 53, 466–478.
Gillette, R., 1974. Oil and gas resources: did the USGS gush too high? Science 185 (4146), 127–130.
Gleick, J., 1987. Chaos – Making a New Science. Penguin Books, NY, 352 pp.
Goff, J.A., Jordan, T.H., 1988. Stochastic modeling of seafloor-morphology: inversion of sea beam data for second-order statistics. J. Geophys. Res. 93, 13589–13608.
Gonzalez, M.H., Bukacek, R.F., Lee, A.L., 1967. The viscosity of methane. SPE J. 7 (1), 75–79. SPE 1483-PA. http://dx.doi.org/10.2118/1483-PA.
Gordey, S.P., Makepeace, A.J., 1999. Yukon bedrock geology. In: Gordey, S.P., Makepeace, A.J. (Eds.), Yukon Digital Geology. Geological Survey of Canada Open File D3826 and Exploration and Geological Services Division, Yukon, Indian and Northern Affairs Canada, Open File 1999-1(D).
Gosh, D.P., 2007. Wet H2S cracking problem in oil refinery processes – material selection and operation control issues. In: The 2007 Tri-service Corrosion Conference, Denver, Colorado, USA. December 3–7, 2007.
Grace, J., 1991. When are we really going to run out of oil and gas?—statistical, political, economic, and geoscientific issues. In: Coburn, T.C., Schuenemeyer, J.H. (Eds.), Co-conveners, Invited Session and Debate on Petroleum Reserves Estimation, 150th Annual Meeting American Statistical Association, Mathematical Geology, vol. 23, no. 3, pp. 403–442.
Gray, I., 1987. Reservoir engineering in coal seams: part I-the physical process of gas storage and movement in coal seams. SPERE 28–34.
Greaves, M., Xia, T., 2004. Downhole upgrading of Wolf Lake oil using THAI/CAPRI processes – tracer tests. Prepr. Pap. Am. Chem. Soc., Div. Fuel Chem. 49 (1), 69.
Grechka, V., Tsvankin, I., 1998. 3-D description of normal moveout in anisotropic inhomogeneous media. Geophysics 63, 1079–1092.
Gringarten, A.C., Ramey Jr., H.J., October 1973. The use of source and green's functions in solving unsteady-flow problems in reservoirs. SPE Paper 3818 SPE J. 13 (5), 285–296.
Gringarten, A.C., 1984. Interpretation of tests in fissured and multilayered reservoirs with double-porosity behavior. Theory Pract., JPT 36 (4). SPE-10044-PA.
Groeger, A., Bruhn, R.L., 2001. Structure and geomorphology of the Duchesne graben, Uinta Basin, Utah, and its enhancement of a hydrocarbon reservoir. Am. Assoc. Pet. Geol. Bull. 85, 1661–1678.
Groenenboom, J., Snieder, R., 1995. Attenuation, dispersion, and anisotropy by multiple scattering of transmitted waves through distributions of scatters. J. Acoust. Soc. Am. 98, 3482–3492.
Grout, M.A., 1991. Cleats in coalbeds of southern Piceance Basin, Colorado-correlation with regional and local fracture sets in associated clastic rocks. In: Schwochow, S., Murray, D.K., Fahy, M.F. (Eds.), Coalbed Methane of Western North America. Rk. Mt. Assoc. Geol, pp. 35–47.
Guellal, S., Cherruault, Y., 1995. Application of the decomposition method to identify the distributed parameters of an elliptical equation. Math. Comput. Model. 21 (i), 51–55.
Guevara, E.H., Mukhopadhyay, P.K., 1987. Source rock potential and oil source correlation, Permian (Leonardian) strata, central Spraberry trend, Midland Basin, Texas—preliminary study. Am. Assoc. Pet. Geol. Bull. 71, 561–562.
Guevara, E.H., Tyler, N., 1991. The Spraberry trend, Midland Basin, Texas—Development, innovation, and reserve growth potential in a mature giant [abs.]. Am. Assoc. Pet. Geol. Bull. 75, 586.

Guevara, E.H., 1988. Geological Characterization of Permian Submarine Fan Reservoirs of the Driver Waterflood Unit, Spraberry Trend, Midland Basin, Texas. University of Texas at Austin, Bureau of Economic Geology. Report of Investigations 172, 44 pp.

Gunn, R.D., August 31, 1992. Presentation of AAPG Oil Resource Base Estimates before Department of Energy Oil Resource Base Assessment Panel. Austin, Texas.

Gunn, E.A., 1996. Hierarchical planning processes in forestry: a stochastic programming-decision analytic perspective. In: Martell, D.L., Davis, L.S., Weintraub, A. (Eds.), Workshop on Hierarchical Approaches to Forest Management in Public and Private Organizations, Proceedings of a Workshop. Toronto, Canada, May 25–29, 1992. Canada Forest Service, Petawa National Forestry Institute, Information Report PI-X-124, pp. 85–97.

Gutschick, R.C., Sandberg, C.A., 1991. Late Devonian history of the Michigan Basin. In: Catacosinos, P.A., Daniels Jr., P.A. (Eds.), Early Sedimentary Evolution of the Michigan Basin: Geological Society of America Special Paper 256, pp. 181–202.

Hacquebard, P.A., Donaldson, J.R., 1969. In: Dapples, E.C., Hopkins, M.E. (Eds.), Carboniferous coal deposition associated with flood-plain and limnic environments in Nova Scotia, Environments of Coal Deposition, Geological Society of America Special Paper, Boulder, vol. 114, pp. 143–191.

Hacquebard, P.A., 2002. Potential coalbed methane resources in Atlantic Canada. Int. J. Coal Geol. 52, 3–28.

Hadlari, T., Gal, L.P., Zantvoort, W.G., Tylosky, S.A., Allen, T.L., Fraser, T.A., Lemieux, Y., Catuneanu, O., 2009a. Chapter 7 – upper Devonian to Carboniferous strata I – imperial formation play. In: Pyle, L.J., Jones, A.L. (Eds.), Regional Geoscience Studies and Petroleum Potential, Peel Plateau and Plain: Project Volume. Northwest Territories Geoscience Office and Yukon Geological Survey, NWT Open File 2009-02 and YGS Open File 2009-25, pp. 290–364.

Hadlari, T., Lemieux, Y., Tylosky, S.A., Zantvoort, W., Catuneau, O., 2009b. Slope and submarine fan turbidite facies of the Upper Devonian Imperial Formation, northern Mackenzie Mountains. NWT. Bull. Can. Pet. Geol. 57, 129–208.

Hall, S., October 2000. The last of the Neanderthals. National Geography.

Hamada, G.M., 2009. Petrophysical properties evaluation of tight gas sand reservoirs using NMR and conventional openhole logs. Open Renewable Energy J. 2, 6–11.

Hamalainen, R.P., Mantysaari, J., 2001. A dynamic interval goal programming approach to the regulation of a lake-river system. J. Multicrit. Decis. Anal. 10 (2), 75–86.

Hamblin, A.P., 2006. The "Shale Gas" Concept in Canada: A Preliminary Inventory of Possibilities. Geological Survey of Canada Open File 5384, 103 pp.

Hamilton, D.S., Holtz, M.H., Ryles, P., Lonergan, T., Hillyer, M., 1998. Approaches to identifying reservoir heterogeneity and reserve growth opportunities in a continental-scale bed-load fluvial system—Hutton Sandstone, Jackson field, Australia. Am. Assoc. Pet. Geol. Bull. 82, 2192–2219.

Hamblin, A.P., 2006. The "Shale Gas" concept in Canada: a preliminary inventory of possibilities. Geological Survey of Canada Open File 5384.

Hammes, U., Kerans, C., Lucia, F.J., 1996. Reservoir heterogeneity in karst-related reservoirs—Lower Ordovician Ellenburger Group, West Texas. In: Stoudt, E.L. (Ed.), Precambrian-Devonian Geology of the Franklin Mountains, West Texas—Analogs for Exploration and Production in Ordovician and Silurian Karsted Reservoirs in the Permian Basin: West Texas Geological Society, 1996 Guidebook.

Han, D.K., Yang, C.Z., Zhang, Z.Q., Lou, Z.H., Chang, Y.I., 1999. Recent development of enhanced oil recovery in China. J. Pet. Sci. Eng. 22, 181–188.

Hancock, P.L., 1985. Brittle microtectonics: principles and practices. J. Struct. Geol. 7, 437–458.

Handford, C.R., 1981. Sedimentology and genetic stratigraphy of dean and Spraberry formations (Permian), Midland Basin, Texas. Am. Assoc. Pet. Geol. Gists Bull. 65, 1602–1616.

Handy, L.L., Amaefule, J.O., Ziegler, V.N., Ershaghi, I., October 1982. Thermal stability of surfactants for reservoir applications. Soc. Petrol. Eng. J. 371–381. June.

Haneda, H., Komai, T., Yamamoto, Y., 2000. The effect of CO_2-air mixture compositions on the formation and dissociation of CO_2 hydrate. In: Holder, G.D., Bishnoi, P.R. (Eds.), Gas Hydrates, Challenges for the Future, Ann. N.Y. Acad. Sci., 912, pp. 261–271.

Hanes, J., Shepherd, J., 1981. Mining induced cleavage, cleats and instantaneous outbursts in the Gemini seamat Leichhardt Colliery, Blackwater, Queensland. Proc. Aust. Inst. Min. Metall. 277, 17–26.

Hannigan, P.K., Osadetz, K.G., Dixon, J., Bird, T., 2000. Petroleum Resource Assessment of the Kandik Basin, Yukon Territory, Canada. Oil and Gas Resources Branch, Department of Economic Development, Government of the Yukon.

Hannigan, P.K., Morrow, D.W., MacLean, B.C., 2011. Petroleum resource potential of the northern mainland of Canada (Mackenzie Corridor). Geol. Surv. Can. Open. File 6757, 271 pp.

Hannigan, P.K., 2000. Petroleum Resource Assessment of the Bonnet Plume Basin, Yukon Territory, Canada. Oil and Gas Resources Branch, Department of Economic Development, Government of the Yukon.

Hansen, W.B., Long, G.I.W., 1991a. Bakken production and potential in the U.S. and Canada—can the fairway be defined? In: Hansen, W.B. (Ed.), Geology and Horizontal Drilling of the Bakken Formation: Montana Geological Society, 1991 Guidebook, pp. 69–88.

Hansen, W.B., Long, G.I.W., 1991b. Horizontal and Vertical Prospects in the Bakken Formation, Williston Basin. In: Hansen, W.B. (Ed.), Geology and horizontal drilling of the Bakken Formation: Montana Geological Society. Guidebook, pp. 151–163.

Hanson, B., Grattan, S.R., Fulton, A., 1993. Agricultural Salinity and Drainage: A Handbook for Water Managers. University of California Irrigation Program, University of California, Davis, CA.

Harasimowicz, M., Orluk, P., Zakrzewska-Trznadel, G., Chmielewski, A., June 2007. Application of polyimide membranes for biogas purification and enrichment. J. Hazard. Mater. 144 (3), 698–702. ISSN 0304-3894.

Harrison, T.M., Copeland, P., Kidd, W.S.F., Yin, A., 1992. Raising Tibet. Science 255, 1663–1670.

Hascakir, B., Acar, C., Demiral, B., Akin, S. Microwave assisted gravity drainage of heavy oils (IPTC-12536). In Proceedings of International Petroleum Technology Conference, Kuala Lumpur, Malaysia, December 3–5, 2008.

Hatch, J.R., Gluskoter, H.J., Lindahl, P.C., 1976. Sphalerite in coals from the Illinois Basin. Econ. Geol. 71, 613–624.

Hatcher, D.B., Tussing, A.R., November 17, 1997. Long reserves lives sustain prospects for independents in the U.S. lower 48. Oil Gas. J. 95 (46), 49–59.

Haugen, G., Ursin, B., 1996. AVOA analysis of a vertically fractured reservoir underlying shale. 66th Ann. Intern. Mtg., Soc. Expl. Geophys. Expanded Abstracts, 1826–1829.

Hayes, B.J.R., Dunn, J., 2012. Deep Subsurface Saline Aquifer Characterization, Central Mackenzie Valley. Northwest Territories. Northwest Territories Geoscience Office. NWT Open File 2012–07.

Hayes, B.J., Hayes, M.C., 2004. Exploration assessment of tight gas plays, northeastern British Columbia. In: Proceedings, Canadian Society of Petroleum Geologists Annual Conference, 1 CD-ROM.

Hayes, B.J.R., Conrad, L., Dahlin, A., Daniel, S., 2009. Designing the deep basin development entity. In: Drawing and Expanding the Boundaries of DE#2 for Enhanced Exploitation of Alberta's Deep Basin Tight Gas Resource. Proceedings, Canadian Society for Unconventional Gas Annual Convention 2009, Calgary, Alberta, 1 CD-ROM.
Hayes, B.J., 2003. Exploration Assessment of Tight Gas Plays, Northeastern British Columbia. B.C. Ministry of Energy, Mines and Petroleum Resources. Petroleum Geology Open File 2003-3, 61 pp.
Hayes, B.J., 2005. Comparative Evaluation of Tight Gas Play Opportunities, Western Canada Sedimentary Basin. Petrel Robertson Consulting Ltd non-exclusive industry report.
Hayes, B.J., 2009. Evolution of Tight Gas Sandstone Plays and Production, Western Canada Sedimentary Basin. Search and Discovery Article #10182.
Hayes, B.J.R., 2010. Northwest Territories Unconventional Natural Gas Scoping Study. Northwest Territories Geoscience Office. NWT Open File 2010–03.
Hayes, B.J.R., 2011a. Regional Characterization of Shale Gas and Shale Oil Potential, Northwest Territories. NWT Open File 2011–08. Northwest Territories Geoscience Office.
Hayes, B.J.R., 2011b. Regional Characterization of Shale Gas and Shale Oil Potential. Northwest Territories. Luncheon presentation to Canadian Society for Unconventional Resources. www.csur.com.
Hayes, B.J.R., 2012. Petroleum Resource Assessment of Whitehorse Trough. Yukon Geological Survey, Yukon, Canada. Miscellaneous Report MR-6.
Hayes, B.J.R., Archibald, H.B., 2012. Scoping study of unconventional oil and gas potential, Yukon. Yukon Geological Survey, Miscellaneous Report 7, 100 p.
Helmold, K.P., Loucks, R.G., 1985. Origin of high-permeability reservoirs in the upper Minnelusa Sandstone (Permian), Powder River Basin, Wyoming and Montana [abs.]. Am. Assoc. Pet. Geol. Bull. 69, 264.
Henry, M.E., Hester, T.C., 1995. Anadarko Basin province (058). In: Gautier, D.L., Dolton, G.L., Takahashi, K.I., Varnes, K.L. (Eds.), 1995 National Assessment of United States Oil and Gas Resources. U.S. Geological Survey. DDS–30, CD-ROM.
Henry, J.D., 1982. Stratigraphy of the Barnett Shale (Mississippian) and associated reefs in the northern Fort Worth Basin. In: Martin, C.A. (Ed.), Petroleum Geology of the Fort Worth Basin and Bend Arch Area. Dallas Geological Society, pp. 157–177.
Hentz, T.F., Seni, S.J., Wermund Jr., E.G. (Eds.), 1997. Atlas of northern Gulf of Mexico gas and oil reservoirs. In: Pliocene and Pleistocene reservoirs, vol. 2. The University of Texas at Austin Bureau of Economic Geology, Gas Research Institute, Department of Energy, and Minerals Management Service plus 3 plates and CD-ROM, 78 pp.
Hettema, M., Papamichos, E., Schutjens, P., 2002. Subsidence delay: field observations and analysis, oil and gas science and technology. Rev. IFP 57 (5), 443–458.
Hickin, A., Ferri, F., Ferbey, T., Smith, I.R., 2010. Preliminary Assessment of Potential Hydraulic Fracture Sand Sources and Their Depositional Origin, Northeast British Columbia. Geoscience Reports 2010, BC. Ministry of Energy, Mines and Petroleum Resources, p. 35–92.
Hill, D.G., Nelson, C.R., Summer 2000. Gas productive fractured shales: an overview and update. Gas Tips 4–13.
Hill, R.J., Jarvie, D.M., Zumberge, J., Henry, M., Pollastro, R.M., 2007. Oil and gas geochemistry and petroleum systems of the Fort Worth Basin. Am. Assoc. Pet. Geol. Bull. 91, 445–473.
Hill, R., 1963. Elastic properties of reinforced solid: some theoretical principles. J. Mech. Phys. Solid 11, 352–357.

Him, A., Jiang, M., Sapin, M., Diaz, J., Nercessian, A., Lu, Q.T., Lepine, J.C., Shi, D.N., Sachpazi, M., Pandey, M.R., Ma, K., Gallart, J., 1995. Seismic anisotropy as an indicator of mantle flow beneath the Himalayas and Tibet. Nature 375, 571–574.

Hirasaki, G.J., Lawson, J.B., April 1985. Mechanism of foam flow through porous media: apparent viscosity in smooth capillaries. Soc. Petrol. Eng. J. Apr.: 176-190.

Hite, D.M., November 17, 1997. A native corporation evaluates potential of Alaska's Kandik area. Oil Gas. J. 89–93.

Hobbs, B.F., Meier, P.M., 2000. Energy Decisions and the Environment: A Guide to the Use of Multicriteria Methods, International Series in Operations Research & Management Science. Kluwer Academic Publishers, Norwell, MW.

Hoenig, A., 1979. Elastic moduli of a non-randomly cracked body. Int. J. Solids Struct. 15, 137–154.

Hoffmann, A., Mackowiak, J.F., Górak, A., Haas, M., Löning, J.-M., Runowski, T., Hallenberger, K., January 2007. Standardization of mass transfer measurements: a basis for the description of absorption processes. Chem. Eng. Res. Des. 85 (A1), 40–49.

Hojka, K., Dusseault, M.B., Bogobowicz, A.D., 1993. Analytical solutions for transient thermoplastic stress fields around a wellbore during fluid injection into permeable media. J. Can. Pet. Technol. 32 (4), 49–57.

Holditch, S.A., June 2006. Tight Gas sands. JPT 58 (6), 86–94.

Holmgren, C.R., Morse, R.A., 1951. Effect of free gas saturation on oil recovery by waterflooding. Trans. AIME 192, 135–140.

Holtz, M.H., Kerans, C., 1992. Characterization and categorization of west Texas Ellenburger reservoirs. In: Candelaria, M.P., Reed, C.L. (Eds.), Paleokarst, Karst Related Diagenesis and Reservoir Development—Examples from Ordovician-devonian Age Strata of West Texas and the Mid-continent. Permian Basin Section Society of Economic Paleontologists and Mineralogists Publication 92–33, pp. 197–206.

Holtz, M.H., 2008. Summary of sandstone Gulf Coast CO_2 EOR flooding application and response (SPE-113368). In: Proceedings of SPE/DOE Symposium on Improved Oil Recovery, Tulsa, OK, USA, April 20–23, 2008.

Hongmin, Y., Baoquan, Y., Guorui, X., Jiexiang, W., Shao, R.R., Weimin, L., Liang, X., Haitao, G. Air foam injection for IOR: from laboratory to field implementation in ZhongYuan oilfield China (SPE-113913). In Proceedings of SPE/DOE Symposium on Improved Oil Recovery, Tulsa, OK, USA, April 20–23, 2008.

Hopkins, C.W., Rosen, R.L., Hill, D.G. Characterization of an induced hydraulic fracture completion in a naturally fractured Antrim shale reservoir. Paper SPE 51068 Presented at the 1998 SPE Eastern Regional Meeting, Pittsburgh, Pennsylvania, November 9–11.

Horikawa, M.S., Rossi, M.L., Gimenes, M.L., Costa, C.M.M., da Silva, M.G.C., 2004. Chemical absorption of H2S for biogas purification. Braz. J. Chem. Eng. 21 (3), 415–422. July–September 2004.

Hossain, M.E., Islam, M.R., 2009. An Advanced Analysis Technique for Sustainable Petroleum Operations. VDM Publishing Ltd., Germany, 750 pp.

Hossain, M.E., 2008. Experimental and Numerical Investigation of Memory Based Complex Rheology and Rock/Fluid Interactions (Ph.D. dissertation). Dalhousie University, Halifax, Nova Scotia, Canada, 793 pp.

Houseman, G., England, P., 1986. Finite strain calculations of continental deformation, 1, methods and general results for convergent zones. J. Geophys. Res. 91, 3651–3663.

Houseman, G., England, P., 1993. Crustal thickening versus lateral expulsion in the Indian-Asian continental collision. J. Geophys. Res. 98, 12233–12249.

Houze, O.P., Horne, R.N., Ramey Jr., H.J., September 1988. Pressure transient response of an infinite conductivity vertical fracture in a reservoir with double porosity behavior. SPE Form. Eval. 510–518.

Houze, O., Tauzin, E., Artus, V., et al., 2010. The Analysis of Dynamic Data in Shale Gas Reservoirs - Part 1. Company report. Kappa Engineering, Houston, Texas, USA.
Hovanessian, S.A., Fayers, F.J., March 1961. Linear water flood with gravity and capillary effects. Soc. Pet. Eng. J. 32–36.
Howarth, R.B., Winslow, M.A., 1994. Energy use and CO_2 emissions reduction: Integrating pricing and regulatory policies. Energy 19, 855–867.
Howell, D.G., 1995. Kandik Basin. In: National Oil and Gas Assessment: Central Alaska Province, Report. United States Geological Survey Central Energy Team. http://geology.bgsu.edu/Yacobucci/Geol_415_515/Whitehouse.htm.
Huang, J., Griffiths, D.V., Wong, S.-W., 2010. Characterizing Natural Fracture Permeability from Mud Loss Data. Accepted for Publication in the SPE Drilling and Completion (submitted 2010).
Hubbert, M.K., Rubey, W.W., 1959. Mechanics of fluid-filled porous solids and its application to overthrust faulting. Geol. Soc. Am. Bull. 70, 115–166.
Hubbert, M.K., 1974. U.S. energy resources, a review as of 1972, part 1. In: A National Fuels and Energy Policy Study: U.S. 93rd Congress, Senate Committee on Interior and Insular Affairs, No. 93-40 (92-75), 267 pp.
Hudson, J.A., Liu, E., Crampin, S., 1996. Transmission properties of a plan fault. Geophys. J. Int. 125, 559–566.
Hudson, J.A., 1980. Overall properties of a cracked solid. Math. Proc. Camb. Phil. Soc. 88, 371–384.
Hudson, J.A., 1981. Wave speeds and attenuations of elastic waves in material containing cracks. Geophys. J. R. Astr. Soc. 64, 133–150.
Hudson, J.A., 1986. A higher order approximation to the wave propagation constants for a cracked solid. Geophys. J. R. Astr. Soc. 87, 265–274.
Hughes, B.L. and Sarma, H.K. Burning Reserves For Greater Recovery? Air Injection Potential In Australian Light Oil Reservoirs (SPE-101099). In Proceedings of SPE Asia Pacific Oil & Gas Conference and Exhibition, Adelaide, Australia, September 11–13, 2006.
Hunt Jr., J.L., Burgess, G., 1995. Depositional styles from Miocene through Pleistocene in the north-central Gulf of Mexico: an historical reconstruction. Gulf Coast Assoc. Geol. Soc. Trans. 45, 275–284.
Hunt, J.L., Rester, S., 2000. Reservoir characterization during underbalanced drilling. In: A New Model. SPE 59743 Presented at the 2000 SPE/CERI Gas Technology Symposium, Calgary, Alberta, Canada, 3–5 April. http://dx.doi.org/10.2118/59743-MS.
Hunt, E.H., 1938. Developments in rocky mountain region in 1937. Am. Assoc. Pet. Geol. Bull. 22, 677–693.
Hussen, C., Amin, R., Madden, G., Evans, B., 2012. Reservoir simulation for enhanced gas recovery: an economic evaluation. J. Nat. Gas Sci. Eng. 5, 42–50.
Hvitved, J., 2002. Sewer Processes: Microbial and Chemical Process Engineering of Sewer Networks. CRC Press, Florida, USA, ISBN 1-56676-926-4.
Ikelle, L.T., Yung, S.K., Daube, F., 1993. 2-D random media with ellipsoidal auto correlation functions. Geophysics 58, 1359–1372.
Imlay, R.W., 1980. Jurassic Paleobiogeography of the Conterminous United States in its Continental Setting. U.S. Geological Survey Professional Paper 1062, 134 pp.
Islam, M.R., Chakma, A., 1990. Comprehensive physical and numerical modeling of a horizontal well. In: SPE Paper No. 20627, Proc. of the SPE Annual Conf. and Exhib., New Orleans.
Islam, M.R., Chilingar, G.V., 1995. Three-dimensional numerical simulator for microbial enhanced oil recovery. Scientia Iranica 2 (1), 55–64.

Islam, M.R., Farouq Ali, S.M., 1990. New scaling criteria for chemical flooding experiments. J. Can. Petrol. Technol. 29 (1), 29–36.

Islam, M.R., Wadadar, S.S., Bansal, A., 1991. Enhanced oil recovery of Ugnu Tar Sands of Alaska using electromagnetic heating with horizontal wells (SPE-22177). In Proceedings of International Arctic Technology Conference, Anchorage, AK, USA, May 29–31.

Islam, M.R., Erno, B.P., Davis, D., 1992. Hot water and gas flood equivalence of in situ combustion. J. Can. Petrol. Technol. 31 (8), 44–52.

Islam, M.R., Chhetri, A.B., Khan, M.M., 2010. Greening of Operations, Petroleum Operations. Scrivener Wiley, 852 pp.

Islam, M.R., Mousavizadeghan, H., Mustafiz, S., Abou-kassem, J.H., 2010. Reservoir Simulation: Advanced Approach. Scrivener-Wiley, 468 pp.

Islam, J.S., Zatzman, G.M., Mughal, M.A.H., Islam, M.R., 2014. A Comprehensive theory of time, mass, energy, and human thought material. J. Nat. Sci. Sust. Technol. 8 (1), 181–215.

Islam, J.S., Zatzman, G.M., Mughal, M.A.H., Islam, M.R., 2014a. A Knowledge-based cognitive model. J. Info. Intell. Knowl. 6 (1), 1–30.

Islam, M.R., Islam, J.S., Zatzman, G.M., Rahman, M.S., Khan, M.M., 2014b. God, god's particle, god's wrath and force of nature: delinearized history of mass, energy and time. Adv. Sust. Pet. Eng. Sci. 6 (4).

Islam, J.S., Zatzman, G.M., Mughal, M.A.H., Islam, M.R., 2015. Implications of a comprehensive material balance equation on detection of the causes of medical disorders, part 1: elimination of paradoxes. J. Charact. Dev. Novel Mater. 7 (1).

Islam, M.R., 1993. Route to chaos in chemically enhanced thermal convection in porous media. Chem. Eng. Commun. 124, 77–95.

Islam, M.R., 2000. Energy state of the art 2000. In: Zhou, C.Q. (Ed.), Int. Sci. Serv. Charlottesville, VA, 21 pp.

Islam, M.R., 2001. Emerging technologies in monitoring of oil and gas reservoirs. In: SPE Paper No. 68804, Proc. SPE Western Regional Conference, Bakersfield, California, March 26–30.

Islam, M.R., 2002. Emerging technologies in subsurface monitoring of petroleum reservoirs. In: SPE 69440, SPE Latin American and Caribbean Petroleum Engineering Conference, Buenos Aires, Argentina, 25–28 March.

Islam, M.R., 2006a. Comprehensive mathematical modelling of horizontal wells. In: Paper No. 69, Proc. of the 2nd European Conf. on the Mathematics of Oil Recovery, Latitudes Camargue, Arles, France, 11–14 September.

Islam, M.R., 2006b. Computing for the information age. In: Proc. of the 36th International Conference on Computers and Industrial Engineering, Keynote Speech, Taipei, Taiwan, 20–23 June.

Islam, M.R., 2014. "A novel reserve assessment technique for unconventional gas reservoirs", MPC-2014, paper no. 38, Istanbul, 14-17 September.

Israel, S., Colpron, M., Roots, C., Fraser, T., 2012. Overview of Yukon Geology. Yukon Geological Survey website. www.geology.gov.yk.ca/overview_bedrock_geology.htm.

Jack, T.R., Stehmeier, L.G., Ferris, F.G., Islam, M.R., 1991. Microbial selective plugging to control water channeling. In: Donaldson (Ed.), Microbial Enhancement of Oil Recovery – Recent Advances. Elsevier Science Publishing Co., New York, pp. 433–440.

Jackson, K., McQuilkin, M., Pedersen, P.K., Meyer, R., 2011. Preliminary observation on stratigraphy and hydrocarbon potential of middle to Upper Cretaceous strata, Eagle Plain Basin, northern Yukon. In: MacFarlane, K.E., Weston, L.H., Relf, C. (Eds.), Yukon Exploration and Geology 2010. Yukon Geological Survey, pp. 125–134.

Jadhunandan, P.P., Morrow, N.R., 1995. Effect of wettability on waterflood recovery for crude-oil brine rock systems. SPE Reservoir Eng. 10, 40–46.

James, S.W., 1989. Diagenetic history and reservoir characteristics of a deep Minnelusa reservoir, Hawk Point field, Powder River Basin, Wyoming. In: Coalson, E.B. (Ed.), Rocky Mountain Reservoirs—1989: Rocky Mountain Association of Geologists, pp. 81–96.

Jarvie, D.M., Claxton, B.L., Henk, F., Breyerd, J.T., 2001. Oil and shale gas from the Barnett Shale, Fort Worth Basin, Texas [abs.]. Am. Assoc. Pet. Geol. 10, A100. Abstracts with Program.

Jarvie, D.M., Hill, R.J., Ruble, T.E., Pollastro, R.M., 2007. Unconventional shale gas systems—The Mississippian Barnett Shale of north-central Texas as one model for thermogenic shale-gas assessment. Am. Assoc. Pet. Geol. Bull. 91, 475–499.

Jayakumar, R., Sahai, V., Boulis, A., 2011. A better understanding of finite element simulation for shale gas reservoirs through a series of different case histories. In: Paper SPE 142464 Presented at the SPE Middle East Unconventional Gas Conference and Exhibition, Muscat, Oman. http://dx.doi.org/10.2118/142464-MS.

Jeffries, D.J., Farrier, D.R., 1985. Asymptotic results for eigenvector methods. Proc. Inst. Elec. Eng. F. 132, 589–594.

Jimenez, J. The field performance of SAGD projects in Canada (IPTC-12860). In Proceedings of International Petroleum Technology Conference, Kuala Lumpur, Malaysia, December 3–5, 2008.

Jin, Y., McNutt, M.K., Zhu, Y.S., 1994. Evidence from gravity and topography data for folding of Tibet. Nature 371, 669–674.

Jochen, J.E., Lancaster, D.E. Reservoir Characterization of an Eastern Kentucky Devonian shales well using a naturally fractured, layered description. SPE Paper 26192, presented at the 1993 SPE Gas Technology Symposium, Calgary, Alberta, Canada, 28–30 June.

John, D.R., Reid, B.G., 2000. A literature analysis of the WAG injectivity abnormalities in the CO_2 process. In: SPE/DOE Improved Oil Recovery Symposium. Society of Petroleum Engineers, Tulsa, Oklahoma. Paper no. SPE-59329.

Johnson, D.H., DeGraaf, S.R., 1982. Improving the resolution of bearing in passive sonar arrays by eigenvalue analysis. IEEE Trans. Acoust. Speech Signal Process. ASSP-30, 638–647.

Johnson, R.C., Roberts, S.B., 2003. The Mesaverde total petroleum system, Uinta-Piceance Province, Utah and Colorado. Chapter 7. In: USGS Uinta-Piceance Assessment Team, Compilers, Petroleum Systems and Geologic Assessment of Oil and Gas in the Uinta-Piceance Province, Utah and Colorado: U.S. Geological Survey Digital Data Series DDS–69–b, Version 1, CD-ROM.

Jorgensen, S.D., James, S.W., 1988. Integration of stratigraphic high resolution dipmeter data into the development of the Minnelusa "B" sand reservoir in Hawk Point field, Campbell County, Wyoming. In: Diedrich, R.P., Dyka, M.A.K., Miller, W.R. (Eds.), Eastern Powder River Basin–Black Hills: Wyoming Geological Association, 39th Annual Field Conference, Guidebook, pp. 105–116.

Kaiser, E., 1908. Erlaetergunen zur geol. Karte von Preussen. Blatt Bruehe, p. 44.

Kalathingal, P., Kuchinski, R., 2010. The role of resistivity image logs in deep natural gas reservoirs. In: Paper SPE 131721 Presented at the SPE Deep Gas Conference and Exhibition, Manama, Bahrain, January 24–26, 2010. http://dx.doi.org/10.2118/131721-MS.

Kapdi, S.S., Vijay, V.K., Rajesh, S.K., Prasad, R., 2007. Biogas scrubbing, compression and storage: perspective and prospectus in Indian context. Renew. Energ. ISSN: 0960-1481 30 (8), 1195–1202. July 2005.

Kardolus, C.B., van Kruijsdijk, C.P.J.W., 1997. Formation testing while underbalanced drilling. In: Paper SPE 38754 Presented at the 1997 SPE Annual Technical Conference and Exhibition, San Antonio, Texas, USA, 5–8 October. http://dx.doi.org/10.2118/38754-MS.

Kasino, R.E., Davies, D.K., 1979. Environments and diagenesis, Morrow sands, Cimarron County (Oklahoma), and significance to regional exploration, production and well

completion practices. In: Hyne, N.J. (Ed.), Pennsylvanian Sandstones of the Mid-Continent. Tulsa Geological Society Special Publication 1, pp. 169–194.
Kasper, D.L., Hendry, H.E., Renaut, R.W., 1992. Stratigraphy and diagenesis of the Bakken Formation, west central Saskatchewan [abs.]. Am. Assoc. Pet. Geol. Bull. 76, 62.
Kaufman, G.M., Balcer, Y., Kruyt, D., 1975. A probabilistic model of oil and gas discovery. In: Haun, J.D. (Ed.), Methods of Estimating the Volume of Undiscovered Oil and Gas Resources. American Association of Petroleum Geologists Studies in Geology No. 1, pp. 113–142.
Kaufman, G.M., 1963. Statistical Decision and Related Techniques in Oil and Gas Exploration. Prentice-Hall, Englewood Cliffs, NJ, 307 pp.
Kaufman, G.M., 1965. Statistical Analysis of the Size Distribution of Oil and Gas Fields, in Third Symposium on Petroleum Economics and Evaluation. Society of Petroleum Engineers, 109–124.
Kaufman, G.M., 1993. Statistical issues in the assessment of undiscovered oil and gas resources. Energy J. 14 (1), 183–215.
Kazemi, H., 1968. Pressure transient analysis of naturally fractured reservoirs with uniform fracture distribution. In: Paper SPE 2156A Presented at the 1968 43rd Annual Fall Meeting, Houston, Texas, 29 September–2 October.
Kendall, P.F., Briggs, H., 1933. The formation of rock joints and the cleat of coal. Proc. R. Soc. Edinb. 53, 164–187.
Kerans, C., Holtz, M.H., Tyler, N., 1989. Contrasting styles of reservoir heterogeneity in Ellenburger Group carbonates, west Texas [abs.]. In: Cunningham, B.K., Cromwell, D.W. (Eds.), The Lower Paleozoic of West Texas and Southern New Mexico—Modern Exploration Concepts. Permian Basin Section Society of Economic Paleontologists and Mineralogists Publication 89–31, p. 131.
Kerans, C., 1988. Karst-controlled reservoir heterogeneity in Ellenburger Group carbonates of west Texas. Am. Assoc. Pet. Geol. Bull. 72, 1160–1183.
Ketata, C., Satish, M., Islam, M.R., 2005a. Stochastic evaluation of rock properties by sonic-while-drilling data processing in the oil and Gas Industry. In: In the Proceedings of 2005 International Conference on Computational & Experimental Engineering & Science, Hyderabad, India. December 1–6, 2005.
Ketata, C., Satish, M., Islam, M.R., 2005b. Knowledge-based optimization of rotary drilling system. In: In the Proceedings of 2005 International Conference on Computational & Experimental Engineering & Science, December 1–6, 2005, Hyderabad, India.
Ketata, C., Satish, M.G., Islam, M.R., 2006a. The meaningful zero. In: Proc. 36th Conference on Computer and Industries, Taiwan, June.
Ketata, C., Satish, M.G., Islam, M.R., 2006b. Cognitive work analysis of expert systems design and evaluation in the oil and gas industry. In: Proc. 36th Conference on Computer and Industries, Taiwan, June.
Ketata, C., Satish, M.G., Islam, M.R., 2006c. Multiple Solution Nature of Chaos Number-Oriented Equations. CIMCA-2006, November, Sydney, Australia.
Ketata, C., Satish, M.G., Islam, M.R., 2006d. The Meaningful Infinity. CIMCA-2006, Sydney, Australia, November.
Khan, M.I., Islam, M.R., 2006a. True Sustainability in Technological Development and Natural Resources Management. Nova Science Publishers, NY, USA.
Khan, M.I., Islam, M.R., 2007a. The Petroleum Engineering Handbook: Sustainable Operations. Gulf Publishing Co., Houston, USA, 461 pp.
Khan, M.I., Islam, M.R., 2007b. True Sustainability in Technological Development and Natural Resources Management. Nova Science Publishers, New York, USA, 381 pp.

Khan, M.M., Zatzman, G.M., Islam, M.R., 2008. The formulation of a comprehensive mass and Energy balance equation. In: ASME International Mechanical Engineering Congress and Exposition, Boston, MA, 2–6 November.

Kim, K.S., van Stone, R.H., 1995. Hold time crack growth analysis at elevated temperatures. Eng. Fract. Mech. 52 (3), 433–444.

Kim, K.Y., Wrolstad, K.H., Aminzadeh, F., 1993. Effects of transverse isotropy on P-wave AVO for gas sands. Geophysics 58, 883–888.

King, G.E., 2010. Thirty years of gas shale fracturing: what have we learned?. In: Paper SPE 133456 Presented at the 2010 SPE Annual Technical Conference and Exhibition, Florence, Italy, 19–22 September. http://dx.doi.org/10.2118/133456-MS.

Klemme, H.D., Ulmishek, G.F., 1991. Effective petroleum source rocks of the world—Stratigraphic distribution and controlling depositional factors: American Association of Petroleum Geologists Bulletin. v. 75, 1809–1851.

Klett, T.R., 2003. Graphic comparison of reserve-growth models for conventional oil and gas accumulation. In: Dyman, T.S., Schmoker, J.W., Verma, M.K. (Eds.), Geologic, Engineering, and Assessment Studies of Reserve Growth. U.S. Geological Survey Bulletin 2172–F (accessed 09.08.07.). http://pubs.usgs.gov/bul/b2172-f/.

Kline, M., 1972. Mathematical Thought from Ancient to Modern Times. Oxford University Press, New York, USA, 1238 pp.

Kneissl, W., 2001. Reservoir characterization whilst underbalanced drilling. In: Paper SPE/IADC 67690 Presented at the SPE/IADC Drilling Conference, Amsterdam, The Netherlands, 27 February–1 March. http://dx.doi.org/10.2118/67690-MS.

Kocurek, G., 1981. Significance of interdune deposits and bounding surfaces in eolian dune sands. Sedimentology 28, 753–780.

Kocurek, G., 1988. First-order and super bounding surfaces in eolian sequences—bounding surfaces revisited. Sedimentary Geol. 56, 193–206.

Koefoed, 0, 1955. On the effect of Poisson's ratios of rock strata on the reflection coefficients of plane waves. Geophys. Prospect. 3, 381–387.

Kolh, A., Nielsen, R., 1997. Gas Purification. Gulf Publishing Company, Houston, Texas, USA, ISBN 978-0-88415-220-0.

Kopaska-Merkel, D.C., Mann, S.D., 1993. Upward shoaling cycles in Smackover carbonates of southwest Alabama. Gulf Coast Assoc. Geol. Soc. Trans. 43, 173–181.

Kosters, E.C., Bebout, D.G., Seni, S.J., Garrett, C.M., Brown Jr., L.F., Hamlin, H.S., Dutton, S.P., Ruppel, S.C., Finley, R.J., Tyler, N., 1989a. Atlas of Major Texas Gas Reservoirs. University of Texas, Bureau of Economic Geology, Austin, TX, 32–51.

Kosters, E.C., Bebout, D.G., Seni, S.J., Garrett, C.M., Brown Jr., L.F., Hamlin, H.S., Dutton, S.P., Ruppel, S.C., Finley, R.J., Tyler, N., 1989b. Atlas of Major Texas Gas Reservoirs. University of Texas, Bureau of Economic Geology, Austin, TX, 136–139.

Kosters, E.C., Bebout, D.G., Seni, S.J., Garrett, C.M., Brown Jr., L.F., Hamlin, H.S., Dutton, S.P., Ruppel, S.C., Finley, R.J., Tyler, N., 1989c. Atlas of Major Texas Gas Reservoirs. University of Texas, Bureau of Economic Geology, Austin, TX, 94–97.

Kosters, E.C., Bebout, D.G., Seni, S.J., Garrett, C.M., Brown Jr., L.F., Hamlin, H.S., Dutton, S.P., Ruppel, S.C., Finley, R.J., Tyler, N., 1989d. Atlas of Major Texas Gas Reservoirs. University of Texas, Bureau of Economic Geology, Austin, TX, 59–62.

Kralovic, P., 2011. North American Natural Gas Market Dynamics: Shale Gas Plays in North America – A Review. Canadian Energy Research Institute Study No. 123.

Krauss, L.M., Dawkins, R., 2013. A Universe from Nothing: Why there is Something Rather than Nothing. Atria Books, 240 pp.

Krystinik, L.F., Blakeney, B.A., 1990. Sedimentology of the upper Morrow Formation in eastern Colorado and western Kansas; in, Morrow Sandstones of Southeast Colorado and Adjacent Areas. In: Sonnenberg, S.A., Shannon, L.T., Rader, K., von

Drehle, W.F., Martin, G.W. (Eds.), The Rocky Mountain Association of Geologists, Special Paper, pp. 37–50.
Krystinik, L.F., 1990a. Development geology in eolian reservoirs. In: Fryberger, S.G., Krystinik, L.F., Schenk, C.J. (Eds.), Modern and Ancient Eolian Deposits—Petroleum Exploration and Production: Rocky Mountain Section, S.E.P.M. Society for Sedimentary Geology, pp. 13–21 to 13–2.
Krystinik, L.F., 1990b. Diagenesis in ancient eolian sandstone. In: Fryberger, S.G., Krystinik, L.F., Schenk, C.J. (Eds.), Modern and Ancient Eolian Deposits—Petroleum Exploration and Production: Rocky Mountain Section, S.E.P.M. Society for Sedimentary Geology, pp. 14–21 to 14–14.
Kubik, W., Lowry, P., 1993. Fracture identification and characterization using cores, FMS, CAST, and borehole camera: Devonian shale, Pike County, Kentucky. In: Paper SPE 25897 Presented at the 1993 SPE Rocky Mountain Regional/Low Permeability Reservoirs Symposium, Denver, CO, USA, 12–14 April. http://dx.doi.org/10.2118/25897-MS.
Kucuk, F., Sawyer, W.K., 1979. Modeling of Devonian shale gas reservoir performance. In: Proceedings of Third Eastern Gas Shales Symposium, Morgantown, West Virginia, (1979) 1–3 October.
Kucuk, F., Sawyer, W.K., 1980. Transient flow in naturally fractured reservoirs and its application to Devonian gas shales. In: Paper SPE 9397 Presented at The1980 Annual Fall Technical Conference and Exhibition of the Society of Petroleum Engineers of AIME, Dallas, Texas, 21–24 September.
Kugler, R.L., 1993. Depositional texture-dependent and independent diagenetic control of petrophysical properties, Norphlet Sandstone, onshore and offshore Alabama [abs.]. Am. Assoc. Pet. Geol. Bull. 77, p. 1591.
Kulander, B.R., Dean, S.L., 1990. Fractured core analysis. AAPG Stud. Explor. 8, 88.
Kulander, B.R., Dean, S.L., 1993. Coal-cleat domains and domain boundaries in the Allegheny Plateau of West Virginia. AAPG Bull. 77, 1374–1388.
Kulander, B.R., Dean, S.L., Williams, R.E., 1980. Fracture trends in the Allegheny Plateau of West Virginia. map WV-11, scale 1: 250,000 West Virginia Geol. Econ. Surv.
Kuuskraa, V.A., Wicks, D.E., Thurber, J.L. Geologic and reservoir mechanisms controlling gas recovery from the Antrim shale. Paper SPE 24883 Presented at the 1992 SPE ATCE, Washington, D.C., 4–7 October.
Kuuskraa, V.A., Koperna, G., Schmoker, J.W., Quinn, J.C., 1998a. Barnett shale rising star in fort Worth Basin. Oil Gas. J. 96 (21), 67–76.
Kuuskraa, v.A., Schmoker, J.W., Dyman, T.S., June 8, 1998b. Diverse gas plays lurk in gas resource pyramid. Oil Gas. J. 123–130.
Kvenvolden, K.A., Kastner, M., 1990. Gas hydrate of the Peruvian outer continental margin. In: Suess, von Heune (Eds.), Ocean Drilling Program Scientific Results, vol. 112, pp. 517–526.
Kvenvolden, K.A., May 1993. Gas hydrates – geological perspective and global change. Rev. Geophys. 173–187.
Laffez, P., Abbaoui, K., 1996. Modelling of the thermic exchanges during a drilling: resolution with Adomian's decomposition method. Math. Comput. Model. 23 (10), 11–14.
Lai, C.H., Bodvarsson, G.S., Tsang, C.F., Witherspoon, P.A. A new model for well test data analysis for naturally fractured reservoirs. Paper SPE 11688 Presented at the 1983 California Regional Meeting, Ventura, California, 23–25 March.
Lake, L.W., 1989. Enhanced Oil Recovery. Prentice Hall, 550 pp.
Lancaster, D.E, McKetta, S.F, Hill, R.E., Guidry, F.K., Jochen, J.E. Reservoir evaluation, completion techniques, and recent results from barnett shale development in the fort Worth Basin. Paper SPE 24884 Presented at the 1992 SPE ATCE, Washington, D.C., 4–7 October.

Lancaster, D.E., McKetta, S., Lowry, P.H., 1993. Research findings help characterize fort Worth Basin's barnett shale. Oil Gas. J. 91 (10), 59–64.

Lane, L.S., Snowdon, L.R., Obermajer, M., 2010. Rock-Eval/TOC and Oil Show Analyzer Data for Selected Yukon Borehole Samples. Geological Survey of Canada Open File 6652, 17 pp.

Larsen, L., Nilsen, F., 1999. Inflow predictions and testing while underbalanced drilling. In: Paper SPE 56684 Presented at the 1999 SPE Annual Technical Conference and Exhibition, Houston, Texas, USA, 3–6 October. http://dx.doi.org/10.2118/56684-MS.

Larson, Eric D., Clair, Joshua R. St., Sumner, Whitney A., Bannister, Roger A., Proenza, Cathy, October 29, 2013. Depressed pacemaker activity of sinoatrial node myocytes contributes to the age-dependent decline in maximum heart rate. PNAS 110 (44), 18011–18016.

Laubach, S.E., Tremain, C.M., 1991. Regional coal fracture patterns and coalbed methane development. In: Roegiers, J.-C. (Ed.), Proc. 32nd U.S. Symp. Rock Mech. Balkema, Rotterdam, pp. 851–859.

Laubach, S.E., Tremain, C.M., Ayers Jr., W.B., 1991. Coal fracture studies: guides for coalbed methane exploration and development. J. Coal Qual. 10, 81–88.

Laubach, S.E., Tyler, R., Ambrose, W.A., Tremain, C.M., Grout, M.A., 1992. Preliminary map of fracture patterns in coal in the western United States. Wyo. Geol. Assoc. 43, 253–267.

Laubach, S.E., Schultz-Ela, D.D., Tyler, R., 1993. Analysis of compaction effects on coal fracture patterns, Upper Cretaceous Rock Springs Formation, southwestern Wyoming. Mt. Geol. 30, 95–110.

Laubach, S.E., Tyler, R., Schultz-Ela, D.D., Tremain, C.M., 1994. Tectonic analysis of weakly defomed foreland areas [abs]. In: AAPG Ann. Conv. Prog., Denver, CO, p. 149.

Laubach, S.E., et al., 1998. Int. J. Coal Geol. 35, 175–207.

Lavender, A.R., Hill, N.R., 1985. P-SV resonances in irregular low velocity surface layers. Bull. Seism. Soc. Am. 75, 847–864.

Lavoie, D., Hamblin, A.P., Theriault, R., Beaulieu, J., Kirkwood, D., 2008. The Upper Ordovician Utica Shales and Lorraine Group Flysch in Southern Quebec: Tectonostratigraphic Setting and Significance for Unconventional Gas. Geological Survey of Canada Open File 5900, 54 pp.

Lavoie, D., Pinet, N., Dietrich, J., Hannigan, P., Castonguay, S., Hamblin, A.P., Giles, P., 2009. Petroleum resource assessment, Paleozoic successions of the St. Lawrence Platform and Appalachians of eastern Canada. Geol. Surv. Can. Open File 6174.

Lavrov, A., Tronvoll, J., 2004. Modeling Mud Loss in fractured formations. In: Paper 88700 Presented at the 2004 11th Abu Dhabi International Petroleum Exhibition and Conference, Abu Dhabi, U.A.E., 10–13 October. http://dx.doi.org/10.2118/88700-MS.

Law, B.E., Curtis, J.B., 2002. Introduction to unconventional petroleum systems. Am. Assoc. Pet. Geol. Bull. 86, 1851–1852.

Law, B.E., 1976. Large-scale compaction structures in the coal-bearing Fort Union and Wasatch Formations, northeast Powder River basin, Wyoming. In: Landon, R.B., urry III, W.H., Runge, J.S. (Eds.), Geology and Energy Resources of the Powder River, vol. 28. Wyo. Geol. Assoc., pp. 221–229.

Law, B.E., 1993. The relation between coal rank and cleat spacing: implications for the prediction of permeability in coal. In: Proc. Int. Coalbed Methane Symp., II, pp. 435–442.

Law, B.E., 2002. Basin-centered gas systems. Am. Assoc. Pet. Geol. Bull. 86 (11), 1891–1919.

Lawn, B.R., Wilshaw, T.T., 1975. Fracture of Brittle Solids. Cambridge Univ. Press, Cambridge, 204 pp.

Leckie, D.A., Potocki, D.J., visser, K., 1991. The lower Cretaceous Chinkeh formation: a frontier-type play in the Liard Basin of western Canada. Am. Assoc. Pet. Geol. Bull. 74, 1324–1352.

Leckie, D.A., Potocki, D.J., 1998. Sedimentology and petrography of marine shelf sandstones of the Cretaceous Scatter and Garbutt formations, Liard Basin, northern Canada. Bull. Can. Pet. Geol. 46, 30–50.

Lee, P.J., Wang, P.C.C., 1986. Evaluation of petroleum resources from pool size distribution. In: Rice, D.D. (Ed.), Oil and Gas Assessment Methods and Applications. American Association of Petroleum Geologists Studies in Geology No. 21, pp. 33–42.

Lee, P.J., Wang, P.C.C., 1990. An Introduction to Petroleum Resource Evaluation Methods: Canadian Society of Petroleum Geologists Course Notes. Geological Survey of Canada Contribution number 51789, 108 pp.

Lee, W.J., Wattenbarger, R.A., 1996. Gas Reservoir Engineering. SPE Textbook Series.

LeFever, J.A., 1991. History of oil production from the Bakken Formation, North Dakota. In: Hansen, W.B. (Ed.), Geology and Horizontal Drilling of the Bakken Formation: Montana Geological Society, 1991 Guidebook, pp. 3–18.

Letourneau, J., 1991. Petroleum hydrogeology of the Jean Marie Member, northeastern British Columbia. In: Proceedings, Canadian Society of Petroleum Geologists Annual Convention, p. 92.

Levander, A., England, R.W., Smith, S.K., Hobbs, R.W., Goff, J.A., Holliger, K., 1994. Stochastic characterization and seismic response of upper and middle crustal rocks based on the Lewisian gneiss complex, Scotland. Geophys. J. Int. 119, 243–259.

Levin, S.A., 1998. Resolution in seismic imaging: is it all a matter of perspective? Geophysics 63, 743–749.

Levine, J.R., 1993. Coalification: the evolution of coal as source rock and reservoir rock for oil and gas. In: Law, B.E., Rice, D.D. (Eds.), Hydrocarbons from Coal, AAPG Studies in Geology, vol. 38, pp. 39–78.

Levson, v., Pyle, L., Fournier, M., 2012. Identification of Potential Silica Sand Deposits in the Northwest Territories. NWT Open File 2012-6.

Li, Y., Vodacek, A., Kremens, R.L., Ononye, A.E., Tang, C., 2005. A hybrid contextual approach to wildland fire detection using multispectral imagery. IEEE Trans. Geosci. Remote Sens. 43, 2115–2126.

Liétard, O., Unwin, T., Guillot, D.J., Hodder, M.H., 1999. Fracture width logging while drilling and drilling mud/loss-circulation-material selection guidelines in naturally fractured reservoirs. SPE Drill. Complet. 14 (3), 168–177. SPE 57713-PA. http://dx.doi.org/10.2118/57713-PA.

Link, C.M., Bustin, R.M., 1989. Organic maturity and thermal history of Phanerozoic strata in northern Yukon and northwestern District of Mackenzie. Bull. Can. Pet. Geol. 37, 266–292.

Link, C.M., Bustin, R.M., Snowdon, L.R., 1989. Petroleum source potential and depositional setting of Phanerozoic strata in northern Yukon and northwestern District of Mackenzie. Bull. Can. Pet. Geol. 37, 293–315.

Littel, R.J., Filmer, B., Versteeg, G.F., Van Swaaij, W.P.M., 1991. Modelling of simultaneous absorption of H_2S and CO_2 in alkanolamine solutions: the influence of parallel and consecutive reversible reactions and the coupled diffusion of ionic species. Chem. Eng. Sci. 46 (9), 2303–2313.

Liu, E., MacBeth, C., Point, C., Hudson, J.A., 1996. The effective elastic compliance of fractured rock. 66th Ann. Intern. Mtg., Soc. Expl. Geophys. Expanded Abstracts, 1842–1845.

Livio, M., 2005. The Equation that couldn't Be Solved: How Mathematical Genius Discovered the Language of Symmetry. Simon & Schuster, New York, 353 pp.

Lloyd, J.M., Ragland, P.C., Ragland, J.M., Parker, W.C., 1986. Diagenesis of the Jurassic Smackover formation, Jay field, Florida. Gulf Coast Assoc. Geol. Soc. Trans. 36, 201–211.
Lock, B.E., Broussard, S.W., 1989. The Norphlet reservoir in Mobile Bay—origin of deep porosity. Gulf Coast Assoc. Geol. Soc. Trans. 39, 187–194.
Long, P.E., Wood, B.J., 1986. Structures, textures and cooling histories of Columbia River basalt flows. Geol. Soc. Am. Bull. 97, 1144–1155.
Long, D.G.F., 2005. Sedimentology and hydrocarbon potential of fluvial strata in the Tantalus and Aksala formations, northern Whitehorse Trough, Yukon. In: Emond, D.S., Lewis, L.L., Bradshaw, G.D. (Eds.), Yukon Exploration and Geology 2004. Yukon Geological Survey, pp. 167–176.
Lore, G.L., Batchelder, E.C., 1995. Using production-based plays in the northern Gulf of Mexico as a hydrocarbon exploration tool. Gulf Coast Assoc. Geol. Soc. Trans. 45, 371–376.
Lore, G.L., Brooke, J.P., Cooke, D.W., Klazynski, R.J., Olson, D.L., Ross, K.M., 1996. Summary of the 1995 Assessment of Conventionally Recoverable Hydrocarbon Resources of the Gulf of Mexico and Atlantic Outer Continental Shelf. Minerals Management Service OCS Report MMS 96-0047, 41 pp. plus appendices.
Lore, G.L., Ross, K.M., Bascle, B.J., Nixon, L.D., Klazynski, R.J., 1999. Assessment of Conventionally Recoverable Hydrocarbon Resources of the Gulf of Mexico Atlantic Outer Continental Shelf. OCS Report MMS 99-0034. http://www.boem.gov/uploadedFiles/BOEM/About_BOEM/BOEM_Regions/Alaska_Region/Resource_Evaluation/MMS-99-0034.pdf.
Lore, G.L., Marin, D.A., Batchelder, E.C., Courtwright, W.C., Desselles Jr., R.P., Klazynski, R.J., 2001. 2000 assessment of Conventionally Recoverable Hydrocarbon Resources of the Gulf of Mexico and Atlantic Outer Continental Shelf as of January 1, 1999. Minerals Management Service. Outer Continental Shelf Report MMS 2001-087 also available at: www.gomr.mms.gov.
Lore, G.L., 1992. Exploration and Discoveries, 1947–1989: An Historical Perspective. Minerals Management Service OCS Report MMS 91-0078, 87 pp.
Lore, G.L., 1995. An exploration and discovery process model: an historic perspective, Gulf of Mexico Outer Continental Shelf. In: Simakov, K.V., Thurston, D.K. (Eds.), 1994 Proceedings: International Conference on Arctic Margins, September 6–10, 1994, Magadan, Russia, pp. 306–313.
Lorenz, J.C., Sterling Jenny, L., Schechter, D.S., Whigham, C.L., Jensen, J.L., March 2002. 2002, Natural fractures in the Spraberry Formation, Midland basin, Texas: The effects of mechanical stratigraphy on fracture variability and reservoir behavior. AAPG Bulletin V. 86 (No. 3), 505–524.
Loucks, R.G., Anderson, J.H., 1982. Depositional facies, diagenetic terranes, and porosity development in the Lower Ordovician Ellenburger Dolomite, Puckett field, west Texas. In: Roehl, P.O., Choquette, P.W. (Eds.), Carbonate Petroleum Reservoirs. Springer-Verlag, New York, pp. 21–38.
Loucks, R.G., Handford, C.R., 1992. Origin and recognition of fractures, breccias, and sediment fills in paleocave-reservoir networks. In: Candelaria, M.P., Reed, C.L. (Eds.), Paleokarst, Karst Related Diagenesis and Reservoir Development—Examples from Ordovician-Devonian Age Strata of West Texas and the Mid-Continent. Permian Basin Section, Society of Economic Paleontologists and Mineralogists Publication 92–33, pp. 31–44.
Loucks, R.G., Dodge, M.M., Galloway, W.E., 1984. Regional controls on diagenesis and reservoir quality in lower Tertiary sandstones along the lower Texas Gulf Coast. In: McDonald, D.A., Surdam, R.C. (Eds.), Clastic diagenesis: American Association of Petroleum Geologists Memoir 37, pp. 15–46.
Lowey, G.W., Long, D., 2006. Summary of Rock-Eval data for the Whitehorse Trough, Yukon: implications concerning the hydrocarbon potential of a frontier basin. In:

Emond, D.S., Bradshaw, G.D., Lewis, L.L., Weston, L.H. (Eds.), Yukon Exploration and Geology 2005. Yukon Geological Survey, pp. 207–218.
Lowey, G.W., Long, D.G.F., Fowler, M.G., Sweet, A.R., Orchard, M.J., 2009. Petroleum source rock potential of Whitehorse Trough: a frontier basin in south-central Yukon. Bull. Can. Pet. Geol. 57 (3), 350–386.
Lowey, G.W., 2009. A summary of Rock-Eval data for the Bonnet Plume Basin, Yukon: implications for a previously unrecognized oil play. In: Weston, L.H., Blackburn, L.R., Lewis, L.L. (Eds.), Yukon Exploration and Geology 2008. Yukon Geological Survey, pp. 173–186.
Lowey, G.W., 2010. The Bonnet Plume Basin, Yukon, Canada: A Previously Unrecognized Oil Play. Search and Discovery Article #10228.
Peters & Co. Ltd, 2008. Montney report – significant potential upside but... beware of variability. In: Analyst Report Industry clients.
Peters & Co. Ltd, 2010. Unconventional Cardium Play Update. Analyst report for industry clients.
Lucia, F.J., 1969. Lower Paleozoic history of the western Diablo platform of west Texas and south-central New Mexico. In: Seewald, K.O., Sundeen, D.M. (Eds.), The Geologic Framework of the Chihuahua Tectonic Belt. West Texas Geological Society, Guidebook, pp. 39–56.
Lucia, F.J., 1995. Lower Paleozoic cavern development, col- lapse, and dolomitization, Franklin Mountains, El Paso, Texas. In: Budd, D.A., Saller, A.H., Harris, P.M. (Eds.), Unconformities and porosity in carbonate strata, Association of Petroleum Geologists Memoir, 63, pp. 279–300.
Lucia, F.J., 2007. Carbonate Reservoir Characterization: An Integrated Approach. Springer, 332 pp.
Lynch, F.L., 1996. Mineral/water interaction, fluid flow, and Frio Sandstone diagenesis—evidence from the rocks. Am. Assoc. Pet. Geol. Bull. 80, 486–504.
Lynn, H.B., Bates, C.R., Simon, K.M., van Dok, R., 1995a. The effects of azimuthal anisotropy in P-wave 3-D seismic. 65th Ann. Intern. Mtg., Soc. Expl. Geophys. Expanded Abstracts, 723–730.
Lynn, H.B., Simon, K.M., Bates, C.R., 1996. Correlation between P-wave AVOA and S-wave traveltime anisotropy in a naturally fractured gas reservoir. Lead. Edge 15, 931–935.
Mabres, A., Freire, D., Bevilacqua, L., 2003. Eliminación de Sulfuro de Hidrógeno en Emisiones Gaseosas Atmosféricas. In: 5tas. Jornadas de Preservación de Agua, Aire y Suelo en la Ind. del Petróleo y del Gas, Buenos Aires, Argentina, November 5–7, 2003.
Mace, J.E., Amerhein, C., 2001. Leaching and reclamation of a soil irrigated with moderate SAR waters. Soil. Sci. Soc. Am. J. 65, 199–204.
Macedo, R., 2011a. Producers expected to focus on drilling as Duvernay land rush abates. Dly. Oil Bull. December 19, 2011.
Macedo, R., 2011b. No clear picture yet on Horn River Exshaw well: Quicksilver. Dly. Oil Bull. November 8, 2011.
MacGowan, D.B., Jiao, Z.S., Surdam, R.C., Miknis, F.P., 1993. Normally pressured vs. abnormally pressured compartments in sandstones in the Powder River Basin, Wyoming—a comparative study of the Muddy Sandstone and the Minnelusa Formation. In: Stroock, B., Andrew, S. (Eds.), Jubilee Anniversary Field Conference: Wyoming Geological Association, 50th Annual Field Trip, Guidebook, pp. 281–295.
MacKenzie, J.J., 1996. Oil as a Finite Resource: When is Global Production Likely to Peak? World Resources Institute, Washington, D.C., 22 pp.
Mackowiak, J.F., Górak, A., Kenig, E.Y., 2009. Modelling of combined direct-contact condensation and reactive absorption in packed columns. Chem. Eng. J. 146, 362–369.

Macquarie Equities Research, 2010. Going straight to the 'source' – a spotlight on the emerging Canadian shale oil plays. In: Analyst report for industry clients.. Available at http://resourceinvestingnews.com/files/2010/06/Macquarie-Equities-Research.pdf.

Macrae, J.C., Lawson, W., 1954. The incidence of cleat fracture in some Yorkshire coal seams. Trans. Leeds. Geol. Assoc. 6, 224–227.

Mahony, J., November 23, 2011. Alberta Bakken drawing more junior interest. Dly. Oil Bull.

Majidi, R., Miska, S.Z., Yu, M., Thompson, L.G., 2008a. Quantitative analysis of mud losses in naturally fractured reservoirs: the effect of rheology. In: Paper SPE 114130 Presented at the 2008 Western Regional and Pacific Section AAPG Joint Meeting, Bakersfield, California, USA, 31 March–2 April. http://dx.doi.org/10.2118/114130-MS.

Majidi, R., Miska, S.Z., Yu, M., Thompson, L.G., 2008b. Modeling of drilling fluid losses in naturally fractured formations. In: Paper SPE 114630 Presented at the 2008 SPE Annual Technical Conference and Exhibition, Denver, Colorado, USA, 21–24 September.. http://dx.doi.org/10.2118/114630-MS.

Mallick, S., Frazer, L.N., 1991. Reflection/transmission coefficients and azimuthal anisotropy in marine seismic studies. Geophys. J. Int. 105, 241–252.

Mallick, S., Craft, K.L., Meister, L.J., Chambers, R.E., 1998. Determination of the principal directions of azimuthal anisotropy from P-wave seismic data. Geophysics 63, 692–706.

Mamora, D., Seo, G., 2002. Enhanced gas recovery by carbon dioxide sequestration in depleted gas reservoir. In: SPE Paper 77347, Presented at the SPE Annual Technical Conference and Exhibition held in San Antonio, Texas.

Mancini, E.A., Benson, D.J., 1980. Regional stratigraphy of Upper Jurassic Smackover carbonates of southwest Alabama. Gulf Coast Assoc. Geol. Soc. Trans. 30, 151–165.

Mancini, E.A., Mink, R.M., Bearden, B.L., Wilkerson, R.P., 1985. Norphlet Formation (Upper Jurassic) of south-western and offshore Alabama—environments of deposition and petroleum geology. Am. Assoc. Pet. Geol. Bull. 69, 881–898.

Mancini, E.A., Mink, R.M., Bearden, B.L., 1986. Integrated geological, geophysical, and geochemical interpretation of Upper Jurassic petroleum trends in the eastern Gulf of Mexico. Gulf Coast Assoc. Geol. Soc. Trans. 36, 219–226.

Mancini, E.A., Mink, R.M., Tew, B.H., Bearden, B.L., 1990. Natural gas plays in Jurassic reservoirs of southwestern Alabama and the Florida panhandle area. Gulf Coast Assoc. Geol. Soc. Trans. 40.

Manrique, E. Enhanced oil recovery (EOR): trends, risks, and rewards. Presented at the ACI Optimising EOR Strategy 2009, London, UK, March 11–12, 2009.

Marchaim, U., 1992. Biogas Processes for Sustainable Development. FAO, Rome, Italy, ISBN 92-5-103126-6.

Markert, J.C., Al-Shaieb, Z.A., 1984. Diagenesis and evolution of secondary porosity in upper Minnelusa sandstones, Powder River Basin, Wyoming. In: McDonald, D.A., Surdam, R.C. (Eds.), Clastic Diagenesis: American Association of Petroleum Geologist Memoir 37, pp. 367–390.

Marrett, R., Allmendinger, R., 1991. Estimates of strain due to brittle faulting: sampling of fault populations. J. Struct. Geol. 13, 735–738.

Marrett, R., 1996. Aggregate properties of fracture populations. J. Struct. Geol. 18, 169–178.

Marrett, R., 1997. Permeability, porosity, and shear-wave anisotropy from scaling of open fracture populations. In: Fractured Reservoirs: Descriptions, Predictions, and Applications. Rk. Mt. Assoc. Geol.

Martin, G.W. (Ed.), 1990. Morrow Sandstones of South-East Colorado and Adjacent Areas: Rocky Mountain Association of Geologists—Morrow Sandstone, pp. 1–8.

Martini, A.M., Budal, J.M., Walter, L.M., Schoell, M., September 1996. Microbial generation of economic accumulations of methane within a shallow organic-rich shale. Nature 12.
Martiniuk, C.D., 1991. Regional geology and petroleum potential of the Bakken Formation, southwestern Manitoba. In: Hansen, W.B. (Ed.), Geology and Horizontal Drilling of the Bakken Formation. Montana Geological Society Guidebook, pp. 43–67.
Martinsen, R.S., 1997. Stratigraphic controls on the development and distribution of fluid-pressure compartments. In: Surdam, R.C. (Ed.), Seals, Traps, and the Petroleum System, American Association of Petroleum Geologists Memoir 67, pp. 223–241.
Marzano, M.S., Pense, G.M., Andronaco, P., 1988. A comparison of the Jurassic Norphlet Formation in Mary Ann field, Mobile Bay, Alabama, to onshore regional Norphlet trends. Gulf Coast Assoc. Geol. Soc. Trans. 38, 85–100.
Mast, R.F., Dingler, J., 1975. Estimates of Inferred and Indicated Reserves for the United States by State, in Geological Estimates of Undiscovered Oil and Gas in the United States. U.S. Geological Survey Circular 725, 78 pp.
Mast, R.F., Dolton, G.L., Crovelli, R.A., Root, D.H., Attanasi, E.D., Martin, P.E., Cooke, L.W., Carpenter, G.B., Pecora, W.C., Rose, M.B., 1989. Estimates of Undiscovered Conventional Oil and Gas Resources in the United States—A Part of the Nation's Energy Endowment. U.S. Department of the Interior, U.S. Geological Survey, and Minerals Management Service, 44 pp.
Masters, J.A., 1979. Deep Basin gas trap, western Canada. Am. Assoc. Pet. Geol. Bull. 86 (11), 1851–1852.
Masters, J.A. (Ed.), 1984. Elmworth – Case Study of a Deep Basin gas field. American Association of Petroleum Geologists Memoir 38, 316 pp.
Mattar, L., 2008. Production analysis and forecasting of shale gas reservoirs: case history-based approach. In: Paper SPE 119897 Presented at the SPE Shale Gas Production Conference, Fort Worth, Texas, USA. http://dx.doi.org/10.2118/119897-MS.
Matthews, M.D., Perlmutter, M.A., 1994. Global cyclostratigraphy—an application to the Eocene Green River Basin. In: DeBoer, P.L., Smith, D.G. (Eds.), Orbital Forcing and Cyclic Sequences. International Association of Sedimentologists Special Publication 19, pp. 459–481.
Mavor, M.J., Logan, T.L., 1994. Recent advances in coal gas-well open-hole well completion technology. JPT 46, 587–593.
Max, M.D., 2003. Natural Gas Hydrate. Kluwer Academic Publishers, The Netherlands.
Mazzotti, A., 1991. Amplitude, phase and frequency versus offset applications. Geophys. Prospect. 39, 863–886.
McAdam, K., 1993. The Jean Marie Member in the Hossitl-ring Area, British Columbia. BC Ministry of Energy, Mines and Petroleum Resources, Petroleum Geology Branch. Special Paper 1993-1, 6 p.
McBeth, I.H., Reddy, K.J., Skinner, Q.D., 2003a. Chemistry of coalbed methane product water in three Wyoming watersheds. J. Am. Water Res. Assoc. 39, 575–585.
McBeth, I.H., Reddy, K.J., Skinner, Q.D., 2003b. Chemistry of trace elements in coalbed methane product water. J. Water Res. 37, 884–890.
McBride, E.F., Land, L.S., Mack, L.E., 1987. Diagenesis of eolian and fluvial feldspathic sandstones, Norphlet formation (Upper Jurassic), Rankin County, Mississippi and Mobile County, Alabama. Am. Assoc. Pet. Geol. Bull. 71, 1019–1034.
McCabe, P.J., 1988. Depositional Environments of Coal and Coal-bearing Strata. Canadian Society of Petroleum Geologists Short Course, 66 pp.
McCabe, P.J., 1998. Energy resources—cornucopia or empty barrel? Am. Assoc. Pet. Geol. Bull. 82, 2110–2134.
McCall, K.R., Guyer, R.A., Zhu, L., Boitnott, G.N., Hilbert Jr, L.B., Plona, T.J., June 1996. Experimental determination of the linear and nonlinear dynamic moduli of rock from

quasistatic measurements. In: Aubertin, M., Hassani, F., Mitri, H. (Eds.), Proceedings, 2nd North American Rock Mechanics Symposium: NARMS'96. Quebec, Canada, 1921. Balkema, Rotterdam, Netherlands, pp. 147–154.

McCrossan, R.G., 1969. An analysis of size frequency distribution of oil and gas reserves of western Canada. Can. J. Earth Sci. 6 (2), 201–211.

McCulloch, C.M., Deul, M., Jeran, P.W., 1974. Cleats in bituminous coalbeds. U.S. Bur. Min. Rep. Invest. 7910, 23.

McCulloch, C.M., Lambert, S.W., White, J.R., 1976. Determining cleat orientations of deeper coalbeds from overlying coals. U.S. Bur. Min. Rep. Invest. 8116, 25.

McDonald, D.A., Surdam, R.C. (Eds.), Clastic Diagenesis. American Association of Petroleum Geologists Memoir 37, pp. 15–45.

McDouglas, J.G., Howard, M.G., 1989. Advances in borehole imaging with second generation CBIL (circumferential borehole imaging log) borehole televiewer instrumentation. In: 28th Annual Conference Proceedings. Ontario Petroleum Institute Inc., Lamberth, Ontario, Canada paper 12.

McFarlan Jr., E., Menes, L.S., 1991. Lower cretaceous. In: Salvador, A. (Ed.), The Geology of North America, the Gulf of Mexico Basin, vol. J. Geological Society of America, pp. 181–204.

McGookey, D.P., 1975. Gulf Coast Cenozoic sediments and structures: an excellent example of extra-continental sedimentation. Gulf Coast Assoc. Geol. Soc. Trans. 25, 104–120.

McHenry, H.M., 2009. Human Evolution. In: Ruse, Michael, Travis, Joseph (Eds.), Evolution: The First Four Billion Years, pp. 256–280.

McKelvey, V.E., 1972. Mineral resource estimates and public policy. Am. Sci. 60, 32–40.

McLean, K., McNamara, D., 2010. Fractures interpreted from acoustic formation imaging technology: correlation to permeability. In: Proceedings, Thirty-sixth Workshop on Geothermal Reservoir Engineering, California, USA, 31 January–2 February 2011. SGP-TR-191.

McManus, D.A., 1959. Stratigraphy and depositional history of the Kearny Formation (Lower Pennsylvanian) in western Kansas (Unpublished Ph.D. dissertation). University of Kansas, Lawrence, 150 pp.

Medeiros, F., Ozkan, E., Kazemi, H., 2006. A semianalytical, pressure-transient model for horizontal and multilateral wells in composite, layered, and compartmentalized reservoirs. In: Paper SPE 102834 Presented at the SPE Annual Technical Conference and Exhibition, San Antonio, Texas, USA. http://dx.doi.org/10.2118/102834-MS.

Megill, R., May 1993. Business side of geology: new data unearths wider view. Am. Assoc. Pet. Geol. Explor. 22–23.

Meisner, J., Demirmen, F., 1981. The creaming method: a Bayesian procedure to forecast future oil and gas discoveries in mature exploration provinces. J. R. Stat. Soc. 144 (1), 1–13.

Meissner, F.F., 1978. Petroleum geology of the Bakken formation, Williston Basin, North Dakota and Montana. In: Estelle, D., Miller, R. (Eds.), Economic Geology of the Williston Basin: Montana Geological Society 1978 Symposium, pp. 207–227.

Meissner, F.F., 1991. Petroleum geology of the Bakken formation Williston Basin, North Dakota and Montana. In: Hansen, W.B. (Ed.), Geology and Horizontal Drilling of the Bakken Formation: Montana Geological Society, 1991 Guidebook, pp. 19–42.

Melancon, J.M., Bacigalupi, S.M., Kinler, C.J., Marin, D.A., Prendergast, M.T., 1995. Estimated proved oil and gas reserves, Gulf of Mexico Outer Continental Shelf, December 31, 1994. In: Minerals Management Service OCS Report MMS 95-0050, 56 pp.

Mercer, R.F., 1974. Liberated, produced, recycled or contamination. In: SPWLA 15th Annual Logging Symposium, 2–5 June, 1974.

Merle, H.A., Kentie, C.J.P., van Opstal, G.H.C., Schneider, G.M.G., September 1976. The Bachaquero study – a composite analysis of the behavior of a compaction drive/solution gas drive reservoir. SPE JPT 1107–1115.

Miller, B.M., Thomsen, H.L., Dolton, G.L., Coury, A.B., Hendricks, T.A., Lennartz, F.E., Powers, R.B., Sable, E.G., Varnes, K.L., 1975. Geological estimates of undiscovered oil and gas in the United States. U.S. Geol. Surv. Circ. 725, 78.

Miller, M.A., Jenkins, C.D., Rai, R.R., 2010. Applying innovative production modeling techniques to quantify fracture characteristics, reservoir properties, and well performance in shale gas reservoirs. In: Paper SPE 139097 presented at the SPE Eastern Regional Meeting, Morgantown, West Virginia, USA. http://dx.doi.org/10.2118/139097-MS.

Miller, B.M., 1986. Resource appraisal methods: choice and outcome. In: Rice, D.D. (Ed.), Oil and Gas Assessment Methods and Applications. American Association of Petroleum Geologists Studies in Geology No. 21, pp. 1–23.

Millheim, K.K., Cichowicz, L., February 1968. Testing and analyzing low-permeability fractured Gas wells. J. Pet. Technol. 193–198.

Milliken, K.L., Land, L.S., Loucks, R.G., 1981. History of of burial diagenesis determined from isotopic geochemistry. Frio Formation, Brazario County, Texas Am. Assoc. Pet. Geol. Bull 65, 1397–1413.

Milliken, K.L., Mack, L.E., Land, L.S., 1994. Elemental mobility in sandstones during burial—Whole-rock chemical and isotopic data, Frio Formation, south Texas. J. Sediment. Res. A64, 788–796.

Minerals Management Service, 1996. An assessment of the undiscovered hydrocarbon potential of the Nation's Outer Continental Shelf. In: Minerals Management Service OCS Report MMS 96-0034, 40 pp.

Minerals Management Service OCS Report MMS 97-0019, 207 pp. plus appendices. In: Dunkel, C.A., Piper, K.A. (Eds.), 1997. 1995 National Assessment of United States Oil and Gas Resources—Assessment of the Pacific Outer Continental Shelf Region.

Minitab, 2000. MINITAB 13.1. Minitab, Inc, State College, PA.

Mink, R.M., Mancini, E.A., 1995. Upper Jurassic and Lower Cretaceous oil reservoirs of the updip basement structure play: southwest Alabama. In: John, C.J., Byrnes, M.R. (Eds.), Gulf Coast Association of Geological Societies Transactions, vol. 45, pp. 441–448.

Molinda, G.M., Mark, C., 1996. Rating the strength of coal mine roof rocks. In: Aubertine, M., Hassani, F. (Eds.), Proceedings of the 2nd North Am. Rock Mech. Symp. Balkema, Rotterdam, pp. 413–422. N.

Molnar, P., Tapponnier, P., 1975. Cenozoic tectonics of Asia: effects of a continental collision. Science 189, 419–426.

Momper, J.A., Williams, J.A., 1979. Geochemical exploration in the Powder River Basin. Oil Gas. J. 129–134.

Monahan, P., 1999. Stratigraphy and Potential Hydrocarbon Objectives of Mississippian to Lower Cretaceous Strata in the Eastern Liard Basin Area. British Columbia Ministry of Energy, Mines, and Mineral Resources Report, 13 pp.

Montes, M., Legorburu, I., Garetto, T., 2008. Eliminación de Emisiones Atmosféricas de COVs por catálisis y adsorción. CYTED, Madrid, Spain, ISBN 978-84-96023-64-2.

Montgomery, S.L., Morgan, C.D., 1998. Bluebell field, Uinta Basin—Reservoir characterization for improved well completion and oil recovery. Am. Assoc. Pet. Geol. Bull. 82, 1113–1132.

Montgomery, S.L., Schechter, D.S., Lorenz, J.C., 2000. Advanced reservoir characterization to evaluate carbon dioxide flooding, Spraberry trend, Midland Basin, Texas. Am. Assoc. Pet. Geol. Bull. 84, 1247–1273.

Montgomery, S.L., Jarvie, D.M., Bowker, K.A., Pollastro, R.M., 2005. Mississippian Barnett Shale, Fort Worth Basin, north-central Texas: gas-shale play with multi-trillion cubic foot potential. Am. Assoc. Pet. Geol. Bull. 89, 155–175.

Moody, J.D., September 16, 1974. Reported in West J., (Ed.), U.S. Oil-Policy Riddle: How Much is Left to Find? Oil Gas J., p. 25–28.

Moore, C.H., Druckman, Y., 1981. Burial diagenesis and porosity evolution, upper Jurassic Smackover, Arkansas and Louisiana. Am. Assoc. Pet. Geol. Bull. 65, 597–624.

Moore, C.H., 1984. The upper Smackover of the Gulf Rim—depositional systems, diagenesis, porosity evolution and hydrocarbon production. Gulf Coast Assoc. Geol. Sci. Trans. 34, 283–307.

Moore, M.C., 2010. Policy and regulation in the face of uncertainty: the case of disruptive markets for unconventional natural gas in North America. In: Paper CSUG/SPE 138167 Presented at the Canadian Unconventional Resources & International Petroleum Conference, Calgary, Alberta, Canada, 19–21 October 2010. http://dx.doi.org/10.2118/138167-MS.

Morel, J.A., Bursk, P.H., Finch, J.W., 1986. A seismic-stratigraphic analysis of North Hollingsworth field, Fall River County, South Dakota. In: Noll, J.H., Doyle, K.M. (Eds.), Rocky Mountain Oil and Gas Fields: Wyoming Geological Association 1986 Symposium, pp. 257–271.

Morgan, C.D., 1997. Second field demonstration of completion techniques in a (DOE class 1) fluvial-dominated deltaic lacustrine reservoir, Uinta Basin, Utah [abs.]. Am. Assoc. Pet. Geol. 81, 85.

Moridis, G.J., Blasingame, T.A., Freeman, C.M., 2010. Analysis of mechanisms of flow in fractured tight-gas and shale-gas reservoirs. In: Paper SPE 139250 Presented at the SPE Latin American and Caribbean Petroleum Engineering Conference, Lima, Peru. http://dx.doi.org/10.2118/139250-MS.

Moritis, G., 2008. Worldwide EOR Survey. Oil Gas J. 2008 (106), 41–42, 44–59.

Morrell, G., Dietrich, J., 1993. Evaluation of the hydrocarbon prospectivity of the Old Crow Flats area of the northern Yukon. Bull. Can. Pet. Geol. 41 (1), 32–45.

Morton, R.A., Jirik, L.A., Galloway, W.E., 1988. Middle-upper Miocene depositional sequences of the Texas coastal plain and continental shelf: geologic framework, sedimentary facies, and hydrocarbon plays. In: The University of Texas at Austin Bureau of Economic Geology Report of Investigations No. 174, 40 pp.

Moslow, T.F., 2005. Geological nature and characterization of "tight" gas reservoirs in the Western Canada Sedimentary Basin and comparison to U.S. Rocky Mountain Basins. In: Proceedings, American Association of Petroleum Geologists Annual Convention, p. A98.

Mourzenko, V.V., Thovert, J.F., Adler, P.M., March 1995. Permeability of a single fracture; validity of the Reynolds equation. J. Phys. II Fr. 5, 465–482.

Mousavizadegan, S.H., Mustafiz, S., Rahman, M., 2007. The adomian decomposition method on solution of non-linear partial differential equations. J. Nat. Sci. Sustain. Technol. 1 (1), 115–131.

Muche, H., Zimmermann, H., 1985. The Purification of Biogas, ISBN 3-528-02015-6. Munich, Germany.

Mudford, B., Lundegard, P., Lerche, I., 1995. Timing of hydrocarbon generation and accumulation in fault-bounded compartments in the Norphlet Formation, offshore Alabama. Mar. Pet. Geol. 5, 549–558.

Mufti, A.A., Tadros, G., Jones, P.R., 1997. Field assessment of fibre-optic Bragg grating strain sensors in the confederation Bridge. Can. J. Civ. Eng. 24 (6), 963–966.

Murray, G.E., 1961. Geology of the Atlantic and Gulf Coastal Province of North America. Harper and Brothers, New York, 692 pp.

Murray, G.H., 1968. Quantitative fracture study—Sanish pool, McKenzie County, North Dakota. Am. Assoc. Pet. Geol. Bull. 52, 57–65.
Mustafiz, S., Islam, M.R., 2005. Adomian decomposition of two-phase, two-dimensional non-linear PDEs as applied in well testing. In: 4th Int. Conf. on Computational Heat and Mass Transfer, Proc. of 4th ICCHMT, Paris-Cachan, 17–20 May.
Mustafiz, S., Islam, M.R., 2008. State of the art of reservoir simulation. Pet. Sci. Tech. 26 (11–12), 1303–1329.
Mustafiz, S., Belhaj, H., Ma, F., Islam, M.R., 2005a. A new two phase extension of modified Brinkman formulation for fluid flow through porous media. In: ASME International Mechanical Engineering Congress and Exposition (IMECE), Orlando, Florida, USA, November.
Mustafiz, S., Belhaj, H., Ma, F., Satish, M., Islam, M.R., 2005b. Modeling horizontal well oil production using modified Brinkman's model. In: Proc. ASME International Mechanical Engineering Congress and Exposition (IMECE), Orlando, Florida, USA, November.
Mustafiz, S., Biazar, J., Islam, M.R., 2005d. An Adomian decomposition solution to the modified Brinkman model (MBM) for a 2-dimensional, 1-phase flow of petroleum fluids. In: Proc. CSCE 33rd Annual Conference, Toronto, Canada, 2–4 June.
Mustafiz, S., Moussavizadeghan, H., Islam, M.R., 2008. "Adomian Decomposition of Buckley Leverett Equation with Capillary Terms". Pet. Sci. Tech. 26 (15), 1796–1810.
Mustafiz, S., Mousavizadegan, H., Islam, M.R., 2008. The effects of linearization on solutions of reservoir engineering problems. Pet. Sci. Tech. 26 (11–12), 1224–1246.
Myal, F.R., Frohne, K.-H., 1992. Drilling and early testing of a sidetrack to the slant-hole completion test well: a case study of Gas recovery research in Colorado's Piceance Basin. In: Paper SPE 24382 Presented at the SPE Rocky Mountain Regional Meeting, Casper, Wyoming, USA, May 18–21, 1992. http://dx.doi.org/10.2118/24382-MS.
Naami, A.M., Catania, P., Islam, M.R., 1999. Numerical and experimental modeling of viscous fingering in two-dimensional consolidated porous medium. In: CIM Paper No. 118, CIM Conference, Regina, October.
Najurieta, H.L., July 1980. A Theory for Pressure Transient Analysis in Naturally Fractured Reservoirs. JPT, p. 1241.
National Energy Board, 2000. Petroleum Resource Assessment of the Eagle Plain, Yukon Territory, Canada, 74 pp.
National Energy Board, 2001a. Petroleum Resource Assessment of the Liard Plateau, Yukon Territory, Canada, 63 pp.
National Energy Board, 2001b. Petroleum Resource Assessment of the Whitehorse Trough, Yukon Territory, Canada, 59 pp.
National Energy Board, 2009. A Primer for Understanding Canadian Shale Gas. Energy Briefing Note, 23 pp.
National Energy Board, British Columbia Ministry of Energy and Mines, and Oil and Gas Commission of British Columbia, 2000. Analysis of Horizontal Gas Well Performance in British Columbia. Joint technical report, 27 pp.
National Petroleum Council (NPC), September 1992. Report of the Reserves Appreciation Subgroup of the Source and Supply Task Group: 1992 National Petroleum Council Natural Gas Study, Washington, D.C., 169 pp.
National Research Council, 1991. Undiscovered Oil and Gas Resources: An Evaluation of the Department of the Interior's 1989 Assessment Procedures. National Academy Press, Washington, D.C., 108 pp. plus appendices.
Nehring, R., January 1981. The Discovery of Significant Oil and Gas Fields in the United States. The Rand Corp.. R-2654/1-USGS/DOE, 236 pp.

Neidell, N.S., Cook, E.E., 1986. Seismic Method for Identifying Low Velocity Subsurface Zones. US Patent 4 571 710.
Nelson, R.A., 1985. Geologic Analysis of Naturally Fractured Reservoirs. Gulf Pub. Co., Houston, 320 pp.
Nelson, R.A., 2001. Geologic Analysis of Naturally Fractured Reservoirs, second ed. Gulf Publishing Co. Book Division, Houston.
Nemat-Nasser, S., Oranratnachai, A., 1979. Minimum spacing of thermally induced cracks in brittle solids. J. Energ. Res. Tech. 34–40.
Ng, M.C., Aguilera, R., July 1999. Well test analysis of horizontal wells in bounded naturally fractured reservoirs. J. Can. Petrol. Technol. 38 (7), 20–24.
Nguyen, T.V., 1986. Experimental Study of Non-Darcy Flow through Perforations. paper SPE 15473.
Nickelsen, R.P., Hough, V.D., 1967. Joints in the Apalachian Plateau of Pennsylvania. Geol. Soc. Am. Bull. 78, 609–629.
Norbeck, J.H., 2010. Identification and Characterization of Natural Fractures while Drilling Underbalanced (MSc thesis). Colorado School of Mines, Golden, CO.
Norbeck, J.H., 2012. Identification and Characterization of Natural Fractures while Drilling Underbalanced (M.S. thesis). Department of Engineering, Colordao School of Mines, 85 pp.
Norford, B.S., 1996. Ordovician and Silurian. In: The Geology, Mineral and Hydro-carbonPotential of Northern Yukon Territory and Northwestern District of Mackenzie. Geological Survey of Canada Bulletin 422, pp. 119–162.
Norris, D.K., Hopkins, W.S., 1977. The Geology of the Bonnet Plume Basin, Yukon Territory. Geological Survey of Canada Paper 76–8, 20 pp.
Norris, D.K., 1996a. Devonian. In: The Geology, Mineral and Hydrocarbon Potential of Northern Yukon Territory and Northwestern District of Mackenzie, Geological Survey of Canada Bulletin 422, pp. 163–200.
Norris, D.K., 1996b. Triassic. In: The Geology, Mineral and Hydrocarbon Potential of Northern Yukon Territory and Northwestern District of Mackenzie, Geological Survey of Canada Bulletin 422, pp. 253–265.
Norris, D.K., 1997. The Geology and Mineral and Hydrocarbon Potential of Northern Yukon Territory and Northwestern District of Mackenzie. Geological Survey of Canada Bulletin 422, 401 pp.
Nouri, A., Vaziri, H., Islam, M.R., 2002. A New theory and methodology for modeling sand production. Energy Sources 24 (11), 995–1008.
Nouri, A., Vaziri, H., Kuru, E., Islam, R., 2006. A comparison of two sanding criteria in physical and numerical modeling of sand production. J. Pet. Sci. Eng. 50, 55–70.
NPC, September 2003. Balancing Natural Gas Policy. National Petroleum Council, 34.
O'Connel, R., Budianksy, B., 1976. Seismic velocities in dry and saturated cracked solids. J. Geophys. Res. 79, 5412–5426.
Ohaeri, C.U., Vo, D.T., 1991. Practical solutions for interactive horizontal well test analysis. In: Paper SPE 22729 Presented at the Presented at the 1991 Annual Technical Conference and Exhibition, Dallas, Texas, 6–9 October.
Oil and Gas Directorate, 1995. Kandik Basin. Chapter 3, Northern Yukon. In: Morrell, G.R. (Ed.), Petroleum Exploration in Northern Canada: A Guide to Oil and Gas Exploration and Potential, Indian and Northern Affairs Canada, pp. 49–53.
Okiongbo, K.S., 2011. Effective Stress-Porosity Relationship above and Within the Oil Window in the North Sea Basin. Res. J. Appl. Sci. Eng. Technol. 3 (1), 32–38.
Oldenburg, C.M., Pruess, K., Benson, S.M., 2001. Process modeling of CO_2 injection into natural gas reservoirs for carbon sequestration and enhanced gas recovery. Energy Fuels 15 (2), 293–298.

Oldenburg, C.M., 2003. Carbon dioxide as cushion gas for natural gas storage. Energy Fuels 17 (1), 240–246.
Olorode, O.M., 2011. Numerical Modeling of Fractured Shale-Gas and Tight-Gas Reservoirs Using Unstructed Grids (MSc. thesis). Texas A&M University.
Olson, J.E., Pollard, D.D., 1989. Inferring paleostresses from natural fracture patterns. New Meth. Geol. 17, 345–348.
Olson, J.E., 1993. Joint pattern development: effects of subcritical crack growth and mechanical crack interaction. J. Geophys. Res. 98, 12251–12265.
Ong, A., 2008. The human and ethical living. In: Kenway, J., Fahey, J. (Eds.), Globalizing the Research Imagination. Routledge, London, pp. 87–100.
Onishi, T., Katoh, K., Takabayashi, K., Uematsu, H., Okatsu, K., Wada, Y., Ogata, Y., 2007. High pressure air injection into light oil reservoirs: experimental study on artificial ignition. In Proceedings of IEA Collaborative Project on Enhanced Oil Recovery 28th Annual Workshop and Symposium, Vedbæk, Denmark, September 4–7.
OPEC, 2011. Share of World Oil Reserves 2010. OPEC.
Osadetz, K.G., Chen, Z., Bird, T.D., 2005a. Petroleum Resource Assessment, Eagle Plain Basin and Environs, Yukon Territory, Canada. Yukon Geological Survey, Open File 2005-2, Geological Survey of Canada Open File 4922, 88 pp.
Osadetz, K.G., MacLean, B.C., Morrow, D.W., Dixon, J., Hannigan, P.K., 2005b. Petroleum Resource Assessment, Peel Plateau and Plain, Yukon Territory, Canada. Geological Survey of Canada Ope. n File 4841, 76 pp.
Osmond, J.C., 1992. Greater natural buttes gas field, Uintah County, Utah. In: Fouch, T.D., Nuccio, V.F., Chidsey Jr., T.C. (Eds.), Hydrocarbon and Mineral Resources of the Uinta Basin, pp. 143–163. Utah: Utah Geological Association, Guidebook 20.
Overbey, W.K., Yost, L.E., Yost, A.B., 1988. Analysis of natural fractures observed by borehole video camera in a horizontal well. In: Paper SPE 17760 Presented at the 1988 SPE Gas Technology Symposium, Dallas, June 13–15.
Ozkan, E., Ohaeri, U., Raghavan, R., June 1987. Unsteady flow to a well produced at a constant pressure in a fractured reservoir. SPE Form. Eval. 279–300, 186–200.
Paige, S., 1912. Description of the Llano and Burnet Quadrangles [Texas]. U.S. Geological Survey Atlas of the United States. Llano-Burnet folio 183, 16 pp.
Palmer, I., Khodaverian, M., Vaziri, H., Wang, X., 1996. Mechanics of openhole cavity completions in coalbed methane wells. In: Aubertin, M., Hassani, F., Mitri, H. (Eds.), Proc. 2nd North Am. Rock Mech. Symp. Balkema, Rotterdam, pp. 1089–1096.
Papay, J., 1999. Improved Recovery of Conventional Natural Gas. Published in the German Magazine Erdoel, Erdgas, Kohle; Part I: Theoretical Discussion of Recovery Methods. No. 6, pp. 302–308, January 1999 and Part II: Results of a Pilot Test; No. 7–8, pp. 354–355, July–August 1999.
Parrish, J.T., Curtis, R.L., 1982. Atmospheric circulation, upwelling, and organic-rich rocks in the Mesozoic and Cenozoic Eras. Palaeogeogr. Palaeoclimatol. Palaeoecol. 40, 31–66.
Passey, Q.R., Creaney, S., Kulla, J.B., Moretti, F.J., Stroud, J.D., 1990. A practical model for organic richness from porosity and resistivity logs. AAPG Bull. 74, 1777–1794.
Patz, M.J., Reddy, K.J., Skinner, Q.D., 2004. Chemistry of coalbed methane discharge water interacting with semi arid ephemeral stream changes. J. Am. Water Resour. Assoc. 40, 1247–1255.
Paul, R.J., 1993. Seismic detection of overpressure and fracturing: an example from the Qaidam Basin, People's Republic of China. Geophysics 58, 1532–1543.
Pawlowicz, J.G., Anderson, S.D.A., Rokosh, C.D., Beaton, A.P., 2009a. Mineralogy, Permeametry, Mercury Porosimetry and Scanning Electron Microscope Imaging of the

Banff and Exshaw Formations; Shale Gas Data Release. Energy Resources Conservation Board. ERCB/AGS Open File Report 2008–13.
Pawlowicz, J.G., Anderson, S.D.A., Rokosh, C.D., Beaton, A.P., 2009b. Mineralogy, Permeametry, Mercury Porosimetry and Scanning Electron Microscope Imaging of the Colorado Group; Shale Gas Data Release. Energy Resources Conservation Board. ERCB/AGS Open File Report 2008–14.
Pearl, J., 2009. Causality: Models, Reasoning, and Inference. Cambridge University Press, ISBN 0-521-77362-8 (2000, 2nd edition 2009).
Pearson, D.G., Brenker, F.E., Nestola, F., McNeill, J., Nasdala, L., Hutchison, M.T., Matveev, S., Mather, K., Silversmit, G., Schmitz, S., Vekemans, B., Vincze, L., March 13, 2014. Hydrous mantle transition zone indicated by ringwoodite included within diamond. Nature 507, 221–224. http://dx.doi.org/10.1038/nature13080.
Pedersen, T.F., Calvert, W.E., 1990. Anoxia vs. productivity—What controls the formation of organic-carbon-rich sediments and sedimentary rocks? Am. Assoc. Pet. Geol. Bull. 74 (4), 454–466.
Pedrosa, Jr., Oswaldo, A., Aziz, K., November 1986. Use of a hybrid grid in reservoir simulation. SPE 13507 SPE Reserv. Eng. 611–621.
Perez, M., Gibson, R., 1996. Detection of Fracture Orientation Using Azimuthal Variation of P-Wave AVO Responses: Barinas field (Venezuela), 66th Ann. Internat. Mtg., Soc. Expl. Geophys., Expanded Abstracts, 1353–1356.
Perez, M., 1997. Detection of Fracture Orientation Using Azimuthal Variation of P-Wave AVO Response (M.S. thesis). Mass. Inst. of Technol., Cambridge, 95 pp.
Peters, E.J., Flock, D.L., April 1981. The onset of instability during two-phase immiscible displacement in porous media. Soc. Pet. Eng. J 249–258.
Peterson, F., 1988. Pennsylvanian to Jurassic eolian transportation systems in the western United States. Sediment. Geol. 56 (1/4), 207–260.
Petrel Robertson Consulting Ltd. (PRCL), 2005. Regional Geological and Geophysical Assessment, Central Mackenzie Valley, NWT and Eagle Plain, Yukon Territory. Non-exclusive industry report, 250 pp.
Petrel Robertson Consulting Ltd. (PRCL), 2009. Exploration Evaluation of the Southern Northwest Territories and Southeastern Yukon. Non-exclusive industry report, 288 pp.
Petrel Robertson Consulting Ltd. (PRCL), 2010a. Assessment of Canada's Natural Gas Resources Base. Report for Canadian Society for Unconventional Gas. www.csur.com/images/news/2011/Petrel_GIP2010_web.pdf.
Petrel Robertson Consulting Ltd. (PRCL), 2010b. Horn River Basin Aquifer Characterization Project – Geological Report. Geoscience BC, Report 2010-11, 192 pp.
Petta, T.J., Rapp, S.D., 1990. Appleton field—U.S.A. Gulf of Mexico Basin, Alabama. In: Beaumont, E.A., Foster, N.D. (Eds.), Structural Traps IV—tectonic and Nontectonic Fold Traps: American Association of Petroleum Geologists Atlas of Oil and Gas Fields and the Treatise of Petroleum Geology, pp. 299–318.
Pfeifer, January 17, 2014. Methane Hydrates Could Be Energy of the Future, Innovation in Energy.
Pieraccini, M., Bicci, A., Mecatti, D., Macaluso, G., Atzeni, C., 2009. Propagation of Large Bandwidth Microwave Signals in Water. IEEE Trans. Antennas Propag. 57 (11), 3612–3618.
Pigage, L., 2009. Yukon Table of Formations, vol. 3.2. Yukon Geological Survey and Oil and Gas Resources Branch.
Pitman, J.K., Sprunt, E.S., 1986. Origin and distribution of fractures in lower Tertiary and Upper Cretaceous rocks, Piceance Basin, Colorado, and their relation to the occurrence of hydrocarbons. In: Spencer, C.W., Mast, R.F. (Eds.), Geology of Tight Gas

Reservoirs. American Association of Petroleum Geologists Studies in Geology 24, pp. 221–233.
Pitman, J.K., Anders, D.E., Fouch, T.D., Nichols, D.J., 1986. Hydrocarbon potential of nonmarine Upper Cretaceous and lower Tertiary rocks, eastern Uinta Basin, Utah. In: Spencer, C.W., Mast, R.F. (Eds.), Geology of Tight Gas Reservoirs. American Association of Petroleum Geologists Studies in Geology 24, pp. 235–252.
Pizzaro, J.O.S., Trevisan, O.V., October 1990. Electrical heating of oil reservoirs: numerical simulation and field test results. J. Pet. Tech. 1320–1326.
Pointon, K., Langan, M., 2002. Distributed Power Generation Using Biogas Fuelled Microturbines. Available from: www.berr.gov.uk/files/file14934.pdf (accessed 26.04.11).
Pollard, D.D., Aydin, A., 1988. Progress in understanding jointing over the past century. Geol. Soc. Am. Bull. 100, 1181–1204.
Pollard, D.D., Segall, P., 1987. Theoretical displacements and stresses near fracture in rock: with applications to faults, joints, veins, dikes, and solution surfaces. In: Atkinson, B.K. (Ed.), Fracture Mechanics of Rock. Academic Press, Orlando, pp. 277–350.
Pollard, D.D., Segall, P., Delaney, P., 1982. Formation and interpretation of dilatant echelon cracks. Geol. Soc. Am. Bull. 93, 1291–1303.
Pollastro, R.M., Schenk, C.J., 1991. Origin and diagenesis of clay minerals in relation to sandstone paragenesis—An example in eolian dune reservoirs and associated rocks, Permian upper part of the Minnelusa Formation, Powder River Basin, Wyoming. Am. Assoc. Pet. Geol. Bull. 75, 1136–1137.
Pollastro, R.M., Hill, R.J., Ahlbrandt, T.A., Charpentier, R.R., Cook, T.A., Klett, T.R., Henry, M.E., Schenk, C.J., 2004. Assessment of Undiscovered Oil and Gas Resources of the Bend Arch–Fort Worth Basin Province of North-Central Texas and Southwestern Oklahoma, 2003. U.S. Geological Survey Fact Sheet 2004–3022, 2 pp.
Pollastro, R.M., Jarvie, D.M., Hill, R.J., Adams, C.W., 2007. Geologic framework of the mississippian barnett shale, Barnett-Paleozoic total petroleum system, Bend arch, fort Worth Basin, Texas. Am. Assoc. Pet. Geol. Bull. 91, 405–436.
Pollastro, R.M., 2007. Total petroleum system assessment of undiscovered resources in the giant Barnett Shale continuous (unconventional) gas accumulation, Fort Worth Basin, Texas. Am. Assoc. Pet. Geol. Bull. 91, 551–578.
Porat, B., Friedlander, B., 1988. Analysis of the asymptotic relative efficiency of the music algorithms in resolving plabe waves in noise. IEEE Trans. Acoust. Speech Signal Process. ASSP-36, 532–544.
Potential Gas Committee (PGC). Biennially, 1971–1995. Potential Supply of Natural Gas in the United States (As of December 31). Potential Gas Agency, Colorado School of Mines.
Potter, J., Richards, B.C., Cameron, A.R., 1993. The petrology and origin of coals from the Lower Carboniferous Mattson Formation, southwestern District of Mackenzie, Canada. Int. J. Coal Geol. 34, 113–140.
Poulton, T.P., Braun, W.K., Brooke, M.M., Davies, E.H., 1993. Jurassic. Subchapter 4H. In: Stott, D.F., Aitken, J.D. (Eds.), Sedimentary Cover of the Craton in Canada, Geological Survey of Canada, Geology of Canada, No. 5, pp. 321–357.
Poulton, T.P., 1996. Jurassic. In: The Geology, Mineral and Hydrocarbon Potential of Northern Yukon Territory and Northwestern District of Mackenzie. Geological Survey of Canada Bulletin 422, pp. 267–299.
Power, M., 1992. Lognormality in the observed size distribution of oil and gas pools as a consequence of sampling bias. Math. Geol. 24 (8), 929–945.
Powers, M.C., 1967. Fluid-release mechanisms in compacting marine mudrocks and their importance in oil exploration. Am. Assoc. Petrol. Geol. Bull. 51, 1240–1254.

Prather, B.E., 1992. Evolution of a Late Jurassic carbonate/evaporite platform, Conecuh embayment, northeastern Gulf Coast, U.S.A. Am. Assoc. Pet. Geol. Bull. 76, 164–190.

Price, N.J., 1966. Fault and Joint Development in Brittle and Semi-brittle Rock. Pergamon Press, New York, 176 pp.

Price, L., 2000. Causes, characteristics, and implications of pervasive oil-induced horizontal fracturing in Bakken source system reservoir rocks [abs.]. Am. Assoc. Pet. Geol. Bull. 84, 1244.

Pulham, A., 1999. Reservoir characterization—Sedimentology, trace fossils, biostratigraphy, and permeability. The Outcrop 48 (2), 4–5.

Puri, R., Evanoff, J.C., Brugler, M.L., 1991. Measurement of coal cleat porosity and relative permeability characteristics. SPE Gas. Technol. Symp. 93–104.

Puryear, C.I., Castagna, J.P., 2008. Layer-thickness determination and stratigraphic interpretation using spectral inversion: Theory and application. Geophysics 73 (2), R37–R48. http://dx.doi.org/10.1190/1.2838274.

Putnam, P.E., Christensen, S.L., May 31–June 4, 2004. McMurray formation SAGD (steam-assisted gravity drainage) reservoirs in Northeastern Alberta: comparative architecture and performance. In Proceedings of CSPG-Canadian Heavy Oil Assoc.-CWLS Joint Conf. (ICE2004), Calgary, AB, Canada.

Pyle, L.J., Allen, T.L., Fraser, T.A., Zantvoort, W.G., 2008. Petroleum Play Data for Upper Devonian and Carboniferous Siliciclastics (Imperial and Tuttle Formations), Mackenzie Corridor. Geological Survey of Canada Open File 5894, 1 CD-ROM.

Pyle, L.J., Gal, L.P., Lemiski, R.T., 2011. Measured Sections and Petroleum Potential Data (Conventional and Unconventional) of Horn River Group Outcrops – Part 1, NTS 96D, 96E, and 106H, Northwest Territories. Northwest Territories Geoscience Office, NWT Open File 2011-09, 116 pp. and Microsoft Excel (.xlsx) files.

Qian, Z., Xu, L., Li, Z., Li, H., Guo, K., 2010. Selective absorption of H_2S from a gas mixture with CO_2 by aqueous *N*-methyldiethanolamine in a rotating packed bed. Ind. Eng. Chem. Res. 49 (16), 6196–6203. In: Goodman, R.E., Tsang, C.-F. (Eds.), Fractured and Jointed Rock Masses. Balkema, Rotterdam, pp. 431–434.

Rader, K., von Drehle, W.F., Martin, G.W. (Eds.), 1990. Morrow Sandstones of Southeast Colorado and Adjacent Areas: Rocky Mountain Association of Geologists—Morrow Sandstone, pp. 51–58.

Rader, K., 1990. In: Sonnenberg, S.A., Shannon, L.T. (Eds.), Petrography of Morrow Sandstones in Southeast Colorado, Southwest Kansas, and Northwest Oklahoma.

Rahman, M.A., Mustafiz, S., Biazar, J., Islam, M.R., 2007. "Experimental and numerical modeling of a novel rock perforation technique". J. Franklin Institute 344 (5), 777–789.

Raistrick, A., Marshall, C.E., 1939. The Nature and Origin of Coal and Coal Seams. English Univ. Press, London, 282 pp.

Ramalho, J., Elliott, D., Francis, P., Medeiros, R.S., 2009. Tight Gas Reservoir Exploitation with Underbalanced Drilling Technology. Paper IPTC 14109 Presented at the 2009 International Petroleum Technology Conference, Doha, Qatar, 7–9 December. http://dx.doi.org/10.2523/14109-MS.

Ramalho, J., 2006. Underbalanced drilling. In: The Reservoir, An Integrated Technology Approach. Paper SPE 103576 Presented at the 2006 SPE Russian Oil and Gas Technical Conference and Exhibition, Moscow, Russia, 3–6 October. http://dx.doi.org/10.2118/103576-MS.

Ramírez, M., 2007. Viabilidad de un proceso para la eliminación conjunta de H2S y NH3 contenido en efluentes gaseosos. Universidad de Cádiz, Cádiz, Spain.

Ramos, A.C.B., Davis, T.L., 1997. 3-D AVO analysis and modeling applied to fracture detection in coalbed methane reservoirs. Geophysics 62, 1683–1695.

Ransohoff, T.C., Radke, C.J., May 1988. Mechanisms of Foam Generation in Glass-Bead Packs. SPE Reservoir Eng. 3 (2), 573–585.
Rascoe, B., Adler, F.J., 1983. Permo-Carboniferous hydrocarbon accumulations, mid-continent, U.S.A. Am. Assoc. Pet. Geol. Bull. 67, 979–1001.
Ratner, M., Tiemann, M., 2014. An Overview of Unconventional Oil and Natural Gas: Resources and Federal Actions. Congressional research service report, January 23.
Raymond James Ltd., 2008. Canadian energy Sector – Horizontal Wells Make "Monumental Montney" Economic. Analyst report for industry clients.
Reed, C.J., Leyendecker, C.L., Khan, A.S., Kinler, C.J., Harrison, P.F., Pickens, G.P., 1987. Correlation of Cenozoic sediments—Gulf of Mexico Outer Continental Shelf, part 1: Galveston area, offshore Texas, through Vermilion area, offshore Louisiana. Minerals Management Service OCS Report MMS 87-0026, 35 pp. plus appendices.
Reimer, J., 1994. Speculations concerning reservoir diagenesis and sourcing of Upper and Middle Devonian gas fields, northeastern British Columbia: a hydrogeological perspective. Can. Soc. Pet. Geol. Reserv. 21 (10), 1–2.
Rhodes, J.A., Maxwell, G.B., 1993. Jurassic stratigraphy of the Wiggins arch, Mississippi. In: Sartin, A.A., Barrett, M.L., Rosen, N.C. (Eds.), Transactions of the 43rd Annual Convention, Gulf Coast Association of Geological Societies Transactions, 43, pp. 333–344.
Rice, D.D., Clayton, J.L., Pawlewicz, M.J., 1989. Characterization of coal-derived hydrocarbons and source rock potential of coal beds, San Juan basin, New Mexico and Colorado, USA. Int. J. Coal Geol. 13, 597–626.
Rice, C.A., Ellis, M.S., Bullock Jr., J.H., 2000. Water Co-produced with Coalbed Methane in the Powder River basin, Wyoming: Preliminary Compositional Data. Open File Report 00-372. US Geological Survey, Denver, CO.
Richards, E.W., Gunn, E.A., 2000. A model and tabu search method to optimize stand harvest and road construction schedules. For. Sci. 46 (2), 188–203.
Richards, P.G., Menke, W., 1983. The apparent attenuation of a scattering medium. Bull. Seismol. Soc. Am. 73, 1005–1022.
Carboniferous; Subchapter 4E Richards, B.C., Bamber, E.W., Higgins, A.C., Utting, J., 1993. In: Stott, D.F., Aitken, J.D. (Eds.), Sedimentary Cover of the Craton in Canada, Geological Survey of Canada, Geology of Canada, 5, pp. 202–271.
Richards, B.C., Bamber, E.W., Utting, J., 1996. Upper Devonian to Permian. In: The Geology, Mineral and Hydrocarbon Potential of Northern Yukon Territory and northwestern District of Mackenzie Geol. Surv. Can. Bull. 422, 201–251.
Richardson, R.J.H., et al., 1988. An Evaluation of the Coal Reseoruces of the Ardley Coal Zone to a Depth of 400 M, in the Alberta Plans Area. Alberta Research Council Open File Report OFR 1988-02, 96 pp.
Ricketts, B.D., 1984. A Coal Index for Yukon and District of Mackenzie, Northwest Territories. Geological Survey of Canada Open File 1115, 8 pp.
Ricketts, B.D., 1988. The Monster Formation: A Coastal Fan System of Late Cretaceous Age, Yukon Territory. Geological Survey of Canada Paper 86-14, 27 pp.
Rickman, R., Mullen, M., Petre, E., Grieser, B., Kundert, D., 2008. A practical use of shale petrophysics for stimulation design optimization: All shale plays are not clones of the Barnett Shale. In: Annual Technical Conference and Exhibition. Society of Petroleum Engineers. SPE 11528.
Rokosh, C.D., Pawlowicz, J.G., Berhane, H., Anderson, S.D.A., Beaton, A.P., 2009. What Is Shale Gas? An Introduction to Shale Gas Geology in Alberta. Energy Resources Conservation Board/Alberta Geological Survey Open File Report 2008-08, 26 pp.
Romeo, L.M., Escosa, J., Bolea, I., 2006. Postcombustion CO_2 Sequestration. Universidad de Zaragoza, Zaragoza, Spain.

Root, D.H., Attanasi, E.D., 1993a. A primer in field-growth estimation. In: Howell, D.G. (Ed.), The Future of Energy Gases: U.S. Geological Survey Professional Paper 1570, pp. 547–554.

Root, D.H., Mast, R.F., 1993. Future growth of known oil and gas fields. Am. Assoc. Pet. Geol. Bull. 77 (3), 479–484.

Root, D.H., 1981. Estimation of inferred plus indicated reserves for the United States. In: Dolton, G.L., Carlson, K.H., Charpentier, R.R., Coury, A.B., Crovelli, R.A., Frezon, S.E., Khan, A.S., Lister, J.H., McMullin, R.H., Pike, R.S., Powers, R.B., Scott, E.W., Varnes, K.L. (Eds.), Estimates of Undiscovered Recoverable Conventional Resources of Oil and Gas in the United States: U.S. Geological Survey Circular 860, pp. 83–87.

Rose, W., May.2000. Myths about later-day extensions of Darcy's law. J. Pet. Sci. Eng. 26 (1–4), 187–198.

Ross, D.J.K., Bustin, R.M., 2008. Characterizing the shale gas resource potential of Devonian-Mississippian strata in the Western Canada Sedimentary Basin: application of an integrated formation evaluation. Am. Assoc. Pet. Geol. Bull. 92, 87–125.

Ross, E., July 27, 2011. Shale play the target of ConocoPhillips NWT acquisition. Dly. Oil Bull.

Rottenfusser, B., Ranger, M.A., May 31–June 4, 2004. Geological comparison of six projects in the Athabasca oil sands. In Proceedings of CSPG-Canadian Heavy Oil Association-CWLS Joint Conf. (ICE2004), Calgary, AB, Canada.

Royden, L., Burchfiel, B.C., King, R.W., Wang, E., Chen, Z., Shen, F., Liu, Y., 1997. Surface deformation and lower crustl flow in eastern Tibet. Science 276, 788–790.

Royden, L., 1996. Coupling and decoupling of crust and mantle in convengent orogens: Implications for strain partitioning in the crust. J. Geophys. Res. 101, 17679.

Ruble, T.E., Lewan, M.D., Philp, R.P., 2001. New insights on the Green River petroleum system in the Uinta Basin from hydrous pyrolysis experiments. Am. Assoc. Pet. Geol. Bull. 85, 1333–1372.

Ruckelshaus Institute of Environment and Natural Resources (RIENR), 2005. Water Production from Coalbed Methane Development in Wyoming: A Summary of Quantity, Quality and Management. Final Report to Wyoming State Governor. Retrieved from http://www.uwyo.edu/enr/ienr/cbm.asp.

Ruger, A., Tsvankin, I., 1997. Using AVO for fracture detection: analytic basis and practical solutions. The Lead Edge 16, 1429–1434.

Ruger, A., 1995. P-wave reflection coefficients for transversely isotropic media with vertical and horizontal axes of symmetry. Expanded Abstracts 65th Ann. Intern. Mtg., Soc. Expl. Geophys 278–281.

Ruger, A., 1998. Variation of P-wave reflectivity with offset and azimuth in anisotropic media. Geophysics 63, 935–947.

Rutherford, S.R., Williams, M.W., 1989. Amplitude-Versus-Offset variations in gas sands. Geophysics 54, 680–688.

Ryder, R.T., Fouch, T.D., Elison, J.H., 1976. Early Tertiary sedimentation in the western Uinta Basin, Utah. Geological Soc. Am. Bull. 87, 496–512.

Saghir, M.Z., Islam, M.R., 1999. Viscous fingering during miscible liquid-liquid displacement in porous media. Int. J. Fluid Mech. Res. 26 (4), 215–226.

Saghir, Z., Chaalal, O., Islam, M.R., 2000. Experimental and numerical modeling of viscous fingering. J. Pet. Sci. Eng. 26 (1–4), 253–262.

Saghir, M.Z., Vaziri, H., Islam, M.R., 2001. Heat and mass transfer modeling of fractured formations. J. Comput. Fluid Dyn. 15 (4), 279–292.

Sain, K., 2012. Gas – Hydrates – a probable solution to India's Energy Crisis. Int. J. Earth Sci. Eng. 5 (2) (Editorial note).

Saller, A.H., Moore, B.R., 1986. Dolomitization in the Smackover formation, Escambia County, Alabama. Gulf Coast Assoc. Geol. Soc. Trans. 26, 275–282.

Sandvol, E., Ni, J., Kind, R., Zhao, W., 1997. Seismic anisotropy beneath the southern Himalayas-Tibet collision zone. J. Geophys. Res. 102, 17813–17823.
Sanfillippo, F., Brignoli, M., Santarelli, F.J., Bezzola, C., 1997. Characterization of conductive fractures while drilling. In: Paper SPE 38177 Presented at the 1997 SPE European Formation Damage Conference, the Hague, The Netherlands, 2–3 June. http://dx.doi.org/10.2118/38177-MS.
Sanford, J., September 16, 2008. Making Bakken – how new technology is getting oil – and money – from prairie stone. Can. Bus.
Sarma H., Das, S., March 15–18, 2009. Air injection potential in Kenmore oilfield in Eromanga Basin, Australia: a screening study through thermogravimetric and calorimetric analyses (SPE-120595). In Proceedings of SPE Middle East Oil and Gas Show and Conference, Bahrain, The Kingdom of Bahrain.
SAS (SAS Institute), 2000. SAS Proprietary Software Release 9.0. SAS Institute, Raleigh, NC.
Sayers, C.M., Ebrom, D.A., 1997. Seismic traveltime analysis for azimuthally anisotropic media: theory and experiment. Geophysics 62, 1570–1582.
Sayers, C.M., Kachanov, M., 1995. Microcrack-induced elastic wave anisotropy of brittle rocks. J. Geophys. Res. 100, 4149–4156.
Sayers, C.M., Rickett, J., 1997. Azimuthal variation in AVO response for fractured gas sands. Geophys. Prospect. 45, 165–182.
Schenk, C.J., Pollastro, R.M., January 2002. Nat 0113-01 Gas Production in the United States. U.S. Geological Survey Fact Sheet, 0113–01.
Schenk, C.J., Richardson, R.W., 1985. Recognition of interstitial anhydrite dissolution—a cause of secondary porosity, San Andres Limestone, New Mexico, and Upper Minnelusa Formation, Wyoming. Am. Assoc. Pet. Geol. Bull. 69, 1064–1076.
Schenk, C.J., Schmoker, J.W., 1993. Role of halite in the evolution of sandstone porosity, Upper Jurassic Norphlet Formation, Mississippi salt basin. In: Sartin, A.A., Barrett, M.L., Rosen, N.C. (Eds.), Transactions of the 43rd Annual Convention, Gulf Coast Association of Geological Societies Transactions, 43, pp. 357–362.
Schenk, C.J., Schmoker, J.W., Scheffler, J.M., 1986. Characterizing Permian upper Minnelusa sandstone reservoirs, Powder River Basin, Wyoming, for enhanced oil recovery [abs.]. Am. Assoc. Pet. Geol. Bull. 70, 1054.
Schenk, C.J., 1990. Overview of eolian sandstone diagenesis, Upper Jurassic Denkman Sandstone Member of the Norphlet Formation, Mississippi and Alabama. In: Fryberger, S.G., Krystinik, L.F., Schenk, C.J. (Eds.), Modern and Ancient Eolian Deposits—Petroleum Exploration and Production. Rocky Mountain Section, S.E.P.M. (Society for Sedimentary Geology), pp. 20–21.
Schenk, C.J., 1995a. Mississippi-Louisiana salt basins province (049). In: Gautier, D.L., Dolton, G.L., Takahashi, K.I., Varnes, K.L. (Eds.), 1995 National Assessment of United States Oil and Gas Resources. U.S. Geological Survey DDS–30, CD-ROM.
Schenk, C.J., 1995b. Western gulf Coast province (047). In: Gautier, D.L., Dolton, G.L., Takahashi, K.I., Varnes, K.L. (Eds.), 1995. National Assessment of United States Oil and Gas Resources. U.S. Geological Survey DDS–30, CD-ROM.
Schlumberger Oilfield Glossary, October 22, 2006. Vitrinite Reflectance. http://www.schlumberger.com.
Schmidt, R., 1979. Multiple emitter location and signal parameter estimation. In: Proc. RADC Spectral Estimation Workshop (Rome, NY), pp. 243–258.
Schmidt, R., 1986. Multiple emitter location and signal parameter estimation. IEEE Trans. Antennas Propag. AP-34, 276–280.
Schmoker, J.W., Hester, T.C., 1983. Organic carbon in Bakken formation, United States portion of Williston Basin. Am. Assoc. Pet. Geol. Bull. 67, 2165–2174.
Schmoker, J.W., 1995. Williston Basin province (031). In: Gautier, D.L., Dolton, G.L., Takahashi, K.I., Varnes, K.L. (Eds.), 1995 National Assessment of United States Oil and Gas Resources. U.S. Geological Survey DDS–30, CD-ROM.

Schmoker, J.W., 1996. A resource evaluation of the Bakken Formation (Upper Devonian and Lower Mississippian) continuous oil accumulation, Williston Basin, North Dakota and Montana. Mt. Geol. 33, 1–10.

Schoenberg, M., Sayers, C.M., 1995. Seismic anisotropy of fractured rock. Geophysics 60, 204–211.

Schultz-Ela, D.D., Yeh, J., 1992. Predicting fracture permeability from bed curvature. In: Tillerson, J.R., Wawersek, W.R. (Eds.), Proc. 33rd U.S. Symp. Rock Mechanics.

Schutter, S.R., 2003. Hydrocarbon occurrence and exploration in and around igneous rocks. In: Petford, N., McCaffrey, K.J.W. (Eds.), Hydrocarbons in Crystalline Rocks. Geological Society Special Publication 214. London: Geological Society, pp. 7–33.

Scotia Capital Canada, 2011. Exploration of the Alberta Bakken: A Resource Play Mosaic in the Making? (Analyst report for industry clients).

Scott, A.R., Kaiser, W.R., Ayers Jr, W.B., 1991. Composition, distribution, and origin of Fruitland Formation and Pictured Cliffs Sandstone gases, San Juan Basin, Colorado and New Mexico. In: Schwochow, S.D., Murray, D.K., Fahy, M.F. (Eds.), Coalbed Methane of Western North America, Rocky Mountain Assoc. Geol. Field Conf, pp. 93–109.

Scott, A.R., Kaiser, W.R., Ayers Jr., W.B., 1994. Thermogenic and secondary biogenic gases, San Juan Basin, Colorado and New Mexico-implications for coalbed gas producibility. AAPG Bull. 78, 1186–1209.

Scott, A.R., Tyler, R., Comer, J.B., 1996. Evaluating changes in maximum horizontal stress in the Piceance Basin, Colorado, on the basis of timing of cleat development [abs.], South-Central Section. Geol. Soc. Am. Abst. Prog. 28, 63.

Secor, D.T., 1965. Role of fluid pressure in jointing. Am. J. Sci. 263, 633–646.

Segall, P., 1984. Formation and growth of extensional fracture sets. Geol. Soc. Am. Bull. 95, 454–462.

Selley, R.C., 1998. Elements of Petroleum Geology (Chapter 6), second ed. Academic Press, London. pp. 268–269.

Seni, S.J., Desselle, B.A., Standen, A., 1994. Scope and construction of a gas and oil atlas series of the Gulf of Mexico: examples from Texas offshore lower Miocene plays. Gulf Coast Assoc. Geol. Soc. Trans. 44, 681–690.

Seni, S.J., Brooke, J., Marin, D., Kazanis, E., 1995. Chronostratigraphic hydrocarbon plays and depositional styles in the northern Gulf of Mexico. Gulf Coast Assoc. Geol. Soc. Trans. 45, 519–527.

Seni, S.J., Hentz, T.F., Kaiser, W.R., Wermund Jr., E.G., 1997. In: Atlas of Northern Gulf of Mexico Gas and Oil Reservoirs. Miocene and older reservoirs, vol. 1. The University of Texas at Austin Bureau of Economic Geology, Gas Research Institute, Department of Energy, and Minerals Management Service, 199 pp. plus 3 plates and CD-ROM.

Serra, K., Reynolds, A.C., Raghavan, R., 1982. New pressure transient analysis methods for naturally fractured reservoirs. In: Paper SPE 10780 Presented at the 1982 SPE California Regional Meeting, San Francisco, California, 24–26 March.

Shanley, K.W., Cluff, R.M., Robinson, J.W., 2004. Factors controlling prolific gas production from low-permeability sandstone reservoirs: Implications for resource assessment, prospect development, and risk analysis. Am. Assoc. Pet. Geol. Bull. 88 (8), 1083–1121. http://dx.doi.org/10.1306/0325040305.

Shapiro, R., Zatzman, G.M., Mohiuddin, Y., 2007. Towards understanding disinformation. J. Nat. Sci. Sustain. Technol. 1 (3), 471–504.

Shebl, M.A., 1995. The impact of reservoir heterogeneity on fluid flow in the Tensleep Sandstone of the Bighorn Basin. In: Jones, R.E. (Ed.), Resources of Southwestern Wyoming. Wyoming Geological Association, 1995 Guidebook.

Sheldon, R.P., 1967. Long-distance migration of oil in Wyoming. Mt. Geol. 4, 53–65.
Shen, F., Toksoz, M.N., 1998. Scattering characteristics in heterogeneous fractured reservoirs from waveform estimation. 68th Ann. Intern. Mtg., Soc. Expl. Geophys. Expanded Abstracts.
Shen, F., Zhu, X., Toksoz, M.N., 1997. Anisotropy of aligned fractures and P-wave azimuthal AVO response. Expanded Abstracts, 2001–2004 67th Ann. Intern. Mtg., Soc. Expl. Geophys.
Shen, F., 1998. Seismic Characterization of Fractured Reservoirs (Part I) Crustal Deformation in Tibetian Plateau (Part II) (Ph.D. dissertation). Massachusetts Institute of Technology.
Shepherd, J., Creasy, J.W., Huntington, J.F., 1981. Geometric and dynamic analysis of some faults in the western coalfield of New South Wales. Proc. Aust. Inst. Min. Metall. 279, 19–32.
Shi, Z., Wang, X., 2007. Comparison of Darcy's Law, the Brinkman Equation, the Modified N-s Equation and the Pure Diffusion Equation in PEM Fuel Cell Modeling. COMSOL Conference proceeding, Boston.
Shuey, R.T., 1985. A simplification of the Zoeppritz equations. Geophysics 50, 609–614.
Shurr, G.W., Ridgley, J.L., 2002a. Unconventional shallow biogenic gas systems. Am. Assoc. Pet. Geol. Bull. 86, 1939–1969.
Shurr, G.W., Ridgley, J.I., 2002b. Unconventional shallow biogenic Gas systems. AAPGA Bull. 86 (11), 1939–1969.
Sierra, R., Tripathy, B., Bridges, J.E., Farouq Ali, S.M., 2001. Promising progress in field application of reservoir electrical heating methods (SPE-69709). In: Proceedings of SPE International Thermal Operations and Heavy Oil Symposium, Margarita Island, Venezuela, March 12–14, 2001.
Sigsby, R.J., 1976. Paleoenvironmental analysis of the big Escambia Creek–Jay–Blackjack Creek field area. Gulf Coast Assoc. Geol. Soc. Trans. 26, 258–278.
Skempton, A.W., 1970. The consolidation of clays by gravitational compaction. Q. J. Geol. Soc. London 125, 373–411.
Smith, M.G., Bustin, R.M., 1998. Production and preservation of organic matter during deposition of the Bakken Formation (Late Devonian and Early Mississippian), Williston Basin. Palaeogeogr. Palaeoclimatol. Palaeoecol. 142, 185–200.
Smith, M.G., Bustin, R.M., 2000. Late Devonian and Early Mississippian Bakken and Exshaw black shale source rocks, western Canada sedimentary basin—a sequence stratigraphic interpretation. Am. Assoc. Pet. Geol. Bull. 84, 940–960.
Smyth, M., Buckley, M.J., 1993. Statistical analysis of the microlithotype sequences in the Bulli seam, Australia, and relevance to permeability for coal gas. Int. J. Coal Geol. 22, 167–187.
Snow, D.T., 1965. A parallel Plate model of fractured permeable Media (Ph.D. dissertation). University of California, Berkeley.
Snowdon, L.R., Price, P.R., 1994. Rock-Eval/TOC Data for Three Wells in the Kandik Basin, Western Yukon Territory. Geological Survey of Canada Open File 2899, 24 pp.
Snowdon, L.R., 1988. Petroleum Source Potential and Thermal Maturation Reconnaissance in Eagle Plain, Yukon Territory. Geological Survey of Canada Open File 1720, 115 pp.
Society of Petroleum Engineers, 1987. Definitions for oil and gas reserves. J. Pet. Technol. 577–578. Solution of transient energy equation in rocks subjected to laser.
Sonnenberg, S.A., Shannon, L.T., Rader, K., von Drehle, W.F., 1990. Regional structure and stratigraphy of the Morrowan Series, southeast Colorado and adjacent areas. In: Sonnenberg, S.A., Shannon, L.T., Rader, K., von Drehle, W.F., Martin, G.W. (Eds.), Morrow sandstones of southeast Colorado and adjacent areas: Rocky Mountain Association of Geologists Guidebook, pp. 37–50.

Sonnenberg, S.A., 1985. Tectonic and sedimentation model for Morrow sandstone deposition, Sorrento field area, Denver Basin, Colorado. Mt. Geol. 22, 180–191.

Sorokhtin, O.G., Chilingar, G.V., Khilyuk, 2007. Global Warming and Cooling. Elsevier, 313 pp.

Sparks, D.P., McLendon, T.H., Saulsberry, J.L., Lambert, S.W., 1995. The effects of stress on coalbed methane reservoir performance, Black Warrior Basin. U.S.A. SPE 30734 SPE Ann. Tech. Conf. 339–351.

Spears, D.A., Caswell, S.A., 1986. Mineral matter in coals: cleat mineral and their origin in some coals from the English Midlands. Int. J. Coal Geol. 6, 107–125.

Spencer, C.W., 1995a. Uinta-Piceance Basin Provice (020), USGS report.

Spencer, C.W., 1995b. Uinta-Piceance province (020). In: Gautier, D.L., Dolton, G.L., Takahashi, K.I., Varnes, K.L. (Eds.), 1995 National Assessment of United States Oil and Gas Resources. U.S. Geological Survey DDS–30, CD-ROM.

Sperr, J.T., 1991. Exploration models for Bakken reservoirs. In: Hansen, W.B. (Ed.), Geology and Horizontal Drilling of the Bakken Formation. Montana Geological Society, 1991 Guidebook, pp. 143–149.

Spivey, J.P., Semmelbeck, M.E., 1995. Forecasting long-term gas production of dewatered coal seams and fractured gas shales. In: Paper SPE 29580 Presented at the 1995 Rocky Mountain Regional/Low Permeability Reservoirs Symposium, Denver, Colorado, 20–22 March.

Sposito, S., Mattigod, S.V., 1977. On the chemical foundation of the sodium adsorption ratio. Soil. Sci. Soc. Am. J. 41, 323–329.

Squarek, J., Dawson, M., July 24, 2006. Coalbed methane expands in Canada. Oil Gas. J. 37–42.

Srugoa, P., Rubinstein, P., January 2007. Processes controlling porosity and permeability in volcanic reservoirs from the Austral and Neuquén basins, Argentina. AAPG Bull. 91 (1), 115–129.

Stach, E., Mackowsky, M.T.H., Teichmueller, M., Taylor, G.H., Chandra, D.R., 1982. Coal petrology, Gebruder Borntraeger. Stuttgart–Berlin 5–86.

Standenes, S., 1995. Seismic Monitoring – A Tool for Improved Reservoir Management. SEISMIC 95, London.

Steinfeld, G., Sanderson, R., 1998. Landfill Gas Cleanup for Carbonate Fuel Cell Power Generation. National Renewable Energy Laboratory, Morgantown, West Virginia, USA.

Stell, J.R., Brown, C.A., 1992. Comparison of production from horizontal and vertical wells in the Austin Chalk, Niobrara, and Bakken plays. In: Schmoker, J.W., Coalson, E.B., Brown, C.A. (Eds.), Geological Studies Relevant to Horizontal Drilling—Examples from Western North America: Rocky Mountain Association of Geologists, pp. 67–87.

Stephen, R.A., 1984. Finite-difference seismograms for laterally varying marine models. Geophys. J. R. Astr. Soc. 79, 185–198.

Stephen, R.A., 1988. Lateral heterogeneity in the upper oceanic crust at DSDP Site 504. J. Geophys. Res. 93, 6571–6584.

Stewart, G., Ascharsobbi, F., 1988. Well test interpretation for naturally fractured reservoirs. In: Paper SPE 18173 Presented at the 1988 Annual Technical Conference and Exhibition, Houston, Texas, 2–5 October.

Stewart, P.R., September 1970. Low-permeability gas well performance at constant pressure. J. Pet. Technol. 1149–1156.

Stickey, D.C., 1993. New forces in international energy law: a discussion of political, economic, and environmental forces within the current international energy market. Tulsa J. Comp. Int. Law 1 (1), 95.

Stone, D.S., 1967. Theory of Paleozoic oil and gas accumulation in Big Horn Basin, Wyoming. Am. Assoc. Pet. Geol. Bull. 51, 2056–2114.

Stoneley, R., 1982. The structural development of the Wessex Basin: implications for exploration. Proc. Ussher Soc. 8, 1–6.

Story, C., 1998. Norphlet geology and 3-D geophysics— Fairway field, Mobile Bay, Alabama. The Lead Edge 17 (2), 243–248.

Stott, D.F., 1982. Lower Cretaceous Fort St. John Group and Upper Cretaceous Dunvegan Formation of the Foothills and Plains of Alberta, British Columbia, District of Mackenzie and Yukon Territory. Geol. Surv. Can. Bull. 328, 124 pp.

Stratigraphic distribution and controlling depositional factors. AAPG Bull. 75 (12), December 1991, 1809–1851.

Streltsova, T.D., 1982. Well pressure behavior of a naturally fractured reservoir. In: Paper SPE 10782 Presented at the 1982 California Regional Meeting, San Francisco, California, 24–26 March.

Strumm, W., Morgan, J.J., 1996. Aquatic Chemistry, Chemical Equilibria and Rates in Natural Waters, third ed. Wiley, Canada. p. 591.

Sundaram, N.S., Sarwar, M., Bang, S.S., Islam, M.R., 1994. Biodegradation of anionic surfactants in the presence of petroleum contaminants. Chemosphere 26 (6), 1253–1261.

Svärd, T., 2004. Adsorption of Hydrogen Sulfide at Low Temperature. Lund University, Lund, Sweden.

Swanson, D.C., 1979. Deltaic Deposits in the Pennsylvanian Upper Morrow Formation of the Anadarko Basin, 1. Tulsa Geological Society Special Publication. , pp. 115–168.

Sweeney, J.J., Gosnold, W.D., Braun, R.I., Burnham, A.K., 1992. A Chemical Kinetic Model of Hydrocarbon Generation from the Bakken Formation. Lawrence Livermore National Laboratory Report UCRL-ID-112038, Williston Basin, North Dakota, 57 pp.

Takahashi, H., Abe, H., 1989. Fracture mechanics applied to hot, dry rock geothermal energy. In: Atkinson, B.K. (Ed.), Fracture Mechanics of Rock. Academic, San Diego, California, pp. 241–276.

Tang, Y., Jenden, P.D., Teerman, S.C., 1991. Thermogenic methane formation in low-rank coals – published models and results from laboratory pyrolysis of lignite. In: Manning, D.A.C. (Ed.), Organic Geochemistry—Advances and Applications in the Natural Environment. Manchester University Press, Manchester, pp. 329–331.

Tapponnier, P., Molnar, P., 1976. Slip-line field theory and large-scale continental tectonics. Nature 264, 319–324.

Taylor, K.C., Hawkins, B.F., Islam, M.R., 1990. Dynamic interfacial tension in surfactant-enhanced alkaline flooding. J. Can. Petrol. Technol. 29 (1), 50–55.

Teng, L., Mavko, G., 1996. Fracture signatures on P-wave AVOZ. 66th Ann. Intern. Mtg., Soc. Expl. Geophys 1818–1821. Expanded Abstracts.

Teramoto, T., Takabayashi, K., Onishi, T., Okatsu, K., September 5–8, 2005. Air Injection EOR in highly water saturated oil reservoir. In Proceedings of IEA Collaborative Project on Enhanced Oil Recovery 25th Annual Workshop and Symposium, Stavanger, Norway.

Terwilliger, P.L., Wilsey, L.E., Hall, H.N., Bridges, P.M., Morse, R.A., 1951. An experimental and theoretical investigation of gravity drainage performance. Trans. AIME 192, 285–296.

Tew, B.H., Mink, R.M., Mancini, E.A., Mann, S.D., Kopaska-Merkel, D.C., 1993. Geologic framework of the Jurassic (Oxfordian) Smackover Formation, Alabama and panhandle Florida coastal waters and adjacent federal waters area. Gulf Coast Assoc. Geol. Soc. Trans. 43, 399–411.

Thompson, L.G., Manrique, J.L., Jelmert, T.A., 1991. Efficient Algorithms for Computing the Bounded Reservoir Horizontal Well Pressure Responses. In: Paper SPE 21827 Presented at 1991 Rocky Mountain Regional Meeting and Low-Permeability Reservoirs Symposium, Denver, CO, April 15–17.

Thomsen, L., 1986. Weak elastic anisotropy. Geophysics 51, 1954–1966.

Thomsen, L., 1988. Reflection seismology over azimuthally anisotropic media. Geophysics 53, 304–313.

Thomsen, L., 1993. Weak anisotropic coefficients. In: Castagna, J., Backus, M. (Eds.), Offset Dependent Reflectivity, Soc. Expl. Geophys.

Thomsen, L., 1995. Elastic anisotropy due to aligned cracks in porous rock. Geophys. Prospect. 43, 805–829.

Tidwell, V.C., Robert, J.G., 1995. Laboratory investigation of matrix imbibition from a flow fracture. Geophys. Res. Lett. 22, 1405–1408.

Ting, F.T.C., 1977. Origin and spacing of cleats in coal beds. J. Press. Vessel Technol. Trans. ASME 99, 624–626.

Toksoz, M.N., Charette, E.E., 1990. Effects of random heterogeneities on seismic waves: Implications for determination of reservoir heterogeneities. In: Presented at Soc. Expl. Geophys. Japan Int. Conf. On Geotomography.

Tondeur, D., Teng, F., 2008. Carbon capture and storage for greenhouse effect mitigation. In: Letcher, T.M. (Ed.), Future Energy: Improved, Sustainable and Clean Options for Our Planet. Elsevier, China, pp. 305–330. ISBN 978-0-08-054808-1.

Tortike, W.S., Farouq Ali, S.M., 1987. A framework for multiphase nonisothermal fluid flow in a deforming heavy oil reservoir. In: SPE 16030, Presented at the 9th SPE Symposium on Reservoir Simulation, San Antonio, TX.

Tremain, C.M., Laubach, S.E., Whitehead, N.H., 1991. Coal fracture (cleat) patterns in Upper Cretaceous Fruitland Formation, San Juan Basin, Colorado and New Mexico: implications for exploration and development. In: Schwochow, S., Murray, D.K., Fahy, M.F. (Eds.), Coalbed Methane of Western North America. Rk. Mt. Assoc. Geol., pp. 49–59.

Treybal, R.E., 1996. Operaciones de Transferencia de Masa, second ed. Mc Graw Hill, Mexico, ISBN 968-6046-34-8.

Tristone Capital Inc., 2008. Montney Python – The Holy Grail of Canadian Tight Gas. Analyst report for industry clients.

Tromp, P.L., Cardinal, D.F., Steidtmann, J.R., 1981. Stratigraphy and depositional environments of the "Leo sands" in the Minnelusa Formation, Wyoming and South Dakota. In: Reid, S.G., Miller, D.D. (Eds.), Energy Resources of Wyoming: Wyoming Geological Association, 32d Annual Field Conference, Guidebook, pp. 11–22.

Trotter, J.F., 1984. The Minnelusa revisited. In: Goolsby, J., Morton, D. (Eds.), The Permian and Pennsylvanian Geology of Wyoming: Wyoming Geological Association, 35th Annual Field Conference, Guidebook, pp. 127–151.

Tsvankin, I., 1995. Body wave radiation patterns and AVO in transversely isotropic media. Geophysics 60, 1409–1425.

Tsvankin, I., 1996. P-wave signatures and notation for transversely isotropic media: an overview. Geophysics 61, 467–483.

Tsvankin, I., 1997. Reflection moveout and parameter estimation for horizontal transverse isotropy. Geophysics 62, 614–629.

Turcotte, D.L., 1992. Fractals and Chaos in Geology and Geophysics. Cambridge Univ. Press, Cambridge, 221 pp.

Turta, T., Singhal, A.K., Sim, S.S.K., Hawkins, B.F., 2007. Basic Investigations on Enhanced Gas Recovery (EGR) by Gas/Gas Displacement. CIPC 2007-124, June 2007.

Tyler, N., Gholston, J.C., 1988. Heterogeneous Deep-Sea Fan Reservoirs, Shackelford and Preston Waterflood Units, Spraberry Trend, West Texas. University of Texas at Austin, Bureau of Economic Geology. Report of Investigations 171, 38 pp.

Tyler, N., Gholston, J.C., Guevara, E.H., 1997. Basin morphological Controls on Submarine-Fan Depositional Trends—Spraberry Sandstone, Permian Basin, Texas. Bureau of Economic Geology, Geological Circular 97–6, 43 pp.

U.S. Bureau of Mines and U.S. Geological Survey, 1980. Principles of a Resource/Reserve Classification for Minerals. U.S. Geological Survey Circular 831, 5 pp.
U.S. Energy Information Administration, 2009. Natural Gas. Web page at. www.eia.doe.gov/oil_gas/natural_gas/info_glance/natural_gas.html.
U.S. Energy Information Administration, 2010. Summary: U.S. Crude Oil, Natural Gas, and Natural Gas Liquids Proved Reserves 2009. www.eia.gov/oil_gas/natural_gas/data_publications/crude_oil_natural_gas_reserves/cr.html.
U.S. Energy Information Administration, 2011a. World Shale Gas Resources: An Initial Assessment of 14 Regions outside the United States. www.eia.gov/analysis/studies/worldshalegas/.
U.S. Energy Information Administration, 2011b. Review of Emerging Resources: U.S. Shale Gas and Shale Oil Plays. www.eia.gov/analysis/studies/usshalegas/.
U.S. Geological Survey, March 26, 1974. USGS Releases Revised U.S. Oil and Gas Resource Estimates. U.S. Geological Survey news release, 5 pp.
U.S. Geological Survey, 1975. Geological Estimates of Undiscovered Recoverable Oil and Gas Resouces in the United States. U.S. Geological Survey Circular 725.
U.S. Geological Survey, 1981. Estimates of Undiscovered Recoverable Conventional Resources of Oil and Gas in the United States. U.S. Geological Survey Circular 860.
U.S. Geological Survey, 1995. 1995 National Assessment of United States Oil and Gas Resources. U.S. Geological Survey Circular 1118, 20 pp.
U.S. Department of Interior, Bureau of Land Management, 2003. Final Environmental Impact Statement and Proposed Plan Amendment for the Powder River Basin Oil and Gas Project, vols. 1–3. Buffalo Field Office, Buffalo, WY.
U.S. Department of Interior, Geological Survey, 1986. Hydrology of Area 50, Northern Great Plains and Rocky Mountain Coal Provinces, Wyoming and Montana. Water Resources Investigations Open-File Report 83–545.
USGS, 2008. Geologic controls on the growth of petroleum reserves chapter I of geologic, engineering, and assessment studies of reserve growth. U.S. Geol. Surv. Bull. 2172-I.
Utting, J., Zonneveld, J.P., MacNaughton, R.B., Fallas, K.M., 2005. Palynostratigraphy, lithostratigraphy, and thermal maturity of the Lower Triassic Toad and Grayling, and Montney Formations of western Canada, and comparison with Coeval Rocks of the Sverdrup Basin, Nunavut. Bull. Can. Pet. Geol. 53, 5–24.
Vail, P.R., 1987. Seismic stratigraphy interpretation using sequence stratigraphy, part 1, seismic stratigraphic procedure. In: Bally, A.W. (Ed.), Atlas of Seismic Stratigraphy: American Association of Petroleum Geologists Studies in Geology 27, vol. 1, pp. 1–10.
Vairogs, J., Hearn, C.L., Dareing, D.W., Rhoades, V.W., September 1971. Effect of rock stress on Gas production from low-permeability reservoirs. J. Pet. Technol. 1161–1167.
Van der Meer, L.G.H., 2005. Carbon dioxide storage in natural gas reservoirs. Oil Gas Sci. Technol. 60 (3), 527–536.
Van Krevelen, D.W., 1981. Coal. Elsevier, Amsterdam, 514 pp.
Van Waggoner, J.C., Posamentier, H.W., Mitchum, R.M., Vail, P.R., Sarg, J.F., Louitit, T.S., Hardenbol, J., 1988. An overview of the fundamentals of sequence stratigraphy and key definitions. In: Wilgus, C.K., et al. (Eds.), Sea-Level Changes: An Integrated ApproachSociety of Economic Paleontologists and Mineralogists Special Publication 42, pp. 39–45.
Vaziri, H.H., Xiao, Y., Islam, R., Nouri, A., 2003. Numerical modeling of seepage-induced sand production in oil and gas reservoirs. J. Pet. Sci. Eng. 36, 71–86.
Ver Steeg, K., 1942. Jointing in coal beds of Ohio. Econ. Geol. 37, 503–509.
Verga, F.M., Carugo, C., Chelini, V., Maglione, R., De Bacco, G., 2000. Detection and characterization of fractures in naturally fractured reservoirs. In: Paper SPE 63266 Presented at the 2000 Annual Technical Conference and Exhibition, Dallas, Texas, USA, 1–4 October. http://dx.doi.org/10.2118/63266-MS.

Verma, M.K., 2003. Modified Arrington method for calculating reserve growth—A new model for United States oil and gas fields. In: Dyman, T.S., Schmoker, J.W., Verma, M.K. (Eds.), Geologic, Engineering, and Assessment Studies of Reserve Growth: U.S. Geological Survey Bulletin 2172–D. http://pubs.usgs.gov/bul/b2172-d/ (accessed 09.08.07.).

Viberg, M., Ottersten, B., 1991. Sensor array processing based on subspace fitting. IEEE Trans. Signal Process. 39, 1110–1121.

Vilotte, J.P., Diagnieres, M., Madariaga, R., 1982. Numerical modeling of intraplate deformation: simple mechanical models of continental collision. J. Geophys. Res. 87, 10709–10728.

Wadadar, S.S., Islam, M.R., 1994. Numerical Simulation of Electromagnetic Heating of Alaskan Tar Sands Using Horizontal Wells. J. Can. Petrol. Technol. 33 (7), 37–43.

Wade, W.J., Moore, C.H., 1993. Jurassic sequence stratigraphy of southwest Alabama. Gulf Coast Assoc. Geol. Soc. Trans. 43, 431–443.

Walsh, J., Ross, C., Smith, M., Harper, S., Wilkins, W., 1988. Biogas Utilization Handbook. Georgia Tech Research Institute, Atlanta, Georgia, USA, ISBN 0962464740.

Walsh, W., Adams, C., Kerr, B., Korol, J., 2006. Regional "Shale Gas" Potential of the Triassic Doig and Montney Formations, Northeastern British Columbia. BC Ministry of Energy, Mines and Petroleum Resources, Resource Development and Geoscience Branch. Petroleum Geology Open File 2006-02, 19 pp.

Wang, D., Cheng, J., Wu, J., Wang, G., September 29–October 2, 2002. Experiences learned after production of more than 300 million barrels of oil by polymer flooding in daqing oil field (SPE-77693). In Proceedings of SPE Annual Technical Conference and Exhibition, San Antonio, TX, USA.

Wang, Z., Peden, J.M., Lemanczyk, R.Z., 1994. Gas kick simulation study for horizontal wells. In: Paper IADC/SPE 27498 Presented at the 1994 IADC/SPE Drilling Conference, Dallas, Texax, USA, 15–18 February. http://dx.doi.org/10.2118/27498-MS.

Wark, K., Warner, C.F., 1981. Air Pollution: Its Origin and Control. Harper and Row Publishers, New York, New York, USA.

Warn, G.F., Sidwell, R., 1953. Petrology of the spraberry sands of west Texas. J. Sediment. Pet. 23, 67–74.

Warpinski, R.N., Kramm, R.C., Heinze, R.J., Waltman, K.C., Comparison of single and dual-array microseismic mapping techniques in the barnett shale. Paper SPE 95568 Presented at the 2005 Annual Technical Conference and Exhibition, Dallas, Texas, 9–12 October.

Warren, J.E., Root, P.J., 1963. The behavior of naturally fractured reservoirs. SPE J. 3 (3), 245–255. http://dx.doi.org/10.2118/426-PA. SPE 426-PA.

Watson, A.T., Gatens III, J.M., Lee, W.J., Rahim, Z., 1989. An analytical model for history matching naturally fractured reservoir production data. In: Paper SPE 18856 Presented at the 1989 Production Operations Symposium, Oklahoma City, Oklahoma, 13–14 March.

Wattenbarger, R.A., El-Banbi, A.H., Villegas, M.E., Maggard, J.B., 1998. Production analysis of linear flow into fractured tight gas wells. In: Paper SPE 39931 Presented at the 1998 SPE Rocky Mountain Regional/Low-permeability Reservoirs Symposium and Exhibition, Denver, Colorado, USA, 5–8 April. http://dx.doi.org/10.2118/39931-MS.

Wax, M., Kailath, T., 1985. Detection of signals by information theoretic criteria. IEEE Trans. Acoust. Speech Signal Process. ASSP-33, 378–392.

Wax, M., Ziskind, I., 1989. On unique localization of multiple sources by passive sensor arrays. IEEE Trans. Acoust. Speech Signal Process. ASSP 37, 996–1000.

Wazwaz, A.M., El-Sayed, S.M., 2001. A new modification of Adomian decomposition method for linear and nonlinear operators. Appl. Math. Comput. 122 (3), 393–405.

Wazwaz, A.M., 2001. A new algorithm for calculating Adomian polynomials for nonlinear operators. Appl. Math. Comput. 111 (1), 33–51.
Website 1, NY government report, New York's oil and natural gas history – a long story, but not the final chapter. http://www.dec.ny.gov/docs/materials_minerals_pdf/nyserda2.pdf.
Website 2, http://www.windows2universe.org/our_solar_system/planets_table.html.
Website 3, http://pubs.usgs.gov/fs/fs021-01/fs021-01.pdf.
Website 4, https://www.transformsw.com/wp-content/uploads/2013/05/Shale-Gas-Reservoirs-Similar-yet-so-different-2010-RMAG-DGS-3D-Symposium-Roth.pdf.
Webster, R.W., 1984. Petroleum source rocks and stratigraphy of the Bakken Formation in North Dakota. In: Woodward, J., Meissner, F.F., Clayton, J.L. (Eds.), Hydrocarbon Source Rocks of the Greater Rocky Mountain Region: Rocky Mountain Association of Geologists, pp. 57–81.
Weeks, L.G., 1958. Fuel reserves of the future. Am. Assoc. Pet. Geol. Bull. 42 (2), 431–441.
Wei, Y., Economides, M.J., 2005. Transverse hydraulic fractures from a horizontal well. In: Paper SPE 94671 Presented at the 2005 SPE Annual Technical Conference and Exhibition, Dallas, Texas, USA, 9–12 October. http://dx.doi.org/10.2118/94671-MS.
Welge, H.J., 1952. A simplified method for computing oil recovery by gas or water drive. Trans. AIME 195, 91–98.
Welte, D.H., Stoessinger, W., Schaefer, R.G., Radke, M., 1984. Gas generation and migration in the Deep Basin of western Canada. In: Master, J. (Ed.), Elmworth – Case Study of a Deep Basin Gas Field. American Association of Petroleum Geologists Memoir 38, pp. 35–47.
WEO, September 2013. World Energy Resources. World Energy Organization report.
Wharton Jr., J.B., 1948. Isopachous maps of sand reservoirs. Am. Assoc. Pet. Geol. Bull. 32 (7).
Wheeler, D.M., Scott, A.J., Coringrato, V.J., Devine, P.E., 1990. Stratigraphy and depositional history of the Morrow Formation, southeast Colorado and southwest Kansas. In: Sonnenberg, S.A., Shannon, L.T., Rader, K., von Drehle, W.F., Martin, G.W. (Eds.), Morrow Sandstones of South-East Colorado and Adjacent Areas: Rocky Mountain Association of Geologists—Morrow Sandstone, pp. 9–36.
White, D.A., Gehman, H.M., 1979. Methods of estimating oil and gas resources. Am. Assoc. Pet. Geol. Bull. 63 (12), 2183–2192.
White, D., Colpron, M., Buffett, G., Roberts, B., 2006. Structural constraints for oil and gas assessment in the Whitehorse Trough: new results from seismic profiling. In: Emond, D.S., Bradshaw, G.D., Lewis, L.L., Weston, L.H. (Eds.), Yukon Exploration and Geology 2005. Yukon Geological Survey, pp. 315–323.
White, D.A., 1980. Assessing oil and gas plays in facies-cycle wedges. Am. Assoc. Pet. Geol. Bull. 64 (8), 1158–1178.
White, D.A., 1993. Geologic risking guide for prospects and plays. Am. Assoc. Pet. Geol. Bull. 77 (12), 2048–2061.
Wilkinson, D.S., Maire, E., Embury, J.D., 1997. The role of heterogeneity on the flow and fracture of two-phase materials. Mater. Sci. Eng. A. Struct. 233, 145–154.
Williamson, I.A., 1967. Coal Mining Geology. Oxford Univ. Press, Oxford, 266 pp.
Wilson, G.V., 1975. Early differential subsidence and configuration of the northern Gulf Coast Basin in southwest Alabama and northwest Florida. Gulf Coast Assoc. Geol. Soc. Trans. 25, 196–206.
Winker, C.D., 1982. Cenozoic shelf margins, northwestern Gulf of Mexico Basin. Gulf Coast Assoc. Geol. Soc. Trans. 32, 427–448.
Witherspoon, P.A., Wang, J.S.Y., Iwai, K., Gale, J.E., 1980. Validity of cubic law for fluid flow in a deformable rock fracture. Water Resour. Res. 16 (6), 1016–1024.
Wo, S., Whitman, L.D., Steidtmann, J.R., April 14–16, 2009. Estimates of potential CO_2 demand for CO_2 EOR in Wyoming Basins (SPE-122921). In Proceedings of SPE Rocky Mountain Petroleum Technology Conference, Denver, CO, USA.

Woodrow, C.K., Elliott, D., Pickles, R., Weatherford, D.B., 2008. One Company's first exploration UBD well for characterizing low permeability reservoirs. In: Paper SPE 112907 Presented at the 2008 SPE North Africa Technical Conference and Exhibition, Marrakech, Morocco, 12–14 March. http://dx.doi.org/10.2118/112907-MS.

Worrall, D.M., Snelson, S., 1989. Evolution of the northern Gulf of Mexico, with emphasis on Cenozoic growth faulting and the role of salt. In: Bally, A.W., Palmer, A. (Eds.), The Geology of North America—An Overview, The Geology of North America, v. A. Geological Society of America, pp. 97–137.

Wright, C.J., Gallun, R.A., 2008. Fundamentals of Oil & Gas Accounting, fifth ed. PennWell Books. p. 750, ISBN 978-1-59370-137-6.

Wu, R.S., Aki, K., 1990. In: Scattering and Attenuation of Seismic Waves, vol. 132. Birkhauser, Basel.

Wyoming Department of Environmental Quality, 2001. SAP: Water Quality Rules and Regulations (Chapter 1). Department of Environmental Quality and Water Quality Division, Cheyenne, WY.

Wyoming Department of Environmental Quality, 2006. NPDES Permitting, Water Quality Division. In: Wyoming Geological Association Symposium, vol. II, pp. 243–472. Retrieved 01.11.06 from: http://deq.state.wy.us/wqd/npdes/.

Wyoming Geological Association, 1981. Powder River Basin Oil and Gas Fields.

Xiong, H., Holditch, S.A., 2006. Will the blossom of unconventional natural gas development in North America be repeated in China?. In: Paper SPE 103775 Presented at the 2006 SPE International Oil & Gas Conference and Exhibition, Beijing, China, 5–7 December. http://dx.doi.org/10.2118/103775-MS.

Xu, R., 1981. In: Geology, Geological History and Origin of Qinghai-Xizang Plateau, vol. 1. Since Press, Beijing, pp. 139–144.

Yacobucci, M.M. Field Trip to the Devonian Dundee Formation, OH, Whitehouse.

Yang, Y., Aplin, A.C., 2004. Definition and practical application of mudstone porosity-effective stress relationships. Petrol. Geosci. 10, 153–162.

Yin, X.C., Liu, X.H., 1994. Investigations of fracture instability in crack growth for several metals – part I: experimental results. Int. J. Fract. 69 (2), 123–143.

Zaman, M., Bjorndalen, N., Islam, M.R., 2004. Detection of precipitation in pipelines. Pet. Sci. Technol. 22 (9–10), 1119–1141.

Zaman, M., Agha, K.R., Islam, M.R., 2006. Laser based detection of paraffin in crude oil samples: numerical and experimental study. Pet. Sci. Technol. 24 (1), 7–22.

Zatzman, G.M., Islam, M.R., 2005. Natural gas pricing. In: Mokhatab, S., Speight, J.G., Poe, W.A. (Eds.), Handbook of Natural Gas Transmission and Processing. Elsevier Inc.

Zatzman, G., Islam, M.R., 2006. Gas energy pricing. In: Mokhatab, S., Poe, W., Speight, J. (Eds.), Gas Engineering Handbook. Elsevier, p. 672.

Zatzman, G.M., Islam, M.R., 2007. Truth, consequences and intentions: the study of natural and anti-natural starting points and their implications. J. Nat. Sci. Sustain. Technol. 1 (3), 169–174.

Zatzman, G.M., Islam, M.R., 2007a. Truth, consequences and intentions: the study of natural and anti-natural starting points and their implications. J. Nat. Sci. Sustain. Technol. 1 (2), 169–174.

Zatzman, G.M., Islam, M.R., 2007b. Economics of Intangibles. Nova Science Publishers, 431 pp.

Zatzman, G.M., Khan, M.M., Chhetri, A.B., Islam, M.R., 2007. A delinearized history of time and its roles in establishing and unfolding knowledge of the truth. J. Nat. Sci. Sustain. Technol. 1 (4), 489–533.

Zatzman, G.M., Khan, M.M., Islam, M.R., 2008. The Use of Newly Formulated Comprehensive Mass and Energy Balance Equation to Develop an Atmospheric Model. paper ID: 588–301, WSEAS Transaction.

Zatzman, G., 2012. Sustainable Energy Pricing: Nature, Sustainable Engineering, and the Science of Energy Pricing. Wiley Scrivener, 608 pp.

Zekri, A.Y., Mustafiz, S., Chaalal, O., Islam, M.R., 2008. "The Areal Sweep Efficiency of the First Contact Miscible Displacements: An Experimental Approach". Pet. Sci. Tech. 26 (17), 2033–2047.

Zeng, F., Zhao, G., 2010. The optimal hydraulic fracture geometry under non-Darcy flow effects. J. Pet. Sci. Eng. 72, 143–157.

Zhong, L., Islam, M.R., October 1995. A new microbial plugging process and its impact on fracture remediation. In: SPE Paper no. 30519, Proc. SPE Annual Conference and Exhibition, Dallas, TX.

Zhong, L., Siddiqui, S., Islam, M.R., 1999. A standard method for testing acidization-foam-forming surfactants. Energy Sources 21 (1/2), 191–213.

Zhu, X., 1997. 3-D Numerical Modeling of the P and SV Wave Reflections from Fractured.

Zuber, M.D., Williamson, J.R., Hill, D.G., Sawyer, W.K., Frantz, J.H., September 20–October 2. A comprehensive reservoir evaluation of a shale Gas reservoir – the New Albany shale. Paper SPE 77469, Presented at the 2002 Annual Technical Conference and Exhibition, San Antonio, TX.

INDEX

Note: Page numbers followed by "f" and "t" indicate figures and tables respectively

D

E

FSC
www.fsc.org
MIX
Papier | Fördert
gute Waldnutzung
FSC® C083411

Kolibri 360 GmbH
Ferdinand-Jühlke-Straße 7
99095 Erfurt, Deutschland
produktsicherheit@kolibri360.de